SCIENTIFIC INQUIRY AND NATURE OF SCIENCE

Science & Technology Education Library

VOLUME 25

SCOPE

The book series *Science & Technology Education Library* provides a publication forum for scholarship in science and technology education. It aims to publish innovative books which are at the forefront of the field. Monographs as well as collections of papers will be published.

The titles published in this series are listed at the end of this volume.

Scientific Inquiry and Nature of Science

Implications for Teaching, Learning, and Teacher Education

Edited by

L.B. Flick
Illinois Institute of Technology,
Oregon State University,
Corvallis,
OR, U.S.A.

and

N.G. Lederman
Illinois Institute of Technology,
Chicago,
IL, U.S.A.

KLUWER ACADEMIC PUBLISHERS
DORDRECHT / BOSTON / LONDON

A C.I.P. Catalogue record for this book is available from the Library of Congress.

ISBN 1-4020-2671-4 (HB)
ISBN 1-4020-2672-2 (e-book)

Published by Kluwer Academic Publishers,
P.O. Box 17, 3300 AA Dordrecht, The Netherlands.

Sold and distributed in North, Central and South America
by Kluwer Academic Publishers,
101 Philip Drive, Norwell, MA 02061, U.S.A.

In all other countries, sold and distributed
by Kluwer Academic Publishers,
P.O. Box 322, 3300 AH Dordrecht, The Netherlands.

Printed on acid-free paper

Printed in the Netherlands.

TABLE OF CONTENTS

LAWRENCE B. FLICK & NORMAN G. LEDERMAN

INTRODUCTION

A renewed stress on scientific inquiry and nature of science are what distinguish current reform documents in science education from previous efforts. Unfortunately, classroom teachers, as well as teacher educators, remain uncertain about the specific attributes of scientific inquiry and nature of science, let alone their integration into current science instruction and curricula. Although intimately related, scientific inquiry and nature of science are different constructs. The purpose of this text is to help clarify both the theoretical and practical aspects of inquiry and nature of science as well as provide some guidance relative to their inclusion in science teaching, teacher education, and research. Consequently, the text contains chapters that are dedicated solely to inquiry, solely to nature of science, and to the interaction of inquiry and nature of science. To introduce these two complex ideas, we use the term *inquiry*, and variously the terms *learning (teaching) about inquiry* and *learning(teaching) about science* in our discussion of the *nature of science*.

Not since the introduction of the term "hands-on" into science education have teachers and scholars in the field been so dominated by a single concept. The concept capturing the imagination of so many people is, *inquiry*. The contemporary stimulus prompting the use of the term in popular and scholarly discourse was the publication of the *National Science Education Standards* (NRC, 1996). Publication of this document was part of a larger movement to generate national standards for science subject matter, science teaching, and assessment, among other concerns. But as ubiquitous as the standards movement has become, it can not be the only or, perhaps, even the main reason that the term *inquiry* has captured so much attention. With so much being written on the subject of teaching science as inquiry, there is always the tendency for the quantity of discourse to dilute the meaning of key terms. As usage flows between popular and professional literature, concepts are applied in an increasing number of settings and to increasingly diverse instances. It is important for a profession to periodically take a reading on the directions major ideas have been taken. This book examines *inquiry* and *nature of science*, two major and interrelated foci of the reforms in science education. The first concept is *inquiry*. The term has taken on three different meanings in the context of the National Science Education Standards. Inquiry stands for a fundamental principle of how modern science is conducted. Inquiry refers to a variety of processes and ways of thinking that support the development of new knowledge in science. In addition to the *doing* of science, inquiry also refers to knowledge *about* the processes scientists use to develop knowledge, that is the *nature of science itself*. Thus, inquiry is viewed as two different student outcomes, ability to *do* scientific processes and knowledge *about* these processes. Finally, inquiry is viewed as a teaching approach

L.B. Flick and N.G. Lederman (eds.) Scientific Inquiry and Nature of Science, ix-xviii.

that can be used to teach students the traditional subject matter of the sciences. The logic here is that students will best learn science if they learn using a reasonable facsimile of the processes scientists follow. Closely related to inquiry is *nature of science*, a term that has created as much confusion as inquiry. As a consequence of the conventions of what is considered acceptable scientific inquiry, as well as the fact that inquiry is performed by humans, the knowledge produced necessarily has certain characteristics that limit and delimit its applications and ontological status.

The *National Science Education Standards* used the term inquiry as the label for a core principle of science. Simply stated that principle is that knowledge about the world derives from human efforts to systematically gather and interpret observations that become evidence for or against explanations and theory through collaboration, discussion, and debate. This general principle implies that scientific knowledge about the world is subject to the interpretation and reinterpretation of a body of evidence in a fluid environment of scientific ideas and theory. Schwab (1962) in analyzing the nature of science contrasted "stable enquiry", were scientific principles define problems, with "fluid enquiry", where principles are treated as problems themselves. This contrast highlights two complementary activities of science, constructing bodies of evidence related to existing theory and constructing new theory. The principle of disciplined inquiry is at the center of both forms of investigation. On the basis of this central principle the National Science Education Standards made "teaching science as inquiry" a core principle for science education.

When translated into classroom curriculum, instruction, and assessment, teaching science as inquiry and teaching about the nature of science have resonated with core principles in teaching science. These core principles derive from work on understanding the nature of student thinking in complex tasks such as those required by inquiry-oriented tasks. Throughout the last century the principle that science should be taught "from the beginning...studied (though not exclusively) by direct contact with the environment" (United States Bureau of Education, 1893) has received periodic reinforcement. However understanding the cognitive demands of complex tasks, such as posing and investigating problems in science classrooms, has developed more recently. Educational research examining the skills of expert learners involved with complex tasks resonates closely with teaching science as inquiry.

Teaching scientific inquiry and teaching about the nature of scientific inquiry means finding ways for engaging students in investigative activity and also teaching appropriate ways of thinking that support development of meaning. Active student involvement that prompts the use of relevant intellectual skills goes to the core of what teachers strive to do in any subject. Educators have long been interested in a better understanding of how to stimulate student thought. Teaching science as inquiry poses a particularly important and difficult instance of this goal. Addressing scientific problems through inquiry requires that students link a scientific purpose with scientific procedures that lead to a conclusion supported by reasoned argument (Reiser et al. 2001). The cognitive skills needed for this kind of task are beyond what most students are capable of doing without direct teacher involvement. Understanding how to support student development and use of relevant cognitive skills is a broad area of research by cognitive and educational psychologists as well as science educators that is relevant to teaching science as inquiry.

Forming and maintaining a sense of scientific purpose in the process of designing and/or engaging in investigative activity, requires a reflective state of mind. The nature of complex tasks, such as inquiry, require the learner to track their own progress, use of cognitive resources, and awareness of when error checking or correction is needed. Reflection is also required for students to understand the nature of the scientific work they are engaged in, in other words the nature of science itself. Educators and scholars over recent decades have become increasingly interested in metacognitive capabilities that support learners in this kind of complex task. Prominent areas of work have included studying from texts, expository writing, and scientific reasoning (Brown et al., 1983). Indeed, it is this alternation of *doing* and *reflecting* that provides students with opportunities to develop their inquiry skills as well as an understanding about what they have done. In the language of the standards, students will learn to *do* inquiry as well as learn *about* inquiry, or about the nature of science itself.

Early work in teaching science as inquiry (see Welch, 1981; Harms & Yager, 1980) was not successful. Specifically, low achieving students seemed to be further disadvantaged by instruction focusing on inquiry. Hope for improvement of this situation comes from important advances in knowledge about the kinds of cognitive skill and knowledge needed for inquiry-type tasks and how to scaffold student thinking. The premises underlying recent efforts to teach scientific inquiry are derived from research suggesting that with appropriate instructional scaffolding, students can learn cognitive skills for learning how to learn. By extension, students can be guided in developing the metacognitive skills necessary for following the course of investigating a scientific problem. In the process of pursuing investigations, students can also develop knowledge of science and knowledge about science. These advances are accomplished by explicit teaching of cognitive and metacognitive processes in the context of learning science and participating in investigations. Explicit approaches to teaching these skills particularly benefit low-achieving and/or disadvantaged students (White & Frederickson, 1998).

As students gain skills for conducting an investigation, they can also be guided to consider how the scientific enterprise functions from a larger perspective. A metacognitive awareness necessary for guiding and shaping procedures and constraining the relationship between data and conclusions, can be extended to consider the context within which the work takes place. Science does not operate outside of the influences of culture, politics, and society. As the student's world grows so should their understanding of how science, its people and ideas, fit into it. It is only at this level of thought about science and scientific knowledge that students reach the full intent of science education envisioned by current reformers. It is one thing to be able to focus on a scientific question, for example where does salt go when dissolved in water, and quite another to recognize that question as part of a much larger process of building scientific knowledge.

Recognizing the nature of inquiry as disciplined observation and interpretation under the constraints of a stated problem is a major accomplishment. Recognizing that inquiry is a way of thinking and a way of knowing encompassing the work of scientists over hundreds of years is a second major accomplishment. We must not underestimate the scale of these two achievements. The framers of a new science education want to broaden the way students see the world to include a scientific perspective. A favorite story in science is an analogy suggesting the scale of this change in thinking. We picture Newton pondering the scientific problems of force and motion. A round apple, reminiscent of the moon hanging in the sky, falls from a tree and strikes him on the head triggering an insight. The forces operating on earth also operate on the scale of the heavens (Bronowski, 1973, p. 222). Science and our perception of the world is forever changed. This is the goal we have for students. The nature of scientific work as investigative processes, bookish knowledge, and discipline is limited, it involves much more. It is creativity always influenced by the culture, politics, and social values of the time. Jacob Bronowski was an acute observer of the development of science and the interplay among ideas and people and culture. His thoughts on the practice of science are the stuff that breathe life into abstract principles and disciplined procedures. "Every theory is based on some analogy, and sooner or later the theory fails because the analogy turns out to be false. A theory in its day helps to solve the problems of the day" (p. 140). Historical precursors to modern scientific work, alchemy and astrology, were attempts to use human life as an analogy for how the world works. This turns out to be a false analogy. But like the early thinkers that pre-date modern science, our students themselves are stargazers and fascinated by transforming materials.

The focus on inquiry as critical to the development of scientific literacy provides students within a framework within which they can better understand the nature and limitations of the knowledge produced as part of the scientific body of knowledge. Such understandings are critical, especially when we quickly come to realize that it is unreasonable to assume that our citizenry will make decisions about scientifically and technologically-based issues by running to the garage to conduct authentic investigations. More realistically, experiences with inquiry provide our students (and citizens) with foundational experiences from which they can reflect on the nature and limits of scientific knowledge and claims. It is based upon this knowledge that the general citizenry will derive meaning and research conclusions concerning knowledge claims. This is the value of *nature of science*. There are numerous lists and definitions that one can find related to nature of science. However, the empirical literature and the *National Science Education Standards* typically use the phrase to refer to the characteristics of the knowledge as directly derived from how the knowledge is produced. That is, the nature of scientific inquiry has implications for the knowledge produced. Again, although disagreements exist among philosophers, scientists, and educators there is virtually no disagreement that scientific knowledge is a) subject to change (tentative); b) partly derived from human imagination, creativity and imagination; c) necessarily derived from a combination of observation and inference; d) embedded within a social and cultural context; and e) at least partially derived from the empirical world.

It is critical for us to do more than avoid debates about nature of science by rising to a level of generality where disagreements do not exist. As educators, it is absolutely critical that we carefully consider what aspects of nature of science are accessible to school-aged students and what aspects make sense for *all* students to know. It is not, for example, fruitful to insist that middle school students should come to understand that there are no observations, but really only inferences.

Clearly, scientific inquiry and nature of science are the foundations of current conceptions of what it means to be scientifically literate. Clearly, promoting students' understandings of these complex and abstract ideas and processes require types of teaching, learning, and assessment much different from what has been typically observed in our classrooms over the past century. The authors presented in this text have attempted to address, from varied perspectives, the educational issues surrounding attention to inquiry and nature of science in our K-12 classrooms, as well as teacher education.

Organization of the Book

Bybee has, perhaps, influenced the inclusion of inquiry and nature of science in the science curriculum at the policy level more than any science educator. His chapter sets a concrete foundation for the text by clearly defining the different perspectives on the meaning of inquiry and its relationship to nature of science within teaching and curriculum. Further, he does an excellent job of clarifying the importance of inquiry and nature of science to science teaching and learning, and the role of science curriculum in the current socio-cultural climate. The implications provided with respect to teacher education clearly follow. Overall, this chapter provides an lens through which to view the more specific perspectives of the following chapters.

Part I examines three educational contexts that pose significant challenges to reforming science and mathematics teaching. In order to understand the broad implications of teaching inquiry and the nature of science to the broadest possible population of students, DeBoer sets an historical foundation for the book. In particular, he provides an analysis of the history of scientific inquiry as a goal of science education. Specific attention is given to the varying rationales for the inquiry approach within curriculum with respect to current day socio-historical perspectives. One completes this chapter with a clear understanding that the inclusion of inquiry and nature of science has been advocated for a long period of time and the rationale has not always been limited to the value of these topics to an understanding of science.

Haberman examines teaching children of poverty in urban settings. These environments pose some of the most difficult educational environments in our system. For the teachers that not only survive in these environments but help students thrive, Haberman describes how they address three of the central questions of in our field: What knowledge is of most worth? How does learning best occur? and What is the purpose of learning? Teaching science and inquiry, in the hands of skilled teachers, can play an important role in promoting real thinking and real learning in the toughest of schools.

Lederman and Stefanich take an extremely innovative approach to their discussion of traditional issues related to inclusive instruction. Their discussion provides an excellent primer for those not well-versed in the literature on inclusive instruction, but then takes matters several steps further by clearly showing how an instructional approach focused on inquiry and nature of science is quite consistent with what the literature on inclusive science instruction has been recommending for years. Additionally, the authors provide an innovative perspective in their description of how the experiences of students with disabilities can be used to enhance inquiry-oriented and nature of science instruction for *all* students.

Novak and Krajcik provide a *tour de force* of learning technologies and their potential to impact student understanding of science through inquiry. Emphasis is placed on technologies as a cognitive tools that, in concert with the teacher, provide a variety of scaffolding for student inquiry. They present examples of learning technologies whose features support the development of integrated understandings through access to a wide variety of information, capabilities for multiple forms of representation, and channels of communication that break the barriers of typical classrooms. Each category of technology is discussed in the context of research critically examining of the role technology can play in science teaching and learning. Across the breadth of this chapter, the reader gains a sense of future technological innovation through tantalizing examples of the integration of technologies. Computer probes, hand-held computers, and web resources connect students to the environment, students to experts, students to each other.

In Part II, authors take readers through issues in teaching and learning scientific inquiry from the elementary grades to teacher education. Metz begins with the obvious that the work of professional scientists and children learning science in elementary classrooms is fundamentally different. However, she raises a central, penetrating question of how problematic are these differences for leading children to understand science as inquiry? By examining the goal structures of scientific work and of exemplary elementary curricula, she challenges readers to re-examine the capabilities of children and the intended goals of elementary science. Contrary to the design of most curricula in the US, Metz argues that it is possible for less to become more if we "understand the nature of scientists' knowledge and how they use it." Finally, drawing on recent research in classrooms, she challenges elementary science educators to recognize the fundamental role of discourse in examining competing ideas in science and to scaffold student engagement in such discourse as a central task of science learning.

Magnusson, Palincsar, and Templin present an argument for the central importance of communication within and among scientific communities. They apply the central tenets of this argument to heuristic for guided inquiry in elementary classrooms. The chapter uses data from classroom students to illustrate how their heuristic can be used by teachers to encourage use of and advance student skills in classroom conversation that transitions into scientific discourse. A key feature of the basic argument instantiated in the heuristic is that scientists operate in two arenas of discourse. In "workbench" science, scientists engage in informal communication and argument along with creative speculations. In formal publication, scientists must respect formal modes of presentation and argument. The guided discovery heuristic helps teachers structure classrooms as a scientific community.

Prescriptions for teaching science in high school has a long history of emphasizing lab work and viewing science as part of real world experiences. In sketching some of these historical antecedents to teaching science as inquiry, Flick highlights the more contemporary view that to accomplish an understanding and use of inquiry, teachers are charged with teaching and supporting the use of appropriate cognitive skills. This is a radical shift in teaching responsibilities for many teachers. Flick outlines the complementary nature of research in cognitive and developmental psychology and research in science classrooms both emphasizing the importance of cognitive supports for engaging in challenging, inquiry tasks. The chapter includes discussion by teachers of their efforts to incorporate inquiry in their classrooms based on a year-long professional development program.

Abell, Smith, and Volkman examine the roles of teacher educators as they present scientific inquiry in university courses. To do this, they lead us through three different contexts where teachers or pre-service teachers are learning about the nature of science inquiry. The contexts are (a) a physics content course specifically designed for pre-service elementary education majors, (b) an undergraduate elementary science methods course "Teaching Subject Matter to Diverse Learners", and (c) a graduate level methods course for practicing teachers at various levels. The analysis describes a variety of roles for science teacher educators that include the teacher who pushes students to generate evidence-based arguments, who "tells" scientific explanations, who scaffolds instruction for inquiry, and who assesses and evaluates student learning. Each of these create a different kind of instructional tension that challenges both instructors and students.

Holliday confronts a long-standing issue related to the broad interpretation of inquiry teaching in science, the issue of explicit versus implicit teaching practices as they apply to inquiry-related instruction. While recognizing that science teaching in general and inquiry in particular require varied forms of teaching, many authors have, perhaps unintentionally, given the impression that inquiry implies an implicit or indirect approach with students. Holliday prompts us to consider the instructional challenges of delivering the content and spirit of scientific inquiry to all students.

Part III takes up the design of curriculum that fosters students opportunity to learn inquiry and the design of assessment tasks the provide students opportunities to demonstrate their knowledge of inquiry. Sherin, Edelson, and Brown present a novel approach to analyzing curriculum that is highly relevant to the design of inquiry-oriented materials. Task-structured curricular designs are contrasted with content-structured designs. The analysis specifically highlights structural differences that compare discipline structure with problem-centered structures. Sherin and colleagues deepen the analysis by noting that some curricula, such as prior-conceptions-driven models, map out both target concepts and pathways that students may traverse while achieving intended understanding. It may be more difficult for task-structured curricula to provide an account of student learning than content-structured curricula. That having been said, they examine empirical and theoretical sides of the question of whether task-structured curricula can achieve their learning goals.

Elementary classrooms have been a perennial challenge to the establishment of a consistent and sound science curriculum. Abell and McDonald explore the operation of inquiry-oriented science in two elementary classrooms and conjecture that an inquiry-based science curriculum could be a model of reform-minded, inquiry-based curriculum and instruction across elementary school subjects. They argue for a view of elementary science that avoids both activity-driven curricula that emphasizes the action over content and text-driven curricula the emphasizes passive reading of content over meaningful activity. Their integrated vision is elaborated through two descriptions of classroom instruction one based on fourth-grade, teacher-developed curriculum and another based on a fifth-grade, published curriculum from the National Science Resources Center.

Champagne and Gummer focus on the challenges faced by classroom teachers as they work to provide student opportunity to learn what is intended by state implementations of national standards. Specifically they focus on understandings about the nature of scientific inquiry and how those understandings are communicated to teachers through school system documents. If teachers are charged with structuring opportunities to learn for students, then how clear and coherent is the language of these documents and how well aligned are state tests with the stated content? They take us through an in-depth analysis of the New York State documents describing the Standards for Science and Science Inquiry and develop the contrasting images of scientific inquiry they present to teachers.

The educational challenges of teaching and learning about the nature of science is the focus of Part IV. In his introductory chapter, Lederman provides a concrete foundation for the perspectives on nature of science evident in national reforms and the empirical research on the topic. Upon completion of the chapter the reader should be clear on the often misunderstood distinction between nature of science and scientific inquiry as well as be cognizant of the numerous overlaps. The major thrust of the chapter is to provide readers with an organizational template for the successful integration of nature of science into current curriculum and classroom instruction.

Richard Duschl is perhaps the most influential contemporary science educator for promoting the importance of nature of science and the use of history of science to teach to that end. Duschl does an excellent job of reviewing the relationship between history of science and nature of science instructional outcomes. Importantly, he discusses both the effective use of history as well as the abuse of history as a venue for teaching nature of science. Few would argue that history of science could not be a fruitful way to promote student understandings of nature of science. However, Duschl makes it quite clear that inclusion of history of science is not a universal panacea, but rather it can be abused just as any other curriculum focus.

The value of students' experiences with authentic scientific inquiry as a means to promote understandings of nature of science is the focus of this "cutting edge" chapter by Schwartz and Crawford. Over the years, most educators have accepted, untested, the claim that students come to understand nature of science naturally through experiences with inquiry. Hence, the popularity of providing both teachers and students with experiences working with active scientists. In what probably represents one of the few contemporary reviews of research on the assumed value of research experiences, Schwartz and Crawford clarify what the research actually says and they also provide clear guidance on how such experiences can be useful.

Shipman is one of handful of science professors that have consistently made an effort to promote students' understanding of nature of science and scientific inquiry while they are learning college level astronomy. He does an excellent job of describing the "history" of college level science instruction and how, in recent years, there has been a shift in instructional approach. Shipman is quite realistic in his discussion and does not recommend an overnight and total reorganization of college level instruction. Rather, his well-crafted discussion and instructional examples carefully considers the logistical constraints of college level instruction which make it quite a different matter than what is possible in K-12 instruction.

As an companion to Shipman's chapter, Abd-El-Khalick reviews the literature of college students' understandings of nature of science. He brings the most recent research to bear in his discussion of what is known, what is assumed, and what can be done to alleviate the perceived problems. In ever case, what is provided to the reader is thoroughly supported by current research. Although one would expect research support for recommendations to be the norm, the literature on nature of science and scientific inquiry is replete with unsubstantiated recommendations.

Bell probes the various aspects of nature of science as advocated throughout the literature on the topic and then grapples with the hard questions of developmental appropriateness. Teachers often are concerned about how capable their students are of understanding some of the abstract ideas labeled as nature of science. Bell provides incisive, research-based guidance on teaching nature of science and what is possible to teach younger students. He makes a strong argument for the possibility of teaching most nature of science aspects to students often thought of as too young for such curriculum topics.

How to Use the Book

The purpose of this book is to help clarify both the theoretical and practical aspects of inquiry and nature of science as well as provide some guidance relative to their inclusion in science teaching, teacher education, and research. Certainly the authors do not share a common point of view and the increased clarity we seek does not imply a simplification of these complex constructs. But discussion and debate that should follow from reading these chapters will stimulate deeper understanding. Researchers will find the contributions to theory offered by the authors to be valuable for framing studies to examine the two central constructs of the book in finer detail. In this sense the book will be useful as a scholarly resource and for readings in graduate courses. The careful examination of issues of practice offer new approaches and insights to science educators who work with teachers in their courses and in the field. Selected chapters will provide a foundational resource for teachers in workshops and in-service programs conducted by science specialists in districts or state departments of education. Ultimately it is our hope that the book as a whole will support clearer thinking about and reflection on inquiry and the nature of science as it applies to student learning.

REFERENCES

Bronowski, J. (1973). *The ascent of man*. Boston, MA: Little Brown & Company.

Brown, A. L., Bransford, J. D., Ferrara, R. A. & Campione, J. C. (1983). Learning, remembering, and understanding. In Flavell, J. H. & E. M. Markman. *Cognitive Development, Vol. 3* of P. H. Mussen (Ed.). *Handbook of child psychology, 4th Edition*. New York, NY: John Wiley & Sons.

Brown, A. L. & Campione, J. C. (1990). Interactive learning environments and the teaching of science and mathematics. In M. Gardner, J. G. Greeno, F. Reif, A. H. Schoenfeld, A. DiSessa, & E. Stage (Eds.). Toward a scientific practice of science education. Hillsdale, NJ: Lawrence Erlbaum Associates, Publishers.

Gardner, M., Greeno, J. G., Reif, F., Schoenfeld, A. H., DiSessa, A, & Stage, E. (1990). Toward a scientific practice of science education. Hillsdale, NJ: Lawrence Erlbaum Associates, Publishers.

Harms, N. & Yager, R. (Eds.). (1980). What research says to the science teacher (Vol. 3). Washington DC: National Science Teachers Association (ERIC Document Reproduction Service No. ED 205367).

Reiser, B., Tabak, I., Sandoval, W. A., Smith, B. K., Steinmuller, F., Leone, A. J. (2001). BGuILE: Strategic and conceptual scaffolds for scientific inquiry in biology classrooms. In S. M. Carver & D. Klahr (Eds.). *Cognition and instruction: Twenty-five years of progress*. Medwah, NJ: Lawrence Erlbaum Associates, Publishers.

Schwab, J. J. (1962). The teaching of science as enquiry. In J. J. Schwab & P. F. Brandwein, *The teaching of science*, Cambridge, MA: Harvard University Press.

United States Bureau of Education (1893). Report of the committee on secondary school studies. Washington, D.C.: Government Printing Office.

Welch, W. W. (1981). Inquiry and school science. *Science Education, 65*, 33-50.

White, B. Y., & Frederickson, J. (1998). Inquiry, modeling, and metacognition: Making science accessible to all students. *Cognition and Instruction.* 16 (1), 3-118.

CHAPTER 1

RODGER W. BYBEE

SCIENTIFIC INQUIRY AND SCIENCE TEACHING

SCIENTIFIC INQUIRY AND SCIENCE TEACHING

Using the Internet, a science teacher provided students access to hundreds of ultraviolet (UV) photographic images of the outer atmosphere of Earth. Upon examining the images, students discovered numerous dark "holes" in the outer atmosphere. The teacher posed a general question for the students--What could possibly explain the "dark holes"? The teacher asked the students to propose explanations for the dark holes, and then afterward, directed them to support their explanations with scientific knowledge.

After several days of study and research, the students were divided on their explanations about the origin of these atmospheric holes. One group proposed that the holes are caused by numerous small comets that enter Earth's outer atmosphere from elsewhere in the solar system. A second group argued that the holes are not natural phenomena but the result of several kinds of electronic noise.

- Are these students engaged in scientific inquiry?
- Are the teaching strategies appropriately described as inquiry?

As you think about your response to these questions try to clarify the criteria you used to define what counts as inquiry-oriented science teaching; and, by extension, what is not inquiry-oriented science teaching. Did you center on the role of students working from the photographic images? How did you consider the teacher's role? Specifically, the fact that the teacher (and not the students) posed the general question.

Did you judge this as an interesting activity, but not inquiry-oriented? Perhaps you used hands-on investigations as essential criteria for inquiry; and analyzing photographic images is not "hands-on."

Perhaps more information would help you answer the questions about scientific inquiry and science teaching. As students analyzed the images and completed their library research for scientific knowledge, they proposed explanations such as the following. Group one argued that the holes are caused by comets. Comets are composed mainly of ice that vaporizes when it enters Earth's atmosphere, creating a cloud of water vapor. The water vapor absorbs UV light from the Sun, which reflects off the Earth and back toward the spacecraft sensors, creating the dark holes seen in the images.

Based on the same UV photographic images and opportunities for library research, the second group proposed a different explanation: The holes result from different kinds of electronic noise. The group reasoned as follows: Cosmic rays can

L.B. Flick and N.G. Lederman (eds.) Scientific Inquiry and Nature of Science, 1-14.

interact with data systems on the spacecraft and cause a specified number of elements (pixels) of an image to be dark. Further, lightning near the receiver and other sources of electronic noise can interfere with the transmission, causing the Earth-based computer to interpret the images as being dark.

Does this information change your original response to the questions: Are the students engaged in scientific inquiry? Are the teaching strategies appropriately described as inquiry? If you changed your view, how do you justify the change? If you maintained your original position, why do you think it is the best response?

I designed this introductory discussion to engage your ideas about scientific inquiry and science teaching, so there is no need to reach closure on these questions. Rather, the intention is to open the discussion about the implications of a contemporary view of scientific inquiry for teaching.

A discussion of scientific inquiry and teaching strategies brings together several important domains of science education. First, one must consider the essential features of scientific inquiry, since those features may be applicable to learning outcomes and teaching strategies. Second, the topic requires recognizing what we know about how students learn science. Finally, the aforementioned domains must be synthesized into the science curriculum in a way that enhances the learning outcomes and is understandable and usable by classroom teachers. Discussion of these three domains forms the core of this chapter.

SCIENCE: A WAY OF EXPLAINING THE NATURAL WORLD

How scientists know and explain the natural world and what they mean by explanation and knowledge are both directly related to the processes, methods, and strategies by which they develop and propose explanations. What are the basic elements that underlie science as a way of knowing and explaining the natural world? Succinctly, scientific knowledge of the natural world must be based on observations and experimental data, that explanations about the way the world works must be evaluated against empirical evidence.

Many science textbooks and some teaching instill the notion that science proceeds as a prescribed method. Students learn that scientists begin with observations. Based on their observations, scientists state a hypothesis. Often the hypothesis takes the form of an "If-Then" proposition. That is, the hypothesis has a predictive quality, which can be challenged or confirmed through further observation of experimentations. If the observations or experiments confirm the prediction, the hypothesis endures and investigations continue. The scientific method, as presented, is logical, objective, and impersonal. Textbooks and teaching leave students with the view that all of science proceeds in much the same way. To the degree that students and the public have any sense of inquiry and the nature of science, this is it. Unfortunately, this view is neither entirely correct nor absolutely wrong. Perhaps some scientists might conduct their work in this manner. Certainly they report their work this way at scientific meetings and in journals.

History offers numerous examples where scientists did not proceed as described. Einstein, for example, did not begin with observations and experiments. A singular approach to a scientific method gives no recognition to the diversity of approaches among scientific disciplines. Compare, for example, particle physics and ecology, or one might even consider differences within a discipline. Evolutional biologists consider population changes during vast time scales, while molecular biologists focus on understanding the function of cells in their current evolutional form.

Although differences in the concepts that form a discipline and the questions that guide inquiries are important, there is a set of common goals for any scientific inquiry. All scientists are trying to improve their understandings and explanations about the natural world. Further, they agree that using empirical data reasoning in constructing explanations, avoiding bias, and presenting explanations for skeptical review are all "rules" of the scientific process. Note that this set of commonly agreed upon rules shifts from the aforementioned perspective of using the correct method to a view that centers on constructing explanations through the use of data, imagination and logic, and public review. Science is a human activity. Scientists may share some fundamental assumptions such as the essential place of data, the importance of logic, and the need to base their explanations on evidence, but scientists vary in their talents, imagination, intuition, and courage. Given the same experimental results, two individuals or groups of scientists may form different explanations (such as the students in the introduction). Some scientists may provide a very safe and secure explanation for experimental results, while another will give a bold and courageous explanation. This discussion leads to the conclusion that we need to provide students with a broader view of scientific inquiry and the nature of science.

INQUIRY: A CONTENT GOAL AND TEACHING METHODS

One educational lesson of the standards-based reform, including the National Science Education Standards (NSES), has been to differentiate the aims, goals, and outcomes of education from the means, methods, and techniques of teaching. Logic suggests that one begin designing programs and recommending teaching methods by first identifying the educational outcomes, then implementing the appropriate means for achieving the outcomes, and assessing the degree to which students have attained them. This is a simple and obvious point, but one that seems clouded and obscure in many educational discussions, especially ones about inquiry in the science classroom.

The NSES includes both the scientific perspectives, discussed in the content standards "Science as Inquiry" and the teaching perspective, discussed in the "Science Teaching Standards." Before continuing the discussion, an example of the standards will help clarify this presentation of inquiry.

- As a result of activities in grades 5-8, all students should develop:
- Abilities necessary to do scientific inquiry.
- Understandings about scientific inquiry.

This is the actual standard for grades 5-8. The standards for K-4 and 9-12 are similar. The standard also includes a clarification and definition of what is meant by abilities and understandings. Table 1 includes summary statements of the abilities and understandings for "Science as Inquiry." Appendix A provides a complete narrative for "Science as Inquiry."

Table 1. Science as Inquiry, Grades 5-8

Abilities Necessary to Do Scientific Inquiry	Understandings about Scientific Inquiry
• Identify questions that can be answered through scientific investigation.	• Different kinds of questions suggest different kinds of scientific investigations.
• Design and conduct a scientific investigation.	• Current scientific knowledge and understanding guide scientific investigations.
• Use appropriate tools and techniques to gather, analyze, and interpret data.	• Mathematics is important in all aspects of scientific inquiry
• Develop descriptions, explanations, predictions, and models using evidence.	• Technology used to gather data enhances accuracy and allows scientists to analyze and quantify investigation results.
• Think critically and logically to make the relationships between evidence and explanation.	• Scientific explanations emphasize evidence, have logically consistent arguments, and use scientific principles, models, and theories.
• Recognize and analyze alternative explanations and predictions.	• Science advances through legitimate skepticism.
• Communicate scientific procedure and explanations.	• Scientific investigations sometimes result in new ideas and phenomena for study, generate new methods or procedures for investigation, or develop new techniques to improve the collection of data.
• Use mathematics in all aspects of scientific inquiry.	•

For a different view, namely, inquiry as teaching strategy, Table 2 summarizes the key ideas from the Science Teaching Standards. The first standard A refers to an inquiry-based program for students. Here, in order to be internally consistent, the referent must be the standards for "Science as Inquiry" because that is the content or outcome of teaching and learning. The teaching standards express various other factors associated with that goal; that is, facilitating, learning, assessing, providing time, space, and resources, developing communities of learners, and designing school science programs. Combining content standards (Table 1) with the teaching standards (Table 2) describes the goals of inquiry in the science classroom and the various means of achieving those goals.

Table 2. Science Teaching Standards—Key Ideas

Key Ideas
A. Teachers of science plan an inquiry-based science program for their students.
B. Teachers of science guide and facilitate learning.
C. Teachers of science engage in ongoing assessment of their teaching and of student learning.
D. Teachers of science design and manage learning environments that provide students with time, space, and resources needed for learning science.
E. Teachers of science develop communities of science learners that reflect the intellectual rigor of scientific inquiry and the attitudes and social values conducive to science learning.
F. Teachers of science actively participate in the ongoing planning and development of the school science program.

The vision of inquiry presented in the NSES has the dimensions just described. For example, early in the NSES one can find this statement:

> Scientific inquiry refers to the diverse ways in which scientists study the natural world and propose explanations based on the evidence derived from their work. Inquiry also refers to the activities of students in which they develop knowledge and understanding of scientific ideas, as well as an understanding of how scientists study the natural world. (NRC, 1996, p. 23.)

This quotation affirms the earlier point of grounding inquiry in science and the work of scientists. It also presents an educational connection; namely, that inquiry has something to do with the activities of students, and by implication, the strategies of teaching. The quotation reminds us that science teaching also has something to do with developing students' understanding of scientific ideas.

Engaging students in activities based on the inquiry as presented in the NSES should contribute to greater:

- understanding of scientific concepts,
- appreciation of "how we know" what we know in science,
- understanding of the nature of science,
- development of skills necessary to become independent inquirers about the natural world, and
- disposition to use the skills, abilities, and attitudes associated with science (NRC, 1996, p. 105).

To conclude this section, two specific aims of science teaching stand out in this discussion of inquiry. As a result of their educational experiences, students should develop an understanding of the defining qualities of science as a way of knowing and explaining the natural world. A second aim complements this goal of science teaching; namely, as a result of their experiences, students should develop some cognitive abilities and manipulative skills associated with scientific inquiry. Related to these aims that center on inquiry and the nature of science are associated aims of understanding science concepts and an appreciation of inquiry and the nature of science.

LEARNING: HOW STUDENTS FORM SCIENTIFIC EXPLANATIONS

In this discussion, *explanations* are used in a distinct way. Before clarifying this particular use of the term, supposedly, when one asks for an explanation, one might have several different things in mind. If a student is late for science class, the teacher might ask him to explain why he is late. Here, explanation is used as a justification. Take another situation: a student might ask for an explanation of the mathematics involved in a scientific investigation. Here, explanation is used as a procedure or how to do something. In contemporary society, many people say they have an explanation for something when, in fact, they express a personal opinion. A variation on the latter occurs when individuals indicate they have a theory about something. For most instances, the individual does not have a theory (or explanation) in the scientific use of the term. Rather, the individual has a personal explanation.

Use of the term explanation here is more distinct. Consider a situation where you are told that the number of alcohol-related deaths decreased in a certain part of the state in which you live. You might reasonably consider why this change occurred and ask for an explanation. You are asking for an account of why something happens. This is the meaning of explanation proposed in the following discussion. To be specific, explanation is an account of why something happens or continues to happen. Further, the explanatory account uses data and is a reasoned proposition based on data.

Explanations are central to knowing about the natural world. This assumption applies to science and to students learning science. Explanations are ways to learn about what is unfamiliar by relating it to what is already familiar. Although this statement presents a paradox, resolution of the paradox lies in our continuing quest to understand how people learn (Bransford et al., 1999).

A recent synthesis of research by Bransford et al, on how people learn suggests the following:

- students build their scientific understanding on what they already know and believe;
- students formulate new scientific knowledge by modifying and refining their current concepts and adding new concepts to what they already know;
- understanding science is more than knowing facts; it involves placing and retrieving facts in a conceptual framework;
- learning is mediated by the social environment in which learners interact with others;
- the ability to apply knowledge to novel situations (transfer of learning) is affected by the degree to which students learn science with understanding in a variety of contexts; and
- effective learning requires that students take control of their own learning through reflection and self-assessment (Bransford et al., 1999).

Our understanding of learning presented in the National Research Council report *How People Learn* challenges educators, in particular curriculum developers, classroom teachers, and assessment specialists to reconsider textbooks, teaching, and tests. Considering the theme of scientific inquiry and the nature of science, educators have some obligation to recognize students' current conceptions (i.e., misconceptions) about inquiry and the nature of science. For example, middle-school students use personal experiences as evidence to justify explanations. When asked to use evidence as support for an explanation, students evaluate evidence incorrectly and apply it inconsistently, or not at all (Kuhn et al., 1988).

Most high-school students have difficulty correctly analyzing arguments. They accept inadequate sample size, confuse correlation with causation, and do not recognize significant and insignificant differences (Jungwirth & Dreyfus, 1990, 1992). (For further information on students' understanding, see: Ogborn, Kress, Martins, & McGillicuddy, 1996; Driver, Leach, Miller, & Scott, 1996; Lederman, 1992.)

This research on students' understanding and abilities of inquiry and learning suggests that educators should use, as the basis for curricular design, a conceptual framework for scientific inquiry and the nature of science. The National Science Education Standards (NRC, 1996) and Benchmarks for Science Literacy (AAAS, 1993) establish such a framework. The Atlas for Science Literacy (AAAS, 2001) provides further background for curricular design in the area of inquiry and the nature of science. Further, contemporary understanding of how students learn has clear implications for science teaching.

TEACHING: USING STRATEGIES THAT UNITE SCIENCE AND LEARNING

This discussion leads to the second part of the title—the theme of science teaching. Here, we confront the problem of how to structure science teaching so it 1) accommodates the goals of developing the abilities and understandings of scientific inquiry, 2) enhances the learning of science concepts, and, 3) recognizes the realities of classrooms and schools.

In the section *Inquiry: A Content Goal*, it was implied that students should learn the content and abilities of scientific inquiry as described in the national standards. Here, it is proposed that how much scientific inquiry students learn is directly influenced by how they are taught. The key idea, of course, is how they are taught. Most educators have not addressed issues of teaching in a clear, direct, and meaningful way. Arguably, one need is for a systematic, coherent, and focused approach to teaching. Although strategies may vary, the orientation and sequence of lessons should systematically apply our understanding of student learning. The lessons should be coherent in that they have orderly and logical parts that contribute in instructional sequence and are understandable and usable by classroom teachers. Finally, the lessons should clearly focus on the learning outcomes of inquiry and/or science concepts.

An instructional model with five stages, the 5Es—engage, explore, explain, elaborate, and evaluate—structures the learning experience for students. This instructional model is congruent with contemporary learning theory that suggests students learn best when they are allowed to develop their understanding of concepts over time. The focus is on teaching for meaning and understanding, and on reducing, but not eliminating, the emphasis on memorizing facts, principles, and the definitions of vocabulary words.

Contemporary learning theory suggests that students come to learning situations with knowledge and explanations for their world. As students form their understanding, they link new information with the knowledge they bring to the learning experience. Therefore, it is important that curricula and teachers provide well-articulated learning experiences that will help students make the connections between new concepts and the knowledge they already have.

Meaningful learning does, however, take time. If students are truly to understand science and develop useful skills, they cannot simply read, memorize, and recite isolated bits of information and vocabulary words. They must take time to wrestle with new ideas, to discuss their ideas with their classmates and teachers, to collect data and use it data to draw conclusions, to practice skills, and finally relate what they are learning to the world around them. The 5E instructional model is summarized in Figure 1.

ENGAGEMENT: This phase of the instructional model initiates the learning task. The activity should (1) make connections between past and present learning experiences, and (2) anticipate activities and focus students' thinking on the learning outcomes of current activities. The student should become mentally engaged in the concept, process, or skill to be explored.

EXPLORATION: This phase of the teaching model provides students with a common base of experiences within which they identify and develop current concepts, processes, and skills. During this phase, students actively explore their environment or manipulate materials.

EXPLANATION: This phase of the instructional model focuses students' attention on a particular aspect of their engagement and exploration experiences and provides opportunities for them to verbalize their conceptual understanding, or demonstrate their skills or behaviors. This phase also provides opportunities for teachers to introduce a formal label or definition for a concept, process, skill, or behavior.

ELABORATION: This phase of the teaching model challenges and extends students' conceptual understanding and allows further opportunity for students to practice desired skills and behaviors. Through new experiences, the students develop deeper and broader understanding, more information, and adequate skills.

EVALLUATION: This phase of the teaching model encourages students to assess their understanding and abilities and provides opportunities for teachers to evaluate student progress toward achieving the educational objectives.

Figure 1. The 5E instructional model

TOWARD AN UNDERSTANDING OF INQUIRY IN SCIENCE CLASSROOMS

The opening segment in which a science teacher used UV photographic images and had students develop explanations for dark holes provides the context for this discussion. At the conclusion of that example, it was asked if students were engaged in scientific inquiry. It also was asked if the teaching strategies could be appropriately described as inquiry.

Inquiry as a teaching strategy should capture the spirit of scientific investigation and the development of knowledge about the natural world. As an approach to teaching and learning, the essential features of inquiry should not necessarily be defined as "activity-based, hands-on" or other approaches to teaching. Too often, such terms serve as the ends; the goals as though doing activities in themselves were the aim and the defining quality of our inquiry approach to teaching. This section outlines several defining features of inquiry in the science classroom and describes continua of classroom experiences that accommodate the essential features of an inquiry approach to teaching and learning. These two aspects of inquiry are based on *Inquiry and the National Science Education Standards* (NRC, 2000).

First, the essential features of inquiry in the science classroom:

- learner engaged in a scientifically oriented question,
- learner gives priority to evidence in responding to the question,
- learner uses evidence to develop an explanation,
- learning connects the explanation to scientific knowledge, and
- learner communicates and justifies the explanation.

Several points are worth noting about these features of inquiry. First, they all center on the learner's mental activity. Second, this activity clearly has a scientific orientation and has as its aim, developing scientific explanations. Third, the learner makes a connection to current scientific knowledge. Finally, there are elements of justification and communication. The essential features highlight the learner but do not describe specific teaching methods.

The second point gets much closer to teaching strategies. However, you will note that it does not describe a specific set of teaching methods that define inquiry. Rather, it continues an emphasis on the learner and expands the essential features to continua of classroom experiences. The following chart (see Figure 2) is an explicit attempt to move our thinking beyond an either/or position relative to inquiry. For example, inquiry deserves a deeper understanding than—hands-on, kit-based, process-oriented; or even, hands-on, minds-on.

To be clear, the position described here acknowledges the importance of learning-oriented questions, but it does not use this as the exclusive criteria for defining inquiry. The following chart clarifies a variety of experiences on a continuum where some are more teacher-centered and others more student-centered. Critical decisions about what strategies and methods best facilitate the experiences described will have to be left to the professionals who design curriculum materials, and those teachers in classrooms.

The perspective on scientific inquiry and science teaching presented here challenges many current notions of teaching science as inquiry. Many of those notions reveal a typological or a single, preconceived idea about scientific inquiry and teaching strategies. For instance, inquiry is equated with "hands-on" or "student-originated activity." Perspectives such as these often reveal a single teaching strategy or method as the defining characteristics of inquiry. The perspective presented here has several defining characteristics. Inquiry in the science classroom should:

- focus on the learning, more than pedagogy,
- focus on learning outcomes, more than the experience of students, and
- focus on science and learning theory, more than teaching methods.

1. Learner engages in scientifically oriented questions	Learner poses a question	Learner selects among questions, poses new questions	Learner sharpens or clarifies question provided by teacher, materials, or other source	Learner engages in question provided by teacher, materials, or other source
2. Learner prioritizes evidence in responding to questions	Learner determines what constitutes evidence and collects it	Learner directed to collect certain data	Learner given data and asked to analyze	Learner given data and told how to analyze
3. Learner formulates explanations from evidence	Learner formulates explanation after summarizing evidence	Learner guided in process of formulating explanations from evidence	Learner given possible ways to use evidence to formulate explanation	Learner provided with evidence
4. Learner connects explanations to scientific knowledge	Learner independentl y examines other resources and forms the links to explanations	Learner directed toward areas and sources of scientific knowledge	Learner given possible connections	
5. Learner communicate s and justifies	Learner forms reasonable	Learner coached in development	Learner provided broad	Learner given steps and procedures for

explanations	and logical argument to communicate explanations	of communi-cation	guidelines to use and sharpen communica-tion	communica-tion

Figure2. Essential features of classroom inquiry

Source: National Research Council. 2000. *Inquiry and the National Science Education Standards, A Guide for Teaching and Learning*. Washington, DC: National Academy Press, p. 29.

CONCLUSION

Release of the *National Science Education Standards* (NRC, 1996) has engaged professional discussions of inquiry in the science classroom. These discussions have different perspectives that categorized inquiry as: strategies for teaching science, models for learning science, and content for science education. Although the profession has accepted these views, it seems that we have overemphasized the teaching strategies and underemphasized the scientific contemporary learning theory as the basis for including inquiry in school programs.

In this brief essay, some contemporary views of inquiry were challenged and the *National Science Education Standards* were used as the basis for an alternative perspective. The alternative view of inquiry is based in science and provides a definition that incorporates science content and teaching strategies. The perspective has practical implications for teachers. Specifically, a systematic approach to instruction and an expanded view of scientific inquiry and teaching strategies.

REFERENCES

American Association for the Advancement of Science. (1993). *Benchmarks for Science Literacy*. New York: Oxford Press.

American Association for the Advancement of Science. (2001). *Atlas for Scientific Literacy*. Washington, DC: American Association for the Advancement of Science and National Science Teachers Association.

Bransford, J., Brown, A., Cocking, R. (1999). *How People Learn: Brain, Mind, Experience, and School*. Washington, DC: National Academy Press.

Driver, R., Leach, J., Millar, R. & Scott, P. (1996). *Young People's Images of Science*. Philadelphia, PA: Open University Press.

Jungwirth, E. & Dreyfus, A. (1990). Identification and acceptance of a posteriori causal assertions invalidated by faulty inquiry methodology: An international study of curricula expectations and reality. In D. Herget (Ed.), *More History and Philosophy of Science in Science Teaching* (pp. 202-211). Tallahassee, FL: Florida State University.

Jungwirth, E. & Dreyfus, A. (1992). After this, therefore because of this: One way of jumping to conclusions. *Journal of Biological Education, 26*, 139-142.

Kuhn, D., Amsel, E., & O'Loughlin, M. (1988). *The Development of Scientific Thinking Skills*. San Diego, CA: Academic Press.

Lederman, N.G. (1992). Students' and teachers' conceptions of the nature of science: A review of the research. *Journal of Research in Science Teaching. 29*, 331-359.

National Research Council (NRC). (1996). *National Science Education Standards.* Washington, DC: National Academy Press.
National Research Council (NRC). (2000). *Inquiry and the National Science Education Standards: A Guide for Teaching and Learning*. Washington, DC: National Academy Press.
National Science Foundations, Foundations-Volume 2. (1999). *Inquiry: Thoughts, Views, and Strategies for the K-5 Classroom.* Arlington, VA: Division of Elementary, Secondary, and Informal Education, NSF.
Obborn, J., Kress, G., Martins, I. & McGillicuddy, J. (1996). *Explaining Science in the Classroom.* Philadelphia, PA: Open University Press.
Spickler, T. and McCreary, C. (1999). *Making the Case for Teaching Science Using a Hands-On, Inquiry-Based Approach.* Pittsburgh, PA: Bayer Corporation.
St. John, M. (1999). *Wait, Wait! Don't Tell Me! The Anatomy and Politics of Inquiry.* (The 1998 Catherine Malong Memorial Lecture). New York: The City College Workshop Center, City College School of Education.

APPENDIX A
FUNDAMENTAL ABILITIES AND CONCEPTS THAT UNDERLIE "SCIENCE AS INQUIRY" (GRADES 9-12)

This discussion is from the
National Science Education Standards National Research Council 1996

19.1 Abilities Necessary to do Scientific Inquiry

Identify Questions and Concepts that Guide Scientific Investigations. Students should formulate a testable hypothesis and demonstrate the logical connections between the scientific concepts guiding a hypothesis and the design of an experiment. They should demonstrate appropriate procedures, a knowledge base, and conceptual understanding of scientific investigations.

Design and Conduct Scientific Investigations. Designing and conducting a scientific investigation requires introduction to the major concepts in the area being investigated, proper equipment, safety precautions, assistance with methodological problems, recommendations for use of technologies, clarification of ideas that guide the inquiry, and scientific knowledge obtained from sources other than the actual investigation. The investigation may also require student clarification of the question, method, controls, and variables; student organization and display of data; student revision of methods and explanations; student organization and display of data; student revision of methods and explanations; and a public presentation of the results with a critical response from peers. Regardless of the scientific investigation performed, students must use evidence, apply logic, and construct an argument for their proposed explanations.

Use Technology and Mathematics to Improve Investigations and Communications. A variety of technologies, such as hand tools, measuring instruments, and calculators, should be an integral component of scientific investigations. The use of computers for the collection, analysis, and display of data is also a part of this standard. Mathematics plays an essential role in all aspects of an inquiry. For example, measurement is used for posing questions, formulas are used for developing explanations, and charts and graphs are used for communicating results.

Formulate and Revise Scientific Explanations and Models Using Logic and Evidence. Student inquiries should culminate in formulating an explanation or model. Models should be physical, conceptual, and mathematical. In the process of answering the questions, the students should engage in discussions and arguments that result in the revision of their explanations. These discussions should be based on scientific knowledge, the use of logic, and evidence from their investigation.

Recognize and Analyze Alternative Explanations and Models. This aspect of the standard emphasis the critical abilities of analyzing an argument by reviewing current scientific understanding, weighing the evidence, and examining the logic so as to decide which explanations and models are best. In other words, although there may be several plausible explanations, they do not all have equal weight. Students should be able to use scientific criteria to find the preferred explanations.

Communicate and Defend a Scientific Argument. Students in school science programs should develop the abilities associated with accurate and effective communication. These include writing and following procedures, expressing concepts, reviewing information, summarizing data, using language appropriately, developing diagrams and charts, explaining statistical analysis, speaking clearly and logically, constructing a reasoned argument, and responding appropriately to critical comments.

19.2 Understandings about Scientific Inquiry

- Scientists usually inquire about how physical, living, or designed systems function. Conceptual principles and knowledge guide scientific inquiries. Historical and current scientific knowledge influence the design and interpretation of investigations and the evaluation of proposed explanations made by other scientists.
- Scientists conduct investigations for a wide variety of reasons. For example, they may wish to discover new aspects of the natural world, explain recently observed phenomena, or test the conclusions of prior investigations or the predictions of current theories.
- Scientists rely on technology to enhance the gathering and manipulation of data. New techniques and tools provide new evidence to guide inquiry and new methods to gather data, thereby contributing to the advance of science. The accuracy and precision of the data, and therefore the quality of the exploration, depends on the technology used.
- Mathematics is essential in scientific inquiry. Mathematical tools and models guide and improve the posing of questions, gathering data, constructing explanations and communicating results.
- Scientific explanations must adhere to criteria such as: a proposed explanation must be logically inconsistent; it must abide by the rules of evidence; it must be open to questions and possible modifications; and it must be based on historical and current scientific knowledge.
- Results of scientific inquiry—new knowledge and methods—emerge from different types of investigations and public communication among scientists, in communicating and defending the results of scientific inquiry, arguments must be logical and demonstrate connections between natural phenomena, investigations, and the historical body of scientific knowledge. In addition, the methods and procedures that scientists used to obtain evidence must be clearly reported to enhance opportunities for further investigation.

PART I: HISTORICAL AND CONTEMPORARY EDUCATIONAL CONTEXTS

CHAPTER 2

GEORGE E. DEBOER

HISTORICAL PERSPECTIVES ON INQUIRY TEACHING IN SCHOOLS

INTRODUCTION

There is a tendency to treat inquiry teaching as if it were something new and innovative, a recently invented approach to science teaching. But in one form or another, inquiry teaching has been part of the educational landscape at least since the middle of the nineteenth century (Bybee and DeBoer, 1994; DeBoer, 2001). The purpose of this chapter is to review the history of inquiry teaching to clarify the various meanings that this pedagogical approach has had. The term "scientific inquiry" will be used to refer to the general process of investigation that scientists use as they attempt to answer questions about the natural world, and the term "inquiry teaching" will be used to refer to pedagogical approaches that model aspects of scientific inquiry. Inquiry teaching mirrors scientific inquiry by emphasizing student questioning, investigation, and problem solving. Just as scientists conduct their inquiries and investigations in the laboratory, at field sites, in the library, and in discussion with colleagues, students engage in similar activities in inquiry-based classrooms.

Science, as practiced by scientists and as studied in classrooms, is both process and product. It is both a body of richly interconnected observations and interpretations regarding the natural world, and it is a set of procedures and logical rules that guide those observations and interpretations. The same is true of classroom science. Science can be studied for its interconnected concepts, but science can also be practiced in the classroom in ways similar to those used by scientists themselves. Inquiry teaching uses the general processes of scientific inquiry as its teaching methodology. And just as scientists seek to understand the natural world through their investigations, students in inquiry classrooms try to advance their understanding of the principles and methods of science through theirs. It is important to note, however, that inquiry teaching does not require students to behave exactly as scientists do. Scientific inquiry is simply a metaphor for what goes on in an inquiry-based classroom.

In addition to being a model for teaching science, the processes of science themselves can be an object of study, and learning activities can be explicitly organized to teach the nature of science as process. For example, in studying how science is done, students might learn how to control variables in an experimental design; or they might learn what is required to make claims regarding one variable influencing or causing a change in another variable; or they might learn about the

L.B. Flick and N.G. Lederman (eds.) Scientific Inquiry and Nature of Science, 17-35.

importance of mathematics in the development of scientific knowledge. These are all examples of scientific inquiry being an object of study. The fact that the term "inquiry teaching" is used in different ways has sometimes led to confusion about what inquiry teaching is. Does inquiry teaching suggest a pedagogy that is modeled after the investigative nature of science or does it suggest a content to be studied, i.e., the nature of science? In this chapter, the discussion will be limited to inquiry teaching as pedagogy. When scientific inquiry becomes an object of study it falls under the category of the nature of science (NOS), a topic that is being dealt with by other contributors to this volume.

It should also be noted that inquiry teaching and teaching *about* inquiry do not necessarily go hand-in-hand. Just because scientific inquiry is the thing being studied does not mean that inquiry teaching is being practiced. In fact, the nature of scientific inquiry can be taught in very non inquiry-oriented ways. Teachers can provide students with examples of scientific investigations, explain the logic of these studies, and show students how evidence was used to answer the question being raised in the inquiry. Teachers can introduce the language of scientific inquiry by describing the differences between theories, hypotheses, evidence, conclusions, observations, and inferences. All of these things can be accomplished using traditional, teacher-directed methods. Of course, teachers can also introduce the methods of science through inquiry-oriented activities, and students can develop an understanding of the processes of science by engaging in actual scientific inquiries. The important thing to realize is that inquiry teaching and teaching about scientific inquiry are not necessarily one and the same, and using the term *inquiry teaching* for both of them often leads to confusion.

The Nature of Scientific Inquiry

Before beginning this historical overview of inquiry teaching, it is useful to briefly review some of the more general characteristics of scientific inquiry that are related to inquiry teaching as well as some of the reasons inquiry teaching is used as a pedagogical approach. Although it is impossible to provide an exact definition of scientific inquiry or elaborate to any significant degree the nature of science here, it is at least worthwhile to state some of the more general aspects of scientific inquiry, especially as they relate to inquiry teaching. In its most essential form, scientific inquiry is what scientists do when they investigate the natural world. According to the *National Science Education Standards* (National Research Council, 1996): "Scientific inquiry refers to the diverse ways in which scientists study the natural world and propose explanations based on the evidence derived from their work" (p. 23). Inquiry means looking into or investigating something. It is a search, an attempt to find out. According to St. John (1999) "...inquiry entails the perception of depth. It has the quality of penetrating into something, going deeper, so you can see what you haven't been able to see before. When you begin an inquiry, you are deliberately setting out to search for what you don't know" (p. 109). How that search is conducted, the rules concerning what constitutes evidence, the logic that is applied in drawing conclusions, the special techniques that are used to gather information—all this is developed by a community of practicing scientists and described by philosophers of science. At one end of the spectrum, using an inquiry approach to teaching may simply mean modeling the inquisitive, questioning nature of science, and at the other it may mean learning how to apply very detailed procedures and protocols of a particular field of science.

Scientific Inquiry as a Model for Pedagogy

There are many purposes that can be served by having students model scientific inquiry either in a general or in a more focused way. One purpose might be the preparation of future scientists. Prospective scientists need to engage in scientific investigations and gain experience as scientists-in-training. They need to become increasingly familiar with the methodologies used in the various disciplines, with appropriate equipment, and with ways of taking and recording measurements. They need to learn how to collect data, how to analyze it, and how to test it against their predictions. They need to learn what constitutes evidence that will support or refute their predictions. Laboratories, as well as natural field sites, are places where students as potential future scientists can collect data that is important to their investigations. How early this initiation into the work of scientists should begin and how extensive early training in the actual practice of science should be are questions that are open to debate and discussion.

A second purpose that can be served by having students model scientific inquiry is the development of citizens who may not become scientists themselves but who will be autonomous, independent thinkers. As informed citizens they should have an inquisitive and questioning attitude and a faith in their ability to ask important questions and seek answers to those questions, be able to solve problems by drawing together necessary resources, and work alone or with others on projects to see them through to completion. What is needed pedagogically is an open and supportive learning environment where students receive assistance in generating questions that are interesting and important to them and in designing productive strategies for investigating those questions. Student inquiries and investigations may be laboratory based or archival, and they can focus on science-related societal issues or on the natural world itself. Because the main purpose is to encourage autonomy and independence of thought, students need to be given as much practice as possible in generating their own questions and in devising their own strategies for answering those questions. Inquisitive students can find answers to their questions through conversations with their teacher, from print and electronic resources, or from investigations of the natural world. The goal is to develop citizens with an inquiring attitude and the skills needed to search for answers to questions they feel are important.

Inquiry teaching can also be used as a pedagogical tool for accomplishing other goals. For example, it has long been recognized that the direct, hands-on experience that accompanies scientific investigations is an important way to strengthen not only an understanding of the methods of science, but also the content and principles of science. Hands-on experiences provide students with a deeper understanding of the way the world works as well as a way to personally confirm and verify the principles of science. Scientific investigations put the learner in direct contact with the natural world so that the phenomenon in question becomes more real to the student. Scientific inquiry, especially when it involves direct hands-on investigations, has often been promoted as a way to support an understanding of scientific content.

Another way that inquiry teaching can be used indirectly to support other educational goals is by taking advantage of inquiry teaching's motivational power. There is a psychological justification for having students actively involved in hands-on inquiries and investigations known as self-determination theory (Deci, 1975; Deci & Ryan, 1985). According to this theory, individuals have a need for competence, autonomy, and relatedness. When these needs are met, students' intrinsic motivation is increased and their efforts are more effective and meaningful. Inquiry-oriented classrooms give students more control over their own learning. When students can select questions that are interesting to them, when they can be active in the design of investigations, and when they can interact with their classmates in doing the work and in reporting and discussing the results, then they develop a greater sense of control and autonomy, and the activity becomes more enjoyable to them. In addition, children and adolescents have social instincts that are satisfied by working collaboratively with their classmates on problems of mutual interest, and the physical activity that usually accompanies scientific investigations adds to the intrinsic satisfaction that students experience when compared to more passive reception-style learning.

Given the range of benefits that can come from inquiry teaching, it is no surprise that educational leaders have forcefully promoted it as a valuable part of the science program over the years. What is surprising is that inquiry teaching has not met with more success. (See Hurd, et al., 1980; Welch, Klopfer, Aikenhead, and Robinson, 1981; Costenson and Lawson, 1986; and Bybee, 2000). Some have blamed this failure on a misunderstanding of the purposes of inquiry teaching or a misconception concerning exactly what inquiry teaching is, especially given that there are so many ways that students can engage in inquiry activities and so many different reasons that justify its use. Perhaps the reason inquiry teaching has been so difficult to implement is because educators have not paid enough attention to the basic purposes which they want inquiry-oriented classroom activities to serve or the outcomes they wish to achieve. The modest success that inquiry teaching has enjoyed may be due to a failure by educators to pay careful attention to their own educational goals and a failure to employ just those strategies that are most suited to their own particular situations. There is no single method of inquiry teaching (NRC, 2000). The exact way that scientific inquiry is used as a guide to classroom teaching depends on the specific goals and purposes a teacher has. Inquiry teaching is a broad array of approaches that has as its most general characteristic a problem to be solved or a question to be answered. If science educators were clearer about their goals and felt confident in choosing those aspects of inquiry teaching that were best suited to those goals, the many and varied methods of inquiry teaching would be much more widely and successfully used.

But perhaps the most important reason why inquiry teaching has not enjoyed more success is because its essential nature is often misunderstood. Successful inquiry teaching demands a significant intellectual commitment by students and a depth of engagement in the subject they are studying. An inquiry—a question or problem or issue to be investigated—should provide focus, direction, and purpose to a student's work. It should provide the drive and motivation needed for students to move toward deeper understanding of the content. Unfortunately, inquiry teaching has too often been misunderstood as simply having students do things, with students performing activities as if the activity were an end in itself. Activities without intellectual substance are pointless. Keeping students busy with hands-on activities is not enough. When inquiry teaching is identified with low-level activity-based teaching, it is bound to fail.

Finally, it is important to remember that inquiry teaching is not the only way to teach science or even the best way in all circumstances. To use inquiry-based approaches is not a black-and-white, either-or decision. For one thing, classroom activity can vary according to the degree of direction that the teacher provides and the degree of independence the student is given. If the questions students are investigating and the methods they employ are too highly prescribed by the teacher, then the activities may not lead to a genuine desire on the part of students to find something out, and student engagement may suffer. On the other hand, if the teaching is too open-ended, there is the possibility that students will become lost in their investigations and learn little. Questions of how much inquiry teaching and what type of inquiry teaching are appropriate must be answered by individual teachers in the context of the goals they have for their own students, and always with an eye toward the student's level of intellectual engagement. The variety of ways that inquiry teaching can be used to advance student learning are discussed in the National Research Council's publication, *Inquiry and the National Science Education Standards* (NRC, 2000).

HISTORICAL PERSPECTIVES

In what follows a number of the arguments that have been made over the years to justify incorporating aspects of scientific inquiry in the classroom are presented. With this as background, it should be easier for educators to choose those reasons that make the most sense to them and to have their students conduct inquiries and investigations in the ways that are most appropriate for their particular situations. Certainly it is easy to be confused by the interconnectedness of the various goals and approaches related to scientific inquiry, but that challenge can be met by a careful examination of each of these goals and by being careful not to be misled by what are often exaggerated claims of the benefits of inquiry in the classroom.

Nineteenth-Century Efforts to Incorporate Scientific Inquiry into the Classroom

Before the middle of the nineteenth century, the school curriculum was dominated by classical studies. It was not until scientists in Europe and the United States began promoting the value of science for its contribution to intellectual development that it became a regular part of the school curriculum. They argued that science was fundamentally different from the other school subjects because it offered practice in inductive logic. Mathematics and grammar, which were mainstays of the school curriculum, begin with rules and clear logical inferences that follow from those rules. Inductive science, on the other hand, begins with specific detailed observations and moves toward general principles. Scientific knowledge depends on the empirical observations upon which the generalizations are built. To justify its entry into the school curriculum, science was presented as something different from the other subjects already being taught. What was special about science was its basis in observation and inductive reasoning. That meant that students had to learn how to observe the natural world and draw conclusions from those observations. The study of science was justified largely on the basis of its ability to develop the intellect in ways that were fundamentally different from what was usually done in schools. This form of intellectual development was thought to be especially important because science was taking on an increasingly significant role in the modern world and because life in a modern democratic society depended on the independence of mind that was characteristic of science.

A key advocate for science was the prominent British biologist, Thomas Huxley (1825-1895). Huxley, who served as president of the Royal Society, was a staunch defender of Darwinism, popularized science, and a frequent lecturer and essayist on many subjects including the importance of science in the school curriculum.

According to Huxley:

> The great peculiarity of scientific training, that in virtue of which it cannot be replaced by any other discipline whatsoever, is this bringing of the mind directly into contact with fact, and practicing the intellect in the completes form of induction; that is to say, in drawing conclusions from particular facts made known by immediate observation of Nature. ... In teaching him botany, he must handle the plants and dissect the flowers for himself; in teaching him physics and chemistry, you must not be solicitous to fill him with information, but you must be careful that what he learns he knows of his own knowledge. ... And especially, tell him that it is his duty to doubt until he is compelled, by the absolute authority of Nature, to believe that which is written in books. (Huxley, 1899, p. 126-127)

This view of how science should be taught became the justification for the emerging science laboratory. Laboratory instruction and teaching science as a process of investigation received support from another prominent nineteenth century British intellectual, Herbert Spencer (1820-1903). Spencer was primarily a social scientist and philosopher but he also published a two-volume work on biology (Spencer, 1864, 1867). Spencer introduced the term "survival of the fittest" and was best known for applying theories of evolution to the study of society. His most lasting contribution to the field of education was his essay "What Knowledge Is of Most Worth" (Spencer, 1864) in which he argued for the inclusion of science in the school curriculum. According to Spencer, the laboratory should provide the opportunity for students to develop a clear conception of natural phenomena, something that could not be accomplished through book learning alone. In addition to giving precise mental images to go along with the verbal abstractions found in books, the laboratory also provided practice in drawing conclusions from observations, what Spencer called "judgment":

> No extent of acquaintance with the meaning of words can give the power of forming correct inferences respecting causes and effects. The constant habit of drawing conclusions from data, and then of verifying those conclusions by observation and experiment, can alone give the power of judgment correctly. And that it necessitates this habit is one of the immense advantages of science. ...By science, constant appeal is made to individual reason. Its truths are not accepted upon authority alone; but all are at liberty to test them—nay in many cases, the pupil is required to think out his own conclusions. Every step in a scientific investigation is submitted to his judgment. He is not asked to admit it without seeing it to be true. (Spencer, 1864, pp. 88-89)

Spencer believed that placing students in direct contact with natural objects and phenomena and having them draw their own conclusions from those observations had other advantages as well. The generalizations that were discovered by students through their own inquiries would be remembered longer and the process of inquiry would make the learner independent of the authority of the teacher. According to Spencer: "Children should be led to make their own investigations, and to draw their own inferences. They should be *told* as little as possible, and induced to *discover* as much as possible" (pp. 124-125).

Inductive approaches to teaching were also supported by Johann Friedrich Herbart (1776-1841), the German philosopher and educator whose work became popular in the United States toward the end of the nineteenth century. He and his followers in the U.S. (DeGarmo, 1895) believed that the best way for students to develop an understanding of new concepts was by having them discover the relationships between phenomena on their own and by having teachers relate new concepts to the experiences of the learner. Like Spencer, he felt that independent discovery would produce a fuller and more meaningful understanding of concepts. Herbart also believed that inductive teaching was greatly aided by informal conversation among students and teachers. Conversation would give the student an opportunity "to test and to change the accidental union of his thoughts, to multiply the links of connection, and to assimilate, after his own fashion, what he has learned" (Herbart, 1835/1901, p. 56).

Support for the laboratory and for student investigations to develop inductive reasoning abilities also came from Charles Eliot, a chemist and president of Harvard University from 1869 to 1895. In addition to introducing laboratory study into the college curriculum at Harvard, Eliot was also a vigorous advocate for science education in the schools. According to Eliot, science teaching in both schools and colleges should "develop and discipline those powers of the mind by which science has been created and is daily nourished—the powers of observation, the inductive faculty, the sober imagination, the sincere and proportionate judgment" (Eliot, 1898, p. 6).

The laboratory allowed students to observe the world and reason into the nature of things on their own. The direct and independent contact with the natural world that could be accomplished in the laboratory would provide a clear and unbiased view of the world that could not be achieved through book study, and learning how to conduct independent investigations would free individuals from the authority of both the text and the teacher. The specific skills that were needed to accomplish these goals included training of the sense organs, practice in organizing and comparing sense impressions and drawing inferences from them, training in making accurate records of these sense impressions in written form or in memory, and training in the ability to formulate clear logical statements about the conclusions drawn from observation (Eliot, 1898, p. 322).

Eliot's leadership in getting science taught inductively was most evident when he became chair of the prestigious Committee of Ten of the National Education Association (NEA) in 1892. Each of the science groups that were represented there strongly supported the use of the laboratory in science teaching. The purpose of laboratory instruction was to develop the students' reasoning skills and their ability to acquire knowledge independently. Students would come to understand "the reason of things." They would observe changes and processes as well as static facts and they would examine cause and effect relationships (NEA, 1893, p. 213). At all times teachers would keep in mind that the topics they were teaching were for the purpose of exercising the students' personal intellectual powers. The important thing was that students should not be taught dogmatically. They needed to be taught inductively so that they could develop their own ways of seeking knowledge.

Despite these efforts to introduce inquiry-based teaching into the classroom, especially by means of the laboratory, textbook-based methods remained the dominant mode of teaching at the turn of the century. According to one New York State report: "While the laboratory method is almost universally approved by the science teachers, the text-book method prevails in the schools, to such an extent that laboratory work is incidental, inefficient, and in many cases excluded altogether" (University of the State of New York, 1900, p. 706). In a book published in 1902 by Smith and Hall, the authors argued against the textbook approach to school science teaching and offered specific suggestions on how the science program could be strengthened. Smith, speaking for chemistry, felt that the high school course should be laboratory-based, utilize as much independent discovery by students as possible, focus on a meaningful understanding of the facts and principles of chemistry, and be oriented toward practical applications in everyday life.

Studying chemistry would help students develop their ability to think, including their ability to compare, to discriminate, and to reason inductively. The laboratory would be used both as a place for the verification of chemical principles and as a place for independent discovery. Verification labs and practical illustrations would "make the understanding of the law more vivid, the recollection of its content more lasting, and, above all, to show…what the nature of its experimental basis is" (Smith & Hall, 1902, p. 106). The most important feature of the laboratory, however, was that it could be used to place the student in the role of discoverer. Smith cited Armstrong (1898) as the founder of the *heuristic method* in England, an approach where no books or directions from the teacher were used and students raised their own questions as they examined materials that they were given.

But Smith also recognized the impracticality of having students spend all their time in independent discovery activities because of the additional time required, and so he opted for teacher-guided inquiry as the preferred approach. Under this method the teacher raised the questions, provided the materials, offered suggestions about what to look for next, and asked leading questions to move students along the path of discovery. Students still made careful observations and reasoned from those observations, but they did not have the same degree of independence that characterized the heuristic method.

Hall, speaking for physics, likewise offered a range of inquiry-based teaching methods and analyzed the advantages and disadvantages of each. The first he called the true discovery or heuristic approach in which students were given the maximum amount of freedom to explore the natural world on their own. As did Smith, Hall felt that this method required too much time and that students were often not well enough equipped to draw anything but the most superficial conclusions from their investigations. The second method he called the verification approach in which students confirmed scientific facts or principles in the laboratory. Although the approach helped strengthen students' understanding of the concepts of science, Hall felt that the method led to unscientific attitudes because students were too tempted to look for the right answer or consider only the evidence that supported the expected result. The third method was the guided discovery approach, what Hall called inquiry. Using this method, students did not have to discover everything on their own but they did have to seek solutions to questions for which they did not have the answers. In this way they would still be acting as genuine investigators and not simply confirming something that was already known to them.

Inquiry Teaching During the First Half of the Twentieth Century

During the latter half of the nineteenth century the goals of science education were expressed mainly in terms of individual personal development. These goals included having a familiarity with the facts and principles of science essential to living in a scientific age and the mental discipline that comes from practice in inductive reasoning. Being able to draw conclusions independently from evidence would free individuals from a dependence on the intellectual authority of others, something that would serve them well as citizens in a democratic society. But during the first half of the twentieth century, these personal benefits began to take on less importance and science education came to be justified more explicitly in terms of its societal value. Education took on a more pragmatic bent as it sought to address the pressing problems that the rapidly growing country faced--issues related to immigration, urbanization, public health, and other socially-based problems. There was also a growing belief in the importance of child-centered approaches to education that was due in large part to the influence of John Dewey (1902/1990). The trend in all of education was toward more practical work for students both because it would be more interesting to them and because practical studies could address issues having importance in society.

In this changing atmosphere, inquiry-style teaching was now seen as a way to develop the abilities needed to solve specific problems having social significance rather than as a way to discipline the mind through inductive reasoning. Throughout the first half of the twentieth century, Dewey argued that citizens in a democratic society should be inquirers regarding the nature of their physical and social environments and active participants in the construction of society. They should ask questions and have the resources to find answers to those questions, independent of external authority. To prepare them for life in a democracy, formal education needed to give students the skills and dispositions to formulate questions that were significant and meaningful to them, and since there is a shared, collaborative aspect to life in a democratic society, students also needed to develop a capacity for cooperative group inquiry.

Regarding science instruction, Dewey said:

> ...students should be introduced to scientific subject-matter and be initiated into its facts and laws through acquaintance with everyday social applications. Adherence to this method Is not only the most direct avenue to understanding of science itself but as the pupils grow more mature it is also the surest road to the understanding of the economic and industrial problems of present society. (Dewey, 1938, p. 80)

In one of the first formal statements on the importance of engaging students in the solution of real world problems, the science committee of the National Education Association's Commission on the Reorganization of Secondary Education (CRSE) issued a report that said:

> The unit of instruction, instead of consisting of certain sections or pages from the textbook, or of a formal laboratory exercise, should consist of a definite question, proposition, problem, or project, set up by the class or by the teacher. Such a problem demands for its solution recalling facts already known, acquiring new information, formulating and testing hypotheses, and reasoning, both inductive and deductive, in order to arrive at correct generalizations and conclusions. This method calls for an organization in which information, experimental work, and methods of attack, all are organized with reference to their bearings on the solution of the problem. (National Education Association, 1920, p. 52)

Similarly, the physics committee of the CRSE addressed the importance of the laboratory as a place for genuine inquiry rather than as a place to "verify laws," to "fix principles in mind," to "acquire skills in making measurements," or to "learn to be accurate observers."

> With a project or a problem as the unit of instruction and its solution as the motive for work, the pupil should go to the laboratory to find out by experiment some facts that are essential to the solution of his problem.... With such a motive he is more nearly in the situation of the real scientist who is working on a problem of original investigation. He is getting real practice in the use of the scientific method. (National Education Association, 1920, p. 53)

Numerous practical questions also arose around laboratory instruction besides those having to do with its aims and purposes. Studies were conducted to determine if students learned more from laboratory experiences than from teacher-led demonstrations and if laboratory instruction was cost effective. There was also the question of whether the laboratory should precede or follow classroom instruction. In a truly inductive approach, laboratory work is exploratory and therefore should come first, but this often created scheduling problems that could not be easily overcome.

In 1932, the Thirty-First Yearbook Committee of the National Society for the Study of Education (NSSE) analyzed the reasons why the laboratory might be used in science teaching. They identified seven purposes of laboratory work: (1) To develop simple laboratory techniques. (2) To establish for oneself the principles of science already established and accepted. (3) To gain familiarity with the objects of science. (4) To provide illustrations to develop a better understanding of the principles of science. (5) To provide training in the scientific method. (6) To provide scientific training in the solution of the pupil's own problems. (7) To study science problems that the student might have (NSSE, 1932, p. 270).

The committee felt that the best use of the laboratory was for the solution of the students' own science problems. The committee said: "It is the very essence of the original reason for the existence of every laboratory…" (NSSE, 1932, pp. 271-272). But not everyone felt so positively about that aspect of laboratory work. Francis Curtis (NSSE, 1932) felt that the interpretation of data was generally beyond the abilities of most students and he felt that genuine scientific investigations were of little use to most students because these activities focused on the practices of actual research scientists, an unlikely career choice for most students.

During the first half of the twentieth century, there continued to be a strong focus on teaching students the scientific way of thinking. For the most part this was accomplished in the context of problems and projects that were interesting to students and that had social relevance. Because science had been introduced into the school curriculum by nineteenth-century scientists as an inductive, laboratory-based study, much of the discussion during the early part of the twentieth century was on the appropriate use that should be made of the laboratory. Should it be used to strengthen concepts, to verify scientific principles, to teach laboratory techniques, or to provide a place for students to engage in genuine investigations? Many science educators during the early years of the twentieth century felt that the laboratory should be used as a place where students could work on problems of interest to them and that had social, as well as scientific, relevance and importance.

Inquiry Teaching in the Age of Reform

By the 1950s a growing number of scientists, science educators, and industry leaders began to argue that science education had lost its academic rigor and had become intellectually soft. They were concerned about the applied, practical orientation of many science courses and what they saw as an over-emphasis on social relevance and student interest. They said that the primary job of the schools should be the training of disciplined intelligence and the transmission of the cultural heritage. Education had become too student centered and too concerned with practical applications. What was needed was a return to disciplinary rigor. This focus on rigor and the structure of the disciplines became the thrust of a curriculum reform movement that began in the 1950s and lasted throughout the 1960s and into the 1970s. The attention paid to students' intellectual development was reminiscent of a similar thrust during the late nineteenth century, although the reasons given for it were now different. In the nineteenth century the concern was for personal intellectual development, but in the 1950s it was to prepare individuals who might become scientists and to develop a public that was sympathetic to science. Science had become important for national security and economic development. Science teaching was no longer a casual activity. To many, a strong educational program in science was essential for maintaining the security of the country.

Leaders of this movement believed science should be taught as it was practiced by scientists in order to give it the most authenticity possible. Students should be taught the fundamental ideas of the disciplines and they should be taught those ideas through investigations that mirrored the way scientists themselves generated new knowledge. Scientific inquiry would be the model for classroom teaching and learning. Although they were most often referred to as inquiry teaching, other terms were used to describe pedagogical approaches that were modeled after scientific inquiry as well. These included discovery learning, inductive teaching, problem solving, and project learning.

The key difference between this version of inquiry teaching and earlier versions was that it was linked even more closely to actual scientific inquiry in order to make it as intellectually rigorous possible. Learning the structure of the disciplines meant learning the disciplines in the way that scientists understood them, including both the content and the modes of inquiry that were used. Whether learned through the laboratory or through a textbook, the conclusions of science and the evidence that supported those conclusions would go hand-in-hand. For future scientists this would give them the advantage of an early introduction to the logic and methods of their chosen field of work, and for the general public it would give them a truthful and accurate picture of the nature of science and an appreciation for the methods of science. Public understanding and appreciation were of critical importance in a political environment where scientific research had come to depend so much on public funding. The scientific community realized that it needed the support of the public to continue its work.

The individual most often associated with the reform movement's notion of scientific inquiry was Joseph Schwab. To Schwab (1962), scientific content and processes were intimately connected and inseparable. An accurate representation of science required that the principles of science be taught in the context of the evidence on which they were based. Content should be taught in relation to the methods that generated that knowledge. Content could not stand alone and neither could method. In many ways this was the same argument that was made by nineteenth-century scientists who said that students should learn how to reason about science and have opportunities to develop good judgment regarding scientific facts. They should be given practice in drawing their own conclusions from observations and they should verify their conclusions by experiment. Nineteenth-century students were taught not just the conclusions of science but also the reasoning that led to those conclusions. The difference between the nineteenth-century view and what Schwab was proposing is that nineteenth-century educators were focused on the personal intellectual development of the students. Schwab's concern was different. His concern was for the welfare of the nation. In his words:

> A hundred fifty years ago, science was an ornament of a leisurely society. It was still mainly pursued by amateurs and gentlemen. It was a gratuitous activity of the enquiring intellect, an end pursued for its own sake....It is so no longer. Industrial democracy has made science the foundation of national power and productivity. (Schwab, 1962, p. 18)

According to Schwab, the nation faced three important needs. The first was to increase the number of scientists. The second was to develop competent political leaders who could develop policy agendas based on an understanding of science. The third was to educate a public that was sympathetic to the tentativeness of scientific knowledge and the fluid nature of scientific investigation so they would support basic research in the sciences. It was not Schwab's primary objective that students should be able to conduct scientific inquiries themselves but rather that they should understand the nature of scientific inquiry as a dynamic and ongoing activity and that they should understand scientific content in the context of the evidence upon which it was based. As he put it:

> Of the two components—science as enquiry and the activity of enquiring—it is the former which should be given first priority as the objective of science teaching in the secondary school. It is a view of science as enquiry which is necessary if we are to develop the informed public which our national need urgently demands. (1962, p. 72)

Despite the clear distinction that Schwab made between inquiry as content and inquiry as pedagogy, the two were often confused. Reformers wanted students to understand the interconnectedness of the contents and methods of the science disciplines in as intellectually rigorous a way as possible. They believed that this understanding of science as both process and product could be accomplished both through inquiry-based teaching methods and through non-inquiry-based methods. Students might, for example, be led step-by-step through the historical development of a particular scientific idea. They might be shown the way a problem was formulated and solved, how data was collected and analyzed, and how conclusions were drawn. In this way the logic of discovery could be explained directly to the students. But students might also be asked to formulate their own problems and solutions and learn about inquiry by practicing inquiry.

Although Schwab and other educational leaders recognized the usefulness of direct teaching about scientific discovery, they believed that it was more important to have students conduct their own investigations because it promoted deeper intellectual engagement with the content and more meaningful understanding of the nature of scientific inquiry. One such inquiry-based method that Schwab proposed had students analyze historical papers that scientists had written so they could study the logic of discovery and the fluid nature of scientific inquiry. This kind of historical investigation was just one of many inquiry-based methodologies he proposed. He also suggested that students could be asked to analyze their textbook and their teachers' lectures and to always put themselves in the position of evaluating the validity of the claims of others and the adequacy of the evidence they presented. Discussion was seen as a particularly powerful inquiry-based teaching strategy because it required an active engagement with the content and provided a richer and more varied set of ideas to work with rather than the opinions of a single expert.

It needs to be emphasized, however, that Schwab's main interest was not in preparing students to be inquirers into the nature of the physical world. His goal was for students to have the fullest and most complete understanding of science possible, both its content and its methods, so that they would have a firm foundation for further science study if they were to become scientists, and so that they would be sympathetic to the scientific enterprise if they were not destined for science careers. The primary purpose of performing investigations in the classroom was to understand more fully the nature of science, not to learn the skills necessary to conduct scientific work themselves. And it was certainly not his intention that students learn how to apply the methods of science to problems of practical or social concern. This way of conceptualizing inquiry teaching was clearly in opposition to the more functional and practical approaches proposed earlier in the century.

Although probably unintended, linking method so closely with disciplinary content also had the effect of privileging expert knowledge and making scientific method less accessible to the general public. Whereas early twentieth century educators promoted scientific method as something that could be applied to a wide range of scientific and social problems that were within the range of almost anyone to investigate, mid-century reformers saw scientific inquiry as discipline-specific. To really understand scientific method required a deep understanding of the discipline in question. Thus the science courses that were developed during this period were inaccessible to many students because of their conceptual difficulty and theoretical sophistication, and they failed to address the social world of students, their personal interests, or practical concerns. In a major shift from the approach taken during the first half of the twentieth century, student investigations became much more closely tied to the logically organized science content and much less so to phenomena in their everyday experience.

Science Literacy

By the early 1970s, however, the educational focus began to shift from disciplinary study to preparing an enlightened citizenry that would have the skills to function effectively in a scientific world. Although not without its critics, science education for social relevance and democratic participation regained much of the importance it had held during the first half of the twentieth century. The idea of science education for a broad and functional understanding of science came to be referred to as science literacy. This neo-progressive attitude was also represented in newly developed programs in environmental education, values education, humanistic education, and in the science, technology, and society (STS) movement.

In this new intellectual environment, science knowledge and the processes of science were to be used to answer questions that people encountered in their everyday lives. Science was to be practical and useful to people. Science teaching would focus on science as a social and cultural force, on the relationship between science and technology, and on preparing citizens who could use scientific knowledge and processes to solve problems they encountered in everyday living (Hurd, 1970). A position statement on scientific literacy from the National Science Teachers Association (1971) said: "The major goal of science education is to develop scientifically literate and personally concerned individuals with a high competence for rational thought and action" (p. 47). Issues students would investigate might include endangered species, genetic engineering, global warming, nuclear waste disposal, or air and water pollution.

Students would conduct independent investigations and learn to apply the methods of science to problems of personal and societal concern. Of less interest was the study of science as a structured discipline. The discipline-based approaches of the 1960s treated scientific inquiry as fundamentally linked to the disciplines and not as a general method that could be applied to a wide range of scientific and socially-based problems. Students practiced scientific inquiry to acquire an understanding of the disciplines that was as intellectually complete and authentic as possible. In the more socially oriented period that followed, student inquiries were not aimed so much at the basic principles and concepts of science as they were toward science-related issues that had social relevance. The inquiries they engaged in were often more appropriately called problem-solving, or personal and social decision-making activities. In this new intellectual environment, the logic of science and the scientific way of thinking were still important but they were important for solving practical problems that citizens faced in their everyday lives. Science teaching would no longer take place just in the classroom and the laboratory but also in the communities where students lived.

Because democratic citizenship implies social responsibility, education for citizenship meant that students needed to acquire the knowledge and skills that would enable them to analyze science-related social issues and to evaluate alternative solutions for resolving them. They would learn skills of data collection, interpretation, and communication of results by investigating science-related social issues directly (Ramsey, 1997). Some suggested that social responsibility implied using science to transform society. Hofstein and Yager said: "The use of societal issues as organizers for the science curriculum of the 80s has many advantages. First of all, it helps delineate content that can be useful for improving the quality of life…" (1982, p. 542). According to Ramsay (1997), "learners would develop a sense of purpose and control about their use of science knowledge and skills in the democratic process of social change…" (p. 310). And pedagogically it was felt that the sense of purpose and control that comes with problem ownership would prove motivating to students.

Critics of an issues-oriented approach to science teaching said that such an approach lacked substance and did not convey a sense of the structural integrity of science, and because society's problems are always changing, investigating today's problems would not provide students with the knowledge or skills needed to deal with problems in the future (Kromhout and Good, 1983). There was also the question of whether it was appropriate to teach scientific inquiry in the context of science-related social problems. Doing so implied that the methods of science were general and had applicability to a broad range of problems, but mid-century reformers had insisted that scientific inquiry was intimately connected to the content of the science disciplines. The question was not whether the methods of science *could be* generalized to the study of social problems but whether this is the kind of inquiry we want students in our science classes to engage in. Dewey and other early twentieth century educators spoke of general method that had broad application to a wide range of problems that could be studied in the science classroom. Neo-progressives of the 1970s and 1980s agreed with that position, whereas discipline-based reformers of the 1950s and 1960s did not.

Inquiry Teaching Today

By the late 1980s, goal statements in science education included an understanding of science content for its cultural, disciplinary, and intellectual value and for its application to everyday decision-making and problem-solving. And whether it was used to develop students' abilities to solve problems that were personally and socially relevant, for personal intellectual development, or as a motivational device, inquiry teaching had a role to play. Rather than settling on a single approach to science teaching, however, the tendency was to combine all of these goals under the general heading of science literacy.

Published in 1989, Project 2061's *Science For All Americans* (AAAS, 1989) was an attempt to reach a consensus on what students should know to be scientifically literate in the broadest possible sense. The common core of learning was selected on the basis of five criteria: (1) Does the content enhance one's long-term employment prospects and the ability to make personal decisions? (2) Does the content help one to "participate intelligently in making political decisions involving science and technology?" (3) Does the content "present aspects of science, mathematics, and technology that are so important in human history or so pervasive in our culture that a general education would be incomplete without them?" (4) Does the content help people ponder the enduring questions of human existence? (5) Does the content enrich children's lives at the present time regardless of what it may lead to in later life? (pp. *xix-xx*). The goals of personal intellectual development and responsible citizenship are both included in these statements.

When it came to inquiry teaching, the authors of *Science For All Americans* recommended that science teaching should be consistent with the nature of scientific inquiry. Accordingly: "Students need to get acquainted with the things around them—including devices, organisms, materials, shapes, and numbers—and to observe them, collect them, handle them, describe them, become puzzled by them, ask questions about them, argue about them, and then try to find answers to their questions. …Students should be given problems…that require them to decide what evidence is relevant and to offer their own interpretation of what the evidence means" (p. 201). The goal of inquiry teaching is to "help people in every walk of life to deal sensibly with problems that often involve evidence, quantitative considerations, logical arguments, and uncertainty…" (p. *xiv*).

Following soon after the publication of *Science For All Americans*, the National Research Council (NRC) contributed to the advancement of scientific literacy through publication of the *National Science Education Standards* NRC, 1996). The goals for school science identified by the *National Standards* were to prepare students who would be able to:

- experience the richness and excitement of knowing about and understanding the natural world;
- use appropriate scientific processes and principles in making personal decisions;
- engage intelligently in public discourse and debate about matters of scientific and technological concern; and

- increase their economic productivity through the use of the knowledge, understanding, and skills of the scientifically literate person in their careers. (p. 13)

The *National Science Education Standards* is an all-encompassing document that includes a wide range of content and process goals. An important feature of the *National Standards* is its focus on inquiry teaching. In the *National Standards* and the follow-up volume, *Inquiry and the National Science Education Standards* (NRC, 2000), inquiry teaching is described as "a set of interrelated processes by which...students pose questions about the natural world and investigate phenomena; in doing so, students acquire knowledge and develop a right understanding of concepts, principles, models, and theories." It is recommended that "...designers of curricula and programs must be sure that the approach to content, as well as the teaching and assessment strategies, reflect the acquisition of scientific understanding through inquiry. Students will then learn science in a way that reflects how science actually works" (NRC, 1996, p. 214).

In the *Standards*, inquiry teaching is a pedagogical approach that is consistent with the nature of science and that provides useful skills for investigating problems of personal interest or social concern. In both *Science for All Americans* and the *National Science Education Standards*, there was recognition of the importance of inquiry teaching for giving an accurate portrayal of scientific investigation, for contributing to one's personal intellectual development, and for offering a way of thinking that would be used in the solution of everyday problems.

But perhaps the primary justification for using inquiry teaching, particularly in the NRC's publications, is the argument that inquiry teaching is a more effective teaching strategy, that it is more engaging, and that students learn more from inquiry-based approaches to teaching. Much of *Inquiry and the National Science Education Standards* is devoted to a discussion of psychological arguments about inquiry teaching's efficacy, especially as a way to learn the concepts and principles of science. The authors say, for example, that "students are much more likely to understand and retain the concepts that they have learned this way" (p. xiii). They say that the *Standards* "seek to promote curriculum, instruction, and assessment models that enable teachers to build on children's natural, human inquisitiveness" (p. 6). They point out that research on people who have expertise in a field shows that these people have "inquiry procedures available that help them solve new problems efficiently and effectively" (p. 116). Finally they argue that as students "develop their abilities to question, reason, and think critically about scientific phenomena, they take increasing control of their own learning" (p. 120) which is motivating to them.

SUMMARY

Inquiry teaching in science classrooms has had a long and varied history. The rationale for using inquiry teaching has changed according to the educational philosophy of its supporters, but regardless of the arguments used to justify its use, inquiry teaching has remained a significant aspect of science education throughout the years. In the nineteenth century, as science was becoming a regular part of the school curriculum, scientific investigation in the classroom was seen as a way to develop students' inductive reasoning skills, something that the other school subjects could not do. In the early twentieth century, students were encouraged to apply general methods of scientific inquiry to problems of social concern. By the 1950s and 1960s the focus had shifted away from practical and applied problem-solving to a rigorous treatment of the individual scientific disciplines. This was in part for purposes of personal intellectual development but mainly so that personnel needs could be met in technical and scientific fields and so that lay people would have sufficient understanding of science to offer their unqualified support for scientific research. Within a relatively short period of time, however, there was a revisiting of the idea that it was important for students to acquire skills in scientific inquiry so that they could solve problems of personal and social concern and be active, contributing citizens in a democratic society. Finally, by century's end, goal statements in science education recognized the validity of a wide range of arguments favoring inquiry teaching that had been made over the years. These documents also provided new support to inquiry teaching by pointing out the congruence between scientific inquiry and effective student-centered teaching. In addition, a developing focus on constructivist pedagogy was consistent with important aspects of inquiry teaching.

Although the essence of inquiry teaching is not always easy to grasp and implementation has proven difficult, educational leaders have consistently recognized the potential of inquiry-based pedagogies to enhance student learning and to provide students with the skills needed to function effectively in a democratic society. It is important to recognize, however, that there is no single way to think about what inquiry teaching is and no single argument that justifies its use. It is a multifaceted approach to teaching that can be used to accomplish many differing purposes. Understanding the variety of ways that inquiry teaching can be used and the range of meanings it can have should aid educators in moving toward pedagogies that are effective and motivating to students, and that deepen their intellectual engagement with scientific ideas and that give them a better sense of what science is.

REFERENCES

American Association for the Advancement of Science (1989). *Science for all Americans.* New York: Oxford University Press.

Armstrong, H. (1898). On the heuristic method. *Special reports on educational subjects, vol. II.* London: Eyre & Spottiswood.

Bybee, R. (2000). Teaching science as inquiry. In J. Minstrell & E. van Zee (Eds.), *Inquiring into inquiry learning and teaching in science* (pp. 20-46). Washington, DC: AAAS.

Bybee, R. & DeBoer, G. (1994). Research on goals for the science curriculum. In D. Gabel (Ed.), *Handbook on research on teaching and learning* (pp. 357-387). New York: Macmillan.

Costenson, K. & Lawson, A. (1986). Why isn't inquiry used in more classrooms? *The American Biology Teacher, 48* (3): 150-158.

DeBoer, G. (1991). *A history of ideas in science education: implications for practice.* New York: Teachers College Press.

Deci, E. (1975). *Intrinsic motivation.* New York: Plenum Press.

Deci, E. & Ryan, R. (1987). The support of autonomy and the control of behavior. *Journal of Personality and Social Psychology, 53* (6), 1024-1037.
DeGarmo, C. (1895). Herbart and the Herbartians. New York: Charles Scribner's Sons.
Dewey, J. (1990). *The school and society; the child and the curriculum*. Chicago, Illinois: The University of Chicago Press. (Original work published in 1902.)
Dewey, J. (1938). Experience and education. New York: Collier.
Eliot, C. (1898). *Educational reform*. New York: Century.
Herbart, J. (1901). *Outlines of educational doctrine* (C. DeGarmo, Ed.; A. Lange, Trans.). New York: Macmillan. (Original work published in 1835).
Hofstein, A. & Yager, R. (1982). Societal issues as organizers for science education in the 80s. *School Science and Mathematics, 82*, 539-547.
Hurd, P. (1970). *New directions in teaching secondary school science*. Chicago: Rand McNally.
Hurd, P., Bybee, R., Kahle, J., & Yager, R. (1980). Biology education in secondary schools of the United States. *The American Biology Teacher, 42* (7): 388-410.
Huxley, T. (1899). *Science and education*. New York: Appleton.
Kromhout, R. & Good, R. (1983). Beware of societal issues as organizers for science education. *School Science and Mathematics, 83*, 647-650.
National Education Association (1893). *Report of the committee on secondary school studies*. Washington, D.C.: U.S. Government Printing Office.
National Education Association (1920). *Reorganization of science in secondary schools: A report of the commission on the reorganization of secondary education*. (U.S. Bureau of Education Bulletin No. 20). Washington, D.C.: U.S. Government Printing Office.
National Research Council (1996). *National science education standards*. Washington, DC: National Academy Press.
National Research Council (2000). *Inquiry and the national science education standards*. Washington, DC: National Academy Press.
National Science Teachers Association (1971). NSTA position statement on school science education for the 70's. *The Science Teacher, 38*, 46-51.
National Society for the Study of Education. (1932). *A program for teaching science: Thirty-first yearbook of the NSSE*. Chicago: University of Chicago Press.
Ramsey, J. (1997). STS issue instruction: Meeting the goal of social responsibility in a context of scientific literacy. In W. Graber & C. Bolte (Eds.), *Scientific literacy* (pp. 305-330). Kiel, Germany: Institute for Science Education.
St. John, M. (1999). The value of knowing what you do not know. In *Foundations (Volume 2) Inquiry: thoughts, views, and strategies for the K-5 classroom* (pp. 109-111). Arlington, VA: The National Science Foundation.
Schwab, J. (1962). The teaching of science as enquiry. In *The teaching of science* (pp. 1-103). Cambridge, MA: Harvard University Press.
Smith, A. & Hall, E. (1902). *The teaching of chemistry and physics in the secondary school*. New York: Longmans, Green.
Spencer, H. (1864). *Education: Intellectual, moral, and physical*. New York: Appleton.
Spencer, H. (1864, 1867). *Principles of biology*, 2 vols. London: Williams and Norgate.
University of the State of New York (1900). *Report of the committee of nine* (High School Report, Bulletin No. 7). Albany: University of the State of New York Press.
Welch, W., Klopfer, L., Aikenhead, G., & Robinson, J. (1981). The role of inquiry in science education: Analysis and recommendations. *Science Education, 65*, 33-50.

CHAPTER 3

MARTIN HABERMAN

THE SPECIAL ROLE OF SCIENCE TEACHING IN SCHOOLS SERVING DIVERSE CHILDREN IN URBAN POVERTY

Writing this paper has taken me on a journey to a place I have been heading all my life but have resisted getting to. Over the past forty-five years I have dealt with the issue of why children in poverty and children of color are not adequately served by the public schools. During this period I have developed numerous programs for preparing teachers whose students do achieve and do succeed in school, but the number of such teachers has never caught up with the need for them. Even increasing the pools of new teachers in the 43 states where alternative certification programs are offered will not prove equal to the task of providing all children in poverty with the teachers they deserve. Most of the 2.2 million teachers who will be hired in the next decade will be placed with children in poverty. But if those who are recruited do not stay in teaching long enough to become proficient and then remain as career-long teachers, we will simply be perpetuating the present churn of teachers who come and go at the expense of children of color and children in poverty. Just as we import agricultural and domestic workers to do the jobs Americans won't do, we are now increasing the importation of substantial numbers of English- and Spanish-speaking teachers from Asia, Europe, South America, Africa, and the Caribbean. But the ones who survive culture shock and who become culturally sensitive enough to Learn how to teach urban youngsters will be far fewer than those who quit or fail. And the ones who stay and become truly effective (at least three years) will be far fewer than the number of strong insensitives who will stay because they can't get other jobs with equal benefits or because being a teacher enables them to remain in this country and bring their families. By 2010 not having enough effective teachers for children in poverty will have been a problem of sufficient duration to reveal that in-the-box thinking will recruit and prepare only enough effective career teachers to help only some students in specific urban schools (those with special kinds of school leaders) but not enough effective career teachers to impact the schooling of children in poverty nationally.

In preparing for this paper I have been led to rethink the nature of teaching and learning in urban schools, the societal conditions under which these schools function, the politics surrounding them, the ways in which they are funded, the research regarding the achievement gap, and the nature, of the way these schools are

L.B. Flick and N.G. Lederman (eds.) Scientific Inquiry and Nature of Science, 37-53.

administered. I have considered the advocacies of both those with simple-minded proposals as well as those I know to be sophisticated analysts of the problem, noting again and again the difference between those who believe schools shape society and those who understand that schools reflect society. Finally, the question of who benefits from the present miseducation of children in poverty has been my constant companion.

THE PROBLEM

Students in poverty who attend school for 13 years and who graduate are more likely to be miseducated that students who are pushed out. Graduates are exposed and rewarded over longer periods for learning misconceptions that will prevent them from ever participating successfully in the world of work or becoming lifelong learners.

In “The Pedagogy of Poverty vs. Good Teaching” (Haberman, 1991), I argued that poverty schools succinctly answer three grand questions: What knowledge is of most worth? How does learning best occur? and What is the purpose of learning? The three answers demonstrated by schools serving students in poverty are that knowledge is a set of basic skills learned by following directions for the purpose of securing an immediate reward.

In “Unemployment Training: The Ideology of Nonwork Taught in Urban Schools” (Haberman, 1997), I focused on the curriculum and argued that the culture of poverty schools socializes students into an ideology that will prevent them from participating in the world of work. The third and final component of this problem definition was set forth in “Urban Schools:

Day Camps or Custodial Institutions?” (Haberman, 2000). The essential contention here was that experts and researchers, as well as lay constituencies, do not yet recognize that they are using school reform strategies to transform places that are no longer functioning as schools.

For this paper I decided to focus my analysis on issues related to teaching science in the urban middle school because this level is the nexus of the problems of educating urban youngsters. It is in middle school where the pretense that urban school districts can effectively educate children of color in urban poverty is revealed to be a monstrous hoax perpetrated on those least able to demand what is in their own best interests. It is at this level where the teacher’s role changes from the more personal, nurturing elementary teacher in the self-contained classroom to the impersonal subject matter specialist who must deal with 150 students daily; where the school climate is dominated by the war between early adolescent peer groups versus school managers; where suspensions and push-outs increase markedly; where individual help for students below grade level from either teachers or aides decreases; where high stakes testing is introduced as the standard for going on to high school; and where the least creative, most directive, most punitive forms of instruction are conducted by the greatest number of temporary teachers passing through the schools. The urban middle school is a wasteland for the administrators and staff as well as for the children. In my study of prison schools I have found greater concern for individual differences and the needs of the prisoner students than is typical in most urban middle schools. In prison schools the staff recognizes that control and punishments can only go so far—that they will not lead to the internal controls the youngsters will need to function on the outside. In the typical urban middle schools there is still a search for stronger and stronger punitive consequences that will control the youngster—with no regard for what this teaches them about controlling their own behavior in order to function in the world outside.

Early adolescence is the age when society shifts from perceiving children as needing understanding and support to needing control and punishment. Unlike the primary youngsters who were perceived as being cute, young adolescents are perceived as the causes of serious societal problems: gangs, drugs, crime, and the results of uncontrolled sexual activity. The very same children of color and in poverty whose misbehaviors were treated with hugs and encouragement in the primary grades are responded to at age thirteen with fear and foreboding

… particularly the males. We put them in special education or push them out. The urban middle school is not primarily an educational institution at all but a custodial one in which the activities of an age group and population we regard as social dynamite must be contained and restrained.

Urban middle school teachers frequently feel and describe themselves as helpless. "How do you teach eighth-grade social studies (substitute any subject) to second-grade readers?" What do I do with six Inclusion (handicapped) students in a class of twenty-five?" "Do they expect us to be nurses, social workers, and policemen as well as teachers?" "What am I supposed to do with six classes of these kids a day?' The advice of these veteran burnouts to the bushy-tailed neophytes temporarily passing through their schools is clear: "Just cover the material and those who want to learn will. It's all up to them." But teachers who feel powerless can build neither their own feelings of professional self-worth nor the self-concepts of their students. *It takes somebodies to make somebodies. Nobodies don't make somebodies*.

It is typical for middle school teachers to attribute student inadequacies to students' home lives and to the poor teaching the students have received from their previous teachers. In contrast to such teachers there is a small number of star teachers who succeed with the very same children in the very same schools under the very same conditions (Haberman, 1995). For the purpose of this paper I decided to clarify some of my ideas about science teaching by having an in-depth conversation with the best middle school science teacher in a great city school system (Big City). This teacher, "Ms. M." is an individual whom my colleagues and I have Observed over a ten-year period. In responding to my Urban Teacher Selection Interview Ms. M. gave all the answers of a star teacher before she ever began her teaching career. In the district of over 103,000 youngsters in which she teaches, fewer than 50% of middle school youngsters achieve proficiency in any area and as is typical in most urban districts, achievement in science is lower than in any of the four basic curriculum areas. Her students achieve 90% proficiency in science each year in an urban district in which only 27% of the children achieve science proficiency. The university regards Ms. M. as its model mentor of student teachers and interns and has placed many beginners with her. Ms. M. teaches in one of the most difficult middle schools where suspensions are high, teacher turnover is high, and the principals change very year or two. As a teacher of science she had inadequate equipment and facilities, little access to computers, and purchases most of the consumable materials she needs with her own funds. The university has publicly recognized Ms. M. as the outstanding teacher of the 1990—2000 decade—not just in science but in all teacher areas (Haberman, 1999).

Ms. M. is an African American female who decided to become a teacher in her 30s after majoring in biology and trying other careers. She can be described as an individual with very strong commitments to her students and to their learning. She is also an extremely socially conscious person who is well aware of the issues of equity, justice, and the impact of classism and racism on the education of urban youth. In observing her teaching behaviors it is clear that she believes the ultimate values to be preserved are fostering her students' science learnings and encouraging them to function as critical thinkers in their daily living. Consider the conversational interchanges with Ms. M. regarding each of the following activities.

INSTRUCTION

The common assumption is that schools are the solution or the "way out" for children in poverty. For most the reverse is true. The way children and youth are taught and managed in urban poverty schools inculcates a set of values about learning, about the way tolerate to authority, about what knowledge is of most worth, and about how to be successful. This set of values systematically programs students for failure in the world of work and in postsecondary education. Without the skills or desire to participate in the world of work or in training, the graduates of these schools have no basis for any form of adult achievement beyond mindless, low-level jobs at less-than-subsistence wages.

Some examples of almost every form of pedagogy can be found in urban schools: direct instruction, cooperative learning, peer tutoring, individualized instruction, computer-assisted learning, behavior modification, the use of pupil contracts, media-assisted instruction, scientific inquiry, lecture discussion, tutoring by specialists and volunteers, and occasionally even the project method common in progressive education. In spite of this broad range of instructional options, however, there is a typical form of teaching that has become accepted as basic. Indeed, this basic urban style, which encompasses a body of specific teacher acts, seems to become stronger each year since I first noted it in 1959. An urban teacher of the 21st century who did not engage in these basic teacher acts as his/her primary means of instruction would be regarded as deviant. Not performing these acts in an urban school would be considered prima facie evidence of not teaching. The teaching acts that constitute these core functions of urban teaching are:

giving information, asking questions, giving directions, making assignments, monitoring seatwork, reviewing assignments, giving tests, reviewing tests, assigning homework, reviewing homework, settling disputes, punishing noncompliance, marking papers, and giving grades. This basic menu of urban teacher functions characterizes all levels and subjects.

Actually, there are occasions when any one of these 14 acts might be appropriate and even have a beneficial effect. Taken together and performed to the systematic exclusion of other acts, they have become the pedagogical coin of the realm in urban schools. They constitute the pedagogy of poverty, that is, what teachers do, what students expect, and what parents, the school community, and the general public assume teaching to be.

In order to teach science however students must be engaged in a systematic process of inquiry that leads them to question, observe, measure, analyze, and evaluate. While some simply refer to this process as thinking, it is an unnatural act in most urban schools. On the other hand, star teachers incorporate these processes into their daily activities. In this project method the teacher knows the major concepts to be covered in the district's stated curriculum. The teacher and students then together focus on the world around them and develop the major questions for study. While the teacher knows how these questions meet the district's stated curriculum expectations, the students know these questions are relevant to and meaningful in their lives. The questions are then reworked and reworded in ways that make it possible to gather data to answer them. Together, the teacher and students design ways for gathering the data needed. Working in teams, the students then gather and analyze the data. Ultimately, there is a synthesis of the team's findings in some culminating activity, exhibit, or presentation. Teaching that is comprised of such acts requires a great deal of planning, and a great deal of preparation of materials and equipment. It also requires a high level of organizational skills. In addition, such teachers have a high degree of trust in and control over students' behavior. They are not afraid to have students use equipment that is expensive and might be broken, or materials that might be dangerous if misused (the two most common rationalizations of science teachers who teach science from textbooks rather than in laboratories). Only teachers who have complete command of the subject matter, the instructional strategies required to manage various groups engaged in different activities simultaneously, and the relationship skills to connect with and build trust with students can teach science using the project method.

This process is the exact opposite of what usually happens: the school district mandates a test, then buys texts that cover the material to be tested for, then informs the teachers to follow the teachers' manuals accompanying the texts in a given time period. Science teachers in this scheme are no different than reading teachers. Students read the text about various topics, then answer "comprehension" questions. Such test-driven "curriculum development" is not carried out surreptitiously but boasted of openly as "curriculum alignment."

There are four curricula operating in urban schools: the written curriculum of the district stating what students are expected to learn, what the teachers actually teach, what the students actually learn, and what is tested for. The stated curriculum is the broadest encompassing the goals of national professional groups in science as well as stated standards. The second is a much narrower curriculum and reflects what the teachers are personally most comfortable teaching about. The third curriculum is narrower still and reflects those things in which the particular learners have become the most interested. The fourth curriculum (i.e., what is tested for) is, of course, the narrowest but dominant one and *de facto* pronounces the science knowledge of most worth.

How does Ms. M. teach in an urban middle school?

> Martin Haberman: When you first started to teach you used the project method right away. What happened?
>
> Mrs. M.: The kids were confused and I couldn't get them involved right away. They said to me, "Why are you so frustrated? Why are you making your life so difficult? Why don't you make your life easy? You give us the worksheets. We do 'em. You mark them and put them in our folders. And everybody's life is easy.
>
> Martin Haberman: They knew the system and because they liked you they didn't want you to have a hard time.
>
> Ms. M.: After they caught on, though, no one could give them a worksheet again.
>
> Martin Haberman: Another reason for the burnout teachers to dislike you.
>
> Ms. M.: I guess. The kids know the system and will play worksheets with you. But really they'd rather be learning.

The students' stake in maintaining the pedagogy of poverty is of the strongest possible kind: it absolves them of responsibility for learning and puts the burden on the teachers, who must be accountable for making them learn. In their own unknowing but crafty way students do not want to trade a system in which they can hold their teachers responsible for one in which they would make themselves accountable and responsible for what they learn. It would be risky (foolish) for students to swap a "try and make me" system for one that says to the students, "Let's see how well and how much you really can do."

INCLUSION

One of the problems urban teachers face is the large number of inclusion students placed in their classrooms. Private schools and charter schools in urban areas are typically contracting with major metropolitan school districts to take their students with handicapping conditions. This merely adds to the burden that naturally accrues to large systems seeking to serve large populations in poverty where there are higher numbers of students requiring special services. Overlaying the problem is racism. The extremely large number of students of color and students from minority cultural backgrounds who are diagnosed as handicapped cannot be attributed to chance. This situation is blatant in the numbers of African American males designated as deficient in some way. In Big City there are 145 full-time school psychologists assisted by 80 aides and 100 diagnostic teachers. This cadre of 325 "experts" have identified 17,000 out of 103,000 students as having handicapping conditions. In addition there are another 3,000 students in the pipeline waiting to be tested and labeled. This percentage leads the nation and reflects the layers of bureaucracy who benefit from the process of just labeling students. Unfortunately, these numbers also reflect the ignorance of some parents who seek to have their children labeled as handicapped in order to increase their monthly payments from the social security system (SSI).

Typically, each class of 25 students in Big City may have as many as 6 inclusion students. The numbers vary from school to school. I have worked with beginning middle school science teachers who have had over 30 students in a class, including 8 students with handicapping conditions in a room with no running water. Each inclusion student has an Individualized Educational Plan (IEP) on how s/he can best be taught. The regular teacher of science is expected to meet these IEPs in addition to meeting the needs of the students in the rest of the class. In eighth grade in Big City, for example, "normal" would include students who have basic skills that vary from 2nd grade level to high school level. The typical classroom would also have two or three diverse culture groups represented. Consider how Ms. M. deals with this issue of inclusion.

Martin Haberman: How do you deal with inclusion students?

Ms. M.: They do the same things as everyone else.

Martin Haberman: Can they?

Mrs. M.: Sure. It might be necessary to do some things over but they want to be successful like everyone else. I have no problem asking kids to repeat or practice things We all do stuff over all the time.

Martin Haberman Do you have any problems at all with inclusion students?

Ms. M.: We don't have enough computers in the school that are available to my classes. Some of my kids do real well with computers. I have to schedule a few hours a week for my students to have access. That's a problem.

Martin Haberman: Does the fact that you teach science help with inclusion students?

Ms. M.: It sure does. By working in teams they have specific data to collect and things to find out. There's no limit to what an inclusion student or any kid can learn. Some are learning more than "normal" kids. [Project] teaching of science lets it all happen.

Martin Haberman: How do you deal with the fact that the kids you get in any one classroom are so varied in their level of basic skills?

Ms. M.: No desire to learn is more critical than how much they know. Once I get them interested and active they improve their skills levels . . . a lot.

Martin Haberman: Do you read the records and get the recommendations of the psychologists regarding the inclusion students placed in your room?

Ms. M.: Not really. I just make sure I know if there is any physical condition I

should be aware of so that I can help in an emergency. If I listed the biggest problems I face teaching middle school science this one wouldn't even be on the screen.

CONTENT

The issue of content is critical once teachers are committed to being more than test tutors. Consider the following interchange with Ms. M.

Martin Haberman: How do you decide the content to be covered?

Ms. M: (Hands me a pile of curriculum guides containing modules in "Types of Energy," "Microorganisms," "Heat and Temperature," "Land forms," "Living and Nonliving Things.") Each one of these has core content, process outcomes, and key questions.

Martin Haberman: Take me through with an example of what you actually do.

Ms. M.: At the start of the year I give them an overview. Then I give them a choice of what we will study first.

Martin Haberman: This means the district sequence may be ignored.

Ms. M.": Yes. This also makes it difficult for my substitute teachers since each of my six classes may be working on a different unit.

Martin Haberman: Then what happens?

Ms. M.: We sequence the module. Children group themselves with each group trying to include someone who is a good writer, illustrator, reader, math student, etc. They pick who they like but they can be changed. Some individuals may start out without a group and just work near my desk. I had a kid who had difficulty expressing himself but knew a great deal about sharks. He was labeled LD (Learning Disabled). Eventually, he and the others who started out without a group moved away from me and joined a group. They don't want to be embarrassed at having to work with the teacher. They come up for discreet help.

Martin Haberman: What do you do?

Ms. M.: I make sure each group has clear questions, assignments for the members, and is working toward the agreed-upon objectives. About three or four times a week I will help each group as well as individuals and decide who needs reteaching.

Martin Haberman: Give me an example from the "Living Things" module.

Ms. M.: Sure. We raised and answered the question, "What does a typical biology student in our middle school look like?" They started with eye color, height, weight, skin color, lips, hair, head shape, widows' peaks, cleft palates, and moved on to the concept of dominant and recessive traits. Statistically, they compared individuals' traits against the school as a whole and the 6th grade against the 7th and 8th grades. They got into concepts such as homozygote and heterozygote.

Martin Haberman: Did every kid in every group have an assignment?

Ms. M.: Yes, but not necessarily the same assignment. Each student comes up with some testable hypothesis then struggles to propose questions for study.

Martin Haberman: Sounds as if you are teaching more than science content.

Ms. M.: Everything is related to everything else. We measure stuff (math?), write about stuff (language arts?), and learn about the uses of things in the real world (??).

Martin Haberman: So what are you trying to get at?

Ms. M.: My ultimate goal is to give them materials and have them create their own experiments. Then chart and graph their conclusions. Like in the Living Things unit we were just talking about, they came up with lots of questions. 'Do bigger kids have higher pulse rates?" "Do girls have higher pulse rates?" "How high do pulse rates get from running in the hall compared with sitting at your desk?"

Martin Haberman: So you can't guarantee that all kids cover the same material in the same way because they are all doing different things.

Ms. M.: That's right. But they also report to each other and share what they learn. Martin Haberman: How do you explain the fact that by using this method they still do well on the tests?

Ms. M.: It comes down to getting them interested and involved and keeping them asking questions and looking for answers.

Martin Haberman: How does your own background in biology help you in this kind of teaching?

Ms. M.: I know what's important and what to skip.

TEACHING GIRLS

Eighty percent of early childhood and elementary teachers are women. It is not unusual for children to arrive in middle school never having had a male teacher. There are numerous ramifications of this but since our focus here is on the teaching of science it is necessary that this situation be dealt with directly. The emphasis on the need for science teachers having content knowledge typically focuses on high school teachers. State-mandated tests include basic knowledge of the sciences as a requirement for teacher licensure in a particular field taught at the secondary level. This is not the case for teachers of young children. It is assumed that if they have had a methods course in the teaching of science they know basic concepts and are not themselves casualties of their science teaching. But the actual miseducation of children occurs much earlier than high school. By the time students enter high school a new generation of females have become science casualties. It is rare to find teachers of early childhood and elementary grades who have even rudimentary knowledge of basic scientific concepts. (If you ask elementary teachers to explain what causes summer they will tell you, just as they tell the children, that the earth is closer to the sun in the summertime.) In the typical first grade, every morning is spent rehearsing with the children the date, the day of the week, the weather, and the time. After doing this for 180 days most of the children can call the days of the week and even read them, but none can explain what a day is. Many will be able to recognize the words week, month, and year, but they will be unable to explain what they are. They will be able to recognize the words fair, cloudy, rain, snow, hot, cold, daytime, and night time, but they will not be able to explain why these things happen. They will be able to tell time but not know what time is. But these are not examples of the worst damage. The unforgivable miseducation that has occurred is that these children started school wanting to learn these things but have had their natural inquisitiveness had needs to know systematically deflected by their teachers. Their teachers have taught them that it isn't important to know why the sky is blue but to read the word blue. That being able to call and read the names of the months is important but knowing what a month is of no consequence. That reading the words wet and dry and knowing they are opposites is important (that will be tested for). But knowing what these states are is less worthy knowledge. Simply stated, learning to read the names we give things is more valuable than knowing what these things do, what they are made o1 and what causes them. At fundamental, rock bottom, this is what the teachers believe . . . that reading about things is important knowledge required to function in life but that understanding how things work can be an "elective" interest for the children just as it was for the teacher in their teacher education programs.

> Ms. M.'s conversation made it clear that she understood the special needs of the girls in her classroom given the nature of their previous instruction, the nature of the contacts they have had with other women, and the pressures they feel from their peer groups to focus on appearance and acceptance above all else.
>
> Martin Haberman: Is getting the girls involved a problem you work on?
>
> Ms. M.: It's a very important problem.
>
> Martin Haberman: What are some of the things you do?
>
> Ms. M.: One of the first things I do is not make a fuss if they want to leave the room or they think the work would make them sick. They can take the option of not doing it. I say, "If this is too much I will give you a glove. If you are afraid that's ok." If they have an option it really makes a difference. In almost no time they never take these options. They participate. Say they can't do something. I know this will change. Early in the semester I make sure to come in really dressed up with a nice dress and all made up, nice hair and no gloves. I just start doing a dissection of a pig or whatever we are working on... just to connect the thought that "prettiness" has nothing to do with our learning activities. Then I start wearing a coat and encouraging them to do so. My being a woman and an African American helps. They see me doing everything and because they relate to me they all eventually want to do it too. I didn't care about my nails or my hair or my hands when I dug into the work. But I didn't ruin or hurt my appearance.

Martin Haberman: Do you think you are changing their values regarding what girls can do?

Ms. M.: That's exactly what I'm doing. It won't just happen. You have to teach it.

Martin Haberman: What else is special in teaching girls science?

Ms. M.: Girls need to have a lot of discussion before an activity. They are very unsure and even more in the presence of boys. I spend a lot of time asking them what they think. They need a handle on everything before they do it. I have some girls just work with other girls.

Martin Haberman: Why is that?

Ms. M.: For self-confidence. They need confidence. After discussion they need the freedom to do it over again . . . and the time to do it. The girls help other girls and even argue with them. They show themselves they can keep their hair, nails, etc.

Martin Haberman: Peer teaching?

Ms. M.: Right.

Martin Haberman: Is there anything else you do?

Ms. M.: I let the shyest stay with the girls longest until they are ready to work in a mixed group. I allow them to build and test on their own, then work with other girls. They develop and find some interests that build their confidence.

Martin Haberman: Anything else. . . for these really reluctant girls?

Ms. M.: I never say to them, "I didn't hear you. Speak louder."

Martin Haberman: Yes, that's one of the most common mistakes I see in teaching.

Ms. M.: I always say, "I didn't understand you. I have a hearing problem." Once I get them to show some self-confidence they are the best. They have to be able to talk it through. You can't put them down. They have to be able to still be adolescent girls. You must take them on field trips so they can actually see connections. They are into detail: observation, careful, painstaking.

Martin Haberman: Anything else?

Ms. M.: I compliment them for their work and their minds, never for being pretty. I give them compliments like, "you really think like a scientist." Rather than "your hair is nice," or "your dress is cute," etc., I spend a lot of time talking to them about technical careers... that what they know and can do and not how they look is everything.

THE PRINCIPAL AND THE "SYSTEM"

It is quite common for star teachers to bump up against school authorities and rules. The reason for this is that the highest priority in large urban districts is not the learning of the children but the maintenance of the system. Star teachers regard themselves as buffers between the school organization and the best interests of their children. Stars advance and protect the learning of children above all else. Some do this more surreptitiously than others. Ms. M. is more directly honest. The reason these confrontations are inevitable is that stars are succeeding with the very same children that the other teachers are failing with or actually pushing out with suspensions. In order to protect themselves, teachers who feel threatened by the success of stars make it a regular part of their job to complain about them. School principals who are building managers rather than educational leaders inevitably support those following the rules... even if the rules work against the children and their learning. Consider the causes of the friction between Ms. M. and the principal.

Martin Haberman: You seem to do very well with the kids, the parents, and some of the other teachers, but the principal is a problem for you.

Ms. M.: Mr. X (two principals back) was fine. The current lady and I don't see eye to eye.

Martin Haberman: Why is that?

Ms. M.: Well, for one thing she called me into the office and said, "You can't buy anything unless it's approved by me."

Martin Haberman: Why would she say that?

Ms. M.: We needed some materials for experiments and I used my own money rather than filing the requisition forms.

Martin Haberman: Why didn't you file the requisition forms?

Ms. M.: It would take several months if it was approved and the kids wouldn't be able to do the work in the meanwhile. And she wouldn't approve my requests anyway. So the choice is between the kids getting what they need from me or not at all.

Martin Haberman: So you used your own money?

Ms. M.: Yes. I also had parents donate stuff like shovels, thermometers, aquariums, and other stuff. They loved being able to help.

Martin Haberman: What else did you do that angered the principal?

Ms. M.: Just about everything we do... running in the hail, wearing dirty clothes

Martin Haberman: Why were your students running in the hail?

Ms. M.: It was part of the tests they were doing on pulse rate. We did it at times no one was in the halls, but running violates a school rule.

Martin Haberman: How about the dirty clothes?

Ms. M. We were digging in the neighborhood as part of an experiment and having them come in dungarees and baseball caps violated the school dress code. There is also the issue of my wearing tennis shoes rather than high heels ... both for digging and my knee problem.

Martin Haberman: Any more examples.

Ms. M.: She didn't like our fingerprinting project.

Martin Haberman: Anything else?

Ms. M.: I sell the kids pencils in my classroom rather than from the school store because it's easier and more efficient.

Martin Haberman: Didn't you once purchase a large number of safety glasses in September for $.50 each rather than go through procedures and get them in February for $5.00 apiece?

Ms. M.: But that was another principal and there was no problem.

Martin Haberman: Any other reasons the principal doesn't appreciate your work?

Ms. M.: I don't suspend students. The principal recently suspended 25 kids from one class in her new get-tough policy. This doesn't eliminate or control fighting.

Martin Haberman: Anything else?

Ms. M.: When the principal tried to speak at an assembly the kids booed her.

Martin Haberman: Would it be fair to say that you don't have to do or say anything to be in trouble? Your success in teaching and in relating to the kids threaten her. She doesn't see you as an asset who can help the kids learn but as a threat to her authority.

Ms. M.: Yes. She finally called me in and informed me that in the future I would be teaching English.

Martin Haberman: So that if you return to this school next fall you will be teaching English?

Ms. M.: That's right.

Martin Haberman: How do you explain the 27% passing rate in science achievement?

Ms. M.: The people in the central office don't value science... they leave it to the individual school. The individual school doesn't make science teaching a priority. Neither the principal nor the assistant principals know anything about the curriculum in science or care. Because of the low priority there are few consumables.

Martin Haberman: So it's really up to the individual science teacher acting on his/her own?

Ms. M.: Yes, the teachers must be prepared in science and be willing to prepare interesting activities and lots of materials and equipment. It takes a lot of

planning and organization and then a lot of caring about the kids and extra time and effort on the part of the teacher. Some teachers just pick some things out of the book and teach what they feel like. Others feel it is just too much to cover so they don't try.

Martin Haberman: Any other reasons? [As if these were not enough.]

Ms. M.: There is no accountability for anyone—not the teachers, principals, or those in the central office for teaching and learning science.

Martin Haberman: I realize you can't change the world but if our goal was to just help save more kids in the present system what would you suggest?

Ms. M.: Teachers should have to take workshops related to the specific curriculum units they are given to teach. Teachers who just turn on the overheads and lecture or who just play videotapes should be monitored and held accountable. The teachers have to have the kids producing lots of work products, work samples, portfolios. Most of all the teachers have to stop assuming the kids are supposed to show up knowing everything and start teaching them what they need.

Martin Haberman: Anything else?

Ms. M.: From top to bottom, people have to understand that the only way to have power is to give it up.

Martin Haberman: (Long discussion of this concept in the Idylls of the King by Tennyson.) Knowing what you now know, would you do things differently?

Ms. M.: No. The kids come first.

WHAT DOES MS. M. MEAN FOR THE "BIG PICTURE?"

The conversation with Ms. M. continued for several hours. I believe she is typical of other star teachers of middle school science who teach children in poverty and whom I interview on a regular basis. The generalizations I would make based on the population Ms. M. represents are as follows.

Star teachers differ from others in that they perceive their classrooms as laboratories not museums. They seek to not only preserve what is the best of the past but to actively involve their students in finding out new things. This goal is to help their students become producers and not merely consumers of knowledge. In science the students use such production skills to share a common reality and not merely the perceived reality achieved by a student who might produce a piece of art or creative writing. It is only in science where a level of commonly accepted proof must be reached. A student who produces a piece of artwork has certainly gone to a higher level than if s/he had only seen the work of others. But this student does not bear any burden of sharing what s/he thinks he has learned (can prove) with others. The students of star teachers in science ask and answer more questions than their teachers ask of them. This makes science teaching, when it is well done, a completely different paradigm from the activities children typically engage in when they are taught other subjects.

By fostering a paradigm that teaches thinking and reasoning star teachers of science will, to a very great degree, be operating outside or even against the value structure of the school. Schools serving children and youth in poverty are managed by policies and rules based on authority not derived from reason or data. For example, if a middle school adopts a suspension policy for first-time offenses, there is no rationale required other than administrative fiat. No thinking or evidence is required to prove the value of such a policy. It is likely that a star teacher of science in a middle school would discuss an issue like suspension with his/her students, encourage them to raise questions and even do some experimentation whenever an individual begins to think some portion of the universe is placed in jeopardy. It is not an accident that Ms. M.'s students' questions started to irritate other teachers (e.g., "Why don't we raise questions and answer them in your class?") and annoy the principal (e.g., "Will suspended students learn more outside of school?"). The decision to shift Ms. M. to teaching English was not the result of chance.

A second basic understanding we derive from considering the work of star teachers of science is the reminder that exploring the world we live in, rather the teaching of low-level reading skills should be the center of the curriculum. In life, reading is a consequence of learning not the cause. *After* we have generated an interest or a need we read. The present focus in primary grades on reading as if it were a content area rather than a set of skills used for the purpose of learning the real content of the curriculum (i.e., math, science, literature, social studies) is historical refuse. Our first schools were started for children to learn to read the bible and to practice obedience to authority and for no other reasons. The fact that we subsequently became a republic with freedom of religion has not changed much. Currently the hegemony of reading over the curriculum in elementary grades is still maintained by female teachers who are victims of inadequate science teaching doing what they feel comfortable doing, i.e., replicating their own miseducation. The public supports the notion that "reading is basic" because it also reflects their school experiences as well. But when faced with the reality of how creative and interested in the world their children were before they sent them to school and how uninterested in learning their children were by age fifteen, parents understand very clearly that their children were systematically taught to put aside their natural and vital interests in order *to do well in school.*

A third message from Ms. M. and her cohort is that because it is dangerous to the staff's present teaching and maintenance of order in the school, science teaching is encapsulated. It is not allowed to spill over into other areas, either in the ways it is taught or in the nature of the content it is open to considering. Beginning in the primary grades the teaching of science may be done by a specialist for a limited period or two per week. Ms. M. and other star science teachers also typically offer integrated units in which they not only teach science but have the students read, write, compute, and engage in activities that look very much like social studies. ("Why is there more lead in our lawns than in the lawns of rich people on the other side of town?") By limiting the time devoted to science and by separating it from the rest of the curriculum the regular program, which is essentially reading in the elementary school, is left untouched. Also, by having science specialists the regular teachers free themselves from the responsibilities of having to learn anything about science. It is marginalized to just another specialty area such as art, music, and physical education. Encapsulating science in this manner accomplishes devastating results on the children. Star teachers help their students see meaning and relevance in their learning by focusing on the nature of the real world. The most natural and most effective integrated teaching units that children can experience inevitably have science learnings as their core since they deal with the children's reality and their natural concerns. By encapsulating and marginalizing science it is inevitable that the children will not be taught a curriculum of integrated thematic units (i.e., bringing several disciplines to bear) on the solution of the life problems that are most relevant in their lives.

Ms. M. and other star science teachers use information systems in their teaching. Opening the world to their students can only be accomplished by accessing the world. One of the basic reasons the achievement gap between advantaged students and those in poverty continues to grow is their differential in access to and uses of computers. Ms M. and other stars are keenly aware of how race, sex, and economic class control student access and seek to compensate for these differences in their programs and in the opportunities they can open up for their children in related school agencies that offer access.

The issue of how Ms. M. and other stars pursue programs of professional development is also crucial. The teachers who need to be reached with science courses are inevitably individuals who have not majored in science. Unfortunately, universities offer undergraduate courses in science during the day and only on campus. While this may seem to be an unimportant point dealing with logistics it is at the heart of the matter. Universities are, in effect, inaccessible to most of the teaching force. This leaves inservice workshops and other on-the-job mechanisms for reaching teachers, and the evidence is quite clear that such offerings do not reach the typical teacher and do not transform those with fears who are science illiterates.

The result of rethinking the gap between the practices of great science teachers and typical teacher practices is that I believe the gap will widen for children in poverty. Operationally, the more a youngster is dependent on schools for what and how s/he learns the more that student is "disadvantaged." Children and youth from knowledgeable families with means and access have a decreasing dependency on their schools for what they learn. They learn more in spite of not because of their schools. For children in poverty their dependence on their schools as their primary and indeed sole source of learning continues to grow. Selecting and supporting more Ms. M.'s may slow the rate at which the achievement gap widens. But admitting that schools serving children in poverty are a major cause of the problem is still to be acknowledged by educators or the general public. Until that occurs school reform will essentially continue on doing more of the same, harder and faster, in more efficient ways.

REFERENCES

Haberman, M. (1991, December). The pedagogy of poverty vs. good teaching. *Kappan*.
Haberman, M. (1995). *Star teachers of children in poverty*. West Lafayette, IN: Kappa Delta Pi.
Haberman, M. (1997, March). Unemployment training: The ideology of nonwork learned in urban schools. *Kappan*.
Haberman, M. (1999, December). Increasing the number of high-quality African American teachers in urban schools. *Journal of Instructional Psychology*, pp. 131—137.
Haberman, M. (2000, November). Urban schools: Day camps or custodial institutions? *Kappan*.

CHAPTER 4

JUDITH SWEENEY LEDERMAN & GREG P. STEFANICH

ADDRESSING DISABILITIES IN THE CONTEXT OF INQUIRY AND NATURE OF SCIENCE INSTRUCTION

INTRODUCTION

Science educators have a prepare scientifically literate students who understand the concepts, principles, theories, processes of science and limitations of scientific knowledge, and have an awareness of the complex relationships between science, technology, and society (Abd-El-Khalick, Bell, & Lederman, 1998; American Association for the Advancement of Science [AAAS], 1990, 1993; Millar & Osbourne, 1998).

However, if American education is devoted to offer opportunities for *all* students to gain sufficient schooling to help them make life choices and become productive members of society, it is essential that *all teachers* have the knowledge to make appropriate adaptations so that every student, regardless of ability or disability, can become an active participant in the learning process. The complexity of how to educate all students persists, and is further amplified by the current focus on scientific inquiry and nature of science within the curriculum. The degree to which schools are committed to developing opportunities for such learning for all students varies greatly.

Often, when it comes to educating all students the attitude today in a large majority of classrooms is "equal treatment is fair treatment," particularly for students who demonstrate academic achievement at acceptable levels. Many times the students most seriously shortchanged are those with disabilities who have average or above average academic performance. The only services these students receive relate to general accommodations (like translators, assistive access devices, and interpreters). They receive little or no accommodations in science classrooms or laboratories. They are expected to be the "observers" or, if they wish to participate, to become "experts" in providing their own suggestions for accommodation. Fair treatment is not about equal treatment; it is about giving all students what they need to have a successful learning experience. Yet, the perception of "equal treatment" persists. School personnel, the families of school children and even the students themselves generally believe:

- Students are responsible for their own learning.
- When students don't learn, there is something wrong with them.
- Schools are to determine what's wrong, with as much precision as possible, so students can be directed to the tracks, curricula, teachers, and classrooms that match their learning profiles.

L.B. Flick and N.G. Lederman (eds.) Scientific Inquiry and Nature of Science, 55-74.

These attitudes and beliefs must change. Educators must examine their roles and responsibilities in serving the needs of <u>all</u> students in our schools and to focusing on an awareness of unique conditions and needs of students with disabilities. Consider the few elements listed below as needed shifts in current educational practice to more inclusive education for all:

- A shift away from bureaucratic schools structured and organized according to ability, and a shift toward schools structured around student diversity with an instructional program that includes many different ways of organizing students for learning.
- A shift away from teaching approaches that emphasize the teacher as the one who disseminates content that students must retain, and a shift toward teaching approaches that emphasize the role of the learner in creating knowledge, competence, and the ability to pursue further learning.
- A shift away from the school's role of providing educational services and a shift toward a role of providing educational supports for learning.
- A shift away from trying to change or diminish a diverse school culture, and a shift toward valuing diversity.

WHY INCLUSIVE EDUCATION?

Most people's conditioning comes from a multitude of experiences with family and peers, plus observing others in a variety of settings and interacting with others. Interpersonal communication skills are reflected in a person's ability to transmit and receive information from others. But the life experiences for most persons with disabilities are significantly different than those of the general population. Their opportunities to experience and fully adjust to the mores of the majority culture are usually much more limited. In cases of sensory and/or processing deficits, the disabled person may not process experiences in the same way as the general population. The nuances of interpersonal communication are highly sophisticated, and even small deviations can transform feelings of trust and confidence to feelings of skepticism and avoidance.

Many elements of this day-to-day social integration may present special challenges for students with disabilities. During early childhood the quality and extent of interactions with other children and with the greater community outside the family may be significantly affected. During later childhood, group games and team activities are critical for effective social development. Students with disabilities are often excluded because accommodations are necessary to participate as a team member at the same rate or pace of others in the group. Adolescence has some special challenges as individuals investigate interactions involving intimate relationships with a member of the opposite sex. These settings are where adolescents discover interactions that are bonding, interactions that create negative perceptions, and interactions that create appropriate steps to move a relationship from one of acquaintance to one of love and caring. Yet students with disabilities are often excluded from the gatherings and parties where these interactions are played out.

Because of the special challenges of social integration for people with disabilities, the educational environment is an ideal setting to blend together persons with physical, cultural, emotional, and intellectual diversity into a cohesive, mutually supportive group. Group synergism can become more powerful than the collective output of each person acting independently. In a democracy, this is what society is searching for and what businesses seek in their employees. Why shouldn't it be the essence of what we wish to accomplish in schooling? To accomplish these outcomes, the talents and abilities of <u>everyone</u> must be considered. The environment and interactions must be adjusted and modified to meet the capabilities and potentials of each individual. Some things are changing. The personal computer has dramatically helped increase participation for persons with disabilities as new adaptations emerge from improved technology and engineering. This has allowed many persons with disabilities to greatly improve their receptive and expressive communication, possibly reducing the ordinate number of behaviors and interactions they must work through on a daily basis.

The nature of the issues of teaching students with disabilities is complex. Because of that complexity, there is a great need to educate teachers to help them plan substantive science lessons that include adaptations so all students can learn. By looking at the unique ways in which students' reason, their uniqueness can be fostered. All students need opportunities to express themselves, not within a prescribed role, not as an expert, not in accordance with rules and conventions, but as persons with unique skills and talents (Hopfenburg, Levin, Meister, & Rogers as cited in Zorfass, 1991). Teachers must concentrate on the processes of reasoning rather than the products. Such a focus on reasoning skills is clearly consistent with scientific inquiry (both knowing an doing) and understandings of nature of science. Each involve students in thinking about the logic of scientific investigations and the limitations of the knowledge that results.

Learners at all ages develop confidence in their ability to reason through logical relationships. In good teaching, the learners come to believe that outcomes are subject to their own skill and the ingenuity with which they conceptualize. Admittedly it is easier and faster to give answers (Otto, 1991). But if the aim of education is to form one's intelligence rather than to stock one's memory and to produce intellectual exploration rather than mere recall, then students must be presented with a learning environment that stimulates the use of inquiry processes in generating solutions to real-life problems (Zorfass, 1991). Helping students learn about the role of inquiry in science and nature of science requires teachers to develop learning experiences that focus not only on the products of inquiry but also on the characteristics of the knowledge that results from scientific endeavors.

Ryder, Leach, and Driver (1999) refer to the phrase "Nature of Science" as the "knowledge about how scientists develop and use scientific knowledge, how they decide which questions to investigate, how they collect and interpret scientific data, and how they decide whether to believe findings in research journals" (p. 201). Abd-El-Khalick, Bell and Lederman (1998) describe nature of science as meaning the "epistemology of science, science as a way of knowing, or values and beliefs inherent to scientific knowledge and the development of scientific knowledge" (p. 418). Regardless of the definition you wish to use, nature of science is clearly consistent with a focus on inclusive science instruction.

WHY INCLUSIVE EDUCATION?

Legislation from PL 94-142, SB 504, ADA and IDEA 1997 are all initiatives to provide equal opportunities for persons with disabilities to experience the same full and independent life available to the general population. To have the ability to use language and numbers to acquire information and make decisions is a tremendous advantage in modern society. The purpose of education in America is to develop these abilities in all youth, not only those who fit the general norms. To extend special services to those who are not fully developing these skills in a traditional delivery system should be as natural as changing your walking path when there is an obstacle ahead. The essence of decision-making is to modify and adjust in order to obtain a positive outcome. IDEA Amendments of 1997 affect the roles and responsibilities of the regular educator as a member of the IEP Team in several ways:

(1) Regular educators are members of the IEP Team. This requires them to be an active participant in the development, review, and revision of the IEP of students with disabilities served through collaborative measures.

(2) Regular educators will help develop, review, and revise the IEP as member of teams comprised of parents, administrators, and students themselves in collaborative interactions. This requires them to maintain open lines of communication, participate in IEP Team meetings, and implement interventions and adaptations recommended by Team members.

(3) Placement in the regular education classroom with access to the regular classroom and regular curriculum is strongly mandated by law from federal, state, and local level.

This requires the regular educator to adapt and modify classroom expectations to meet the needs and ability levels of the students with disabilities in the regular classroom (IDEA, 1997). It would appear that one of the primary ways a teacher can adjust classroom expectations for students of varied abilities and needs would be through an inquiry-oriented instructional approach. Indeed, science laboratory was seen as way to meet the needs of diverse students (although those with disabilities were not considered) in the early 1900s (Department of the Interior, 1918).

The Individuals with Disabilities Education (IDEA) and Americans with Disabilities Act (ADA) essentially extend equal opportunity for those with disabilities so they can experience the same services and opportunities that have always been accessible to the general population. They are not receiving something extra; they are gaining access to what the general population has taken for granted as being universally available.

The educational environment is an ideal setting to blend together persons with physical, cultural, emotional, and intellectual diversity into a cohesive, mutually supportive group. Group synergism can become more powerful than the collective output of each person acting independently. In a democracy, this is what society is searching for and what businesses seek in their employees. Why shouldn't it be the essence of what we wish to accomplish in schooling? To accomplish these outcomes, the talents and abilities of everyone must be considered. The environment and interactions must be adjusted and modified to meet the capabilities and potentials of each individual.

Lessons that explicitly teach about Nature of Science often provide platforms for all students to share their unique knowledge bases. Although people differ on their lists, there are basic tenets of nature of science that are pertinent and are applicable to K-12 educators' lives and are believed to be accepted across virtually all groups (AAAS, 1993; NRC, 1996;Lederman, 1999). These tenets state that scientific knowledge is (a) tentative (subject to change); (b) empirically based (based on and/or derived from observation of the natural world); (c) subjective (theory-laden); (d) partly the product of human interference, imagination, and creativity (involves invention of explanation); (e) socially and culturally embedded and it necessarily involves a combination of observations and inferences; and (f) necessarily involves observations and inferences. The functions and relationships between theories and laws is a more sophisticated representation of the distinction between observation and inference (Lederman, 1999). It is important to realize that although listed separately these tenets are closely interrelated and should not be examined individually (Bell, Lederman, & Abd-El-Khalick, 2000). These tenets make up the culture of the discipline, science as a way of knowing. Involving students' personal experiences in instruction is an accepted teaching practice to inform and enrich learning. With this type of involvement, students see a connection between the material being presented and their own lives. It is also satisfying for them to feel they have contributed to the learning experience. Often, students with disabilities are left out of these conversations because of the experiential limitations created by their circumstances of their disabilities. However, Nature of Science is a content area in which physically disabled students may have a greater bank of related experiences to add to a classroom discussion than the other students in a class. These students have often experienced years of medication, treatment, and adaptive technology changes. They know first hand the tentative nature of science. They have lived though the politics of the social implications surrounding funding for the research and development these medical advancements. The senses used by a hearing or visually impaired student may lead to a different set of observations and inferences than their classmates. A discussion of these differences can highlight the creativity and subjectivity of science. Giving students the opportunity to share what they know and to feel they have experiences that add to increased understanding and learning validates their worth as contributors to the learning environment and contributes to a positive change in their social status in the science classroom.

Students are dependent upon their science teachers for both curricular and instructional adaptations. The science teacher must assume responsibility for planning, managing and instructing in ways that keep students involved. Successful science teachers help students learn appropriate content, provide experiences where students can demonstrate competency in the application of process skills, and expose their students to social imperatives.

Inquiry is fundamental to developing an understanding of Nature of Science. The 1996 National Science Education Standards have targeted the abilities to understand and perform scientific inquiry as one of the four critical components of science instruction. Inquiry aids students in constructing scientific concepts, metacognition, life-long learning strategies, and in developing an actively curious mindset associated with seeking science knowledge (Haury, 1993; Martin, Sexton, Wagner, & Gerlovich, 1997). Elfin, Glennan, and Reisch (1999) describe science inquiry as being a cluster of values, methods, and activities.

To be challenged, students must relate the challenge from both the familiar and the unfamiliar to their own prior conceptions. Again, two possibilities exist: either students have prior conceptions and possess the capability to address the challenge, or they do not have the intellectual skills and their inability to equilibrate will likely result in frustration and their avoiding future situations perceived as similar. In the first case, conceptual change, through challenge and subsequent mental reorganization will work (Dykstra et al, 1992). In the latter, a knowledge base must be built before using inquiry.

The essence of inquiry can be traced back to Socrates who eloquently used the art of questioning to guide the learning of his students. Inquiry teaching involves drawing upon the intellectual abilities of the learner to solve his/her own problems. In this way, the effective teacher can gain insights into how information is processed and can bring to light both good reasoning and misconceptions that may have developed (Howe & Jones, 1993). John Dewey and Jerome Bruner originally brought the concept of "problem solving through inquiry" into the arena of pedagogical practice.

John Dewey (1910) argued that consciousness of an obstacle is the source of reflective thought. "The origin of thinking is some perplexity, confusion, or doubt" (p. 10). Following this line of thought, a problem in the context of school learning should be considered any situation that causes confusion, perturbation, or perplexity. Bodner (1986) pointed out: "Disequilibration plays an important role in learning. Students need to know that a problem exists before they are willing to accept an explanation" (p. 877). However, there is a "problem" with this idea of using a problem situation itself. How can a curriculum designer or a science instructor get students to become aware that a problem exists, and then elicit a student response using their initiative to solve the problem through their own thinking? If this awareness, of course, comes through disequlibration and confusion, then a discrepant event at the beginning of the instruction seems to provide a good starting point.

If a true picture of science is to be presented in elementary and secondary classrooms, students should be encouraged to invent and imagine. Bruner, the advocate of discovery based on inductive realism, reconsidered his ideas and accepted that "science . . . proceeds by constructing worlds . . . by inventing the facts against which the theory must be tested" (Bruner, 1986, p. 14). In fact, Bruner (1986) not only abandoned the "aboriginal reality," that is, a world existing independently of our interests and knowledge, but he also postulated the existence of two distinctive modes of cognitive functioning-- the narrative and the paradigmatic. The former gives rise to story telling, and the latter to logic and science in general. Although these two modes of thought are irreducible to one another, they are complementary since "many scientific and mathematical hypotheses start their lives as little stories or metaphors" (Bruner, 1986).

From a study conducted by Bell, Lederman, and Abd-El-Khalick (2000), attention to nature of science in a pre-service methods course makes science more interesting. It provides background knowledge necessary for critical thinking/problem solving, provides a more authentic context for learning and understanding science knowledge and its progression, and it provides links for students to make informed decisions in an advanced technological world. Smith and Scharmann (1999) add that an understanding of Nature of Science is "crucial to responsible personal decision-making and effective local and global citizenship" (p. 495). This is supported by the All of the national and state level reforms are calling for the teaching of NOS. However, many science teachers are not familiar with Nature of Science or how to teach it. It was not content included in their own educational backgrounds within K-12 science classes or in their teacher preparation programs. K-12 science textbooks usually provided little more than a brief history of science, chronically identifying a few great scientists and their scientific contributions. But the aspects of science that define it as a discipline, and sets it apart from other disciplines is seldom addressed leaving teachers with few tools to help them understand and teach NOS.

EQUITY ISSUES

Teaching science provides greater flexibility and greater challenges in instructional delivery than any other discipline. As a most basic concept, science can be described as a "search for truth." Within this pursuit of greater understanding is the knowledge (what we know), processes (the tools of the scientist), and attitudes and values (dispositions about the impact and the worth of science). As a broad field, the knowledge base of science is expanding at such a rate that no human being can be expected to know more than a fraction of information. The knowledge base of science contains several elements. Descriptive learning consists of verifiable bits of information; conceptual knowledge develops concepts and generalizations; laws and principles enable us to predict, in some cases with great assurance, about how objects and materials will behave; and theories, provide broad principles to explain a phenomenon. These are all part of the knowledge base in science. The science processes are the skills (observing, classifying, hypothesizing, experimenting, etc.) scientists use as they work to gain a deeper understanding of natural phenomena. The characteristics of the knowledge produced by science processes comprise what we mean by Nature of Science. Finally, the study of attitudes and values is necessary to use science to improve the human enterprise and the quality of life for everyone.

The teaching of science has undergone considerable change during the last century. Early in the 20th century lecture-laboratory was the most common approach. This strategy focused on dissemination of knowledge, both facts and scientific principles, through lecture and sometimes demonstration, followed by laboratory exercises in which students carried out experiments that verified the concepts taught during the teacher-directed instruction.

However, this approach to science teaching was criticized and challenged as the knowledge base expanded and the expectations of the workplace changed. American free enterprise is centered on capturing the unique and creative ideas of the individual to improve production, create new products, and adjust to a changing workplace. These attributes should be reflected in schools, the workplace of children. It is important in the educational process to nurture creative thinking and enlist free expression from students.

Changes in science curriculum led to important reforms in teaching and learning. Following World War II the ideology of science teaching changed to focus more on the actions of the students not as recipients of information, but as active inquirers in the learning process. The initiatives of the federal government as an active facilitator in the improvement of science curriculum following Sputnik led to major educational reforms. In the curriculum reforms developed during the 1960s, students were put in the position of the "original discoverer" and were supposed to discover a pre-existing truth. Many of these innovations are found in current curriculum materials. Students investigate various problems, observe, state and test hypotheses, collect and interpret data. In short, students behave as "junior scientists." The influence of thinkers like Dewey, Piaget, and Bruner led to a shift in the role of the laboratory, where the demonstration and reinforcement of ideas has given way to active investigation by students using the processes of science.

John Dewey systematized the progressive ideas of Jean Jacques Rousseau. Dewey proposed an educational theory that viewed activity and the problem-method as the main avenue for effective learning. He rejected subject matter curriculum, not because of the nature of the subject matter of itself, but because he objected to presenting students with a logical summary of adult experiences. This is why he favored the idea of having students reach their own conclusions by engaging in their own experiences (Dewey, 1910). Thinking, according to Dewey, is a capability people have acquired in the evolutionary process because of the challenges of the natural environment. And this is where he made one of his most important contributions to the nature of human thought. He rejected the idea that thought derives either from contemplation alone or from sensation. Man, he contended, starts thinking when a change in the environment affects his comforts, when circumstances offer different choices to a desired goal, and generally, whenever a problem arises. Therefore, the "problem method" is important to the learning process (Dewey, 1910).

Yet, the most influential factor that brought curricular reform both in Europe and in the USA in the early 1960s was the work of Jean Piaget and Jerome Bruner. They both saw learning as a process, in sharp contrast to the previous behaviorist approach that had viewed learning merely as a product. Learning through activity was the central point of their theories (Piaget, 1970; Bruner, 1963). A renaissance of curriculum development in elementary and secondary school science occurred after Sputnik, largely because of the infusion of federal monies into education and the commitment of the higher education community to support curriculum development efforts. These projects were grounded in contemporary learning theory and sound science. Funding came through the National Science Foundation along with private support. Progressive education was well under way, and new instructional models were developed. The emphases of the new approaches were on students themselves, and their central feature was based on the importance of children's active engagement in their own education.

SIGNATURE FEELINGS

To begin understanding how a school setting can become a supportive, inclusive learning community for all students, educators need insight into why their students feel/react as they do. Signature feelings are one major element of everyone's personal abilities to set and reach goals and to feel part of a community.

Signature feelings are inclinations, often unconscious, that have a significant effect on decisions individuals make about relationships. These feelings are often imprints from one comment or one episode that influence choices concerning interactions. These feelings often result in stereotypes (i.e. you can't trust..., ...has problems with alcohol) and inappropriate generalizations. They affect personal decisions (i.e. the comment overheard from your mother in a telephone conversation with your aunt--"...is a nice girl but she is not very good in mathematics") regarding career choices and goal setting. These signature feelings become carried out in everyday life. They relate to those you greet on the street or who you interact with in social situations and they relate to your academic dispositions (i.e. "I shouldn't take advanced algebra; I'm not very good in math"). They affect whom you make eye contact with, the areas you migrate to, and whom you avoid in a social context.

Persons with disabilities experience many challenges and disappointments resulting from the signature feelings they get from others. This often narrows their opportunities in interpersonal relationships, influences their choices of independence over interdependence, and limits their personal and professional lives. Consistency is critical in nurturing social behaviors. However, persons with disabilities frequently experience inconsistency because adults and even peers often vary their interactions based on prejudices and attitudes in a social context. To perceive trust but experience avoidance can be a de-habilitating interaction for an individual. The result of inconsistent social overtures often creates signature feelings such as not trusting your dispositions of acceptance, or expecting the pain of avoidance if a social context changes. The long term result often becomes to play relationships "close to the heart" rather than confidently reaching out toward collaborative opportunities. Yet skills of confidence and trust are the exact skills that enable people to advance professionally and move through the management hierarchy.

The role of educators is critical in nurturing the interpersonal development of students with disabilities. Each professional in an educational context needs to work vigorously to refocus his/her signature feelings about students with disabilities to become more inclusive and more accepting in their teaching. Part of being a professional means educators have a responsibility not only during work, but during all 24 hours of a day. They must constantly be proactive in modeling inclusive behaviors in their personal lives. Cultural change takes place when those most visible in society model behaviors that improve the quality of life for everyone.

Meeting the social-emotional needs of all students is difficult and some may perceive it as being outside the arena of appropriate educational services. To address these needs one must consider not only the generally accepted norms of mutual cooperation and respect, but also other elements such as etiquette and cultural values. Many students with disabilities (particularly those with physical disabilities, learning disabilities, and attention deficit disorders) have an input deficit or processing deficit that affects their ability to understand the subtle, and sometimes the obvious, messages in an interaction. Individuals with social-emotional anomalies are also more likely to experience inconsistencies in feedback concerning relationships. So, having an opportunity to engage in private interactions can be where those with limited social experiences gain the capacity to feel comfortable and secure in other informal social settings. The social context of the school should be safe and secure for all those present and assistance should be provided to help those who experience social anxieties.

However, most people are generally inclined to ignore the need to nurture appropriate social experiences for persons with disabilities. This inclination becomes more obvious when one looks at social events through another set of glasses. For the past 20 years over 10% of the U.S. school population has been identified as having a disability. This percentage is gradually increasing and will likely approach 15% in the next decade. However, in social occasions the percentage of persons with disabilities is significantly underrepresented. A large percentage of persons who have special life challenges avoid social contexts where their disabilities will affect their interactions. No one appreciates being rejected, and these dispositions are frequently transmitted in group situations, particularly when significant numbers of the people are unfamiliar with each other. Social experiences are frequently much more traumatic and have more long lasting avoidance responses (many times physical) than physical challenges related to disability accommodations.

In general, persons with obvious physical disabilities receive fewer negative overtones than those with "invisible disabilities." In fact, it is not unusual to hear in private conversations people discuss those with "advantaged disabilities." Some people consider "advantaged disabilities" to be deviate maneuvers by guardians or the individuals themselves to use a disability label to leverage additional services or preferential employment conditions. People usually react with surprise, and often disbelief, when they are told that persons with disabilities are not a "protected class" in the work place as are females and minorities. The Individuals with Disabilities Education (IDEA) and Americans with Disabilities Act (ADA) essentially extend equal opportunity for those with disabilities so they can experience the same services and opportunities that have always been accessible to the general population. They are not receiving something extra; they are gaining access to what the general population has taken for granted as being universally available. American schools purport to provide an appropriate and challenging assortment of educational opportunities for all citizens. In the case of students with disabilities, it may often result in a substantive expense, just as medical or legal expenses are accrued if a person encounters an illness, accident, or a miscarriage of justice. Even so, everyone should have a right to pursue happiness and to fully experience the resources made available to the general citizenry.

LEARNED HELPLESSNESS

Schools (educators) appear to be losing large numbers of students who disengage from their educational process and are inclined to become apathetic. The theory of learned helplessness provides some insights into this phenomenon. The phrase was first introduced by Seligman and Maier (1967) and Overmier and Seligman (1967) who reported responses of dogs whose failure experience was an unavoidable shock. This research has been replicated in school settings by Butkowsky and Willows (1980), Dweck (1975), Fowler and Peterson (1981), and Gentile and Monaco (1986,1988).

Gentile and Monaco (1988) described learned helplessness as a group of behaviors commonly noted after exposure to uncontrollable failure experiences. They state:

> After initial exposure to such trauma, the subject tends to increase movement, to increase emotionality as adrenaline increases with the stress associated with discomfort, and to increase motivation to search for solutions which will bring relief. As repeated attempts to gain control do not produce the desired alleviation of suffering, subjects come to believe that the situation is uncontrollable: their fate is unrelated to their behavior. Should such subjects subsequently be placed in a different situation in which their responses might be instrumental in controlling outcomes, they nevertheless cannot perceive the possibility of gaining control due to their past history. (pp. 16-17)

It is likely that learned helplessness may be influenced by changes in physiology of the individual, or vice versa. In studies of rats and humans who have experienced high stress there are higher than normal levels of cortisol and enlarged adrenal glands. The stimulated brain

can produce a cascade of hormones, each influencing specific centers of the brain which can signal physical responses in body organs. One hormone produced in the brain is called corticotrophin-releasing factor (CRF) which stimulates the adrenals to produce cortisol. When released into the bloodstream, the body's own cortisol affects mood, food intake, the sleep-wake cycle, and level of locomotor activity. Furthermore, hormonal stimulation can affect receptor cells and thereby alter the neural architecture -- the hard wiring -- of the brain (Kramer, 1993, pp. 115-117).

Teachers' and parents' lack of understanding of learned helplessness leads to untold suffering by many students afflicted with this condition. These students feel less competent which results in performance deficits unrelated to actual skill deficits (Butkowsky & Willows, 1980, p. 411). Seligman and Maier (1967) noted further that learned helplessness has been found to be de-habilitating to the individual to the point of obstructing his/her performance in school, social settings, and later in life.

From a constructivist point of view, the notion of learned helplessness becomes more meaningful, indeed, for what really matters is what learners bring to the classroom. And what they bring is not only their prior ideas about how the natural world works, but also their fears, anxieties, hopes, and expectations. It is all these that make the difference between failure and success and not just what students believe about electric current, action and reaction, or the greenhouse effect.

Teachers should bear in mind that the central idea of the theory of personal constructs is that the individual "is neither the prisoner of his environment nor the victim of his biography" (Kelly, 1955, p. 560).

RESEARCH SUPPORTING INCLUSIVE SCIENCE TEACHING

Findings in educational research appear to support inclusion as a more desirable alternative than segregated instruction for students with disabilities. Ferguson and Asch (1989) found that the more time children with disabilities spent in regular classes, the more they achieved as adults in employment and continuing education. This held true regardless of gender, race, socioeconomic status (SES), type of disability, or the age at which the child gained access to general education. Research reviews and meta-analyses known as the special education "efficacy studies" (Lipsky & Gartner 1989, p. 19) showed that placement outside of general education had little or no positive effects for students regardless of the intensity or type of disability. In a review of three meta-analyses that looked at the most effective setting for educating students with special needs, Baker, Wang, and Walbert (1994) concluded that "special-needs students educated in regular classes do better academically and socially than comparable students in noninclusive settings" (p. 34). Their review yielded the same results regardless of the type of disability or grade level. It would seem that instruction organized with students in mixed groups (e.g., scientific inquiry) would produce results consistent with the research on inclusive learning.

The reason why many students have supposedly failed to demonstrate critical thinking, creativity, and a number of higher order thinking skills might very well be found not so much in the content of what was supposed to be learned and demonstrated, as in the context of the instruction itself. And the context of instruction is not just the materials, such as springs, thermometers, and voltmeters to be manipulated, but the whole social and physical set up. "Context" is essentially a mental phenomenon. Things "out there" become contextual only when they are invoked, that is, referred to, assumed, or implied in what is communicated. The very act of naming things, or of assuming shared understandings of them, makes their reality for communicators a social and conceptual one, rather than one of simple physical existence in the surrounding world (Edwards & Mercer, 1987, pp. 160-161).

Implicit learning refers to an understanding of nature of science that can be facilitated through process skill instruction, science content coursework, and doing science (Abd-El-Khalick & Lederman, 2000). In other words, it is assumed students learn about science simply by going through the motions. Riley (1979) offers that implicit teaching is an affective goal. Abd-El-Khalick and Lederman (2000) describe the learning of nature of science through implicit means as a by-product of engagement in science-based activities. Teachers expect students to learn this information as a consequence of instruction or as a result of changes in the learning environment, despite the absence of direct reference to the nature of science. Examples of implicit teaching include laboratory tasks, lecture, project tasks (Ryder,

Leach & Driver, 1999) and performance of scientific activities with no reflection on the nature of the activity (Lederman, 1999).

A limitation of student directed learning is as assumption that all students, when challenged with problem situations, will persist in the learning task until they feel a sense of understanding or accomplishment. Many students elect not to persist when the perceive discontent or frustration, others need teacher affirmation to feel successful. A further complication is that the processing skills required in inquiry are not commonplace in the classroom and take considerable time to develop. The application requires a substantial long-term commitment from the teachers that utilize it. We believe the rewards are considerable but the efforts cannot be lighthearted from either the students or teachers.

Simply stated, explicit teaching refers to the direct teaching (not direct instruction) of concepts about the Nature of Science and is a method used to indicate what was taught (Palmquist & Finley, 1997). However, Bell, Lederman, and Abd-El-Khalick (2000) express that explicit teaching does not mean didactic teaching. Ryder, Leach, and Driver (1999) offer an example of explicit teaching which involves using video to initiate a discussion about the ways in which scientific ideas came to be accepted in and outside of scientific communities. Another explicit example would involve giving students partial fossils of trilobites and instructing them to draw the entire organism.

Abd-El-Khalick & Lederman (2000) report that explicit teaching is more effective than implicit teaching. Studies conducted in the late sixties and early seventies focusing on the effectiveness of curriculum that was hands-on and inquiry based did not support that the nature of science can be learned implicitly (Durkee, 1974; Troxel, 1968). Additionally, implicit studies conducted by Scharmann and Harris (1992), ands Spears and Zollman (1997) with the intent of improving conceptions or the nature of science showed no significant gains. However, studies did indicate significant gains in participant understanding of Nature of Science that utilized explicit measures (Akindehin, 1988; Ogunniyi, 1983). Likewise, conceptual change research supports that explicit instruction is necessary to address misconceptions that students hold from both implicit and explicit instruction (Butts, Hoffman, & Anderson, 1993). Akerson, Abd-El-Khalick, & Lederman (2000) determined that explicit reflective activity-based approach to understanding the nature of science was effective in enhancing teachers' views.

There is a good reason to believe that if students are given opportunities to negotiate meanings in a social interaction, the creative invention of concepts, the identification of the relevant concepts, and the development of their relationship might not be an idea so far-fetched. Edwards and Mercer (1987) provided evidence, not just arguments, that "when two people communicate, there is a real possibility that by pooling their experiences together they achieve a new level of understanding beyond that which either had before" (p. 3).

There is evidence that cooperative group learning results in student motivation and academic achievement (Slavin, 1989). Cognitive scientists and science educators emphasize the importance of the social context in the construction of knowledge. Norman (1981) explicitly stated that "human cognition exists within the context of the person, the society, and the culture" (p. 1), and Wheatley (1991) accepted that "knowledge is not something people possess in their heads, but rather something people do together" (p. 11). The idea of a "solo child mastering the world by representing it to himself in his own terms" has also been reconsidered, even by Bruner, the advocate of inquiry learning, who came to recognize that "most learning in most settings is a communal activity, a sharing of the culture" (Bruner, 1986, p. 127).

INSTRUCTION

A first and foremost challenge to teacher education is to familiarize educators and practicing teachers with educational practices that will allow students with disabilities an opportunity to experience a program that is challenging, rewarding, and commensurate with their capabilities. Professional educators must develop proficiency in: 1) how to provide access to knowledge in all aspects of the curriculum; and 2) how to make appropriate adaptations in curriculum content, teaching materials, physical settings, instructional strategies, and assessment instruments for students with disabilities. The challenge to educators is not only to improve teaching, but also to put in place mechanisms that elicit participation and responsibility on the part of the students. This must be characterized by clear vision and coordinated, consistent, and purposeful actions (Stefanich, 1994).

Piaget (1971) stated, "In the majority of cases, an intolerable overloading of the educational program may in the end do harm to both the physical and intellectual health of students, and retard their thinking proportionately to the extent we wish to perfect it" (p. 96). He continued, "When the school requires that the student's effort comes from the student himself rather than being imposed, and that intelligence should undertake authentic work instead of accepting predigested knowledge from outside, it is therefore simply asking that all of the laws of intelligence be respected" (p.159).

Teachers need to acquire the ability to transform their beliefs into classroom practice. They need to have a wide variety of pedagogical routines and approaches concerning organization and management of instruction (Lederman, 1999). Teachers need to focus on specific approaches that they are comfortable with (Abd-El-Khalick & Lederman, 1992) as those approaches can help to transfer knowledge into classroom practice (Lederman & Zeidler, 1987).

Research on instruction provides support for using a variety of methods to deliver instruction. Brophy and Good (1986), in a review of research on instructional methods, reported that the systematic use of a variety of techniques leads to higher levels of student performance than heavy reliance on one technique (p. 342). This is particularly important for instruction directed to students with disabilities. In many cases a teacher cannot discern the limitations of an impairment. Through a variety of instructional approaches, the student is able to bring input from learning experiences in an area of strength to compensate and fill in learning missed because of processing difficulty.

Highly effective teachers adapt and modify. Significant differences can be seen between teachers who teach groups of students and those who look at the uniqueness of each individual. Communication and understanding is needed for all students, but especially for those with disabilities. They do make the life of teachers more difficult and challenging because standard instructional materials are often not appropriate. The type and time of modification must reflect the context of the student's needs.

Research on instruction provides support for using a variety of methods to deliver instruction. Brophy and Good (1986), in a review of research on instructional methods, reported that the systematic use of a variety of techniques leads to higher levels of student performance than heavy reliance on one technique (p. 342). This is particularly important for instruction directed to students with disabilities. In many cases a teacher cannot discern the limitations of an impairment. Through a variety of instructional approaches, the student is able to bring input from learning experiences in an area of strength to compensate and fill in learning missed because of processing difficulty.

Multi-modality instruction is especially critical in helping students with disabilities gain a familiarity with the content material. Scruggs, Mastropieri, Bakken, and Brigham (1993), in presenting suggestions for teaching science lessons to students with disabilities, stated that students with disabilities are likely to encounter far fewer problems when participating in activity oriented approaches to education. The use of multi-modality approaches both in teaching and in assessment has shown positive effects (Cheney, 1989).

Wood (1990) noted that strategies which lend multiple exposures to new terms and concepts enhance opportunities for student mastery. Actual examples or models are especially helpful to students with disabilities. In many cases, these individuals do not have as many experiences as those with greater mobility who constantly receive a variety of exposures to the world by being able to visit places in a community.

Adjustments in the length of assignments and acquiring alternative resources with different levels of reading ability improve student success rates (Lawrence, 1988). To maximize the efficiency of the educational program, all teachers must provide students with assistance in basic skills (Brookover & Lezotte, 1979). If a teacher is to have standards of quality, students must be guided to meet the standard. This is going to typically require resubmitting an assignment until it reflects the desired quality.

For example, in preparing a laboratory report following an experiment, if students are to produce a report of high quality, there must be a process of refinement that requires careful planning with a scope and sequence. The teacher may begin with organization of the collected data, followed by data analysis and establishing inferences and generalizations. Checking arithmetic calculations and organization of the data may require several refinements. Developing a quality narrative demands that the science teacher has a good command of the mechanics of written expression; it may require collaboration with other teachers. Writing refinement may begin with capitalization, then progress to putting a subject and verb in every sentence and then to syntax and clarity of expression. The challenge offered to the student should be attainable. Guidance and encouragement must be provided. Having high expectations does not mean demanding higher standards; it means earning the higher standards by guiding students through the learning process in steps students can master.

This same guidance should be considered as teachers design their objectives and assessments for the nature of science because their instructional intentions significantly affect what occurs in classrooms. The objectives should be thought of as a cognitive instructional outcome (Bell, Lederman, &Abd-El-Khalick, 2000).

CURRENT PROMISING PRACTICES FOR ADAPTING MATERIALS

The use of adapted materials has been reported as being effective with mainstreamed students (Hoover, 1990). Teachers must be willing to accommodate instruction and adjust the learning environment so all students can achieve success and all students can be active participants in the learning process. They must become accustomed to teaching fewer concepts with richer insights, deeper understanding, greater attention to application, and more relationships. Instructors of science methods must model appropriate strategies in their classes if we are to bring about change in current practices.

Studies conducted by Peterson, Marx, and Clark (1978) showed the majority of decisions teachers make are about instructional context and little thought is given to objectives and outcomes. Rosenshine and Stevens (1986), in a meta-analysis of studies on direct instruction, found that more effective teachers maintained a strong academic focus and spent less time on non-academic activities.

Specifically, Olarewaju (1988) found 7th grade students had a better attitude toward an integrated science program in classes guided by instructional objectives. Brophy and Good (1986) reported teachers who plan and organize on a daily basis prior to instruction produce higher levels of student achievement. Prior planning is essential in preparing students for learning. Ausubel (1968) proposed the idea of advanced organizers to help students prepare for learning. By aligning objectives to prior knowledge and experiences, the students have greater opportunity to make personal meaning of the facts and concepts presented in instruction. Shostak (1990) recommended planning the first 5 minutes as "entry" into the lesson. This would include clarification of expectations, reflection using an advanced organizer, and introduction of the lesson topic. In a similar fashion, the last 5 minutes of the lesson should be directed to "closure" to reinforce the key points of the lesson and transfer learning by bringing in applications which relate to age-appropriate experiences.

EVALUATION.

Student evaluations in various forms determine progress and performance by making value judgements on the quality of students' work. Formative evaluation primarily provides feedback to students and guides them in their learning. Effective formative evaluation helps students correct misconceptions, gives the teacher feedback on the effectiveness of instruction and the appropriateness of the curriculum, and helps the teacher match instructional decisions with a student's instructional needs. A type of formative evaluation, immediate feedback is the most helpful to learners. It refers directly to a behavior or assignment that may challenge the students but it is a task possible to perform. Using immediate feedback, students and teachers can check on clarity through observation and questioning, help students unscramble confusion, and help them examine their understanding by performing authentic tasks.

Many students with disabilities will be unable to demonstrate their true level of understanding under traditional testing conditions. Although tests and quizzes are one means of evaluation, they should not be the dominant tool used in evaluation. In many instances, the added difficulty of needing to focus to a large extent on carrying out the mechanics of the response can hinder student performance. Assessment strategies where the student is allowed to share what he/she knows are often more effective. Students with disabilities must be allowed to receive examinations in a comfortable setting and on a timetable that enables them to share what they have learned.

The tests used to evaluate a child's special needs must be racially and culturally nondiscriminatory in the way they are selected and the way they are administered, must be in the primary language or mode of communication of the child, and no one test procedure can be used as the sole determinant of a child's educational program. Accommodations that may be provided include an architecturally accessible testing site, a distraction free space, an alternative location, test schedule variation, extended time, the use of a scribe, sign language interpreter, readers, adaptive equipment, adaptive communication devices, and modifications of the test presentation and/or response format (Thurlow et al., 1993).

The requirements of legislation are clear. Federal and state regulations will add specific details to these regulations. The issue that is less clear is how local school districts plan to implement the law and how educators' reported beliefs and reported practices will impact the implementation of the new law in local schools.

Other evaluations can be used in a science class. Conducting a science investigation monitored by a teacher is an excellent means of assessing student abilities to use science processes. Operational definitions allow students to indicate an understanding of a concept by relating examples or sharing thoughts on how something works.

The use of portfolios and exhibitions allows students to accumulate samples of their work. Paulson, Paulson, and Meyer (1991, p. 60) described portfolios as a personal collection of student work in one or more areas. The portfolio should reflect student participation in selecting the contents, student knowledge criteria for judging merit, and elements of student self-reflection. Through an anecdotal reflection, students can share what they know and what they can do.

A teacher's willingness to examine students' work to ascertain where problems are occurring has been shown to significantly improve student learning. Bloom (1976) and McDonald and Elias (1976) reported that improving the diagnostic ability of teachers results in improving student performance.

Summative evaluation is also an important component in meeting professional responsibilities as a teacher, but must be utilized with discretion. Summative evaluation is valuable in providing feedback on how much information the students have learned and retained. The results of summative evaluation are often used to make placement decisions, to certify students as being competent in a field of knowledge, or to norm and compare students. These measures can provide valuable information, but are generally of little value in helping students learning. Research has shown that student performance decreases when teachers emphasize an evaluation system based on a comparison of classmates' achievements (Brookover & Lezotte, 1979). Evidence has also indicated that increased student absenteeism and higher dropout rates occurred in classrooms with a higher percentage of lower grades, a competitive atmosphere, and arbitrarily high standards (Moos & Moos, 1978; Trickett & Moos, 1974).

SUMMARY AND CONCLUSIONS

The role of the effective teacher is to focus on the science processes and their meanings rather than the product (Martin, 1997), to focus on the students' reasoning skills rather than on accurate direct answers. Always keep in mind that the primary goal of the instructor is to develop higher order thinking skills in the learner, not to disseminate information (Victor & Kellough, 1997; Zorfass, 1991). Teachers must address nature of science or their students will suffer and lag behind those who have been exposed and understand this concept. Reflect on the statement, "If a teacher criticizes a child for inventing the wheel, s/he must be more interested in wheels than invention." It is an understanding of nature of science and scientific inquiry that empower students as citizens in a world with myriad scientifically and technologically-based personal and societal issues.

Teachers need to make adaptations when students are not successfully meeting the demands of the general education setting. Teachers must make adaptations when the learning style or skills of a student do not match the instructional delivery or content objectives (Stainback et al., 1996). Rather than relying on pull-out programs teachers should make carefully designed adaptations in the general education setting.

Collaboration is a supportive system in which teachers utilize the expertise of other educators to solve problems. The existing dual system of regular and special education has fostered definitive boundaries between these two areas with little sharing of expertise and support. Regular classroom teachers often find themselves basically alone in their efforts to serve students with special needs who are placed in their classrooms. This isolation makes teachers more resistant to the changes involved with including special needs students. Perceptions that may interfere with effective collaboration can become ingrained in professional practice. Unless they are brought to the surface, they serve as persistent bottlenecks to collegiality between professionals.

Students with special needs can benefit when adaptations are made in the classroom. Inclusive classrooms can provide rich learning environments for all students. Inclusive classrooms of all kinds provide teachers with the opportunity to design and implement both curricular and instructional adaptations. These adaptations can positively impact student learning. In inclusive settings, where adaptations are made, all children can learn, feel a sense of belonging, and achieve their educational and social goals.

REFERENCES

Abd-El-Khalick, F., Bell, R.L., & Lederman, N.G. (1998). The nature of the science and instructional practice: Making the unnatural natural. *Science Education*, 4, 417-436.

Abd-El-Khalick, F., & Lederman, N.G. (2000). Improving science teachers' conceptions of nature of science: A critical review of the literature. *International Journal of Research in Science Teaching*, 22, 665-701.

Akerson, V.L., Abd-El-Khalick, F., & Lederman, N.G. (2000). Influence of a reflective explicit activity-based approach on elementary teachers' conceptions of nature of science. *Journal of Research in Science Teaching*, 37, 295-317.
Akindehin, F. (1988). Effect of an instructional package on preservice science teachers' understanding of the nature of science and acquisition of science related attitudes. *Science Education*, 73, 73-82.
American Association for the Advancement of Science. (1990). *Science for all Americans*. New York: Oxford University Press.
American Association for the Advancement of Science. (1993). *Benchmarks for science literacy: Project 2061*. New York: Oxford University Press.
Baker, E., Wang, M., & Walberg, H. (1994). The effects of inclusion on learning. *Educational Leadership 52*, (4), 33-35.
Bell, R.L., Lederman, N.G., & Abd-El-Khalick, F. (2000). Developing and acting upon one's conception of the nature of science: A follow-up study. *Journal of Research in Science Teaching*, 37, 563-581.
Brookover, W., & Lezotte, L. (1979). *Changes in school characteristics coincident with changes in student achievement*. East Lansing: Institute for Research on Teaching, College of Education, Michigan State University.
Brophy, J. E., & Good, T. (1986). Teacher behavior and student achievement. In M. Wittrock (Ed.), *Handbook of research on teaching* (3rd ed., pp. 328-375). New York: Macmillan.
Bruner, J. (1986). *Actual minds, possible words*. Cambridge, MA: Harvard University Press.
Butts, D.P., Hoffman, H.M., & Anderson, M. (1993). Is hands-on experience enough? A study of young children's views of sinking and floating objects. *Journal of Elementary Science Education*, 1, 50-64.
Cheney, C. (1989). The systematic adaptation of instructional materials and techniques of problem learners. *Academic Therapy, 25* (1), 25-30.
Department of the Interior (1918). *Cardinal principles of secondary education*. Washington, DC: Government Printing Office.
Durkee, P. (1974). An analysis of the appropriateness and utilization of TOUS with special reference to high ability student studying physics. *Science Education*, 58, 343-356.
Dykstra, D., Boyle, F., & Monarch, I. (1992). Studying conceptual change in learning physics. *Science Education, 76* (6), 615-652.
Edwards, D. & Mercer, N. (1987). Common Knowledge: The Development of Understanding in the Classroom. N4: Methuen.
Eflin, J.T., Glennan, S., & Reisch, G. (1999). The nature of science: A perspective from the philosophy of science. *Journal of Research in Science Teaching*, 1, 107-116.
Ferguson, P., & Asch, A. (1989). Lessons from life: Personal and parental perspectives on school, childhood, and disability. In D. Bicklen, A. Ford, & D. Ferguson (Eds.), *Disability and Society* (pp. 108-140). Chicago: National Society for the Study of Education.
Haury, D. L. (1993). *Teaching science through inquiry. ERIC/CSMEE Digest*. (Report No. EDO-SE-93-4). Washington, DC: Office of Educational Research and Improvement. (ERIC Document Reproduction Service No. ED 359 048).
Howe, A. & Jones, L. (1993). *Engaging children in science*. New York: Macmillan.
Individuals with Disabilities Education Act of 1990. 20 U.S.C. 1400-1485.
Individuals with Disabilities Education Act Amendments of 1997, PL 105-17, 20 U.S.C. 1400-et seq., 105th Congress, 1st session.
Joshua, S., & Dupin, J.J. (1987). Taking into account student conceptions in instructional strategy: An example in physics. *Cognition and Instruction*, 4, 117-135.
Lawrence, P. A. (1988). Basic strategies for mainstreaming integration. *Academic Therapy, 23* (4), 335-349.
Lederman, N.G. (1992). Students' and teachers' conceptions of the nature of science: A review of the research. *Journal of Research in Science Teaching*, 4, 331-359.
Lederman, N.G. (1999). Teachers' understanding of the nature of science and classroom practice: Factors that facilitate or impede the relationship. *Journal of Research in Science Teaching*, 8, 916-929.
Lederman, N.G., & Zeidler, D. (1987). Science teachers' conceptions of the nature of science: Do they really influence teacher behavior? *Science Education*, 71, 721-734.
Lipsky, D., & Gartner, A. (1989). *Beyond separate education: Quality education for all*. Baltimore: Paul H. Brookes.
Martin, D. (1997). *Elementary science methods: A constructivist approach*. Albany, NY: Delmar.

Martin, R., Sexton, C., Wagner, K., & Gerlovich, J. (1997). *Teaching science for all children.* (2nd ed.) Boston: Allyn & Bacon.
Millar, R., & Osborne, J. (Eds). (1998). Beyond 2000: *Science education for the future.* London: King College.
National Research Council. (1996). *National science education standards.* Washington DC: National Academy Press.
Ogunniyi, M.B. (1983). Relative effects of a history/philosophy of science course on student teachers' performance on two models of science. *Research in Science & Technological Education,* 1, 193-199.
Otto, P. (1991). Modeling problem solving inquiry processes. *Journal of Science Teacher Education, 2* (2), 37-39.
Palmquist, B.C., & Finley, F.N. (1997). Pre-service teachers' views of the nature of science during post-baccalaureate science teaching program. *Journal of Research in Science Teaching,* 6, 595-615.
Riley, J.P. (1979). The influence of hands-on science process training on preservice teachers' acquisition of process skills and attitude toward science and science training. *Journal of Research in Science Teaching,* 16, 373-384.
Ryder, J., Leach, J., & Driver, R. (1999). Undergraduate science students' images of science. *Journal of Research in Science Teaching,* 2, 201-218.
Scharmann, L.C., & Harris, W.M., II. (1992). Teaching evolution: understanding and applying the nature of science. *Journal of Research in Science Teaching,* 25, 589-604.
Scruggs, T. E., Mastropieri, M. A., Bakken, J. P., & Brigham, F. J. (1993). Reading vs. doing: The relative effects of textbook-based and inquiry-oriented approaches to science education in special education classrooms. *The Journal of Special Education, 27,* 1-15.
Slavin, R.E. (1989). Students at risk of school failure: The problem and its dimensions. In R.E. Slavin, N.L. Karweit, & N.A. Madden (Eds.), *Effective programs for students at risk* (pp. 3-19). Boston: Allyn and Bacon.
Smith, M.U., & Scharmann, L.C. (1999). Defining versus describing the nature of science: A pragmatic analysis for classroom teachers and science educators. *Science Education,* 83, 493-509.
Spears, J., & Zollman, D. (1977). The influence of structured versus unstructured laboratory on students' understanding of the process of science. *Journal of Research in Science Teaching,* 14, 33-38.
Stainback, W., Stainback, S., & Stefanich, G. (1996). Learning together in inclusive classrooms: What about curriculum? *Teaching Exceptional Children, 28* (3), 14-19.
Stefanich, G. (Dir.). (1994). *A futures agenda: Proceedings of a working conference on science for persons with disabilities.* Cedar Falls, Iowa: University of Northern Iowa.
Thurlow, M.L., Ysseldyke, J.E., & Silverstein, B. (1993). *Testing accommodations for students with disabilities: A review of the literature, Synthesis Report 4.* Washington, D.C.: National Center on Educational Outcomes.
Troxel, V.A. (1968). *Analysis of instructional outcomes of students involved with three sources in high school chemistry.* Washington DC: US Department of Health, Education, and Welfare, Office of Education.
Victor, E., & Kellough, R. D. (1997). *Science for the elementary and middle school.* (8th ed.) Upper Saddle River, NJ: Merrill/Prentice-Hall.
Wood, K. (1990). Meaningful approaches to vocabulary development. *Middle School Journal, 21* (4), 22-24.
Zorfass, J. et al. (1991). Evaluation of the integration of technology for instructing handicapped children (middle school level). Final report of phase II. Washington, DC: Special Education Programs (ED/OSERS). (ERIC Reproduction Service No. ED 342 160).

CHAPTER 5

ANN M. NOVAK & JOSEPH S. KRAJICK

USING TECHNOLOGY TO SUPPORT INQUIRY IN MIDDLE SCHOOL SCIENCE

OVERVIEW

This chapter focuses on the potential of learning technologies to facilitate students' construction of deep understanding of science concepts and process through inquiry. We begin with a discussion of learning technologies including the potential of these tools to assist students toward in-depth and integrated understanding. A discussion of inquiry follows including how inquiry places learners at the center of knowledge building. Intertwined in both sections we discuss the learning principles upon which inquiry is based. In addition, the National Science Education Standards are integrated throughout illustrating how using the various learning technologies meet the Standards. The remainder, as well as the majority of the chapter, is devoted to providing examples of various learning technologies and how they aid learners in developing understandings. Within the entire chapter supporting evidence is provided of research on inquiry and learning technologies to support inquiry.

LEARNING TECHNOLOGIES

Our goal in teaching students science is to help them develop integrated understandings. Integrated understanding occurs when the learner builds meaningful relationships and connections between ideas (Krajcik, Czermiak, Berger, 2002). Understanding is a function of knowledge accumulation and knowledge integration (Linn & Hsi, 2000). It is possible for students to accumulate large numbers of isolated bits of information without creating relationships between those pieces of knowledge or by creating only weak links between ideas thus not demonstrating in-depth or integrated understanding. It is also possible for students to posses a limited amount of knowledge where each piece is well integrated or connected to other knowledge. With integrated knowledge students are able to apply their knowledge to a variety of new situations and contexts as well as to solve problems. Technologies when used as learning tools can be an essential component of an inquiry-based classroom that promotes students towards developing integrated knowledge that allows them to build strong links between ideas. These strong links allow students to transfer their understanding to new contexts.

When technology is used as a tool to promote learning it becomes a learning technology (Krajcik, Blumenfeld, Marx & Soloway, 2000). Learning technologies

L.B. Flick and N.G. Lederman (eds.) Scientific Inquiry and Nature of Science, 75-101.

can be powerful cognitive tools that help teachers' foster inquiry and student learning. They can help transform the science classroom into a learning environment where students actively construct knowledge (Tinker & Papert, 1989; Linn, 1998). They allow students to engage in aspects of inquiry that they would not otherwise be able to do. The National Science Education Standards state:

> A variety of technologies, such as hand tools, measuring instruments, and calculators, should be an integral component of scientific investigations. The use of computers for collection, analysis, and display of data is also part of this standard (p.175).

Computers, software, probes, hand-helds, digital cameras, and the Internet all have the potential to assist students toward in-depth and integrated understanding. They provide dynamic visuals to represent abstract concepts. Learning technologies are tools used by students to ask and investigate "What if?" questions. They allow students to create multiple representations of their understanding. In addition, learning technologies can help students access a variety of information, aid students in collecting various types of scientific data, provide visualization and analysis tools, promote collaboration and sharing of data both within the classroom as well as with other classrooms in the community or at remote locations. These tools may be used to help students create various hypermedia artifacts to represent their understandings (Krajcik et al 2000). They may also help students create models of complex systems (Metcalf-Jackson, Krajcik, Soloway, 2000).

Utilizing learning technologies in an inquiry-based classroom closely emulates how scientists work in the real world. Students can collect and analyze real-time data much like scientists do. They may create computer models to help visualize and explain phenomena. By sharing ideas and data via learning technologies they may create communities of learners. Just as they do with scientists, technologies actively engage students in the learning process by allowing them to interact with phenomena in more meaningful ways to develop more in-depth and integrated understanding of concepts and process.

WHAT IS INQUIRY?

The National Science Education Standards stress that science should be taught through inquiry. "Inquiry into authentic questions generated from student experiences is the central strategy for teaching science" (NEC, p.31, 1996). Classroom environments should be created where students are involved in scientific processes where they investigate various phenomena through observations, measurement, classification, experimentation, making sense of data, and drawing conclusions. This type of atmosphere involves students in similar activities to what scientists engage in and creates an environment for effective and meaningful learning. Students should learn to problem solve, think critically and make decisions within a context of a few broad unifying concepts where they learn the connections between concepts and principles and are able to apply their understandings to new situations. This is in contrast to covering large amounts of material where students learn facts or skills in isolation. Learning theory asserts that for understanding to occur multiple representations of ideas need to be utilized (Blumenfeld, Marx, Patrick, Krajcik and Soloway, 1998). When students make connections between these representations in-depth and integrated learning occurs allowing students to apply their knowledge to new situations.

In an inquiry, project-based classroom students investigate questions rooted in students' real-world experiences (Krajcik, J.S. et. al, 2002; Novak, & Gleason, 2001). These real-world experiences can be framed in meaningful and important questions such as "How clean is the water behind our school?" or "Where does our garbage go?" or "Why should I wear a helmet when I skateboard?" Because they are meaningful and anchored in real-world experiences these driving questions help to contextualize learning for students. These questions present opportunities to draw students in and direct them toward learning that has relevancy to their lives. The driving questions contain within them important science concepts. In the process of investigating these and other sub-questions students develop understanding of science concepts and science process (Schneider, Krajcik, Marx & Soloway, 2001). Students are at the center of knowledge building through their active inquiry into the question. They become a community of learners who are collaborating to answer their question.

In this student-centered classroom individuals actively engage in developing science knowledge. They construct meaning based on their experiences and interactions contextualized in the real world (Krajcik, J.S., Mamlok, R., & Hug, B. 2000). Students engage in various activities and develop multiple representations of their understanding as they engage in the extended inquiry science curricula. Learning technologies, embedded throughout the curriculum, can play an integral role in helping students develop multiple representations. In addition to teacher-designed activities, labs and experiments, an inquiry-based classroom allows for student-designed experiments. In collaboration with peers and with the teacher as a resource, students pursue solutions to their own sub-questions. They research background information using multiple resources, design experimental procedures, make predictions, carry out investigations, analyze data, share their findings and ask new questions. Various learning technologies such as Web Artemis, probes attached to hand-helds, and analysis software, can act as partners to help students more effectively engage in these processes.

Learning technologies utilized within an inquiry-based environment can support a community of learners. They foster student and teacher communication, they help students carry out investigations and they assist students to develop products (Krajcik, Czermiak, Berger, 2002). Through their active engagement with phenomena and through discussions with their peers, teacher, and other knowledgeable people students construct an in-depth and integrated understanding of science content and process. The teacher carefully and masterfully provides scaffolds throughout the curriculum to move students from novices towards an expertise in understanding and problem solving (Krajcik et al., 2002; Reiser, Tabak, Sandoval, Smith, Steinmuller, & Leone, A., 2001). In doing so, the teacher helps build an environment in which students can create a community where learning is continually pushed to the next level. In addition, these technologies can play a powerful role in enhancing this community by increasing both student and teacher motivation by actively engaging students in the learning process.

Dwyer (1994) found that using technology transformed the way in which teachers taught. Teachers transitioned from a teacher-centered classroom, mainly utilizing the lecture method of teaching, to a student-centered classroom where the teacher's role was one of a "collaborator" rather than an "expert". With learning technologies as the catalyst these classrooms were transposed from the teachers as the provider of information to one in which teachers assist students in developing understanding through active engagement with phenomena by asking "What if" questions. The results are more inquiry-based learning environments that align with the National Standards.

LEARNING TECHNOLOGIES - VARIOUS EXAMPLES AND THEIR ROLE IN SUPPORTING INQUIRY

Probes attached to computers

Electronic probes attached to computers allow students to digitally record and graph data thus providing students with scientific laboratory tools. Using probes to collect scientific data is not new. Microcomputer-based laboratories (MBL's) have been in use since the early 1980's. What *is* new is the development of more user-friendly interfaces and easily used yet more sophisticated software for data collection and graphing. There has also been a surge in the variety of probes that are now available for all types of data collection. The cost of such equipment has also decreased making the technology more feasible for schools. In the past, the science teacher who was a "technology nerd" was the one most likely to incorporate the use of probes in the curriculum. With all the advances in probeware the most novice technologically minded teacher can use probes to support students in inquiry.

Similar to activities of scientists, students can use probes and associated software to direct the computer to collect, record, and graph temperature, voltage, relative humidity, light intensity, pH, or dissolved oxygen data, to name a few. These technology tools may offer a fundamentally new way of aiding students' development of science concepts (Mokros & Tinker, 1987). Students can create a real-time graph as they set the probes to continuously collect data over a certain amount of time allowing them to ask their own "What if" questions. Probes enable students to conduct investigations that were difficult if not impossible to perform before the advent of this technology (Krajcik & Layman, 1992). Immediate graphical representations allow students to readily see trends in their data and focus on the concepts they are investigating. This may strengthen students' graphing, science process, and problem solving skills (Linn, Layman, & Nachmias, 1986). The result is better understanding of the concepts. For example, when students want to investigate "Which anti-acid works the best?" as part of a digestion unit[1], they can use new USB technology from PASCO Scientific and attach four pH probes to collect real-time data to investigate their question. Students place pH probes in four similar acidic solutions and start the experiment (See Figure. 1).

[1] This unit was jointly created by Ann Novak and Chris Gleason, Greenhills School, Ann Arbor, MI

Figure 1. Jackie, Cecelia and Alex investigate, "Which anti-acid works the best?"

They observe a quadruple line graph as it is created in real-time. Initially, all four lines are all the same. When students place a different anti-acid in each container the graphs begin to change. Over the course of the next twenty minutes the students' observation of the graph elicits a new question. The group's initial question, "Which anti-acid works the best?" focused solely with "best" as meaning the fastest. While viewing the real-time graph student's notice that a second, very important, factor emerges. Not only is time an important element but effectiveness is also a key element. Certain anti-acids more effectively neutralized the acid, moving the pH of the solution closer to seven. But they may not work the fastest. Students realize that, when deciding which anti-acid works the best they needed to look at both time and effectiveness to neutralize. Figure 2 shows a graph of the first two minutes of the anti-acid experiment. Using DataStudio software (www.pasco.com) the graph is created as the students conduct the experiment.

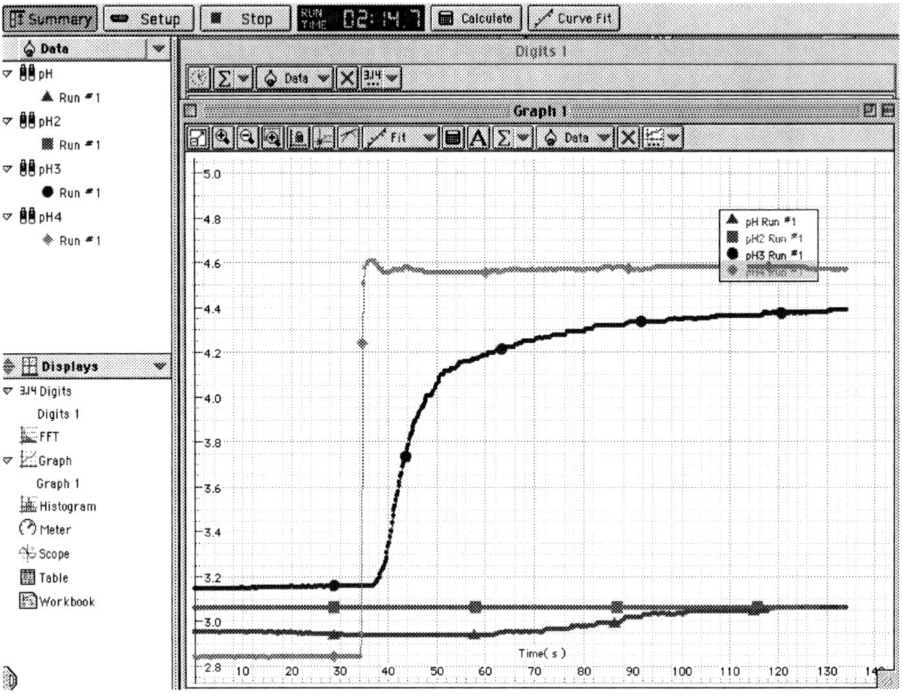

Figure 2. A graph showing four anti-acids as they begin to neutralize an acid

Conducting this experiment, using probes attached to a computer and seeing the graph dynamically visualized in real-time allows students to engage more deeply with the concepts. This learning technology has truly assisted students in inquiry. It allowed them to do something that they otherwise could not have done without it. In addition to collecting real-time data that was simultaneously graphed, it stimulated conversation among group members and encouraged them to critically look at the data, question the objectives of their experiment, and collaboratively revise their question to be more comprehensive.

Another example where MBL's can be used to help students develop understanding in such a way that they could not do without the technology is exploring the concept of evaporation rate as an indicator of relative humidity. In this investigation students attach two temperature probes through interface boxes to the computer. Both probes are covered with towels, placed in water and withdrawn. One probe is placed in a container that has had water swished in it (to increase the moisture content in the bottle). The other is left in the open classroom. As water evaporates from the towel students graphically see the temperature of the thermometer in the classroom dropping significantly while the temperature of the thermometer in the bottle stays relatively unchanged (because little evaporation takes place since the bottle is already filled with water vapor). This is a powerful learning activity that helps students develop understanding of relative humidity. It is one of several activities students engage in when investigating the question, "Why is it so difficult to forecast the weather?"

Students may use pH and conductivity probes to investigate questions about their drinking water. They may determine the boiling and freezing point of water using temperature probes. Motion probes can be utilized for many activities investigating force. All of these examples illustrate how MBL's can be used to foster inquiry. Using software that allows them to make predictions and explanations allows students to understand difficult science ideas (Linn & Hsi, 2000). For a detailed account describing the history of MBL see Bob Tinker's recently written a paper (Tinker, 2000).

Probes Attached to Hand-held Computers

Within the last five years hand-held learning technologies have been developed for schools. A variety of probes can be attached either directly to the hand-held or through a small interface box. These small technology tools become portable laboratories. The advent of these hand-helds opens up a whole new world for contextualizing science and fostering authentic inquiry. Students may now go to where the science is happening rather than simulating the science within the confines of the classroom. The Science Learning in Context (SLIC) project explored just this.

> "Field experiments can occur in real fields and streams, but also in streets, school corridors, home, neighborhoods, subways, workplaces, shopping centers as well as classrooms-any place where interesting phenomena can be observed. ...They make it possible for any place to become a rich context for learning" (Tinker, R. F., & Krajcik, J.S., 2001).

Students can now go to the drinking fountains in the school and test the water as they investigate, "Why does the water on the third floor taste different than the water from the gym?" (Warren, Rosebery & Conant, 1989). They can take these tools outside when investigating "How accurately can I forecast the weather?" and collect real-time temperature, relative humidity and barometric pressure data.

Figure 3. Andrew, Norah and Tom collect weather data using portable mini-computers

Figure 3 shows students collecting weather data outside of their science classroom. These students collect relative humidity, barometric pressure and temperature data over a period of time using portable technologies attached to probes. In addition, they collect wind direction data from homemade wind vanes, wind speed data obtained from charts, daily precipitation and cloud cover data. Each day data is collected, these novice meteorologists compare their own data with that of expert meteorologists by accessing up to date information from the Web. Students can then plot all of their data on one graph over an extended period of time and compare it with data meteorologists have gathered. Students can communicate with each other and the teacher and look for patterns in the data, compare these with the weather conditions for the day, and make connections between the various weather components and their relationship to the weather conditions.[2] Investigating the question "How accurately can I forecast the weather?" helps to contextualize the science and make it meaningful to students. The weather affects the daily lives of students. It will determine if there is soccer practice after school or if weekend plans for a picnic or ski trip will need to be altered. Therefore, understanding weather concepts is important and meaningful to students. A project, inquiry-based approach investigating this driving question and utilizing learning technology tools allows students to construct an integrated understanding of the concepts because they themselves have actively engaged with the phenomena. These novice meteorologists use the same tools that expert meteorologists use. Looking for patterns and relationships among the various weather components from data that they themselves have collected students can now forecast the weather and gain in-depth

[2] This weather unit was developed jointly by Chris Gleason and Ann Novak. Chris and Ann teach 7th/8th grade science at Greenhills School in Ann Arbor, MI.

understanding of weather concepts as well as an understanding as to why it is so difficult to forecast the weather.

Another example of students using portable learning technologies to engage in extended contextualized inquiry is with a stream project. Students can investigate "How clean is the stream behind our school?" with portable hand-helds by using pH, temperature, dissolved oxygen and conductivity (measuring amount of dissolved substances) probes (Novak, & Gleason, 2001). Students may also make qualitative observations such as how much turbidity is in the water, note organic debris in the stream such as dead leaves and sticks, note other substances such as litter, and also record current and recent weather conditions. Combining quantitative data obtained from the probes with various qualitative data, students have a comprehensive picture of the stream. They conduct a longitudinal study, collecting stream data during the fall, winter and spring. They look for patterns and relationships between the various tests during each season. They also look for trends in the data over the entire year. This project [3]exemplifies using learning technologies to foster inquiry because it allows students to construct in-depth knowledge based on a meaningful question with the assistance of portable scientific laboratories. The water near the school eventually flows into the water that is the main drinking source for students and in the watershed that students live. Figure 4 shows a student collecting water quality data using portable technology tools.

Figure 4. Jules measures the amount of dissolved oxygen in the stream using a probe

[3] Ann Novak would like to recognize Chris Gleason and Jay Mahoney, two colleagues from Greenhills School for their collaboration on the water quality unit.

Our students have created initial and final concept maps about water quality. Students' initial concept maps reflect novice understandings of water quality concepts. By contrast their final concepts maps portray rich understandings of the concepts. Research has reported substantial gains in our students' development toward integrated and in-depth understanding (Krajcik & Starr, 2001; Novak & Gleason, 2001).

The weather and water scenarios are only two examples of how hand-helds allow students to go *to* the science and move science beyond the four walls of the classroom. Both show how learning technologies aid students in doing inquiry in ways that they would not otherwise be able to do. They allow students to interact with the phenomena in a more meaningful way to develop deeper more integrated understanding. Bob Tinker and Stephen Bannasch from the Concord Consortium who started to develop probes twenty-five years ago are now developing a new generation of probes for handheld computers. They have developed what they call CCProbe which works across various platforms of computers including Palms and PocketPC handheld computers as well as MacOS, Windows, and Unix desktop operating systems. To learn more about these new developments visit www.concord.org.

Modeling

Modeling is gaining popularity as an effort of science education reform (Gilbert, Boulter & Rutherford, 1998; Gobert & Buckley, 2000). The National Science Education Standards emphasize that students should create models to portray their understanding of various science phenomena (NEC p. 145). Models help students represent complicated or sophisticated phenomena. In creating models students investigate questions, identify relationships between independent and dependent variables and discuss these relationships with peers. Models can help students, just as they help scientists, to both represent their understanding of concepts as well as construct new understanding. Models assist students in developing integrated knowledge by fostering discussions around relationships between variables. We are not suggesting using models to replace experiencing the phenomena, but rather that by creating models, both computer and physical, students can enhance and compliment their experiences with various science phenomena.

Building models is an essential component of what scientist's do. Now students use computer software to participate in this scientific process. They construct computer models based on physical models they create in class. These computer-generated models assist students in making sense of complex, real-world phenomena thus providing a context for deepening understanding of science. In addition, computer models can help overcome some of the limitations of physical models. One advantage of computer models is that time can be compressed. For example, decomposition is a slow process. In physical models of decomposition, days, weeks, and even months may pass before any visible changes can be observed. Building computer models of decomposition allows students to manipulate and test several variables while viewing instantaneous results (Novak & Krajcik, 2002). Creating computer models eliminates ethical concerns when investigating questions that involve living organisms. For example, when asking the questions, "How does acid rain or lack of water affect decomposition?" worms and other decomposers would not be harmed in a computer model. When investigating how communicable diseases are transmitted students can avoid getting sick because they can model the process (Hug & Krajcik, 2002). Another advantage of computer models involves student safety. Students can model the impact of high levels of acid rain on plants without concern of possible dangers if spills occur in class. Computer models allow students to manipulate variables that they might otherwise not be able to do. For example, students could increase the amount of mold using the computer and observe the affect on decomposition. In real life students could not directly increase the amount of mold. For phenomena that are too small for direct observation building computer models are ideal. For example, students cannot directly see bacteria. In computer models where students investigate how germs cause disease they may increase or decrease the amount of bacteria and observe the results. Finally, computer models can be more sophisticated, allow scientists to run more trials, and make it easier to understand how components of a model affect one another.

Model-It is a software program that allows students to create dynamic computer models of complex science systems (Metcalf-Jackson, Krajcik, & Soloway, 2000; GoKnow.com). It is designed especially for learners. It provides users with various supports and an easy to use visual and qualitative interface. Students build models based on questions by building qualitative relationships between identified variables accompanied by detailed descriptions to explain these relationships. As students work together, they dialog about concepts and wrestle with ideas and relationships to create models that accurately portray the relationships between variables. The process of model building assists students to more deeply understand the interrelationships and connectivity of the variables involved within any complex system (Spitulnik, Stratford, Krajcik & Soloway, 1997; Stratford, Krajcik & Soloway, 1998).

The software has prompts that foster dialog. But the teacher is vital in creating an environment that encourages student conversations. S/he can prompt students to think first about what exactly the students want to model and what objects and variables will be needed to accurately portray student understanding of the process. As well, the teacher will need to help students think about what relationships are needed to accurately build the model with emphasis on students having discussions about the causes and effects of those relationships.

Model-it includes three different modes - planning, building, and testing. After determining a driving question for their model, students plan the model by working together to choose the components of their models. These include the physical objects found in the environment and the associated variables.

Figure 5 shows a screenshot of Model-It where students are beginning to plan their model. In this model, these 7th grade students are trying to create a model based on the driving question, "How do people affect temperature and water quality?" The screen shows a palette of objects including people, a stream, weather and roads that students may choose from. In this picture a stream is used as the background and people, roads, factories, and trees are placed on it. These represent the basic components of the model.

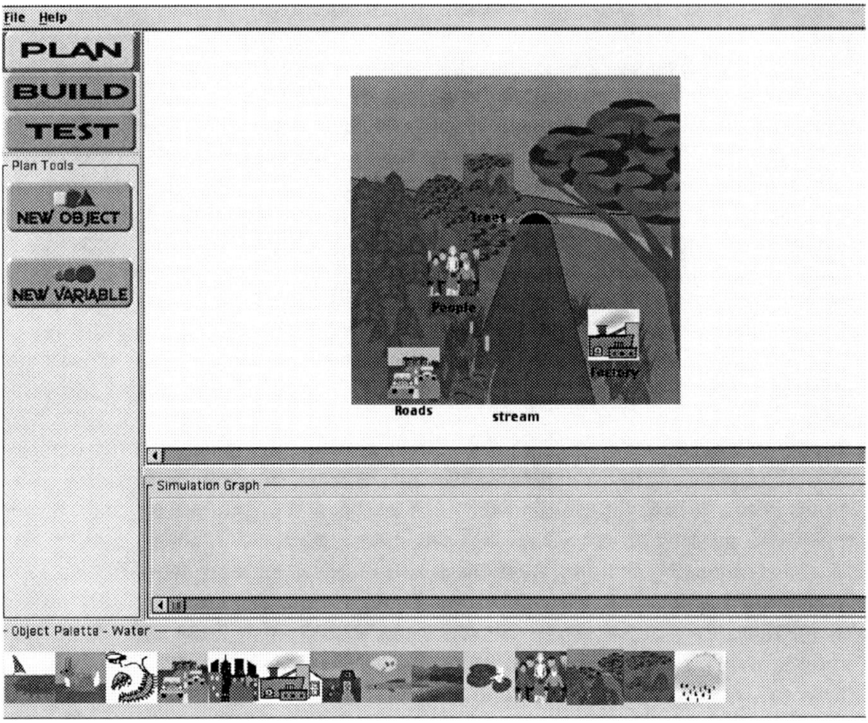

Figure 5. Students choose a stream, people, roads, factories and trees as objects for their model.

Once the question and basic components of the model are chosen students then create one or more variables for each of the objects. These variables should be based on evidence from in-class activities, experiments, text and Internet resources, and discussions with classmates and teachers. Students select a variable button on the screen and a window appears with prompts for students to select an object and create a variable associated with that object. Students also must decide on ranges, either qualitative or quantitative, for each variable. In addition, students must write an explanation, in text, to justify their choices. The variable editor, along with other prompts throughout *Model-it*, fosters discussion among partners to help learners clarify their thoughts and to challenge learners to make connections between ideas. Not only does this encourage collaboration it assists students to thoughtfully engage in the process and build meaningful relationships thus promoting in-depth and integrated knowledge. The prompts that are in Model-It, in our experience, have encouraged high-level discussions among students that we have been unsuccessful at facilitating through small group discussion, guideline sheets and other measures aimed at helping students analyze and find relationships between various components.

After variables for the objects are created then students build relationships between the variables based, again, on evidence from in class activities, experiments, text and Internet resources, and discussions with classmates and teachers. These relationships reflect how one variable affects another variable and are represented as independent and dependent variables.

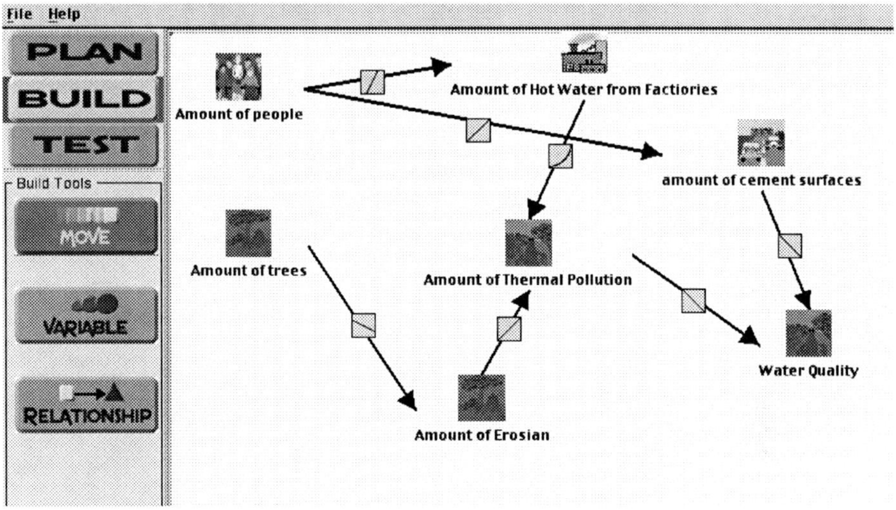

Figure 6. Students build relationships between variables to show how one variable influences another variable

The top portion of the model in Figure 6 illustrates the relationships that the students have created. It shows that as the amount of people increases the amount of cement surfaces increase. And as the amount of cement surfaces increase water quality decreases. Students have already explained, in the variable editor window layered beneath "amount of cement surfaces" that hot roads in the summer heat up rainwater and this hot water can run into streams and cause thermal pollution. As students build the relationships between the variables, new prompts appear that support students in building and explaining their relationships. For instance, from a pull down window options such as 'more and more', 'less and less' and 'about the same' appear that allow students to define their relationships qualitatively. Other prompts support students in writing verbal explanations of their relationships.

The final phase of *Model-it* is the testing phase where students actively predict how one variable will change as another variable is manipulated. Here, students see if the model they have created works the way in which they have planned. Figure 7 shows a screenshot where the thermal pollution model is being tested. In this picture we see meters and graphs.

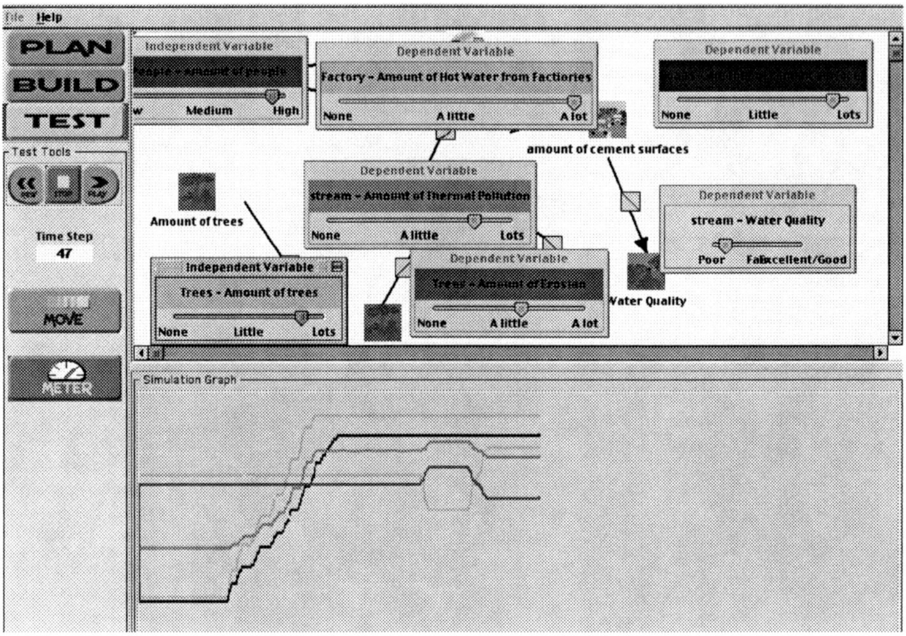

Figure 7. The testing phase of Model-It

As students manipulate the independent variable, for example, the amount of people, the dependent variables, the amount of factories, the amount of thermal pollution and water quality are affected. Students have told the program that as the number of people increases the number of factories manufacturing products will increase. These factories may dump hot water into streams that will result in increased thermal pollution. If thermal pollution increases than water quality will decreased. The graphs automatically change as the variables are manipulated. Amati (2000) further describes how to use Model-It and the benefits of using the tool in urban settings.

Students' models may serve both to help students develop understandings and as assessment tools. Model-It does not provide any content or mechanism to evaluate content. The accuracy of the model relies on the knowledge of the model builders and feedback from discussions with classmates and the teacher. This enhances student learning by placing the responsibility of understanding the concepts with the students. It also actively engages students in the critiquing and feedback loop because students need to rely on the accuracy of their own understanding in order to evaluate their peers' models. If, for example, in the thermal pollution model presented above, as the amount of thermal pollution increased the water quality also increased, rather than decreased, the students (and the teacher) would have to recognize the inaccuracy of this and give suggestions to improve the model. Based on feedback from peers and the teacher, students could then go back to revise and improve their model.

Computer modeling programs, such as *Model-It*, are powerful learning technologies that allow students to create models that they would not otherwise be able to do. In the process they help students more fully engage in inquiry. This is done as they build and test their models, as they critique their models and the models of their peers, and as they revise their models based on this feedback. In all aspects students are actively constructing an integrated understanding of the phenomena of which they are modeling by building models that portray their understanding of the concepts.

World Wide Web

Within the last five years we have seen an explosion of information that students can access from the Web. This can be both a blessing and a nightmare! At its best the Web can be an invaluable learning tool to assist students in doing inquiry because it allows students to gain tremendous insight into questions that they are investigating. Students can go online and obtain up-to-date weather data from local meteorologists and compare this data with data they've just collected with probes during science class. They can view weather fronts and track their changes. They can find data sets that scientists use. They can email scientists to ask them questions. They can communicate with students in classrooms across town or on the other side of the world. Students can research background information when designing their own experiments based on their own questions. Valuable information and activities may be found for any science area students are studying. These are all tremendously valuable uses of the Web. At its worse the Web can seem like an endless jungle of information where students are wondering throughout with a broken compass. Not only can the validity of some of the information be questioned but the appropriateness as well. Students can innocently type in search words and access inappropriate websites (Wallace, Kupperman, Krajcik & Soloway, 2000).

Amidst all the problematic potential of the Web is a huge positive potential to assist students in inquiry. Several very positive uses have been shared above. One of the greatest challenges listed above is to "shrink the jungle" and provide students with a compass when searching the Web but without minimizing the richness and depth of valuable possibilities for various information gathering.

One such fruitful attempt at providing students with support in using the Web is the Artemis Digital Library Project (Hoffman, Wu, Krajcik & Soloway, 2002) (http://www.goknow.com) and the National Science Teachers Association, SciLinks digital library (http://www.scilinks.org) A digital library has pre-selected and pre-approved sites that have been screened for content, validity and reading level of the learner. Using Artemis, students access a research engine that supports students in searching and retrieving information. Students ask driving questions, organize information, save bookmarks or URL's into personal student folders and view past search results. Figure 8 shows a screenshot of the Artemis workspace.

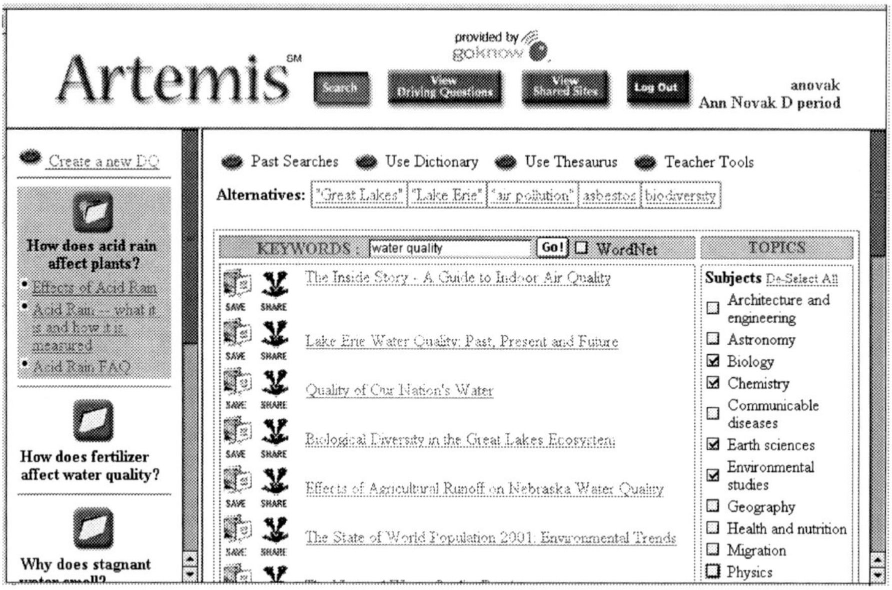

Figure 8. View of the main Artemis workspace

Students can share driving questions with other students who hold Artemis accounts. They may also post insights from articles into a shared folder for others to see. There is a section for students to take notes on articles and save them into their folders. Students may access these notes later as they write up their formal science paper. Once booked marked into their folder, students may access any of their articles from any computer with Internet access. Teachers also have access to all of their student's accounts and are provided with a record of the sites students have visited. Artemis is one vehicle for supporting students to use this learning technology, the Web, for inquiry. Students find it very easy to go back to find articles later.

We find Artemis to be extremely beneficial for several reasons. An essential component of inquiry includes students asking their own questions and designing experiments to investigate these questions. It is vital that students conduct background information searches to clarify questions, gain insights into the concepts associated with their question as well as information that will aid them in experimental design. Artemis facilitates all these levels of investigation. Prior to using Artemis our students would use class time and home time seeking out information often times without finding relevant information. Because Artemis has pre-selected websites, our students are now able to use their time much more efficiently because they can more easily find information that is relevant to their questions. The sharing feature of Artemis is another very worthwhile function. Groups of students are familiar with the questions other groups are exploring. If they find information that is relevant to another group they can post that site through sharing. They can include notes about the article. Figure 9 shows students from three different classes sharing notes. These students are 7th graders in the same school with three different science teachers. This sharing feature helps create a real community of learners. They are designing experiments based on water quality and are using Artemis to look up background information for the experiment. Some are sharing information with their partners. Others have found information that may be useful to other groups. The sharing function of Artemis allows students to share with their own class, classes in their school, or classes in other schools who have Artemis accounts.

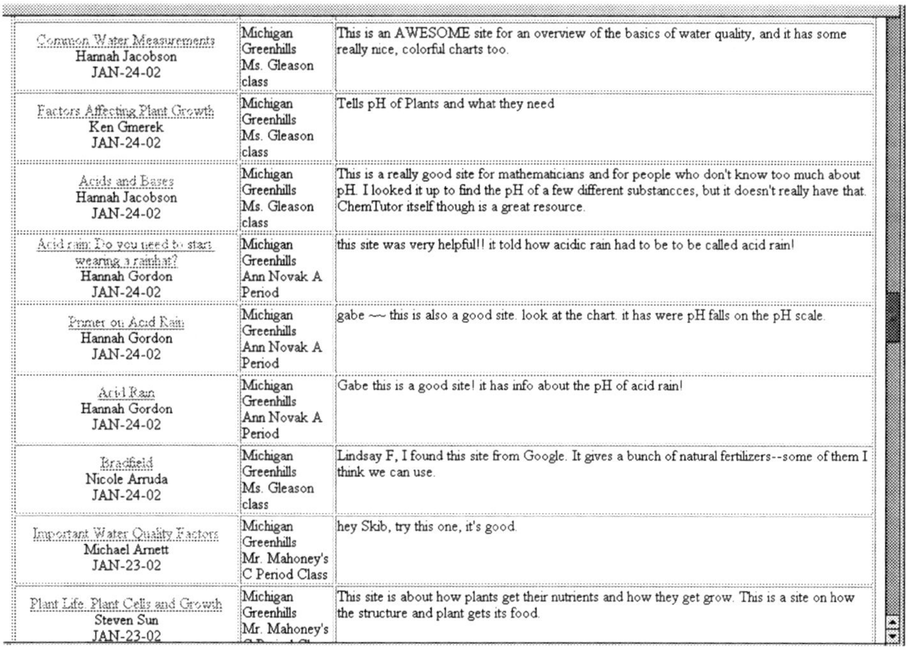

Common Water Measurements Hannah Jacobson JAN-24-02	Michigan Greenhills Ms. Gleason class	This is an AWESOME site for an overview of the basics of water quality, and it has some really nice, colorful charts too.
Factors Affecting Plant Growth Ken Gmerek JAN-24-02	Michigan Greenhills Ms. Gleason class	Tells pH of Plants and what they need
Acids and Bases Hannah Jacobson JAN-24-02	Michigan Greenhills Ms. Gleason class	This is a really good site for mathematicians and for people who don't know too much about pH. I looked it up to find the pH of a few different substancces, but it doesn't really have that. ChemTutor itself though is a great resource.
Acid rain: Do you need to start wearing a rainhat? Hannah Gordon JAN-24-02	Michigan Greenhills Ann Novak A Period	this site was very helpful!! it told how acidic rain had to be to be called acid rain!
Primer on Acid Rain Hannah Gordon JAN-24-02	Michigan Greenhills Ann Novak A Period	gabe ~~ this is also a good site. look at the chart. it has were pH falls on the pH scale.
Acid Rain Hannah Gordon JAN-24-02	Michigan Greenhills Ann Novak A Period	Gabe this is a good site! it has info about the pH of acid rain!
Bradfield Nicole Arruda JAN-24-02	Michigan Greenhills Ms. Gleason class	Lindsay F, I found this site from Google. It gives a bunch of natural fertilizers--some of them I think we can use.
Important Water Quality Factors Michael Arnett JAN-23-02	Michigan Greenhills Mr. Mahoney's C Period Class	hey Skib, try this one, it's good.
Plant Life. Plant Cells and Growth Steven Sun JAN-23-02	Michigan Greenhills Mr. Mahoney's	This site is about how plants get their nutrients and how they get grow. This is a site on how the structure and plant gets its food.

Figure 9. The note sharing section of Artemis

Students report that a component they find most useful is the organizational component of Artemis. Not only can they find and save information from sites in Artemis into their own private account, they may also save information found using other search engines. So all information is readily accessible to them in one area - their own workspace. This is important for several reasons. First, students can take and save notes from any sites into their account. They can then access their Artemis account any place they can gain access to the Web (home, school, Grandma's house, etc.). Second, when they create a bibliography there is never anymore searching for sites. Students often forget to write down bibliographic information as they take notes from articles. When it comes time to create a bibliography they are going back to the computer to retrace their steps to try to find articles. With Artemis this has now become a non-issue.

Hoffman, Wu, Krajcik and Soloway (2002) expand on early attempts to describe how students interact and learn from on-line learning environments like the World Wide Web. Their findings indicate that Artemis does assist students to develop accurate and in-depth understandings if they appropriately use search and assess strategies, resources are thoughtfully chosen and when the teacher provides extensive support and scaffolding. The Web has a vast potential to be a powerful learning technology. But there are many problems associated with it as well. Artemis, and NSTA's SciLinks are programs that successfully provide students with support in utilizing the Web for inquiry.

WISE, Web Integrated Science Environment (http://wise.berkeley.edu) provides another excellent use of the Web. WISE is a web-based learning environment that allows students to examine real-world evidence and analyze current scientific controversies. The various curriculum projects are designed to meet standards (Linn, 1998).

Hypermedia Artifacts

Learning technologies allow students to construct artifacts to portray their understanding of science. Among these are HyperStudio (http://www.hyperstudio.com), PowerPoint by Microsoft Office, and student generated web pages. Students can incorporate text, digital pictures, movies, graphs and more into their documents to create a comprehensive picture of their understanding of various phenomena. For example, students may create hypermedia documents of experiments that they are designing and conducting. These documents illustrate the project from beginning to end becoming living artifacts that represent the steps, both procedural and conceptual, students have taken throughout the process. As a whole they portray student understanding of the science. We provide students with guidelines to conduct their experiments as well as create their hypermedia document. These include text with the question and library research that forms the background of the experiment and which becomes the basis for predictions of the experimental outcome. In addition images captured with a digital camera may be used to drive or clarify student predictions as well as illustrate the experimental set-up. Figure 10 is an example of a procedure from an actual 8th grade project in photosynthesis. The class explored the question, "How do plants get their energy?" The students asked their own sub-questions. This group asked, "Do different types of plants have different rates of transpiration?"

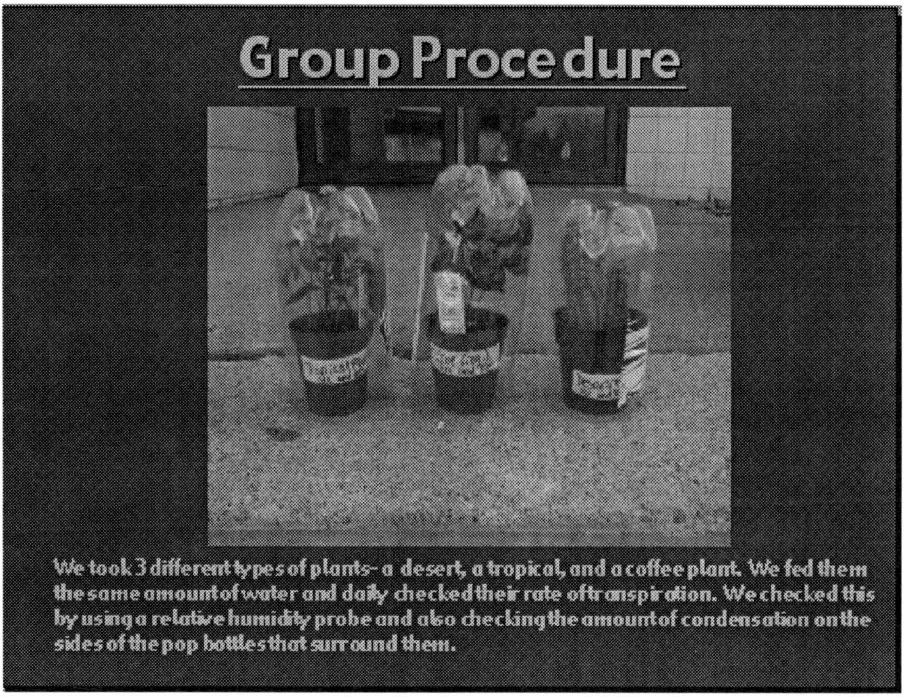

Figure 10. Procedure using PowerPoint from 8th grade science experiment on transpiration rate of various types of plants

This slide conveys to the audience that the students used relative humidity probes to collect data from three different types of plants that were under plastic domes created from pop bottles. Data collection in hypermedia documents may include text with qualitative data, quantitative tables and digital pictures to show sequencing of events or to capture important images of the experiment. Drawings or movies may further convey student understanding. Graphs can be included with discussion of the results where students analyze the data, form conclusions and come up with new questions.

We provide students with guideline sheets that contain clear written expectations for what should be included in an analysis. These include three main components: reporting the results, explaining what those results mean by tying in background information with the data and their own insights and speculations, and comparing their predictions with the results they obtained.

As teachers, we find that students' products, the entire PowerPoint document, provides us with an accurate assessment of the students overall understanding of both the concepts and the scientific process. We also find that we can support students with feedback as they create the PowerPoint document. They are simultaneously conducting the experiment and creating the document over a several week period. We are able to have dialog with student groups on a regular basis. Through our interactions with students we can provide guidance about their thinking of the science and with the process of creating the document that will portray their understanding. This gives students opportunities to rethink and revise their understanding and develop deeper meaning as they experience the phenomena.

Documents that include multiple representations of concepts are especially important in assisting students to develop conceptual understanding of scientific concepts (Kozma, 1991). By building thoughtful connections and constructing meaningful relationships among various representations, students develop more integrated understandings (Linn, 1998).

Figure 11 is an example from one page of an actual 7th grade web site about water quality at a section of a stream they study over the course of a year. This page illustrates the dissolved oxygen (D.O.) level of the stream throughout three seasons, fall, winter, and spring. The students have included a picture of a small waterfall with an explanation that waterfalls can increase D.O. levels. They include a graph of the data collected over the three seasons along with water quality standards. They explain that "fish and other organisms depend on D.O. for their oxygen supply."

Figure 11. 7th grade web page on water quality

One component of all of the above hypermedia examples is the use of digital pictures. Digital pictures can assist students in inquiry in several ways. They can show time lapse data. For instance, students can take snapshots of seeds sprouting to capture the sequence of growth. They can show various stages of an experiment to view later when analyzing the data and sharing the experiment with other students.

They can use digital pictures to support predictions. For example, they can take a digital camera out to the stream. If dissolved oxygen is created in a stream from fast moving water capturing oxygen from the air, students can take pictures of waterfalls with water rushing towards the bottom and use this, in addition to explaining in text, why they are predicting a high dissolved oxygen reading at that location. If decomposers use oxygen and the stream location has enormous amounts of organic waste, dead leaves for example, students can capture the image on cameral to show why they predict low dissolved oxygen at that location. Images of soap bubbles, dead leaves in the water, algae growth, and suspended dirt particles can all be used to help student make predictions about their stream quality. Students can take pictures of their experimental set-ups and well as the results of the experiments. These images are then easily integrated into hypermedia documents to form multimedia presentations (Rivet & Schneider, 2001).

Artifacts like students' web pages, and those created using Hyperstudio or PowerPoint are all examples of hypermedia learning technologies that assist students to create products that portray their understanding.

Hand-held Computer: Applications

There is a whole suite of learning technology application programs for hand-held computers. These have come about within the last five years. Several are discussed here.

Wireless Internet access

Perhaps the most exciting tool that has recently been developed that helps promote inquiry is wireless Internet access. Still in its' infancy stage this technology promises powerful applications both inside the four walls of the classroom and beyond. Students may access the web using a handheld computer without wires or plugs. When a question arises during a classroom discussion, students may instantaneously access the web with their handhelds right from their seats. It is merely the logical extension of the cell-phone generation. Students no longer have to go to the computer lab or move to a classroom computer. In schools where students have handheld devices and wireless internet access students can go on the web, look for information to clarify a question or idea, discuss it as a group and continue the discussion with greater understanding. All of this can happen within minutes. Student can use information from the web to help them make connections between ideas as they occur. In the past teachers might convey to the student that it was a great question, have students look it up for homework and come back the next day to discuss it. By then the momentum may have passed; the window of opportunity may be closed. Or the teacher may say there isn't time and the question or idea simply passes. Now information is only a moment and fingertips away!

Another application of wireless Internet access is to obtain scientific data. In an example discussed earlier in this chapter, students go outside using probes attached to hand-helds to collect weather data. While students are outside collecting data they may use wireless Internet access, within a short range, to instantaneously attain current weather data from the web and compare their real-time data with the data collected by experts. They can be at a nearby stream collecting water quality data and find interesting phenomena that sparks their curiosity as well as influences their stream. For example, students may find algae and want to know more about it. They don't have to wait to go back to the classroom but can look on the web right from the stream. Out in the field they can find pictures that match what they are seeing rather than trying to remember what the algae looked like. Each of the scenarios above illustrates the power of wireless Internet access to help promote inquiry.

Go and Tell

This software combines the use of digital pictures and annotations (http://hi-ce.org/palms) using the Palm and the Palm's digital camera. Students take pictures and write text about the images right on the Palm. They can do this right out in the field. Back in the classroom, students readily upload their files to a computer. This learning technology is very portable and allows the students to go where the science is happening. Students can take pictures and write text about the phenomena they observe.

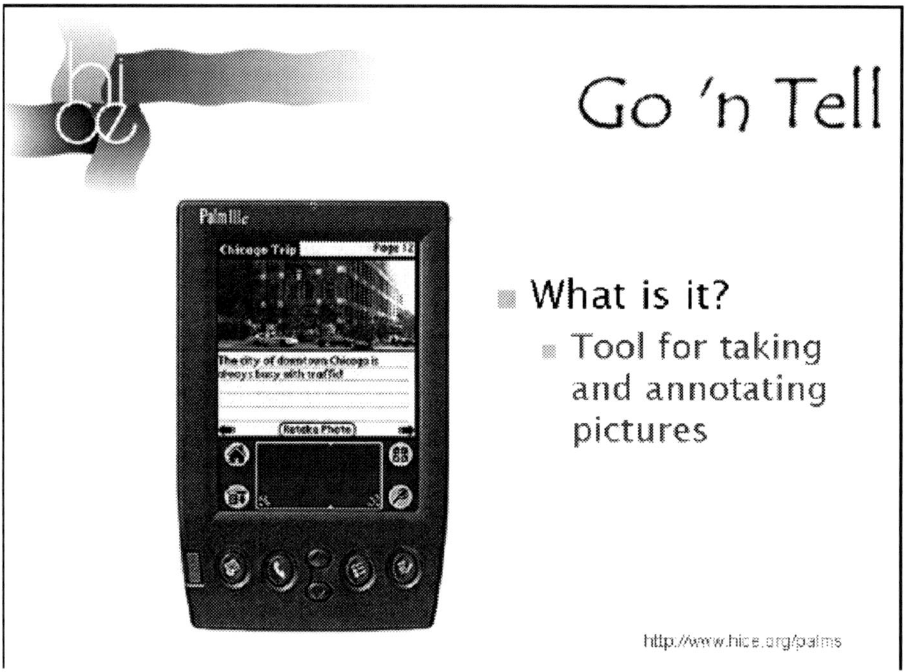

Figure 12. Go 'n Tell software for the palm. Students take pictures and write notes

They can create virtual scrapbooks of fieldtrips. Go 'n Tell can be used to document science experiments. Students can upload their work for printing. In addition they can easily upload their documents to create web pages. Figure 12 shows an image with text created using Go 'n Tell software. It is another fine example of a learning technology that promotes inquiry.

Interactive PicoMap

Concept mapping is a tool invented to help students make connections and see relationships between concepts (Novak & Gowin,1998). Interactive PicoMap (http://hi-ce.org/palms) is a software program that allows students to electronically create concept maps with one added bonus. Two or more students create one concept map together. Each student uses a hand-held. One student invites the others to join the map making process. Together students share ideas and draw concepts and linking words that appear on every participant's concept map. The process encourages student discussions about what to place on the concept map and why. It is truly a collaborative process. Through discussion a consensus must be reached to add concepts and linking words as well as notes in a separate section that "hides" under the concept map. The process of discussion and concept map building facilitates students to make connections between concepts. As a result students develop a much more integrated and in-depth understanding of science. Hand drawn concept maps can look "messy" and be difficult to follow for both students and teacher. PicoMap is an electronic process; it is easy to add, delete, and rearrange concepts and linking words. This put the emphasis on the concepts rather than either trying to decipher what is written, erasing or crossing out sections as new ideas or arrangements make sense, or re-copying concept maps. Concept maps may also be drawn individually to assess student understanding initially, during or at the end of a project.

PicoMap, Go 'n Tell and wireless Internet access are only three software applications of hand-helds. There are many more. These are three examples of learning technologies that truly facilitate students doing inquiry.

Visualization Tools

Visualization tools are software programs that allow students to visualize data collected at both local and remote sites thus enabling them to analyze scientific data. Such programs allow students to explore and manipulate data to see trends not possible without such tools. Features of these programs are similar to those found in powerful, general-purpose visualization environments and are designed specifically for learner use (Edleson, Gordin, & Pea, 1999). *One Sky Many Voices* (Songer, 1996,1998; http:/www.onesky.umich.edu/index.html) and *Kids as Global Scientists* (http://www.onesky.umich.edu/kgs01.html) enable upper elementary and middle school science students to research concepts related to weather and climate. The projects include Internet resources, real-time and near-time data and visualization resources, peers distributed worldwide, and practicing scientists. Students use these programs to see weather fronts in similar ways as meteorologists. Figure 13 shows a sample from the *Kids as Global Scientist* program. In addition to the visualization program, the project also includes activities and an interactive message board that allows students to talk with each other. Research supports these programs as assisting students with positive learning gains (Songer, 1996, 1998).

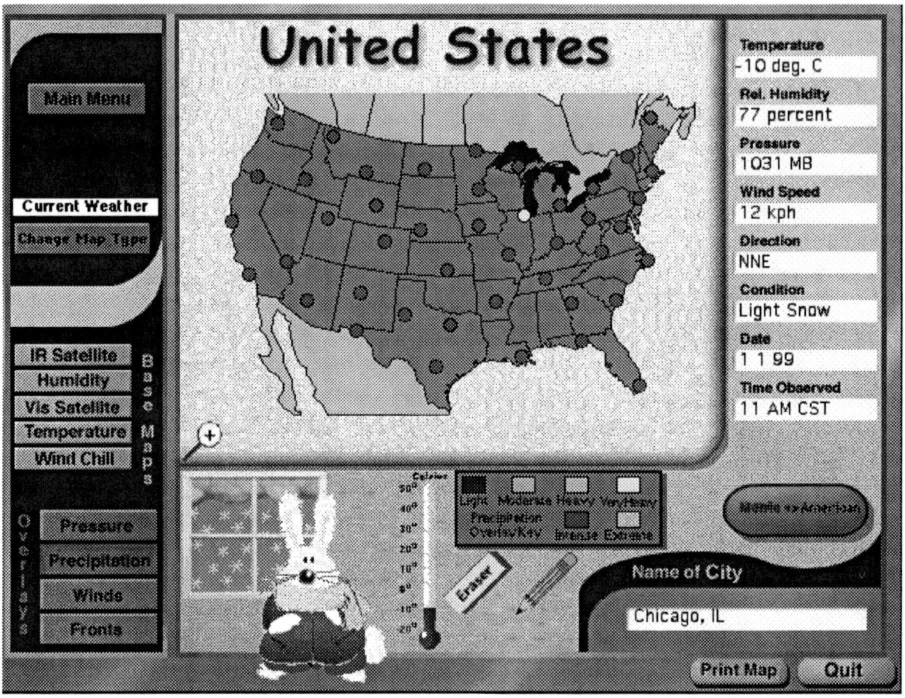

Figure 13. Sample from Kids as Global Scientist Program

The Worldwatcher program (http:/www.worldwatcher.nwu.edu/software.html) allows middle school students to visualize climate patterns in similar ways as climatologists. The FeederWatch program (http://birds.cornell.edu/pfw/) lets students visualize the distribution of different bird populations throughout North America. In addition students may enter data on birds that they themselves are tracking.

Visualization tools are new technologies that assist students in inquiry. Their features create new opportunities for students to use tools similar to scientists that enable them to explore and manipulate data and to see trends not possible without such tools. Such explorations and manipulations of data help create a science classroom reflective of what scientists do and assist students to develop more meaningful understandings (Cohen, 1997).

CONCLUDING COMMENT

Learning technologies serve as powerful tools that assist students in engaging in inquiry. This chapter focused on the potential of these tools to facilitate students to construct deep understanding of science concepts and process through inquiry. The National Education Standards state that inquiry is the central strategy for science teaching and that using a variety of technologies should be an integral component of scientific investigation. Central to inquiry is placing the student at the center of knowledge building with the teacher as a facilitator, collaborator and mentor. Inquiry based classrooms involve students in exploring important and meaningful questions with long-term investigations where a few unifying principles are investigated in contrast to covering large amounts of material where students learn facts or skills in isolation.

Various learning technologies embedded within the curriculum can promote in-depth learning. They allow students to engage in aspects of inquiry that they would not otherwise be able to do. Learning technologies allow students to explore their "What if..?" questions. They allow students to use similar tools and engage in similar activities of scientists. Because less time is needed for gathering and recording data, more time is available for interpreting and evaluating data. The learning technologies discussed in the chapter include microcomputer-based laboratories, probes attached to hand-helds, modeling software, the Web and digital libraries, digital cameras, hypermedia construction tools, applications for hand-helds and visualization tools. Research findings support the use of these tools to assist students in developing in-depth and integrated understanding of science concepts and science process. However, although learning technologies can support students in inquiry, their full advantage will not be released unless teachers carefully support students in the use of these tools.

REFERENCES

Amati, K. . (2000). Model-It. In Minstell, J. Van Zee, E. (Eds.) *Inquiry into inquiry: Science learning and Teaching*, American Association for the Advancement of Science Press, Washington, D.C., pgs. 316 -325.

Blumenfeld, P. C., Marx, R. W., Patrick, H., Krajcik, J. S., & Soloway, E. (1998). Teaching for Understanding. In B. J. Biddle & T. L. Good & I. F. Goodson (Eds.), *International Handbook of Teachers and Teaching* (pp. 819-878). Dordrecht, The Netherlands: Kluwer.

Cohen, K. C. (Editor) (1997). *Internet Links for Science Education: Student-Scientist Partnerships.* Plenum, New York.

Dwyer, D.C. (1994). Apple classrooms of tomorrow: What we've learned. *Educational Leadership 51*, 4–10.

Edelson, D. C., Gordin, D. N., and Pea, R. D. (1999). Addressing the Challenges of Inquiry-based Learning Through Technology and Curriculum Design. *Journal of the Learning Sciences, 8*, 391-450.

Gilbert, J. K., Boulter, C., & Rutherford, M. (1998). Models in explanations, part 1: Horses for courses? *International Journal of Science Education, 20*(1), 83-97.

Gobert, J. D., & Buckley, B. C. (2000). Introduction to model-based teaching and learning in science education. *International Journal of Science Education, 22*(9), 891-895.

Hoffman, J., Wu, H-K, Krajcik, J. S., & Soloway, E. (In press). The Nature of Middle School Learners' Science Content Understandings with the Use of On-line Resources. *Journal of Research in Science Teaching.*

Hug, B., & Krajcik, J.S (2002). Modeling Communicable Diseases. Kendall/Hunt, Dubuque Iowa.

Krajcik, J., Blumenfeld, B., Marx, R. and Soloway. E. (2000). Instructional, Curricular, and Technological Supports for Inquiry in Science Classrooms. In Minstell, J. Van Zee, E. (Eds.) Inquiry into inquiry: Science learning and Teaching, American Association for the Advancement of Science Press, Washington, D.C., pgs. 283 -315.

Krajcik, J.S., Czerniak, C., & Berger, C. (2002). *Teaching Science In Elementary And Middle School Classrooms: A Project-Based Approach*, Second Edition. McGraw-Hill: Boston, MA.

Krajcik, J. S. & Layman, J. W. (1992). Microcomputer-based laboratories in the science classroom. In Lawrenz, F., Cochran, K., Krajcik, J., & Simpson, P. (Editors), Research Matters To the Science Teacher. Manhattan, KS: National Association of Research in Science Teaching.

Krajcik, J. S., Mamlok, R., & Hug, B. (2000). Modern Content and the Enterprise of Science: Science Education in the Twentieth Century. In Corno, L (Ed.), *Education Across a Century: The Centennial Volume. One-hundredth Yearbook of the National Society for the Study of Education*, University of Chicago Press, Chicago, IL.

Krajcik, J. & Starr, M. (2001). Learning Science Content in a Project-based Environment In Tinker, R., & Krajcik, J.S. (Eds), *Portable Technologies: Science Learning in Context*, Netherlands: Kluwer Publishers.

Kozma, R.1991. Learning with media. *Review of Educational Research 6*, 179–211.

Linn, M.C. 1998. The impact of technology on science instruction: Historical trends and current opportunities. In M.C. Linn (Ed.). *International handbook of science education*, Netherlands: Kluwer Publishers.

Linn, M. C., and Hsi, S. (2000). *Computers, Teachers, Peers: Science Learning Partners*. Mahwah, NJ: Lawrence Erlbaum Associates, Inc.

Linn, M. C., Layman J. W. & Nachmias, R. (1986). *Cognitive consequences of microcomputer-based laboratories: Graphing skills development*, Paper prepared for a special issue of the Journal of Contemporary Educational Psychology. Manuscript submitted for publication.

Metcalf-Jackson, S., J.S. Krajcik, and E. Soloway. 2000. Model-It: A Design Retrospective. In M. Jacobson & R. B. Kozma (Eds) *Innovations in Science and Mathematics Education: Advanced Designs for Technologies and Learning*, Lawrence Erlbaum Associates, New York.

Mokros, J.R., and R.F. Tinker. 1987. The impact of microcomputer-based labs on children's ability to interpret graphs. *Journal of Research in Science Teaching 24*, 369–83

National Research Council (1996). *National science education standards*. Washington, D.C.: National Academy Press.

Novak, A.M., & Gleason, C. I., (2001). Incorporating Portable Technology to Enhance an Inquiry, Project-Based Middle School Science Classroom. In R. Tinker, & J. S. Krajcik (Eds.), *Portable Technologies: Science Learning in Context*, Netherlands: Kluwer Publishers.

Novak, A. M. & Krajcik, J.S. (2002). *Modeling Decomposition*. Kendall/Hunt, Dubuque Iowa.

Novak, J., & Gowin, D. B. (1984). *Learning how to learn*. Cambridge: Cambridge University Press.

Reiser, B. J., Tabak, I., Sandoval, W. A., Smith, B. K., Steinmuller, F., & Leone, A. J. (2001). BGuILE: Strategic and conceptual scaffolds for scientific inquiry in biology classrooms. In S. M. Carver & D. Klahr (Eds.), *Cognition and instruction: Twenty-five years of progress* (pp. 263-305). Mahwah, NJ: Erlbaum.

Rivet, A., & Schneider, R. (2001). *Ubiquitous images: Digital cameras to support student inquiry*. Paper presented at the Annual Meeting of the National Association for Research in Science Teaching, St. Louis, MO.

Schneider, R.M., Krajcik, J., Marx, R., & Soloway, E. (2001). Performance of student in project-based science classrooms on a national measure of science achievement. *Journal of Research in Science Teaching, 38,* 821 – 842.

Songer, N. (1997). Can Technology Bring Students Closer to Science? In B.J. Fraser and K. Tobin (Eds.), *International Handbook of Science Education*. Netherlands: Kluwer Publishers.

Songer, N.B., Lee, H.S. and Kam, R. (2002) Technology-Rich Inquiry Science in Urban Classrooms: What are the barriers to inquiry pedagogy? *Journal of Research in Science Teaching 39,* 128-150.

Spitulnik, M. W., Stratford, S., Krajcik, J., & Soloway, E. (1997). Using technology to support student's artifact construction in science. In B.J. Fraser and K. Tobin (Eds.), *International Handbook of Science Education*. Netherlands: Kluwer Publishers.

Stratford, S. J., Krajcik, J., Soloway, E. (1998). Secondary students' dynamic modeling processes: analyzing, reasoning about, synthesizing, and testing models of stream ecosystems. *Journal of Science Education and Technology , 7*, 215-234.

Tinker, R. (1997). *Thinking about science*. http://www.concord.org/library/papers.html Cambridge, MA: Concord Consortium.

Tinker, R. F. (2000). History of Probeware. http://www.concord.org/themes/probeware_history.pdf. Cambridge, MA: Concord Consortium.

Tinker, R. F., & Krajcik, J.S., 2001 (Eds.), *Portable Technologies: Science Learning in Context*, Netherlands: Kluwer Publishers.

Wallace, R. , Kupperman, J., Krajcik, J., and Soloway, E. (2000). Science on the Web: Students On-line in a Sixth Grade Classroom. *Journal of Learning Sciences, 9*, 75-104.

Warren, B., Rosebery, A., & Conant, F. (1989). *Cheche konnen: Science literacy in language minority classrooms.* Paper presented at the First Innovative Approaches Research Project Symposium, Washington, DC.

PART II: TEACHING AND LEARNING SCIENTIFIC INQUIRY

CHAPTER 6

KATHLEEN E. METZ

THE KNOWLEDGE BUILDING ENTERPRISES IN SCIENCE AND ELEMENTARY SCHOOL SCIENCE CLASSROOMS

Analysis of Problematic Differences and Strategic Leverage Points

National curriculum policy documents recommend an emphasis on scientific inquiry throughout the K-12 science curriculum. In this vein, the American Association for the Advancement of Science policy document, *Benchmarks for Scientific Literacy*, contends that students' conceptualization and responsibility over scientific investigations of their own design is fundamental to their understanding of science as a way of knowing. The National Research Council's curriculum document, *The National Science Education Standards*, contends that a focus on inquiry in the teaching of science supports children's learning of scientific concepts, their understanding of the nature of science, and their ability to pursue inquiry of their own. According to this NRC document, "Science as inquiry is basic to science education and a controlling principle in the ultimate organization and selection of students' activities" (p. 105).

However the appropriate relation between the inquiry of professional scientists and the inquiry of children in elementary school science classes is inherently complex. Obviously the knowledge building enterprises in these two contexts will always be fundamentally different. At issue here is where are the differences unnecessary and problematic? How should we use the knowledge building practices of scientists as a "controlling principle" in the design of science curriculum for elementary school children?

ANALYTICAL SCHEMA

Within the confines of this chapter I cannot present a complete comparative analysis of knowledge building enterprises in the communities of scientists and elementary school science classrooms. I focus instead on three aspects of the gulf between these knowledge building enterprises that I view as unnecessarily large and problematic, aspects that suggest key leverage points for educational reform.

Goal-structure constitutes the first aspect of the comparison. In this analysis, I consider such syntactical issues as does the goal-structure tend to be flat or

L.B. Flick and N.G. Lederman (eds.) Scientific Inquiry and Nature of Science, 105-130.

hierarchical; *i.e.*, are the goals largely at the same level or do sub-goals support higher-level goals? I also consider the substance of the goal-structure, contrasting that of scientists and elementary curriculum activities.

Targeted knowledge constitutes the second aspect of the schema. As in the first aspect, I consider both syntactical and substantive facets. First, what is the syntax of the targeted knowledge? More specific, how strong is the emphasis on building connections and hierarchical relations between knowledge elements? Concerning its substance, what is the nature of the knowledge to which the respective enterprise aspires? And what is the epistemic function of this knowledge?

Stance toward competing ideas and the resolution thereof constitutes the third aspect. Here I consider how each enterprise regards and responds to competing ideas or interpretations and, more generally, the function and dynamic of competing ideas in knowledge building.

APPLICATION OF THE SCHEMA TO ONE CASE OF SCIENTIFIC INQUIRY

Prior to a comparative analysis of these two enterprises vis à vis these three facets, I consider their illustration in a particular example of inquiry among professional scientists. I ground my example in a case study of scientific anthropologist Bruno Latour (1999). Latour reports on the collaborative field research of a Brazilian botanist and a French soil scientist (pedologist). It is a single example of the dynamics of a research team, comprised of botanist, pedologist (soil scientist) and geomorphologist, working in the field. Obviously other cases might well reflect other dynamics. My purpose is illustration of the three aspects of the schema by viewing one scientific project through this lens.

Goal Structure

The puzzling phenomenon that has so captured the interest of a Brazilian botanist and consequently a French soil scientist as well is whether the rain forest is advancing or retreating in the area of Boa Vista in Brazil. The goal structure is strongly hierarchical. Discovery constitutes the top-level goal, with such processes of developing hypotheses; developing sampling plans, collecting data, categorizing and analyzing data as sub-goals. It is the top-level goal of discovery that drives, constrains and gives meaning to these secondary goals. Latour's account of how the team discovers the changes between the two ecosystems or, more specifically, constructs a scientific account that is compelling to all reflects multiple cycles of contrasting hypotheses of what is happening, different forms of data collection, competing interpretations of the data based on fundamentally different disciplinary-based interpretative frames. After they are convinced that the rainforest is indeed advancing, the top-level goal shifts to explanation.

More specifically, consider how his case study reflects the hierarchical goal structure of the scientists; *cf.*, how the top-level goal of discovery (of the changing boundary of the ecosystems) drives, constrains and gives meaning to sub-goals in the inquiry process such as observation and categorization and reciprocally how the sub-goals support the goals. For example, the soil scientists fill a special girded box ("pedocomparator") with cubes of soil from locations and depths specified by the protocol, to systematically sample the soil in the Boa Vista area. The purpose of these observations, the sampling plan that frames them, and their interpretation are each subordinate to the top-level goal of discovery (of the nature of changes between the two ecosystems). It is this top-level goal that shapes these processes and gives them meaning:

> At depths below fifty centimeters the soil under the savanna and the soil under the forest appear exactly the same. The hypothesis from infrastructure does not hold. Nothing in the bedrock seems to explain the difference in the superficial horizons –clayey beneath the forest and sandy beneath the savanna. The profile is 'bizarre', and that makes my friends all the more excited. . . (*ibid.*, p.41)

The scientists' investigation of whether or not the rain forest is advancing focuses in on trying to explain this "bizarre" profile and the data that would support them in this endeavor.

Targeted Knowledge

We can examine this same case study in terms of the syntactical and substantive facets of the targeted knowledge structure. In Phase I, the all research team members focused on the single empirical question of the direction of change in the Boa Vista rainforest. Answering this question constituted the top-level targeted knowledge of this phase of the field expedition.

Latour's account richly describes how this top-level targeted knowledge shapes the lower-level targeted knowledge through which each scientist approaches the question: the botanist's focus on changes over time in distribution of plant species, as well as variations in their growth; and the pedologist's focus on the kind of soil in the bedrock at different levels under the rainforest and savanna. For example, in explaining the work of the soil scientist, Latour writes:

> In accordance with the habits of their profession, the pedologists wanted to know whether the bedrock was, at a certain depth, different beneath the forest than beneath the savanna. Here was a simple hypothesis that would have put an end to the controversy between botany and pedology: neither the forest or the savanna is receding, the border that separates them reflects a difference in soil (*ibid*; p. 40).

These discipline-specific research questions allow each scientist to contribute toward their shared top-level goal. Latour analysis' of the scientists' explanatory frameworks also reflects the deep connections between the top-level shared explanatory frameworks of the discipline (on the level of interdependence of organisms, reasoning in terms of organizations of living systems, and adaptation) and explanatory frames of their particular disciplinary focus.

Table 1

Aspect	Analysis
Goal Structure	Hierarchical Top-level goals: Discovery ←--> Build explanatory theory Subgoals: Develop hypotheses > Develop sampling plan > Collect data/ Observe > Analyze > Model. Application: Not part of this team's goal-structure. However the anthropologist studying the team notes their assumption that the knowledge they develop may be applicable by others concerned with the broad spread problem of shrinking rainforests.
Targeted Knowledge	<u>Phase I:</u> Top-level: * Is the rainforest receding or advancing? Secondary: • Changes in the distribution of species of plants & variations in the growth in the Boa Vista area around the boundary of rainforest & savanna • The kind of soil in the bedrock at different levels under the rainforest and savanna areas <u>Phase II:</u> Top-level: * *How* is the rainforest advancing? What biological mechanisms underlay this change? Secondary: • What is the nature of the biological activity that transforms the top 10- 15 cm. from a sandy soil into a clayey-sandy soil? • What is the nature of the activity of the worm population & their impact on the soil environment?
Stance toward competing ideas & resolution thereof	Competing ideas as crucial to the dynamic of the enterprise & its advancement in both discovering how the line between rainforest & savanna is changing & why. Research team composed to incorporate fundamentally different disciplinary perspectives: botanist, soil scientist & geo-morphologist. These different disciplinary perspectives lead to different hypotheses about changing boundaries of rainforest & savanna: the botanist infers the rainforest is advancing; the soil scientist infers it is retreating. The botanist and soil scientist rely on different –and complementary—forms of data to test their competing

	hypotheses. The scientists pose different interpretations of the data, based on their different disciplinary training & perspectives Modeling of the line between rainforest and savanna, integrating data from the two disciplinary perspectives supports the collaborative development of a new theory

Stance Toward Competing Ideas

Competing ideas grounded in their complementary disciplinary perspectives, repeatedly empower the team's progress. Latour's analysis of the team's dynamics documents how competing ideas play a key function in the discovery of changes in the boundary between the two ecosystems, hypotheses they formulate to try to explain the mechanisms of the changes, and the strategies they formulate to test their hypotheses.

Latour describes the botanist's and pedologist's conflicting conjectures, grounded in different data and explanatory schemas:

> Edileusa believes the forest is advancing, but she cannot be certain because the botanical evidence is confused; the same tree may be playing either of two contradictory roles, scout or rear guard. For Armand, the pedologist, at first glance it is the savanna that must be eating up the forest, little by little, degrading the clay soil necessary for healthy trees into a sandy soil in which only grass and small shrubs can survive. Soil goes from clay to sand, not from sand to clay. . l. Soil cannot avoid degradation; if the laws of pedology do not make this clear, then the laws of thermodynamics would. (*ibid*, p. 27)

As the research team prepares the report of this first phase of the expedition, they identify their disciplinary-based differences and the rich argumentation it spurred as crucial to their success. As Latour quotes from the research team's report:

> The interest in this expedition report stems from the fact that, in the first phase of the work, the conclusions of the approaches of botany and pedology appear contradictory. *Without the contribution of the botanical data, the pedologists would have concluded that the savanna is advancing on the forest.* The collaboration of the two disciplines in this case has forced us to ask new questions of pedology. (italics in the original) (Latour, *ibid*; p. 68)

At this end of this first phase of their expedition the team becomes convinced that it is the rainforest that is advancing. They are confident that they have identified the changes in the ecosystem's boundary. The team has developed a possible explanation of the mechanism, an account grounded in the unusual combination of expertise of botanist and pedologist. Developing a well-supported explanation of the mechanism of the change lies ahead, a phase of the research team's work that Latour did not include in his case study. Latour also acknowledges their assumption that eventually what they learn may be applied by others interested in maintaining the rainforest.

COMPARISON OF KNOWLEDGE BUILDING ENTERPRISES OF SCIENTISTS AND CHILDREN IN ELEMENTARY SCHOOL SCIENCE CLASSES

The following sections use the practices of knowledge building among scientists as a foil for examining the practices in elementary science classrooms vis à vis the three aspects of goal-structure, targeted knowledge and stance toward competing ideas and the resolution thereof. I compare how the practices of the elementary school science differ from the practices of the scientific community and examine factors underlying the difference. Obviously there is significant variability in knowledge building practices in both elementary school classrooms and scientific communities. These analyses also consider variability in the knowledge building practices of science elementary school classrooms, to unpack the problematic and the possible.

Goal-Structure

A huge discrepancy frequently exists between the scientists' goal structure and that reflected in the practices of elementary science education classes. I contend that this is one context in which we need to try to modify the extent and form of the differential.

The goal structure of scientists appears somewhat different, from the viewpoint of different intellectual traditions and different scientific disciplines. Largely from the perspective of physics, Reif and Larkin (1991) assert, "the central goal of science is to explain and predict observable phenomena" (p.736). Reif *et al.* contend that a key difference between the goals of science and everyday cognition is "optimal prediction and explanation", as opposed to the prediction and explanation adequate for everyday interaction with the world.

Biologist Ernst Mayr (1997) identifies understanding, prediction and control as goals that philosophers of science have attributed to the scientific enterprise. However he notes that the issue of control may not come up at all in non-applied sciences and prediction plays "a very subordinate role" in many sciences, including the discipline of biology. Mayr explains that in many contexts of biology prediction of the future is impossible, due to the multiplicity of causal factors. More simply, Mayr contends "in most cases, scientists are largely motivated by the simple desire for a better understanding of puzzling phenomena in our world."

In a similar vein, Herbert Simon (2001) views curiosity as driving the discovery that may be followed by verification. Simon argues that, in their analysis of the scientific enterprise, scientists and philosophers of science (but not historians of science) have overemphasized verification to the neglect of discovery. He contends, "Science is concerned with verification as well as discovery, but the former is always kept in the service of the latter."

Donald Stokes' (1997) influential monograph takes an historical and prospective perspective on the relation between control and understanding as aims of science. Through reference to federal policy and funding documents, he tracks assumptions about the relation between the goals of understanding and use. Historically, he finds the two have been viewed largely as disjoint enterprises – with "applied" research divorced from and subsequent to "basic" research. Stokes analyzes research in which the two goal structures are synergistic, as in the work of Pasteur, and argues for the power of the dual goal-structure.

Whether we view control or discovery and explanation as the top-level goals, the goal-structure of science is hierarchical. It is these top-level goals that motivate, shape and give meaning to the sub-goals of observation, categorization, representation and so on. In the words of Karl Popper (1972), "without interests, points of view, problems", the tasks of observation and classification become "absurd" (p. 46).

Consider the goal-structure reflected in the vignette of K-4 science in the recent NRC document, *Inquiry and the National Science Education Standards* (2000). The hypothetical third grade teacher Ms. Flores – presented as a composite of the authors' classroom experiences –has combined her responsibilities to scaffold core biological knowledge with the overarching purpose of engaging children in scientific inquiry that is motivated by their curiosity and desire to investigate particular questions.

The study of the diversity of organisms within the same environment and how the needs of different organisms are met within this same environment constituted the larger context of the unit. A vacant lot comprises the class' field site. In teams of three, the children began by measuring off a square meter and collecting data about what organisms lived there. It was in the context that the children became interested in earthworms:

> During the investigation several students found earthworms in their square meter and became fascinated with earthworm behavior. Some of the other students wanted to know why they did not find earthworms in their study area. Others wanted to know why the worms were different sizes. One student suggested that worms 'liked' to live near some kind of plants and not others, since when she and her dad went fishing they always dug for worms where there was grass. (*ibid.* p. 40)

Ms. Flores supported the children to pursue investigations about of their questions. She used this goal-focused context in which to deepen their understanding of the targeted biological concepts of diversity of organisms, and relation between organisms and their environment, as she teaches the scientific inquiry process *in situ.*

> The NRC curricular document contends that the unit has the "essential features of classroom inquiry":
>
> Her students identified a question of their own interest about earthworms around which to design an investigation. The question derived from their own understanding of the characteristics and environment of earthworms and their curiosity about these animals, and, so the question they chose engaged them thoroughly. As they developed answers to their questions, Ms. Flores helped them to understand that they needed evidence and what the nature of the evidence needed to be. They looked for evidence through their careful observation and what they read in science books. Learning about fair tests increased the likelihood that their evidence would be sound. (*ibid*; pp. 46-47)

In Ms. Flores' classroom, the reasoning processes comprising the scientific inquiry are scaffolded in the context in which they have meaning and purpose[1]. The activity appears to have a robust goal-structure. As in the case of Boa Vista, the questions would motivate, shape and give meaning to the observations and categorization.

This NRC curriculum document includes in their chapter of "Frequently asked questions about inquiry", their response to "Why did the *Standards* choose to leave out the science process skills such as observing, classifying, predicting, and hypothesizing?" (*ibid*; p. 134). The report explains that these processes (sub-goals, from the perspective of the chapter's schema) have not been omitted, but rather embedded in the context of inquiry and elaboration of scientific knowledge, where the students "combine those processes and scientific knowledge to develop their understanding of science" (*ibid*). This stance is supported by Glaser's (1984) model of the bootstrapping relation between one's scientific process and scientific knowledge.

The NRC's K-4 "image of inquiry" bears an intriguing correspondence with the structure of Japanese elementary school science lessons. On the basis of their analysis of science lessons in 10 Japanese fifth grades, Linn, Lewis, Tsucida & Songer (2000) describe an activity structure that reflects a robust goal structure. They identify a sequence of eight components to the structure: "(1) connect lesson to student interest and prior knowledge; (2) elicit student ideas or opinions; (3) plan investigations; (4) conduct investigation; (5) exchange information from investigation; (6) systematically analyze or organize information; (7) reflect and revisit hypotheses or predictions; and (8) connect to next lesson(s). Identify unanswered questions" (pp. 5-6). These Japanese teachers taught the skills needed for scientific inquiry, such as observations, measurement, and categorization, in the context in which they had purpose and meaning.

Particularly at the primary level, this robust goal structure is not representative of most science curricula in the U.S.. The goal-structure in science curricula for young children frequently foregrounds "scientific reasoning skills". Indeed the lower the grade, the more restricted the science process skills tend to be. Additional process skills are added on across the grades until at the end of the elementary school children are designing controlled experiments.

As illustrations of this trend, I examine the science process skills and their manifestation in curricular activities of two of the best elementary science curricula in general use: *(a)* Full Options Science System (FOSS) developed by the Lawrence Hall of Science; and *(b) Science and Technology for Children* (STC) developed by the National Science Resource Center, operated by the Smithsonian Institute and the National Academy of Sciences. I have chosen these because they are deservedly well respected and widely used curricula.

According to the STC curriculum, "in the primary grades, children begin their study of science by observing, measuring and identifying properties. Then they

[1] A much more detailed description of the activity and perceived educative value is included in the document, *Inquiry and the National Science Education Standards* (National Research Council, 2000, pp. 40-48).

move on through a progression of experiences that culminate in grade six with the design of controlled experiments". (National Science Resources Center, 1997, p. iii). The STC curriculum describes observing, measuring, and identifying properties as first grade scientific reasoning skills. STC adds the processes of seeking evidence, and recognizing patterns and cycles in second grade. Identifying cause and effect are added in fourth grade. Consider how these sets of presumed children's scientific reasoning skills are translated into curricular activities. For instance, the *Science and Technology for Children* third grade unit on *Rocks and Minerals* (National Science Resource Center, 1997) summarizes the children's activities as follows:

> Students explore the similarities & difference among rocks; they also study how rocks & minerals are both similar and different. They conduct several tests on minerals and develop a systemic way to record their observations. Finally, students apply the information they have collected to identify the minerals they have been studying by name. These activities introduce students to the way geologists study rocks and minerals. They also help students develop and apply process skills in observing, describing, and recording. (p. 3)

The contribution of the STC curriculum is considerable. The STC curriculum development process clearly benefited from a rich collaboration of scientists, teachers, and teacher educators. The curriculum has richer science content than most of its competitors and is remarkably "teacher friendly" to use. Nevertheless the goal structure appears flatter and lower level than that reflected in the Japanese activity structure or the National Research Council prototype.

In a similar vein, the FOSS curriculum identifies " "Observing, communicating and comparing" as first and second grade "thinking processes". At the third and fourth grade level, FOSS identifies the same processes at the level of "advanced organizing". The curriculum adds "relating" at the fifth and sixth grade level, as instantiated in organizing, comparing, communicating and observing. FOSS relegates inferring to the sixth through eighth grade level.

The FOSS curriculum explicitly grounds the derivation of this schema on strong cognitive stage theory model of children's emergent capabilities:

> The FOSS program is correlated to human cognitive development. The activities are matched to the way students think at different times in their lives. The research that guides the FOSS developers indicates that humans proceed systematically through a predictable, describable sequence of stages of cognitive development. In their early elementary school years, students learn science best from direct experience in which they describe, sort, and organize observations about objects and organisms. Upper elementary students construct more advanced concepts by classifying, testing, experimenting, and determining cause-effect relationships among objects, organisms and systems. " (FOSS, 2002, p. 4)

A related perspective concerning work appropriate for young children in science is reflected in much of the AAAS document, *Benchmark's for Scientific Literacy*:

> "Kindergarten through Grade 2: In the earliest grades, students make observations, collect and sort things, use tools, and build things. They are, for their developmental level, doing science and using technology." (American Association for the Advancement of Science, 1993, p. 31)

A relatively flat and low-level goal structure is also reflected in Vekiri, Baxter and Pintrich's (1998) analysis of talk in three fifth grade science classrooms. Vekiri *et al.* report:

> Most of the teacher questioning, feedback and guidance focused on how to configure the materials, what and how to record in the science notebooks, and what tasks students were to accomplish. However there were fewer instances in which students were asked to use their observations to understand scientific phenomena that were represented in the activities; to identify patterns, examine relations between observations, generate and support predictions, or construct explanations that would account for what their findings indicated. (p. 35)

Vekiri *et al.*'s analysis suggests that the observations are ends in and of themselves, not as sub-goals in supporting the inquiry.

My concern is that goal-structure of science curricula that foregrounds science process skills can be flat and impoverished. The robust purpose of doing science can be lost in the curricular strategy of foregrounding a small number of the "scientific reasoning processes" and designing activities with the strong focus on developing those skills. Given this syntax and substance of a flat and low-level goal-structure, how will children understand the scientific enterprise? What will be their motivation for participating in the exercise?

There is a real danger here, in the words of Karl Popper (1972), that the task may become "absurd". The AAAS science curriculum documents, *Benchmarks for Scientific Literacy*, voices a similar concern:

> Students should be given problems -at levels appropriate to their maturity - that require them to decide what evidence is relevant and to offer their interpretations about what the evidence means. This puts a premium, just as science does, on careful observation and thoughtful analysis ... *However if such activities are not to be destructively boring, they must lead to some intellectually satisfying payoff that students care about.* (American Association for the Advancement of Science, 1993, p. 148) (Emphasis added)

An understanding of the influences underlying this curricular approach and the scope of its extension can shed light on the challenges of changing the practice. This pattern of curriculum development is consonant with the explicit assumption that a complex intellectual practice, such as scientific inquiry, can be effectively taught by decomposition and decontextualization of its component facets. Robert Gagné (1977) argued that educators could parse complex knowledge into component building blocks. They can begin instruction with the simplest building block –in this case, purportedly observation--- and then gradually add on more blocks –such as categorization -- until students are engaging in scientific inquiry, operationalized as the controlled experiment, in sixth grade. More generally, this practice reflects the enduring influence on U.S. education of behavioral psychology, in the form of designing instruction in terms of a sequence of well defined discrete skills.

There are indications of this same trend in how we teach mathematics in the U.S.. In their video-based seminal cross-cultural analysis of mathematics teaching, Stigler and Hiebert (1999) found that "US teachers appear to feel responsible for shaping the task into pieces that are manageable for most students. . . Teachers act as if confusion and frustration were signs that they had not done their job. " (p. 92). Stigler and Hiebert contrast this view with the Japanese teachers' perspective that "students learn best by first struggling to solve mathematics problems. . . . Constructing connections between methods and problems is thought to require time to explore and invent, to make mistakes, to reflect." *(ibid*; p. 91). At issue here is whether our tactic of avoiding the students' temporary struggle by means of dividing up the complex intellectual activity into little pieces, the curriculum may fail to convey the purpose of the authentic activity.

In the context of elementary science education, this particular conceptualization of what pieces should be addressed at what age level, as well as the target image of scientific inquiry, appear to be strongly influenced by interpretations of Piaget's stages of cognitive development. For example, the National Research Council's recent publication *Science for all children: A guide for improving elementary science education in your school district* explicitly grounds what aspects of scientific inquiry children can do at what grades in an interpretation of Piagetian stage theory:

> His [Piaget's] theories still provide basic guidelines for educators about the kind of information children can understand as they move through the elementary school... Through the primary grades, children typically group objects on the basis of one attribute, such as color... The significance of this information for educators is that young children are best at learning singular and linear ideas and cannot be expected to deal with more than one variable of a scientific investigation at a time. . . . . Toward the end of elementary school, students start to make inferences. . . At this stage of development, students are ready to design controlled experiments and to discover relationships among variables." (National Research Council, 1997, pp. 28-29)

Through this lens, the "logic" of experimentation begins to emerge at the end of elementary school. "Concrete operational" capabilities more directly tied to concrete objects – especially observation and categorization --are available early in the elementary school years. The end point is verification, in the form of the controlled experiment. However this strong cognitive stage model of children's emergent capabilities is over simplistic and outdated.

First of all, the field of cognitive development has long rejected the view that cognition advances in such all-encompassing, broad changes in children's capabilities. In the words of seminal cognitive developmental theorist John Flavell (1994), "Virtually all contemporary developmentalists agree that cognitive development is not as general stage-like or grand stage-like as Piaget and most of the rest of the field once thought." (p. 574).

Furthermore analysis of the research literature of children's scientific cognition reveals that children are much more capable than most of the science curricular policy documents assume (Metz, 1995). In short, the assumption that observation, ordering and categorization constitute core scientific reasoning competencies of the elementary school child is not supported by the developmental literature. The research documents these capabilities in much younger children (Markman, 1978; Markman & Callahan; 1983; Markman, Horton, & McLanahan, 1980). The research also documents young children, primary grades and younger, developing and testing informal theories to control and/or understand the world around them (Brewer &

Samarapungavan, 1991; Gopnik & Welman, 1994; Karmiloff-Smith & Inhelder, 1974).

Fortunately the scientific cognition and instruction research literature is examining promising instructional strategies with a more robust goal structure, wherein educators bring scientific inquiry within reach of elementary school children without resorting to decomposition and decontextualization. These include capitalizing on collaborative or, more specifically, socially-distributed expertise to reduce the cognitive demands on the individual learner (Brown and Ashe, Rutherford, Nakagawa, Gordon & Campione, 1993; Resnick , 1991) explicit scaffolding of the higher-level processes of theory formation and testing (Smith, Houghton, Maclin & Hennessey, 1997) and the scaffolding of goal-focused scientific inquiry (Lehrer, Carpenter, Schauble & Putz, 2000; Metz, 2000; Palincsar & Magnusson, 2001; White & Frederickson, 1998).

In summary, in order to convey the rich purpose of the enterprise we need to fundamentally reconfigure the goal-structure underlying the elementary school science curriculum. We need to let such processes as observation, categorization and measurement serve as means, not ends. We need to be careful not to overemphasize the science of verification, but rather foreground the goals of discovery, understanding and explanation.

Knowledge Structure

The distance between the practices of the scientific community and the elementary school classroom vis à vis the targeted knowledge is similarly vast and problematic. In science, the syntax of the knowledge structure is strongly hierarchical, with a relatively small number of ideas regarded as central. The highly structured, strongly hierarchical character of scientific knowledge is reflected in analysis of the current state of knowledge in a discipline, as well as its epistemic aspirations and most highly prized advancements. This syntax is evident in cognitive research examining the problem-solving and knowledge of the expert and, from a different perspective and unit of analysis, descriptions of the knowledge base and aspirations of the discipline as a whole.

In his renowned introduction to physics, Richard Feynman (1995) conveys the tight hierarchical structure of physics knowledge by posing himself the challenge of coming up with a single sentence that will encapsulate the most information about our overall picture of the world. He argues:

> I believe it is the atomic hypothesis (or the atomic fact) that all things are made up of atoms –little particles that move around in perpetual motion, attracting each other when they are a little distance apart, but repelling upon begin squeezed into onto another. In that one sentence, you will see, there is an enormous amount of information about the world, if just a little imagination and thinking are applied. (p. 4)

The cognitive science research of physics cognition documents the highly structured quality of physics expert's knowledge and how they use these connections and super-ordinate explanatory constructs in solving problems (Glaser, 1988; Greeno, 1984; Lesgold, 1988). Physicists / cognitive scientists, Reif and

Larkin (1991), emphasize the centrality of the conceptual structure. Indeed Reif and Larkin characterize science as "conceptual structure enabling numerous predictions." They list maximal generality, parsimony, precision and consistency as the epistemic "requirements" to which the knowledge building enterprise aspires.

Other scientific fields reflect similar knowledge structures and epistemic aspirations. For example, evolution is at the core of biologists' conceptual structure. Evolution and the rich theoretical developments thereof lie at the heart of how biologists think about biodiversity, specialization and adaptation, species and speciation, forces of evolutionary change, and patterns of evolution and extinction.

Although national curriculum policy documents provide a carefully reasoned delineation of core ideas (*cf.*, National Research Council, 1996), U.S. curriculum materials currently available are failing to address the big ideas --even at the middle school and high school level. This strong tendency is reflected in the recent analyses of science curriculum conducted by the American Association for the Advancement of Science and international curriculum comparisons associated with TIMSS (Third International Mathematics and Science Study).

The recently completed AAAS curriculum analyses concluded that all middle school and high school science texts were unsatisfactory in this regard, with the exception of several stand-alone units developed at Michigan State and University of Michigan. Roseman, Kesidou, Stern & Caldwell, authors of the AAAS report, wrote:

> The focus of this effort was to see which textbooks had potential for helping students learn key ideas. But unlike the math textbook study, not *one* of the middle grades science texts evaluated by Project 2061 rated satisfactory. . .
>
> The textbooks covered too many topics and didn't develop any of them well. In addition, the texts included many classroom activities that either were irrelevant to learning key science ideas or didn't help students relate what they were doing to the underlying ideas.

These findings are strongly supported by the comparative analysis of the curriculum in countries included in the Third International Mathematics and Science Study (TIMSS), of particular relevance here given its examination of fourth grade science curricula. Valverde and Schmidt (1998) report that US fourth grade textbooks tend to be longer than those of other countries. Indeed US fourth grade science texts were more than three times longer than the international average (397 pages compared with 125 pages). More significant, the US texts covered many more topics than 75% of the other countries participating in the TIMSS. Valerde and Schmidt comment:

> Does it matter that our textbooks are so comprehensive? Preliminary analyses suggest that it does. This is true because breadth of topics is presented at these textbooks at the expense of depth of coverage. Consequently, our textbooks are limited to perfunctory treatment of subject matter. . . Information collected from the national random sample of teachers in TIMSS indicated that the majority appear to be attempting the Herculean task of covering all the material in the textbook.

Valverde et al.'s analysis indicates that the lack of connection between "facts" reflected in this representation of intended knowledge is frequently manifested in the curriculum itself. Indeed Valverde and Schmidt found that U.S. science curriculum is fundamentally disjointed, with negligible attention to conceptual relations or the structure of the discipline:

> The unfocused curriculum of the United States is also a curriculum of very little coherence. . . U.S. textbooks and teachers present items one after the other from a laundry list of topics from state and local district guides, in a frenzied attempt to cover them all before the year runs out. This is done with little or no regard for establishing the relation between various topics or themes on the list. The loss of these relationships between ideas encourages children to regard these disciplines as no more than disjointed notions that they are unable to conceive of as belonging to a disciplinary whole.

On the basis of his extensive analysis of patterns of performance on the TIMSS in relation to curriculum content, Schmidt identifies these curriculum practices as a key factor in the academic shortcomings of U.S. students.

Further evidence of the liability of this practice of superficial coverage emerges in a large study of predictors of success in the study of physics. Sadler and Tai's (2001) investigated best predictors of performance in college physics. Directly counter to the beliefs of many of their high school physics teachers, Sadler et al. found that:

> Students who had high school courses that spent more time on fewer topics, concepts, problems and labs performed better in college than those who raced through more content in a textbook-centered course." (p.111).

In the words of Phillip Morrison, physicist and author of PSCS, "less is more".

This practice of superficial coverage is deeply problematic. The big ideas and interpretive frames are invariably lost. It perpetuates the image of science as lots of pieces without expectation of coherence. Furthermore, given the bootstrapping relation between the power of one's scientific knowledge and the adequacy of one's scientific reasoning (Glaser & Chi, 1988), superficial coverage handicaps the student's scientific inquiry.

I contend the genesis of this practice is complex and multifaceted. The influential factors include common understandings of scientific knowledge, teachers' beliefs, the sociology of how curriculum frameworks get developed in this country and the forms in which they are articulated.

A common view of scientific knowledge is accumulation of facts or, to use Hammer's (1994) description, a collection of separate pieces with no expectation of coherence. From this perspective, would not teaching more facts mean you were teaching more science? To understand why Phillip Morrison thought "less is more", one needs to understand the nature of scientists' knowledge and <u>how they use it</u>. As physicist William Bragg explained, "The important thing in science is not so much to obtain new facts as to discover new ways of thinking about them" (Bragg, cited in Reif & Larkin, 1992, p. 739).

The value many teachers place on breadth over depth reveals a view of not less is more but more [curriculum] is more [learning]. Both the TIMSS research and Sadler et al.'s study indicate that U.S. teachers believe that they have a responsibility to cover the curriculum by the end of the school year. More specifically, Sadler's work reveals that teachers tend to believe that covering the textbook is key for empowering their students to succeed in more advanced course work. This value is consistent with the epistemological perspective of science as an accumulating set of facts, as opposed to a highly structured interpretative frame.

The practices through which curriculum frameworks become elaborated in this country also contribute to the flat architecture of many objectives with negligible connections or hierarchical organization. Curriculum policy formation is a highly political process, in which many parties have a voice --- most with negligible expertise in science. This process is particularly charged in California, Texas and New York, the largest states with text approval procedures. For example, in their most recent science framework, California acknowledges the many voices that "contributed substantively to the final standards adopted by the State Board of Education, ...including parents, teachers, school administrators, and business and community leaders". President of UC, Richard Atkinson described the new California mathematics framework as a "kitchen sink." I argue so many cooks in the process, many operating from an understanding of science as accumulation of facts, encourages a kitchen sink curricular syntax.

Another influence on the syntax is the expected outcome of the deliberation, a curriculum document that delineates the objectives in the content area for each grade level, at a level of specificity sufficient to define the task of the writers –and to use as the basis to later decide whether or not the texts proposed by the curriculum companies have met the policy's objectives. The practice of conceptualizing the instructional goals of the curriculum in terms of a long list of objectives contributes to neglect of the big ideas and more broadly the incoherence of science curriculum identified in both the TIMSS and AAAS analyses.

There are intriguing counter-examples to this practice. First of all, as beautifully documented in TIMSS curricular analysis, the U.S. is an extreme case of targeted knowledge that is flat and disjointed with little attention to big ideas. AAAS points to Richard Anderson's curriculum units (on Matter and Molecules; Food, Energy and Growth, and Chemistry that Applies) as important existence proofs of more big-idea centered curricula. The influential instructional design work of Ann Brown and Joseph Campione constitutes another example. Brown and Campione view Bruner's principle of organizing curriculum around a "few lithe and beautiful and immensely generative ideas" as a major principle underlying their classroom design work (Brown, Ashe, Rutherford, Nakagawa, Gordon & Campione, 1993). Furthermore Brown and Campione's research indicates that these ideas can function as generative explanatory constructs in children's interpretations of unfamiliar biological phenomena (Ashe & Brown, 1995).

In short, there is strong evidence that the syntax and substance of the elementary science curriculum in the U.S. is superficial, flat and disjointed. The big ideas and explanatory frameworks are not supported by the science curriculum. However there are existence proofs of curricula that do effectively scaffold the big ideas and research indicating that curricula that aspire to depth over breadth are more effective. There is a pressing need for change in this sphere of the syntax and semantics of the targeted knowledge. However given complex factors contributing to the practice, including common understandings of scientific knowledge, teachers' beliefs, the sociology of how curriculum frameworks get developed in this country and the forms in which they are articulated, I envision change in this regard is slow –particularly change that transcends the local level.

Stance toward competing ideas and the resolution thereof

The stance toward competing ideas and how they are resolved constitutes another key difference between the scientific community and that of the elementary school science classroom. I will argue that this gap constitutes another issue fundamental to the challenge of effective science reform, as well as the more general goal of aligning instruction with current knowledge of the socio-cognitive construction and refinement of students' conceptual understandings.

Science is fundamentally an interpretive enterprise. Within the scientific community, competing ideas are manifested in many levels and many contexts. Indeed competing ideas are viewed as a core aspect of doing science and fundamental to the dynamic of theoretical advancements. Driver, Newton, and Osborne (2000) describe the ubiquity of competing ideas and argumentation in the doing of science:

> Practices such as assessing alternatives, weighing evidence, interpreting texts, and evaluating the potential viability of scientific claims are all seen as essential components in constructing scientific arguments (Latour & Woolgar, 1986). In making scientific claims, theories are open to challenge and progress is made through dispute, conflict, and paradigm change. Thus, arguments concerning, for example, the appropriateness of an experimental design, or the interpretation of evidence in light of alternative theories, are seen to be at the heart of science and central to the discourse of scientists (Druker, Chen & Kelly, 1996). Furthermore, the work of scientists also includes argument in the public domain through journals, conferences, and the wider media." (p. 288)

Most elementary school science classrooms and curriculum materials reveal little attention to competing ideas. In elementary school study of science, as well as other subject areas, whereas different ideas may be expressed in turn by different children, there is typically a rapid transition to identify the "right" idea by reference to the authority (of the teacher or text). In science activities, children frequently engage in small group work, where the goal (implicit if not explicit) is consensus. In children's "hands on" science work, there is seldom attention to issues of interpretation. Interpretation of data is viewed as either qualitative eyeballing for trends or application of an algorithm. The place where competing perspectives do appear regularly in many elementary school classrooms is the context of conflict resolution on the playground or "class meeting". Here the teachers and students engage in the articulation and analysis of competing perspectives in order to resolve disagreement and more generally try to maintain a community of social harmony. There is seldom evidence of a teacher and class taking an idea as an object of thought or comparing the merits of competing ideas.

Again there exist informative counter-examples to this trend. Scientific argumentation constitutes a core of the Japanese Itakura instructional method, as reported by Hatano and Inagaki (1991*a*, 1991*b*). In this technique, used in Japan at the elementary and junior high levels, the teacher presents a question and three or four competing hypotheses. The teachers' role is to stay neutral while moderating the discussion; each child's role is to choose the hypothesis (s)he considers most compelling and then to argue for that position while also considering the strength of the competing positions. The conflict is resolved through a combination of the relative persuasiveness of arguments for the different positions and thereafter a critical empirical experiment. In the U.S., the classroom-based work of Lehrer and Schauble constitutes a seminal example of researchers scaffolding teachers to scaffold elementary school children in this kind of argumentation. Lehrer and Schauble's analyses of these classrooms provide compelling evidence that primary grade children can develop and collaboratively consider reasoned arguments about alternative explanatory models about aspects of their world (*e.g.*; why apples change color or how your elbow works) and reasoned alternative interpretations about the data or physical models they design to investigate their ideas (Lehrer, Carpenter, Schauble & Putz; 2001; Penner, Lehrer & Schauble, 1998). Their work involved collaborative professional development and support of the classroom teacher, focusing on the construction, evaluation and revision of models on the part of the teachers and the children.

I conjecture that the minimal place for competing perspectives in the elementary science classroom stems from the interaction of multiple factors, including the teachers' understanding of science, time pressures, and the value on social harmony and conversely the attitude toward argument in everyday discourse. If we view science as "literal and irrevocable truths", to borrow Schwab's (1962) description of this epistemological stance, then what is the place for analyzing competing ideas beyond learning whose views are right? From this epistemological perspective, what would be the rationale to devote limited classroom time to the articulation and comparative analysis of competing ideas?

Different researchers (Burbules & Linn, 1991; Carey, Evans, Honda, Jay & Unger, 1989; Driver, Newton & Osborne, 2000) have argued that current forms of science curriculum investigations both reflect and further contribute to a positivist stance. In the words of Driver and her colleagues:

> Because time and emphasis are not given to such evaluation tasks [thinking about the planning of the experiment, how it might be researched, alternative methods for investigation and how one might choose between them], the main message from much practice work is that 'nature speaks' directly to us from our data and the process of 'making sense', the act of interpretation and the human construction of knowledge are completely overlooked in the priorities given in teaching. (Driver *et al.*, p. 289-290)

Indeed elementary school science curriculum-structured investigations, as well as the templates structuring science fair contributions, typically reflect this remarkably simplistic relation between data and conclusions

Epistemological stance is reflected in the discourse patterns. As linguist Jay Lemke (2001) has documented, the discourse patterns of everyday life and common sense reasoning differ on many aspects from the discourse of science. Another factor in the minimal attention to competing ideas in the elementary school science classroom is the discourse patterns that are familiar to children and elementary school teachers.

In her work on classroom discourse patterns, Elice Forman (1992) investigated the question, "What happens when children are asked to solve scientific tasks in collaboration with a peer?" She interprets the dynamics of these conversations in terms of the challenge of differentiating and appropriately applying everyday and academic discourse. Forman uses Grice's (1975, 1978) four maxims of everyday discourse as a model of how shared meaning is constructed in conversation and, more specifically, how participants in a conversation draw inferences that go beyond that which is said.

Forman explains, according to Grice, "In order for discourse to be mutually understood, speakers and listeners must implicitly agree that contributions to a dialogue are intended to be true (quality), that they are as informative as required and no longer than necessary (quantity), that they are relevant, and that they are expressed in an orderly and unambiguous manner" (*ibid;* p. 147). It is easy to see where these discourse maxims conflict with the discourse maxims of science. In scientific discourse, although in most cases we agree that our colleagues intend their talk to be true, we do not necessarily assume that what our colleagues say is "as informative as required, no longer than necessary", or that it is "relevant" or that it "expressed in an orderly and unambiguous manner".

Forman interprets difficulties children have on the collaborative science task largely in terms of differentiation and coordination of everyday and academic modes of discourse. Forman reports:

> In order to solve collaborative problem solving tasks, students must be able to subordinate their interpersonal needs to dominate their partner by speaking in an authoritative voice to the mutual goal of reaching a consensus about the problem.. .
>
> In my research, I have observed that preadolescent children have a great deal of difficulty differentiating and coordinating modes of debate required by academic discourse and the modes of discourse seen as interpersonally cooperative. Disagreements about task definitions and solutions frequently become competitions focused on issues of authority instead of intellectual debates focused on achieving consensus through the effective use of logic and evidence. (*ibid;* p. 151)

Note that Forman assumed the goal of the academic discourse is consensus. There is significant ecological validity here, in that consensus typically is an implicit if not explicit design criteria for the outcome of small group work in the elementary school classroom. However it conflicts with the norms of scientific discourse. I conjecture that the value placed on coming to consensus in the elementary school classroom stems, at least in part, from the larger value of social harmony and cooperation and discomfort with disagreement.

In my own research in elementary science classrooms over the last five years (in 14 classrooms in 5 different schools), supporting teachers to support children to evaluate competing ideas from a scientific perspective has constituted an enduring challenge. Some teachers –including one at the first grade level-- are effectively engaging their young students in such tasks as comparing the adequacy of alternative representations for a particular purpose, weighing alternative interpretations of data, and comparing the evidence for and against an idea. A large majority of the teachers are slowly developing expertise in these practices.

However I have been startled by several teachers' broad spread use of voting as a way of deciding between competing ideas, competing systems of categorization, or method of empirical investigation. As one second grade teacher who used voting explained, "But they [the children] want to vote because they think the majority is always correct". Another second grade teacher used voting for the children to come to agreement on each step in the class' design of an empirical investigation. She explained, "Kids don't get the opportunity to decide and resolve disagreements because everybody does the deciding for them." These teachers did not appear swayed by their colleague's position that voting was inappropriate to use in science, but useful to help children come to agreement on such social issues as "if you have to step out of line to get something [do] you get your spot back in line?"

I was also startled by how a triad of third graders resolved conflicting interpretations. The children had each been making their own collection of seeds. The intent of the broader enterprise was to have children elaborate their theories of critical seed features for different modes of seed dispersal, theories they would subsequently subject to empirical inquiry (while exploring biodiversity and using structure/ function as an interpretative frame) (Metz, 2002*a*). Their task here was to categorize their seeds according to what they assumed to be the seed's primary mode of dispersal and use this categorization as a basis for reflection on critical seed characteristics for each dispersal mode. However in those instances where they had conflicting ideas, instead of evaluating the reasons underlying different potential categorizations, they decided on the seed's categorization through Rock, Paper, Scissors, a complex version of Eeny, Meeny Miney, Mo. In this case, their teacher interrupted the process and talked with them about why this method was not a scientific way to analyze the competing ideas.

I interpret these observations as reflecting the absence of – or lack of confidence in -- an alternative tactic that they considered more appropriate to the task of choosing between competing ideas. Both voting and Rock, Paper, Scissors relatively expeditiously bring the group to a decision in a manner that the group accepts as equitable and fair. I think they also reflect the strong value on social harmony and the minimizing of disagreement. In short, in these instances it was the "everyday discourse" that prevailed.

Anderson, Holland and Palincsar's (1997) analysis of the workings of a small group in the elementary science classroom further elucidate the complexity of achieving scientific discourse in the elementary school science classroom. The small group work these authors examined occurred within the context of one of the curriculum units lauded by AAAS as grounded in the big ideas. The authors explain that the particular activity videotaped and analyzed in their paper was chosen as "it involved students in using canonical ideas to explain real-world phenomena, and it was designed so that students would encounter (and hopefully overcome) a number of common conceptual difficulties as they constructed their explanations."

However Anderson et al. report that the task functioned as intended for only one of the five children in the targeted group. The authors found that in addition to concerns about the task for which they would be accountable and the quality thereof, attention to interpersonal relations in the form of "the search for power and solidarity" competed with the scientific discourse agenda.

While Anderson *et al.* had structured the outcome of each small group's work as a single artifact (a poster including a picture of the demonstration they had observed, and their observations and explanations thereof), these researchers valued and hoped for articulation of competing ideas in the process of the artifact's construction. However they report that the small group largely avoided conflict by having different children assume responsibility for different aspects of the task; *e.g.*, writing explanations and making the models. The children's system of informal division of the intellectual task "broke down" where two children –one powerful and one weak—were both determined to assume authority for demonstrating the model to the class. The authors' explanation for why they were so dissatisfied with the forms of conflict that did and did not appear provide an intriguing lens onto the status and challenges of considering competing ideas in the elementary school science classroom:

> For us the students' system for avoiding conflicts was as troubling as the conflicts themselves. It was not helpful to Linda, for example, that the mistakes in her explanations went unchallenged until she began to find them herself on Wednesday [third day of the "group work"]. Sometimes we want students to argue. Arguments that combine theory and evidence are essential to the work of scientific communities. . . . Students need to work together to construct the kinds of complex arguments and procedures that are characteristic of functional scientific literacies, but neither children or adults are very good at maintaining equality of participation in communities that are intellectually and socially complex (*ibid*, p. 380).

The research literature reveals the complexity of the teacher placing children into small groups in such a way that will support the "equity of participation" that appears so important in the exchange and collaborative examination of alternative ideas. In accordance with the early recommendations of cooperative learning instructional leaders Johnson and Johnson, the common practice among elementary school teachers consists of grouping a low achieving child, with two or three middle achievers and a high achiever. However, based on her seminal analysis of the research literature concerning children's learning in small groups, Cohen (1994) identifies differences in status, based on perceived academic ability, as well as gender, ethnicity or popularity, as leading to differences in active participation, including the opportunity to voice ideas and to influence the discussion. Cohen's finding that active participation in the small group is an important factor in the child's learning means we need to be particularly concerned with engineering instructional conditions that foster inclusion in the discourse.

I view the huge discrepancy between how the elementary school science classroom and community of scientists regard competing ideas and the resolution thereof as problematic. This approach is not simply a distortion of science. More generally, taking ideas as objects of thought, the careful examination and comparative evaluation of competing ideas is fundamental to the development of understanding. From a socio-cognitive perspective, participating in such analyses is a critical experience for fostering children's more independent engagement in this genre of higher-order thinking.

LEVERAGE POINTS

In their essay on needed research for science education reform, Anderson and Helms (2001) speak of leverage points:

> While a systemic view quickly establishes that there are no simple solutions in the process of reform, one would expect that there are some very strategic leverage points. An old rule of thumb known as the Pareto's principle (named for an Italian economist of an earlier era) says that 80% of the results of an endeavor are produced by only 20% of the input efforts of the endeavor. (p. 5)

I close this chapter by delineating leverage points that I view as key to the reform of elementary science education and briefly describe my own current efforts to address these challenges (Metz, 2000; 2001*a*; 2001*b*, 2002*a*; 2002*b*). The first two concern transformation of curriculum. The last two concern needed foci of teacher professional development that are crucial to supporting the proposed curriculum changes.

Emphasize depth over breadth.

We have strong evidence of the failure of broad superficial coverage. We need to restructure the aspirations of the K-12 curriculum, to focus more deeply on less. In elementary school science, we need to more adequately emphasize the big ideas, how they are related and how they are used as interpretive frames.

My current research involves investigating the power and limitations of elementary children's scientific inquiry, under as optimal conditions as we can design and sustain in the public school classroom. This work is confined to the biology curriculum at the primary grade level. As far as possible within the constraints of the public school classroom, we aim to emphasize depth over breadth. I view depth of knowledge as important, both from its intrinsic importance and the bootstrapping relation between the adequacy of the knowledge and the adequacy of the inquiry (Glaser, 1984). I use the NRC conceptualization of key ideas, but go beyond their grade-specific recommendations as the K-4 ideas do not encompass any explanatory constructs. (NRC's *National Science Education Standards* recommends the three big ideas of *(a)* characteristics of organisms, *(b)* life cycles of organisms and *(c)* organisms and environments as foci for the K-4 life sciences curriculum. We add the ideas of both structure and function of living systems as well as diversity and adaptations of organisms from their list of Grade 5-8 ideas, as these later ideas constitute explanatory structures for making sense of the first set and crucial conceptual tools in formulating explanatory theories.) The curriculum scaffolds these ideas through a process of progressive abstraction. We view supporting teachers to support the children to understand these ideas, the power thereof, and how they are reflected throughout the curriculum as a primary challenge of the curriculum development and corresponding teacher professional development.

Reframe the purpose and form of "hands on" activities

For reasons epistemological, cognitive and affective, we need to foreground discovery and understanding (and perhaps sometimes, in conjunction with these two top-level goals, design and control). We need to let such processes as observation, categorization and measurement serve not as ends but sub-goals. Finally we need be careful not to overemphasize the science of verification to the neglect of the science of discovery.

Scaffolding children's increasingly independent goal-focused inquiry and, conversely avoiding the decomposition and decontextualization of the inquiry process constitutes a major focus on my work. I aim to maintain the robust goal and activity-structure of the authentic activity through a combination of several pedagogical strategies. We capitalize on socially-distributed expertise (Salomon, 1997), in order to reduce the cognitive load of scientific inquiry while maintaining the integrity of the intellectual enterprise. In the vein of Ann Brown's distinction between "blind" and "informed" instruction, we emphasize the metacognitive knowledge layer in the teaching of the tools of inquiry (*e.g.*; Why would I want to use this? When would I want to use this? What might it obscure? Why might I want to use something else?) as this knowledge is key to building independent inquiry. The last phase of each curriculum module consists of pairs of children, *using* their knowledge of the domain and scientific inquiry process to develop, implement and critique a research project with a partner and finally a multi-class conference in which they examine the relations between their research projects. We engineer homogeneous pairings to maximize the probability that each child will have a sense of control and responsibility over the investigation, which is viewed as instrumental in the pursuit of their question. In the research conference, children learn about others' studies, and work with their partner to examine relations between the studies (*e.g.*; which study is most like ours? Do our results agree or not? If not, why might they be different?) and think about future directions (What am I curious about now? What study do I think needs to be done next?) Through these activity structures, we aim to develop a rich sense of science as a way of knowing in the context of a relatively rich understanding of both science content and science process.

Support teachers' capacity to support children to engage in scientific discourse.

A curriculum can structure an activity, but the subsequent scaffolding of student thinking can never be fully scripted. If science classrooms are to transcend the telling of facts, we need to develop teachers' understanding of scientific discourse, especially the bases on which scientists choose between competing ideas.

Teachers' support of the children's scientific discourse has emerged as a critical barrier in successfully empowering the children's scientific discourse and thus constitutes a top-level goal of our curricular and professional development revisions over the last few years. We have used analysis of videos of the project classrooms and the monthly teacher meetings, as well as teacher written feedback and classroom to unpack the challenge this poses for teachers. The project aims to support the teachers empowering of the children's scientific discourse in a number of ways.

Project strategies encompass fostering teachers' understanding of the epistemic enterprise of doing science and, more specifically, the discourse that supports the enterprise. Our tactics include trying to shift the teachers' conceptualization of the epistemic enterprise from simply teaching science content to scaffolding science content in interaction with science process. Without this conceptualization of goal-structure, teachers cannot rationalize taking the time to engage children in science talk. Second, we aim to foster an understanding of the epistemic enterprise itself through case studies of scientists' work framed in terms of thought experiments, as well as extensive curriculum-embedded scaffolding. Third, we engage project teachers in collaborative analysis of classroom discourse and its cognitive, epistemological and affective entailments.[2] They also analyze video cases of teachers who are expert in supporting the discourse. Finally, we focus on developing the teachers' understanding and utilization of the big ideas and powerful explanatory frames of the discipline. All of these elements are crucial to the teacher's support of the scientific discourse.

Transform teachers' understanding of children's capabilities.

The community of elementary school teachers reflects a strong concern with making sure that what they teach and how they teach is "developmentally appropriate." Teachers are understandably strongly resistant to trying new ways of teaching that they believe will exceed their students' developmentally-based capacities. Unfortunately current prevailing assumptions about children's scientific reasoning capabilities significantly underestimate the power of their thinking. While, as John Flavell (1994) notes, "virtually all" cognitive development have come to reject a strong universal stage model such as proposed by Piaget, this advancement has not appeared to have permeated the belief system of elementary school teachers. Changing teachers' understanding of children's capabilities is crucial to achieving their willingness to engage their children in more intellectually substantive inquiry.

I am concerned that low-level expectations about children's capabilities, both in terms of the content they can grasp and the power of the inquiry they can effectively undertake, can become a self-fulfilling prophecy. The curriculum I have written in

[2] For example, the teacher group considered implications of the different strategies they used to decide who has the floor next in a class discussion. Many teachers reported frequently using the random drawing of names, on the grounds that it kept the children alert. However when viewed from a scientific knowledge-building perspective, some teachers argued that the strategy was problematic. The teachers considered alternative strategies that would also more adequately support the collaborative construction of ideas.

connection with my NSF project makes assumptions about children's cognitive capabilities that are at variance with the large majority of elementary school teachers. Raising the bar about the potential power of children's scientific cognition constitutes a key challenge of my work. I have found existence proofs are particularly effective in this regard; *e.g.*, examples of children's written work, their discourse, and video case studies in which teachers closely analyze the power of children's scientific reasoning in realistic classroom settings, in the context of this instructional model and others. We need to help teachers understand the potential power of children's scientific inquiry.

SUMMARY

This chapter compared the knowledge building enterprises in science and elementary school classrooms. I focused on three aspects: *(a)* goal structure; *(b)* targeted knowledge; and *(c)* stance toward competing ideas. I contend that these three aspects represent spheres where the extent of the differences between the enterprises is unnecessary and deeply problematic. For each of these aspects and their interactions, analysis of the practices of scientists can function as a powerful heuristic in the reform of elementary science education. This analysis also reflects the power and complexity of regarding science as inquiry as "a controlling principle in the ultimate organization and selection of students' activities" (National Research Council, 1996, p. 105).

REFERENCES

American Association for the Advancement of Science. (1993). *Benchmarks for scientific literacy*. Washington, D.C.: American Association for the Advancement of Science.

Anderson, C. W., Holland, J. D., & Palincsar, A. (1997). Canonical and sociocultural approaches to research and reform in science education: The story of Juan and his group. *The Elementary School Journal, 97*(4), 359-383.

Anderson, R. D., & Helms, J. V. (2001). The ideal of standards and the reality of school: Needed research. *Journal of Research in Science Teaching, 38*(1), 3-16.

Ashe, D. B., & Brown, A. L. (1995, April 22,1995). *"Otter Fur and Delayed Implantation": Children's Guided Transition from Form-Function Reasoning Towards an Adaptationist Stance.* Paper presented at the Annual Meeting of the American Educational Research Association, San Francisco, CA.

Brewer, W., & Samarapungavan, A. (1991). Children's theories versus scientific theories: Differences in reasoning or differences in knowledge? In R. R. Hoffman & D. S. Palermo (Eds.), *Cognition and the Symbolic Processes: Applied and Ecological Perspectives* (pp. 209-232). Hillsdale, NJ: Erlbaum.

Brown, J.S., Collins, A., & Duguid, P. (1989). Situated cognition and the culture of learning. *Educational Researcher, 18*(1).

Brown, A. L., Ash, D. B., Rutherford, M., Nakagawa, K., Gordon, A., & Campione, J. (1993). Distributed expertise in the classroom. In G. Salomon (Ed.), *Distributed Cognitions: Psychological and Educational Considerations* (pp. 188-228). New York: Cambridge University Press.

Burbules, N. C., & Linn, M. C. (1991.Science education and philosophy of science: Congruence or contradiction? *International Journal of Science Education, 13*(3), 227-241

Carey, S., Evans, R., Honda, M., Jay, E., & Unger, C. (1989). 'An experiment is when you try it and see if it works': A study of grade 7 students' understanding of the construction of scientific knowledge. *11*(Special issue), 514-529.

Cohen, E. G. (1994). Restructuring the classroom: Conditions for productive small groups. *Review of Educational Research, 64*(1), 1-35.
Driver, R., Newton, P., & Osborne, J. (2000). Establishing the norms of scientific argumentation in classrooms. *Science Education, 84*(3), 287-312.
Feynman, R. (1995). *Six easy pieces: Essentials of physics explained by its most brilliant teacher*. Helix Books: Reading, MA.
Flavell, J. H. (1994). Cognitive development: Past, present, and future. In R. D. Parke & P. A. Ornstein & J. J. Rieser & C. Zahn-Waxler (Eds.), *A Century of Developmental Psychology* (pp. 569-588). Washington, D.C.: American Psychological Association.
Forman, E. A. (1992). Discourse, intersubjectivity, and the development of peer collaboration: A Vygotskian approach. In L. T. Winegar & J. Valsiner [Eds.] *Children's development within social context*, Vol. 2: Research and methodology. Lawrence Erlbaum: Hillsdale, NJ. pp. 143-159
FOSS (2001, 1 June 2002). *Foss Grade 1-2 Revisions Edition 2002 - an Overview*, [website]. Full Option Science System.
FOSS (2002). *Teacher Guide: Insects*. Berkeley, CA: Delta Education, Inc.
Gagné, R. (1977). *The Conditions of Learning* (3rd ed.). New York: Rinehart & Winston.
Glaser, R. (1984). Education and thinking: The role of knowledge. *American Psychologist*. 39 (2) 93-104.
Glaser, R., & Chi, M.T.H. (1988). Overview. In R. Glaser & M. T. H. Chi & M. J. Farr (Eds.), *The Nature of Expertise* (pp. xv-xxvii). Hillsdale, NJ: Erlbaum.
Greeno, J., & Simon, H.A. (1984). Problem solving and reasoning. In R. C. Atkinson & R. J. Herrnstein & G. Lindzey & R. D. Luce (Eds.), *Stevens' Handbook of Experimental Psychology* (pp. 589-672). New York: Wiley.
Grice, H. P. (1975). Logic and conversation. *Syntax and Semantics, vol. 3: Speech Acts*, [P. Cole & J. Morgan. Eds.] 41-58. New York: Academic Press.
Grice, H. P. (1978). Further notes on logic and conversation. *Syntax and semantics, vol. 9: Pragmatics* [P. Cole, Ed.] 113-27. New York: Academic Press.
Gopnik, A., & Wellman, H. (1994). The theory theory. In L. A. Hirshfeld & S. A. Gelman (Eds.), *Mapping the Mind: Domain Specificity in Cognition and Culture*. (pp. 257-293). Cambridge: Cambridge University Press. Karmiloff-Smith, A., & Inhelder, B. (1974). If You Want to Get Ahead, Get a Theory. *Cognition, 3*(3), 195-212.
Hammer, D. (1994). Epistemological beliefs in introductory physics. *Cognition and Instruction, 12*(2), 151-183.
Hatano, G. & Inagaki, K. (1991*a*). Motivation for collective comprehension activity in Japanese classrooms. Paper presented at the Annual Meetings of the American Educational Research Association.
Hatano, G. & Inagaki, K. (1991*b*). Sharing cognition through collective comprehension activity. In L. B. Resnick J.M. Levine & S.D. Teasley [Eds.] *Perspectives on socially-distributed cognition*. Washington DC: American Psychological Association. pp. 331-348.
Latour, B. (1999). *Pandora's hope: Essays on the reality of science studies*. Cambridge, Massachusetts: Harvard University Press.
Lehrer, R., Carpenter, S., Schauble, L., & Putz, A. (2000). Designing classroom that support inquiry. In J. Minstrell & E. van Zee (Eds.), *Inquiring into Inquiry Learning and Teaching in Science* (pp. 80-99). Washington, D.C.: American Association for the Advancement of Science.
Lemke, J.L. (2001). Articulating communities: Sociocultural perspectives on science education. *Journal of Research in Science Teaching, 38*(3), 296-316.
Lesgold, A.M., Rubison, H., Feltovich, P.J., Glaser, R., Klopfer, D., & Wang, Y. (1988). Expertise in a complex skill: Diagnosing x-ray pictures. In M. T. H. Chi & R. Glaser & M. J. Farr (Eds.), *The Nature of Expertise* (pp. 311-342). Hillsdale, NJ: Erlbaum.
Linn, M. C., Lewis, C., Tsuchida, I., & Songer, N. B. (2000). Beyond fourth grade science: Why do U.S. and Japanese students diverge? *Educational Researcher, 29*(3), 4-14.
Markman, E. M. (1978). Empirical versus logical solutions to part-whole comparison problems Concerning classes and collections. *Child Development, 49*, 168-177.
Markman, E. M., & Callanan, M. (1983). An analysis of hierarchical classification. In R. J. Sternberg (Ed.), *Advances in the Psychology of Human Intelligence* (Vol. 2, pp. 325-366). Hillsdale, NJ: Erlbaum.

Markman, E. M., Horton, M. S., & Mclanahan, A. G. (1980). Classes and collections: Principles of organization in the learning of hierarchical relations. *Cognition, 8*, 227-241.

Mayr, Ernst. (1997). *This is biology: The science of the living world.* Cambridge, Massachusetts: Belknap Press of Harvard University Press.

Metz, K. (1995). Re-assessment of developmental constraints on children's Science Education. *Review of Educational Research, 65*(2), 93-127.

Metz, K. (2000). *Young children's inquiry in biology: Building the knowledge-bases to empower independent inquiry.* . In J. Minstrell & E. van Zee (Eds.), *Inquiring into Inquiry Learning and Teaching in Science* (pp. 371-404). Washington, D.C.: American Association for the Advancement of Science.

Metz, K. E. (2001*a*). Disentangling immutable from mutable constraints on children's scientific inquiry. Invitational symposium on Cognitive Developmental Research and Instructional Practice. Annual Meetings of the Cognitive Development Society. (Norfolk VA, Oct 26-27).

Metz, K. E. (2001*b*). Elementary school children's statistical reasoning in science. In *Proceedings of the American Statistical Association.* Alexandria VA: American Statistical Association. National Resource Center (1996). *National science education standards.* Washington, D.C.: National Academy Press.

Metz, K.E. (2002*a*) *Children doing science: Investigation of animal behavior.* Unpublished science curriculum module for grades one-three.

Metz, K.E. (2002*a*) *Children doing science: Investigation of plant.* Unpublished science curriculum module for grades one-three.

National Resource Center (1997). *Science for All Children: A Guide to Improving Elementary Science in Your School District.* Washington, D.C.: National Academy Press.

Palincsar, A., & Magnusson, S. J. (2001). The interplay of first-hand investigations to model and support the development of scientific knowledge and reasoning. In S. Carver & D. Klahr (Eds.), *Cognition and Instruction: Twenty-Five Years of Progress* (pp. 151-187). Mahwah, NJ: Lawrence Erlbaum.

Penner, D. E., Lehrer, R. & Schauble, L. (1998). From physical models to biomechanical systems: A design-based modeling approach. *Journal of the Learning Sciences. 7*, 427-449.

Popper, K. (1972). *Conjecture and Refutations: The Growth of Scientific Knowledge* (4th ed.). London, England: Rutledge, Kegan, Paul, Ltd.

Reif, F. & Larkin, J. H. (1991). Cognition in scientific and everyday domains: Comparison and learning implications. *Journal of Research in Science Teaching, 28*(9), 733-760.

Resnick, L. B. (1991). Shared cognition: Thinking as social practice. In L. Resnick & J. Levine & S. Behrend (Eds.), *Socially Shared Cognitions* (pp. 1-19). Hillsdale, NJ: Erlbaum. Rutherford, F.J., & Ahlgren, A. (Eds.). (1990). *Science for All Americans.* New York: Oxford University Press.

Sadler, P. M., & Tai, R. H. (2001). Success in introductory college physics: The role of high school in preparation. *Science Education, 85*(2), 111-136.

Schwab, J. J. (1962). The teaching of science as enquiry. In J. Schwab & P. Brandwein (Eds.), *The Teaching of Science* (pp. 1-103). Cambridge, MA: Harvard University Press.

Simon, H. A. (2001). "Seek and ye shall find": How curiosity engenders discovery. In K. Crowley & C. Schunn & T. Okada (Eds.), *Designing for Science: Implications from Everyday, Classroom, and Professional Settings* (pp. 5-20). Mahwah, NJ: Lawrence Erlbaum.

Smith, S. L., Houghton, C., Maclin, D., & Hennessey, M. G. (2000). Sixth grade students' epistemologies of science: The impact of school experiences on epistemological development. *Cognition and Instruction, 18*(3), 349-422.

Stigler, J. W., & Hiebert, J. (1999). *The teaching gap: Best ideas from the world's teachers for improving education in the classroom.* New York: The Free Press.

Stokes, D. E. (1997). *Pasteur's Quadrant: Basic science and technological innovation.* Washington, D.C.: Brookings Institutional Press.

Valverde, G. A., & Schmidt, W. H. (1998). Refocusing U.S. math and science education. *Issues in Science and Technology Online*

Vekiri, Ioanna, Baxter, Gail, & Pintrich, Paul R. (1998, April 1998). *Scientists invent, discover, and explain; Images of science in elementary hands-on science classrooms.* Paper presented at the 79th Annual Meeting of the American Educational Research Association, San Diego, CA.

White, B. Y., & Frederickson, J. (1998). Inquiry, modeling, and metacognition: Making science accessible to all students. *Cognition and Instruction.* 16 (1), 3-118.

CHAPTER 7

SHIRLEY J. MAGNUSSON, ANNEMARIE SULLIVAN PALINCSAR, & MARK TEMPLIN

COMMUNITY, CULTURE, AND CONVERSATION IN INQUIRY-BASED SCIENCE INSTRUCTION

INTRODUCTION

Professional scientific practice focuses on the production of discoveries about the physical world. Traditional perspectives depict this process as a matter of the diligence and skill of scientists in knowing where and how to observe to uncover the secrets of nature. More contemporary views, however, cast knowledge production as a process of *invention* in which the scientific community ultimately determines whether and what is "discovered." This is a cultural view; that is, it represents scientific practice as thought and activity patterned in particular ways through the social processes of interaction within a community drawn together by shared values and beliefs. If our current desire is that science instruction provide students with opportunities to learn in ways that mirror the activity of actual scientific communities, which underlies the national standard of inquiry-based science teaching, then contemporary views of the nature of science suggest that we need to think in fundamentally different ways about science instruction.

In this chapter, we bring ideas from three different fields of study to describe culture- and community-centered views of scientific practice: contemporary philosophy of science as represented by the work of Pera, social studies of science as represented by the work of Woolgar, and sociocultural perspectives in psychology as represented by the work of Bahktin. We argue that the process of scientific discovery is inextricable from the community of which that scientist is a part, and, hence, the culture of that community. Thus, we argue that teaching and learning science as inquiry (which is the national standard), is also a cultural phenomenon and a community-based endeavor. The purpose of this paper is to represent contemporary views of the nature of scientific activity as community-bound and culturally-based, and then present some ideas about what this view implies for contemporary conceptions of inquiry-based science teaching and learning.

L.B. Flick and N.G. Lederman (eds.) Scientific Inquiry and Nature of Science, 131-155.

SCIENTIFIC INQUIRY AND COMMUNITY

The Methodological Model: A Traditional View

Traditional views of scientific inquiry foreground the individual-as-inquirer, and "official" accounts of knowledge production depict scientists as merely happening upon a discovery by being in the right place at the right time (see Woolgar, 1988 for a more thorough discussion). These representations are rooted in the shift of thinking that occurred in the 16th and 17th centuries when philosophical ways of doing science that harkened back to Aristotle were abandoned. As characterized by Pera (1994), the "founding fathers of modern science" believed that "natural science does not progress through deduction from evident principles but through induction from observations and experiments" (1994, p 129). Sir Francis Bacon, for example, wrote that the purpose of science is "to overcome, not an adversary in argument, but nature in action" (1620, p. 42). Similarly, Galileo wrote that "in the natural sciences the art of oratory is ineffective" because "true and necessary conclusions" will only arise from "sensory experiences and necessary demonstrations" (1953, p. 54). Hooke penned that "the Science of Nature has already been too long made only of the *Brain* and the *Fancy*: it is now high time that is should return to the plainness and soundness of *observations* on *materials* and *obvious* things" (1665, Preface). Finally, Descartes (1628) wrote that "we need a method if we are to investigate the truth of things" and indicated that assiduously following the rules of the method would lead one to "never take what is false to be true" and would "gradually and constantly increase one's knowledge till one arrives at a true understanding of everything within one's capacity" (p. 15).

Pera (1994) has referred to this view of science as the *methodological model*. In this view, scientific research is a game with two players; the inquiring scientist who asks questions and nature who provides the answers. Method is an impartial arbiter in this game, ascertaining whether the game was played well and determining when it is over. Pera notes, "as it is guided or forced by the rules of the arbiter, nature speaks out, and 'knowing' amounts to the scientist's recording of nature's true voice, or *mirroring* its real structure" (Ibid., p. ix).

The point of describing the traditional view and providing specific writings from which it originated, is to suggest its familiarity. Are these not the ideas that still dominate the public's views of science today? Of teachers' views? Is not the discovery learning approach that arose in the 1960s predicated on such views?

The Dialectical Model: A Contemporary View

Despite the familiarity of the methodological model, advancements in various fields brought this model into question over half a century ago. The rise of relativity and quantum theory alone took science from the realm of truth to probability. Moreover, the actual process of discovery from case accounts of scientific activity is far from the commonly depicted action of "uncovering and revealing something which had been there all along" (Woolgar, 1988, p. 55). An accounting of the shift to different views is beyond the scope of this paper, but suffice it to say that many scholars over several decades worked to salvage the methodological model before it succumbed to its own failings. Views of science that took its place include the ideas "that data are theory-laden, that there is no logic of discovery leading from data to cognitive claims, that there is no clear distinction between observational and theoretical concepts, that theories cannot be reduced to their empirical basis, that they are underdetermined by it, and finally that there is no universal method" (Pera, 1994, p. 132).

For some, such ideas are tantamount to a view of scientific knowledge as purely relativistic. But to others, they simply indicate the cultural basis of science; that is, the dependence of the nature of scientific knowledge on the values, beliefs, and standards of a community of practitioners about what is important to know and do, and the norms and conventions that guide how one comes to know. As a case in point, Pera (1994) describes circumstances at the turn of the 17th century that led two scientists – Scheiner and Galileo – who observed the same phenomenon – dark spots when looking at the sun – to draw very different conclusions about the nature of the spots. Scheiner, who assumed the validity of Ptolemy's geocentric theory, came to conclude that the spots were not on the sun but were stars. That conclusion was consistent with a central thesis of Ptolemy's theory: that the heavens were incorruptible. Since the spots were observed to change size and shape, any conclusion that they were on the sun would mean that the sun was not incorruptible. Pera reports that Scheiner (1612) thought it "unseemly and highly unlikely that the spots lie on the surface of the Sun which is a very shiny body," and that his aim was to "free the Sun of the offensive spots" (p. 30). In contrast, Galileo, who found merit in Copernicus' heliocentric theory and rejected Ptolemy's view, concluded that the spots were clouds (giving them a constitution that could account for their sporadic and changing appearance) or perhaps a part of the sun itself. Moreover, Galileo cited the spots as evidence for the need of the scientific community to reject the prevailing theory of Ptolemy and change its ideas. In Galileo's words:

> whether the spots are on the Sun or around the Sun, whether we say they are generative or not, whether we call all these things that vacillate clouds or not, what follows seems certain according to the common opinion of astronomers: that the density and constitution of the heavens as we consider it today can no longer be maintained (Scheiner, 1612, p. 68).

The model that Pera (1994) has developed to depict what he thinks these accounts represent about science, is a game with three players: a scientist or group of scientists, nature, and another group of scientists that debates with the first according to the features of scientific dialectics[1]. In this dialectical model, there is no impartial arbiter; rather, nature responds to a "cross-examination," and knowledge *represents the community's agreement* upon nature's correct answer. Pera notes that such a view should not be interpreted as replacing "objectivity with 'solidarity' or rationality with 'routine conversation' because agreement among members of the community, although not imposed or dictated by nature, is still constrained by it" (p.

ix). He also indicates that science in a dialectical model links rationality "not to certain *properties of theories* fixed by rules, but to the *quality of the arguments* which support the theories" (p. 144). Central to the quality of the argument is the nature of the *objects* and *facts* referred to in the argument, which Pera represents as cultural entities. Objects are defined as "the putative [supposed] reference of a concept about which there is consensus" (p. 160). Examples are cats, sunspots, electrons, genes. Facts are events that represent "a shared state of objects" such as "cats chasing mice, the sun having spots, electrons rotating in their orbits, genes transmitting heredity" (p.163). He argues that "objects and facts guarantee that science is as objective as it is able to be . . . for by constructing facts it also constructs objects . . . [and] objects and facts depend on a consensus over the corresponding concepts and judgments." (p. 161) He goes on to say that, "Science is not objective, however, in the sense that it describes, or makes assertions corresponding to reality in itself, for objects and facts are constructions, not carbon copies, images, or icons of reality" (Ibid.).[2] It is in this sense that scientific practice *invents* rather than discovers nature.

The Dual Community Nature of the Dialectical Model of Science

Studies of the history of science, as well as studies of the nature of current scientific practice, have led philosophers, historians, and sociologists of science to represent scientific inquiry not in terms of individuals, but in terms of communities of practice which provide the motivation, communication, and structure necessary to sustain individual inquiry. As such, a community is *not* a *passive* context for individual knowledge construction; rather, scientific communities enable (and constrain) the production of scientific knowledge. For a contemporary illustration, we present a summary of Woolgar's (1988) discussion of the "discovery" of pulsars by Hewish and Bell working in the Cambridge radio astronomy group.

In the first stage of the discovery process, Hewish and Bell noted unusual data in the form of an anomalous trace on chart recordings of radio signals from space. Early on, these traces were not judged to be worthy of attention; thus, they were not the subject of any further scientific activity. For many scientists, the story may have stopped here. However, at some point enough attention was paid to the anomalous data that they were noted to have some regularity. Hewish and Bell now thought it important to investigate, although their goal was to determine whether the traces were an artifact of interference in the signals. They planned and conducted high speed recordings, the result of which suggested *pulsed* radio emissions. However, there was still much skepticism about the meaning of these data and several researchers in the working group thought that the pulsed signals were spurious. Again, for other scientists, the story might have stopped there, but other members of the Cambridge group became aware of the unusual data and were asked to join Hewish and Bell in investigating the source of the emissions. In hindsight, this was an important turn of events, but at the time it could have been thought of as a fishing expedition, and it certainly delayed the group's ability to publish findings from their work. Curiously, during the next phase of activity, some members of the investigating team thought that the signal might be communication from an extraterrestrial intelligence. That hypothesis led the investigating team to restrict access to information from the project to only those members of the Cambridge group working directly on the investigation. In addition, publication of any results from their work was delayed further as members sought to discount this possibility. Finally, after the group had discounted extraterrestrial intelligence as the source, the investigating team concluded that the signal was from a pulsating radio source and began to do some preliminary investigation of it. Public reporting of the pulsed emissions finally followed, with no mention of the alternative hypothesis that was rejected. Now it was left to the scientific community to determine to what extent there was consensus about this new object (*pulsar*) and fact (*pulsed signal traces on chart recordings*).

Woolgar's account of the discovery of pulsars highlights important features of what he terms a "workbench" community. He argues that at this level, science employs similar problem-solving strategies to those used in many other domains. Workbench science communities typically involve a relatively small group of individuals who work closely with others in on-going collaboration to solve problems of immediate and joint concern. Research documents and laboratory tools within the local working group are artifacts of the workbench community. For the Cambridge radio astronomy group, the discovery process was a crush of immediate problems, most of them technical in nature, which centered on these artifacts. For instance, from the time the anomalous trace was noted until the high speed recordings, time was devoted almost exclusively to tinkering with the laboratory instruments. In the final analysis, what we see in the day-to-day activity of science, is scientists in their workbench community utilizing workbench artifacts to engage in informal speculations and communications to construct knowledge about the world.

Publication of the results represents the point of transition from the workbench community to the professional community arena. The formal re-presentation of the findings by the group was the discovery of a stellar object of immense age. Woolgar reports that tensions and controversy surrounded the "discovery" because other scientists felt that the Cambridge group members were too slow in reporting their data. Members of the Cambridge group countered that they needed to be sure of their discovery before releasing information. In the end, it seems that fear of embarrassment in the professional community drove many of the workbench community's decisions about when and how to inform others. We argue that such situations are quite understandable if one views science as occurring in two types of communities having distinctly different roles which function complementarily. The functioning of the professional community requires that ideas are packaged for maximum comprehension by other community members, and that their import is clearly signaled to indicate the contribution to the community. These ideas are subject to formal criticisms by a widely distributed (geographically) and diverse membership often having competing interests. Such purposes drive these communities to adapt social languages which focus on evolving formal concepts and explanations. Over time, these communities develop a reservoir of agreed-upon knowledge claims and a history of how discoveries occur based on formal accounts offered by community participants. Workbench communities, on the other hand, focus on solving problems as they arise, and therefore develop social languages which primarily encourage informal communications, often at an instrumental level.

What is the significance of the dual community nature of scientific activity?[3] With respect to Pera's dialectical model, it indicates that there are two phases of the game in which the community has to come to agreement regarding nature's "correct" answer: the agreement worked out within the workbench community, followed by the agreement worked out within the corresponding professional community. However, despite the workbench community's best effort to re-present nature's answer for the professional community, the professional community may not accept the answer, and may force a re-examination, reconstruction, or revision of what is considered nature's answer. Thus, we see two very different types of effort in these phases of knowledge production in science. At the workbench, multiple perspectives are fostered and nurtured to create the space for discovery, and innovative and creative thinking may be key to recognizing and constructing the results that are ultimately chosen for formal presentation. For the professional community, however, ideals such as the importance of explanatory power and the coherence of ideas guides how results are re-presented. The focus is on the construction of an argument that bridges powerfully from existing ideas within the formal community to the new ideas, and the presentation of evidence in a way that provides the most convincing backing for the argument.

In the scientific community, the transition from workbench to professional science activity occurs when researchers re-present their results to a broader community of scientists. This process involves making judgments about how to best present one's work. One aspect of the re-presentation involves framing one's results using the social language of the community. Here, we are not simply referring to vocabulary, but to language as a tool, recognizing that the impact of one's ideas will be influenced by one's ability to select appropriate tools for presentation and to use them skillfully. Communication in this phase of activity must serve broader goals because the communication that initially interrelated workbench participants, now must interrelate community members who may be quite distant in time and space. Thus, it is important to carefully select one's tools of expression so that the significance of one's work is maximally signaled to the community. These ideas are consonant with another perspective for thinking about school science learning: sociocultural theory.

LANGUAGE AND COMMUNITY

Sociocultural theorists have found it useful to use the metaphor of "tools" to describe knowledge construction (e.g., Brown, Collins, Duguid, 1989). This characterization contrasts with knowledge conceived of as an abstract entity because tools are embedded in cultural activity. That is, tools must be used to be understood, and "the occasions and conditions for use arise directly out of the context of activities of each community that uses the tool" (Ibid., p. 33). Moreover, tools reflect the views of the community using them, meaning that their appropriate use is a function of having shared understanding of that community; that is, the values, beliefs, norms, and conventions that guide the community's activity (physical and intellectual).

Bakhtin's work makes an even stronger statement about the role of community in human activity by casting language as being community-dependent. He developed this idea by designating "utterance" as the basic unit for the analysis of human language (see Wertsch, 1991a, 1991b). Then, considering that utterances always belong to someone and are always used to address someone, he argued that they can never be analyzed outside of their social context. He pointed out that utterances also interrelate, forming a chain of mutually-aware speech communication, and that they form specific ways of speaking in specific social contexts. For example, the speech patterns of professionals addressing one another at a conference is different from the patterns of speech used by two close friends. These community-based ways of speaking are referred to as "social languages." Wertsch (1991a) uses the following quote from Bakhtin to elucidate these points:

> The word in language is half someone else's. It becomes "one's own" only when the speaker populates it with his [sic] own intention, his [sic] own accent, when he [sic] appropriates the word . . . . Prior to this moment of appropriation, the word does not exist in a neutral and impersonal language (it is not, after all, out of a dictionary that the speaker gets his [sic] words!), but rather it exists in other people's mouths, in other people's concrete contexts, serving other people's intentions: it is from there that one must take the word, and make it one's own. (Bakhtin 1981, pp. 293-294, quoted in Wertsch, 1991, pp. 96).

Bakhtin's analysis emphasizes the role of social context in facilitating individual cognition. Communities, as social contexts, are not passive backdrops against which the cognition of individuals operate. Instead, communities *actively* promote specific forms of community-based dialogue leading to specific genres of speech within that community (see Wertsch, 1991a, 1991b). In this way, language plays a central role in *shaping* the thinking and actions of individual community members toward advancing the work of the community.

We submit that these views do not invalidate our commonly-held conceptions of the practice of science; rather, they resituate them. Thinking in particular ways about particular phenomena, rather than being an outgrowth of careful observation, becomes a product of habits of mind and particular tool use that is informed by the values and beliefs that are appropriated from community involvement. Thus, whereas our familiar notions of science may be rooted in a methodological model, those notions express the values, beliefs, norms, or conventions of particular communities of scientists. The difference is recognizing that scientific (or any other community) knowledge does not automatically arise from independent exploration of the physical world, but is an expression of a particular way of knowing the world that developed through enculturation into particular practices of a community of scientists.

Expressed in a different way, from a sociocultural perspective, learning is viewed as a transformation of participation (physical and intellectual) within a community (Lave, 1991; Rogoff, 1994). As Brown, Collins, and Duguid (1989) state, "situations might be said to *co-produce* knowledge through activity" (emphasis added, p. 32), which means that knowledge is developed as a function of our thoughts and actions in particular contexts, and that the nature of those contexts – that is, the community in which they are embedded – is instrumental to the nature of the knowledge that is produced. Schools are communities, and so are classrooms. Thus, the institution of school already supplies a community context for learning, at least from a social relationship standpoint. In a "traditional" classroom, the bulk of activity consists of students being assigned, completing, and getting feedback relative to "academic tasks" (Doyle, 1986). The issue that is pertinent to this chapter is that such activity does not resemble the practice of any of the communities whose products are typically the targeted understandings of schooling. Thus, despite the community basis of schools, the nature of traditional learning environments does not provide the sort of intellectual context that would support academic learning in particular disciplines. For example, a common activity in high school chemistry or physics is to solve word problems depicting a physical event. In the actual practice of science, one would need to determine what events to observe and what variables to measure and how to measure them before arriving at a point that might resemble solving a word problem, and of course in science, an underlying theoretical frame would be a part of selecting particular events and variables of interest. It is no surprise then that those who have learned problem solving by completing word problems focus on the surface features of the problem rather than underlying principles as is typical of a scientist's approach to problem solving (Champagne, Klopfer, & Gunstone, 1983).

The implications of these views for school science learning is that a more effective environment for the development of scientific knowledge in classrooms is one that embodies cultural elements of the scientific community[4], that guide knowledge production. What Pera's dialectical model tells us is that communication among community members about "nature's correct answer" is a critical part of the production of scientific knowledge, and what the dual community nature of this model suggests is that this conversation happens twice. Having the classroom emulate activity of the scientific community is not new to science education. Indeed, the inquiry-based curriculum development of the 50s and 60s was a major move to support a shift to the learning of science through investigative activity. What this chapter brings to our consideration of enacting such environments, however, is understanding the key role of conversation in the process. In the next section, we present an approach to science instruction that situates physical investigative activity within a rich conversational context, and we discuss key aspects of the role of the teacher in creating such an environment, particularly with respect to supporting students in appropriating the discourse of science.

A GUIDED INQUIRY VIEW OF TEACHING AND LEARNING SCIENCE

We suggest that these views of the nature of science and knowledge production signal two key dimensions of science instruction. First, classrooms need to have learning environments that reflect key elements of the culture of science; that is, the central values, beliefs, norms, and conventions that guide scientific knowledge production. Many of these aspects have been identified in national standards documents, such as in common themes and habits of mind (AAAS, 1989, 1993), or unifying concepts and processes or understandings of science as inquiry (NRC, 1996). In addition, however, is the element pointed out in this chapter: that such environments support not just one community context with its cultural practices, but two. Thus, students need opportunities to learn in an environment that reflects the culture of the workbench science where multiple perspectives and creativity are valued, as well as in an environment that privileges the values and beliefs of the culture of the professional science community where making and evaluating arguments is key. Second, is recognizing the critical role of conversation in learning science, in that part of learning science is appropriating academic scientific discourse (cf. Gee, 1996; Latour, 1987; Lemke, 1990). Thus, in addition to opportunities for physical engagement in investigation (i.e., "hands-on" instruction), students need sustained opportunities to engage in conversation before, during, and after the physical aspects of investigation, both within small groups in which the culture of workbench science dominates with its valuing of multiple perspectives, as well as in a whole class context in which issues of presenting arguments in the form of making knowledge claims and providing data as evidence backing those claims are discussed.

These dimensions are <u>not</u> independent, in fact, they are interdependent. Enacting science instruction as a process of supported engagement in the activity and discourse of science with its attendant language, norms, and conventions, functions to establish and maintain a particular culture. Thus, it is not that a particular classroom culture needs to be established prior to engaging in science instruction, but that science instruction serves to develop the classroom culture in desired ways to lead to the development of targeted scientific knowledge and reasoning. It is useful then, to conceptualize instruction as providing students with opportunities to "try on" scientific activity and discourse, and support them in developing, over time, their facility with such activity and discourse. It is also important to remember that, since the activity and discourse that students experience in the course of scientific inquiry is distinctly different from their everyday activity and conversation and from routine classroom work (Cobb & Yackel, 1996; Driver, Asoko, Leach, Mortimer, & Scott, 1994), students need guidance and support to learn to think and act in different ways.

Appreciating the role of conversation and engagement in scientific discourse is not evident in teacher activity in the science classroom. For example, Newton (1999), reporting on observational research in 34 secondary science classes concluded that: (a) talk was still dominated by exposition and was teacher-led, (b) fewer than half the lessons included deliberative interaction between the teacher and pupils; and when it did occur, it took less than 5% of the lesson time, and (c) little guidance was provided on how to organize interactions; hence, students appeared to experience considerable difficulty with the interpersonal dimensions of classroom discourse. At the elementary level, survey results (Weiss, Banilower, McMahon, & Smith, 2001) indicate that few teachers report having as objectives that students will learn how to communicate ideas effectively or learning to evaluate arguments based on evidence (21% and 8%, respectively), and few teachers routinely ask students to explain concepts to one another or consider alternative explanations (14% and 10%, respectively). Perhaps one reason for the lack of conversation-rich learning environments is a lack of understanding about the nature of the teacher role in such environments. We have been working over 10 years to advance our understanding of inquiry-based instruction, particularly with respect to the role of the teacher, working from a perspective that we refer to as *Guided Inquiry*.

Learning Science in Multiple Community Contexts

Just as scientific knowledge production occurs in phases, Guided Inquiry instruction is conceptualized as occurring in phases. This is not a new idea (e.g., Champagne & Bunce, 1991; Freyberg & Osborne, 1985; Karplus & Their, 1967); however, the difference is that we view particular phases relative to the dominance of a workbench or professional science community context and culture. Figure 1 shows a heuristic that we have developed to depict Guided Inquiry instruction and guide teacher decision-making (Magnusson & Palincsar, 1995; in press). The words in all capital letters represent the *phases* of instruction for *one* cycle of investigation, and a unit of instruction is designed as a *series* of *cycles* of investigation. It is assumed that each cycle begins with engagement around a question, proceeds to investigation from which one derives knowledge claims about the physical world, and ends with reporting of those claims and their associated evidence (typically on poster-size paper), followed by whole class conversation to determine the shared perspective regarding the nature of the physical world, considering the claims that were presented and the extent to which they were backed by convincing evidence or countered by contrary evidence.

The *Reporting* phase is a key element in Guided Inquiry instruction. First, when students know (as do scientists) that they will be responsible for publicly sharing the results of their investigative activity, then that activity becomes influenced by the culture of the context in which it will be shared. This is parallel to the workbench community activity being influenced by the culture of the professional community. Second, this context is a primary opportunity for the enculturation of students relative to the standards and conventions of science (e.g., questioning of one's claim in relation to the data provided as evidence) as well as to come to appreciate the role of a community in setting conventions and standards to support communication and understanding among its members. For example, after a first cycle of reporting claims and evidence, a valuable activity can be to provide opportunity for conversation about conventions and standards for reporting, with the group determining what these should be to enable more effective sharing and greater adherence to the culture of the scientific community. Conventions might include rules of thumb like not using yellow markers to write on posters because yellow is hard to see, or agreements to show only the most critical aspects in drawings so that others do not have difficulty determining what the drawing is trying to communicate. Standard setting might include specifying how drawings will be labeled or agreeing that symbols will be used to represent particular entities in a drawing (e.g., using circuit symbols to show the structure of an electric circuit rather than drawing the circuit elements as they actually appear).

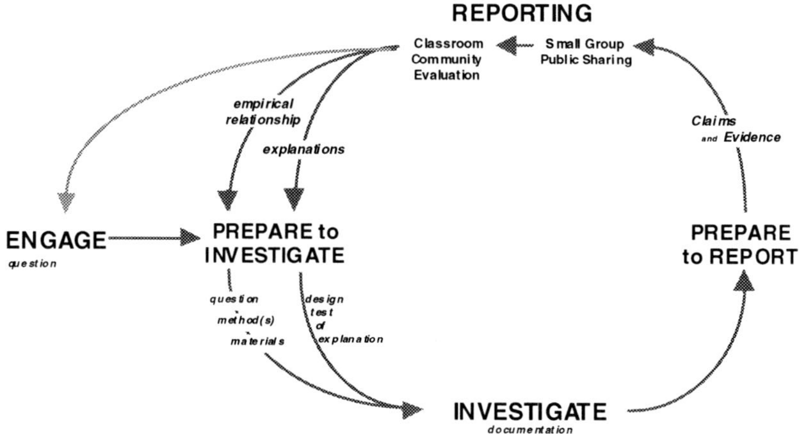

Figure 14. An heuristic representing phases of Guided Inquiry science instruction

Third, this phase provides the most concrete experience relative to the professional science community and culture. It is the context in which the scientific community's expectation is that thinking will converge on a "best answer" to a question, which may have considerable impact on students as they confront the claims and evidence presented by their peers and seek to reconcile those findings with their own. In the scientific community, this is the element that breeds a highly competitive environment, which we do not advocate or desire in classrooms. Thus, a key role for the teacher is to help students see divergent results as a product of different activity (thinking as well as doing), and to remind students that they are learning to "think and act like scientists," which is simply one way of knowing the world.

We think the competitiveness can be moderated by conceptualizing Reporting in two stages. The first stage is small group reporting, which is intended to maintain the culture of workbench science. Thus, it functions as though all the individual student groups are part of one large research group, and they are seeking to inform one another about their independent activity in answering the question or studying the phenomenon of interest. Furthermore, with the workbench culture context, multiple perspectives should be valued. As a result, the teacher should help students appreciate that the range of ideas from the small groups can help each individual or group expand the ways in which it thinks about the question and phenomenon under study. The second phase is whole class conversation about the claims and evidence that have been presented. In this phase, the classroom community evaluates the ideas shared to date, bringing to bear the full measure of the values/beliefs and norms/conventions of the professional community of science. The class determines about which claims it has consensus versus claims that can be rejected or need further investigation. Such conversation, although focusing on claims by groups in the community, if made explicit as a different stage of activity in which different norms come into play, can reduce the personal reaction any individual or group of students may have as the community evaluates their work, while still providing valuable feedback on the extent to which their activity and thinking were consistent with community expectations.

The preparation phases shown in the heuristic – *Prepare to Investigate*, *Prepare to Report* – although key to any instruction, are perhaps different in our view from some conceptions of investigation-based science activity in that they are conceptualized as key opportunities for the teacher to introduce the values, beliefs, norms, and conventions of the scientific community, and provide opportunities for students to try on scientific discourse in a less public forum than the Reporting phase. Thus, it is important for the teacher to seed and shape thinking and action but not be overly corrective so as not to shut down students' autonomous activity. The Reporting phase will provide additional opportunities for students to get feedback about their adherence to scientific norms and conventions, so there is valuable built-in redundancy to support students in becoming enculturated into scientific ways of knowing and doing.

In addition, it is important to recognize that the content of the conversation during these preparatory phases will vary according to the level of inquiry in which students are engaged. Schwab (1962) has defined three levels of inquiry: a *basic* level in which the question and method are known, an *intermediate* level in which only the question is known, and an *advanced* level where even the question is to be determined; in essence, "the student is confronted with the raw phenomenon" (p. 55). Thus, Preparing to Investigate may mean a focus on coming to understand the question and method, and how the method will help address the question (basic inquiry), coming to understand the question and how to figure out a method that will address it (intermediate inquiry), or examining the phenomenon for the purpose of determining the types of questions that are interesting and feasible to ask (advanced inquiry). The teacher's mindfulness about the focus of student activity is key to making the most of the opportunities to introduce and shape student thinking and action relative to the culture of science.

The *Prepare to Investigate* phase is often conducted in a whole class format (although for intermediate or advanced inquiry the teacher may want part of it to involve small group work in which students brainstorm questions or designs of methods), and it is the teacher's role to support the students in thinking through key issues of investigation prior to its occurrence. In contrast, *Preparing to Report* more commonly occurs as a small group activity in which the teacher assists and guides students in determining claims (although it may be efficient and effective at times to work in a whole group to discuss issues of how to analyze the data or what to do with anomalous data). The content of the discussion in Preparing to Report will also vary as a function of the level of inquiry in which students were engaged, such as considering the extent to which method design impacted data collection and analysis, and needs to be publicly shared for full evaluation of claims and evidence.

Notice that three of the four phases of instruction in Guided Inquiry are dominated by conversation and only in the Investigation phase is physical activity dominant. We think this may be a departure from commonly-held notions of inquiry-based science instruction, but one that is consistent with current understanding of knowledge production in the scientific community. For example, the common language of referring to desired science instruction as "hands-on" would suggest the dominance of activity, and even the more recent shift to "hands-on, minds-on instruction" is generally only meant to signal the predisposition necessary to make the most of what one does during activity. Tools like the heuristic in Figure 1 may be important to signaling this different view that is more consistent with how scientific knowledge is actually produced.

Complete attention to the many community- and culture-related issues that are possible to discuss relative to each phase of instruction in guided inquiry is beyond the scope of this chapter. However, a final point is in order. One cycle of investigation in this instructional design provides opportunities for students to work within communities reflecting the workbench and professional science cultures. Nevertheless, one would not expect a single cycle of investigation to result in the desired thinking and action relative to targeted scientific goals for a unit of study. Thus, the teacher's decision-making within any one cycle is also a function of how many cycles of investigation the students have already experienced in the unit of study. During beginning cycles of investigation within a unit (or at the beginning of the school year), the teacher focuses on the most major and basic issues in the conduct of scientific investigation (e.g., what it means to generate a claim, how one thinks about providing evidence for the claim), opportunistically seeding ideas and shaping activity to support students in developing understanding of fundamental issues of scientific investigation and the values, beliefs, norms, and conventions of the scientific community. As cycles progress (either within a unit of study or in units of study undertaken later in the year), the teacher will determine to what extent more sophisticated issues can be taken on, and will shift expectations for student thinking and action to higher levels. In this way, science teaching and learning relative to the culture of science is viewed to be in evolution across the school year, and indeed, across K-12 schooling.

Supporting Students in Appropriating the Discourse of Science

We consider inquiry instruction to be the most sophisticated instruction one can conduct, in no small measure due to the complexities of guiding and shaping conversation to support students in appropriating the discourse of science. This is a complex topic and there is much more to be understood before we have a full accounting of how to think about teaching and learning from this perspective. Considering our points in this chapter that the nature of the discourse is key to the development of scientific knowledge and that the nature of effective inquiry-based instruction is as much or more about engaging in conversation about phenomena as observing them, we present results of studies that we have conducted of classroom discourse that have helped us identify several types of teacher "moves" that are key to supporting students in engaging in such conversation in sustained ways (e.g., Palincsar, Magnusson, and Hapgood, 2001). Results reported in this chapter were of classroom discourse during the Reporting phase of instruction in a unit of study about light that took place in the spring of 1999. Participants were fourth grade students and their teacher from a school in a working class community in the upper midwest with an urban profile (≈ 45% of the student population is African American; ≈ 52% of students receive free/reduced-cost lunch). The teacher, Linda Verhey, was highly experienced (over 25 years), and had been working with us for several years as part of a community of practice of educators seeking to define effective inquiry-based teaching practice relative to the Guided Inquiry orientation to teaching science (Palincsar, Magnusson, Ford, Marano, & Brown, 1998). Table 1 shows the targeted scientific knowledge and reasoning goals for the unit of study involving light.

Table 1. Conceptual Goals for the Unit of Study about Light

SCIENTIFIC CONTENT	• Light can be reflected, absorbed, or transmitted by objects. • There is an inverse relationship between the amount of light reflected from and absorbed by an object: more reflected light, less light absorbed. • Dark or black objects mainly absorb light; light or white objects mainly reflect light. • All objects reflect light. • Light reflects in a particular way: the angle of incoming light equals the angle of the reflected light.
SCIENTIFIC REASONING	▪ Scientists seek to understand why the physical world works in particular ways. ▪ Scientists observe specific aspects of the physical world in order to determine how it works. ▪ Scientists observe carefully and systematically, and record what they observe. ▪ Scientists seek to quantify what they observe to foster accuracy and precision in observation. ▪ Scientists organize their data in particular ways to assist them in identifying relationships.

	▪ Scientists conduct fair and reliable tests to answer a question. ▪ The relationships that scientists identify are presented as knowledge claims to the scientific community. ▪ Knowledge claims are evaluated, and their adequacy is a function of the strength of the evidence supporting them.

Analyses of the studies reported here were designed to reveal the ways in which the teacher encouraged and helped to advance classroom conversation, particularly with respect to reflecting scientific discourse. Three dimensions of teacher activity emerged from these analyses: (1) establishing and maintaining the conversational norms of everyday discourse, (2) working at the intersection of everyday and scientific discourse, and (3) establishing and supporting the norms of scientific practice. We refer to these dimensions as types of teacher "moves" because they indicate the ways in which the teacher engaged with the children to encourage or advance the children's conversation in particular ways. The results presented here come only from the first stage of the Reporting phase[5] in the first cycle of investigation in the unit of study about light; thus, students are quite early in their development of conceptual understanding, and the teacher's primary purpose is to support students in identifying and expressing their ideas and determining to what extent they have similar or different ideas to one another, rather than seeking to shape the conversation in more evaluative ways to have students compare the relative power and merits of different ideas.

Establishing and maintaining the conversational norms of everyday discourse

A primary activity of the teacher during the Reporting phase was to establish norms and conventions for public speaking. This dimension is concerned with the everyday (but critical!) world of interpersonal communication and has little to do with science. In essence, teacher moves in this dimension concern the "etiquette" of reporting, and they communicate about general social conventions that support civil interaction of those who are sharing their claims and evidence (presenters) and those who are listening to the presenters (audience members). From the standpoint of the presenters, aspects of etiquette include: how to stand and face the class in the presentation in order to be easily heard, where to stand relative to one's poster in order to enable others to easily view it to understand what is being presented, and how to speak to increase the chances of being clearly understood (e.g., issues of loudness, diction). From the standpoint of the audience members, aspects of etiquette include: how to indicate attention, and how to engage in conversation with the presenters (e.g., raising one's hand and waiting to be called upon by the presenters). The vast majority of the conversational norms to be observed by presenters were identified within the first ten minutes of the Reporting phase. Thereafter, teacher statements related to this purpose punctuated the discourse as needed, and, in contrast to opening statements directed to students who were presenting, were principally directed to the audience members (e.g., "Those people that have their heads on the desk… I find that extremely rude." [577-581][6] or "How else could you ask that nicely and politely?" [374]).

An important aspect of the teacher's activity in establishing the etiquette of everyday discourse during science instruction was to mediate interpersonal issues. For example, the teacher protected turn-taking, when such protection was warranted: "Let's give Robby a chance to talk" [1037]. In addition, she sometimes deflected comments from an audience member that a presenter experienced as a personal challenge: "You know, he's just wondering if you happen to know, or happen to have an idea" [1021-1023]. We think this latter instance is an interesting case because it raises questions about why the child interpreted a question as a personal challenge rather than as an attempt to simply seek more information. A student's enculturation in other contexts can be a significant factor in his/her willingness or propensity to engage in conversation about thinking. For example, if a child had routinely experienced questioning as a means to belittle or if it was seen as disrespectful, rather than as a routine means to learn more about another's thinking (perhaps it was seen as disrespectful to ask questions about a person's thinking if that information was not offered), it is understandable that the child might misinterpret the meaning of a question. Thus, the teacher's mediation of such situations is critical to providing a context in which student-student conversation will likely occur, and is a prerequisite to students engaging in scientific discourse.

Working at the intersection of everyday and scientific discourse

The second dimension of teacher moves functioned to help students bridge from their everyday discourse to scientific discourse. Three categories of teacher moves observed within this dimension were: a) providing a metascript, b) supporting the articulation of ideas, and c) supporting the collective memory of thinking/activity during the science instruction.

a. Providing a Metascript. This type of move refers to the times that a teacher signals what student thinking is expected to be about; hence, the term metascript. A metascript does not give information about what one should be thinking or saying, like a script would; rather, it provides information regarding what one should be thinking or talking *about*. For example, very early in Reporting the teacher stated:

> "This is what you're supposed to be thinking… what you're sharing with the rest of the scientists in the classroom: your data, what you saw, the claims you made, what thinking you did about those claims. You're also sharing your evidence. The question you should be talking about – How does light behave with solid objects?" [226-234].

Providing a metascript of this type can be an effective way of ensuring that students have some clarity about the contributions they are expected to make to this class-wide conversation (cf., Tharp & Gallimore, 1988). Moreover, metascripts have been observed to be particularly useful in classroom contexts in which children are unaccustomed to playing a prominent role in the discussion (Palincsar, 1986).

A similar move that served to both shape and propel the conversation toward scientific discourse was when Ms. Verhey brokered the conversation for the purpose of "corralling" the class's thinking. This move, like a metascript, signals the need to engage in particular types of thinking. For example, questions such as "Would you agree with that?" [537], " 'A light can be weak or strong'...what do you think about that?" [848-849], and "I'm wondering, can you agree with their claim that 'the light reflects off the mirror'?" [696-698], focused student attention on the examination of the ideas being shared compared to their own, and set the stage for later conversation that would determine about which ideas there were consensus. In another move of this type, Ms. Verhey specifically voiced her conclusion that there was consensus emerging in the conversation: "Great! You're not the only one that thinks that way. That's what they said too. Actually, that's what Barbie's group, I think, said too" [947-953]. In addition, where appropriate, she voiced when there were differences among students' claims so that they would be recognized and could become the focus for later conversation during community reflection upon all of the stated claims: "So, we have a different claim then. Light can go through all kinds of materials. So yours is different from theirs. So we have to add that one. Say it again for me please?" [1117-1120].

Yet another form of discursive move that signaled expectations about thinking occurred when Ms. Verhey noted that differences in thinking were simply part of what happened in the course of building knowledge, and that there was value in agreeing to disagree. For example: "Maybe we haven't changed everybody's thinking and we may not all believe all these things, but right now, these are the claims that we are working with" [801-802], and "Guys, we may disagree with them, but let's let it go" [653].

b. Supporting Students In the Articulation of Their Ideas. This category includes an interesting array of teacher moves that vary in their purpose and sophistication, but each of which as to do with encouraging students' expression of their ideas in a way that fosters scientific discourse. At the most basic level are moves that simply encourage students to express their ideas. For example, in response to one contribution, Ms. Verhey said: "Because you don't think [light] goes wavy, you think it goes [blank], how could you change your claim to say that?" [375]. There were also numerous instances in which Ms. Verhey invited students to diagram their thinking on the board, and then she interpreted the drawing or asked another member of the class to comment on the drawing or draw how their thinking compared.

Revoicing. Another type of move that can be particularly powerful in helping students bridge from everyday to scientific discourse, is known in the sociolinguistic literature as revoicing (O'Connor & Michaels, 1993). Revoicing occurs when the teacher repeats, expands, or reformulates a student's contribution. It serves a broad range of purposes, including: articulating presupposed information, emphasizing particular aspects of a student's contribution, disambiguating terminology, aligning students with positions in an argument or attributing motivational states to students (cf. Forman & Larreamendy-Joerns, 1998).

Several particularly interesting instances in which Ms. Verhey engaged in revoicing are the following: Stefan (questioning a claim that the presenting group made) said, "If it reflects it off the glass tray, would it go through? It went through the glass tray, but I don't get it. If the thing reflected off, then how did it go through?" [354-357]. In expressing his confusion, Stefan opened the door to talking about several of the key ideas in this program of study: light can react with an object in multiple ways (e.g., simultaneously be reflected and transmitted), and there is an inverse relationship among these processes (the more reflected or transmitted, the less absorbed). Ms. Verhey responded by revoicing Stefan's query to advance the conversation toward these ideas: "Stefan, are you having a hard time that light can do two things at once?" [360-361].

In another example of revoicing, Ms. Verhey extended a student's claim, advancing its accuracy: *Sharee*: "Light goes in one path." *Ms. Verhey*: Light goes in one path... goes in a straight path" [525-526]. Ms. Verhey also revoiced for the purpose of raising the level of a question from very specific to more general to increase its power in providing learning opportunities: *S*: "If the ceiling and the wall are made of wood, how come it's not bouncing back off the wall." *T*: Why doesn't [light] just keep bouncing?" *S*: "Yes" [789-792].

Seeding. A final set of examples that we include in this category were instances in which the teacher "seeded" the discussion with useful ideas or information. This type of activity is key to extend students' thinking in particular ways, particularly when important ideas with which students need to work to develop the targeted scientific content and reasoning goals are not likely to be brought up by the student community members.

One example of seeding is when a teacher introduces the need to attend to a particular aspect of the production of scientific knowledge such as determining the cause of a phenomenon. As a case in point, Ms. Verhey prompted the students to consider the mechanism at work considering the behavior of light that they were reporting: "*How* do you think light traps?" [908]. In other cases her questions signaled the potential value of considering the characteristics of the materials: "What else can you tell me about the material? Is there one harder than the other?" [1041-1047], and "And you guys believe only thin material, right? (in response to the claim, "We believe light can go through material" – when each material for which this claim was made was "thin") [858-859]. Finally, in a very opportunistic move that was made possible when one student began to wonder about the amount of light that is transmitted through an object, she queried, "Do you think that when light comes out of the Styrofoam, do you think more light comes out, or less, or equal?" [1203-1205], and subsequently seeded a scheme for quantifying light: "If a 10 went in, how much do you think goes out?" [1209-1210].

Additional examples of seeding include instances in which the teacher introduces language that scientists would specifically use. In Ms. Verhey's teaching, an example was the introduction of the word absorb, which was used to focus students' thinking about the similarities and differences in two students claims: that light was blocked and that light was trapped.

c. Serving As the Collective Memory. The final category of teacher moves to support students in bridging from everyday to scientific discourse is when the teacher serves as the collective memory for the class. The teacher always has privileged information relative to the classroom community — monitoring small group activity during investigation and preparing to report provides the teacher with knowledge of students' data and thinking before it is publicly shared, including information that might not have been planned to be publicly shared. Hence, the teacher is in a position to consider how to use that information to support students in dealing with the challenge of expressing their emerging ideas or the complexities of sorting through the claims and evidence shared during the Reporting phase. For example, in one instance, Ms. Verhey used this information to signal to the class, before they began reporting, that there were differences that should be anticipated across the reports: "And I know that this group doesn't agree with that group and that group doesn't agree with that group" [295-297]. However, Ms. Verhey did not just serve as the collective memory from her privileged position; she also monitored the ongoing conversation to pull on those threads that she believed would be useful to advancing the conversation: "There's another claim out there that we have forgotten" [844].

Establishing and Supporting the Norms of Scientific Practice. Finally, is the dimension of teacher moves that explicitly signal the norms of scientific practice. These moves range from practices in which the teacher privileges certain student activity in order to give it prominence and encourage its appropriation by the community (such as the use of particular language or engagement in particular activity), to interjecting in the conversation in order to press students to dwell more deeply or broadly on critical issues in doing science, such as the evaluation of claims or evidence. In our exploration of this discourse, we were interested in examining features that spoke to the dialectical process as Pera, among others, has characterized it. However, we are aware that the transcript excerpts that we provide as examples will not consistently represent one view of science over another; in fact, the discourse is an interesting blend, with some moves mirroring the dialectical model and others reflecting a more traditional view of the nature of scientific inquiry. Where appropriate, we call attention to this feature of the dialogue.

Privileging Particular Language or Activity. One powerful way in which teachers communicate what is valued is through differentially acknowledging particular language or actions on the part of students. As one example, in the course of class discussion during Reporting, one child reached into his notebook and brought out his data from the first-hand investigation to compare to what was being presented, which Ms. Verhey called attention to, saying: "Oh, I like what you're doing, Robby. You're going to get your materials list!" [955-956]. In addition, there were occasions when she pressed students to come up with language that was more consistent with scientific practice. In one case of this, when a child stated the claim that "Light goes in one path." Ms. Verhey responded, "I'm wondering if we could come up with more of a scientific word than 'goes'...?" A second child responded, "travels?" "Would you agree with that? Okay, Let's switch that – (*writing on the class claims poster*) travels – We just sound more like scientists and that's what we're trying to be" [525-541].

Perhaps the most explicit example of Ms. Verhey reflecting the dialectical nature of scientific practice occurred in an exchange in which she called the class's attention to the role that the classroom community has played in jointly constructing a new claim: "Guys, I want to tell you something that impresses me and that is, they walked up there with three claims. Because of our discussion among the scientists in our classroom, we actually came up with..." Ss: "Four" [558-563]. In another exchange, she reminds the students of the purpose for which they are reporting and notes, "So, that is what Allan is thinking (following Allan illustrating how light from the sun makes it possible for us to see a tree). Does it give you something to think about? That's all we need to do is give you something to think about" [435-437]. Finally, she edits her own recording of students' claims in a way that reflects how the public presentation of scientific ideas obscures the community-bound and cultural basis of scientific practice: "I shouldn't put 'believe' though... I should put, 'light can do three things at one material. Those three things are: reflect, trap, and go through..." [838-842].

In other cases, moves by Ms. Verhey portray a more traditional view of the nature of scientific activity to her students. For example, she reminded the class, "Scientists have to make decisions" [121]. A few lines later she admonished: "Be truthful... only report what you know to be..." *Ss*: "True" [284-285].

Supporting Skepticism and Dissent. If the dialectical process represented in contemporary views of science are going to be possible in classrooms, then clearly there has to be the opportunity for students to express disagreement and skepticism. Indeed, in Ms. Verhey's class, there is evidence that students were invited to assume a skeptical stance. For example, early in Reporting, Ms. Verhey, referring to the upcoming group presentations, advised the class, "They may not convince you" [297-298]. Another case occurred when one student, Bobby, raised several questions for the group that was presenting. When Bobby's questions were met with disgruntlement, Ms. Verhey interjected: "Bobby, you had some really good questions and it's okay that you disagree and your disagreements give us something to..." Bobby: "Worry about?" T: "Not worry, let's not worry about it. Don't worry." [1125-1130]; *Our sense is that Ms. Verhey was expecting the students to supply the words, "think about," rather than "worry about."*]

In another example, Ms. Verhey, transitioning from one group to the next, noted: "Maybe we haven't changed everybody's thinking, and we may not all believe these things, but right now, these are the claims that we are working with" [570-573]. And again, "You might not agree with that everybody. It doesn't sound like everybody did" [545-546].

Informal comparisons of the discourse in Ms. Verhey's class with other classrooms, suggest that her decision-making relative to when to press for consensus, when to acknowledge dissent, and how to use dissent is particularly interesting. We suggest that it is useful for the class to consider a well-defined set of claims; however, it is also the case that students need to be prepared to operate with uncertainty; that is, there will be claims for which there is insufficient evidence or claims that have not yet been articulated clearly enough to receive full consideration of the class's attention. By acknowledging those claims for which there is not sufficient support, Ms. Verhey communicated that the class conversation is a "work in progress." The discourse moves that we describe next are integral to this stance.

Examining the Relationship Between Claims and Evidence. Having generated and interpreted data during the Investigation phase, a key feature associated with the Reporting phase is determining what counts as evidence, and engaging in the critical examination of the relationship between this evidence and the claims the evidence supports, refutes, or calls into question.

Under this category, we examine those instances when Ms. Verhey called the class's attention to the relationship between claims and evidence. At several junctures, she merely elicited a statement of evidence; for example: And what evidence do you have that [light] got trapped?" [733-735]. On occasion, she probed the nature of the evidence: "So, sometimes you believe that [light] does those two things (was transmitted and absorbed). Did you find more than one material that did that? More than ten materials did both those things? Wow! Could you give us some of those numbers[7]?" [978-982]. At other times she prompted the class to evaluate this relationship: "Does that evidence make sense then if they said that they thought [light] goes in a straight path? [401-402].

We also note that Ms. Verhey took advantage of the Reporting phase to challenge students to consider the role that investigative procedures played in generating the data the groups were using as evidence. For example, when class members questioned a particular set of data, Ms. Verhey asked the presenters to clarify the set-up they used in their investigation: "If your light was going like this, was your mirror straight like this? Was it perpendicular, like you learned in math, or did you have it an angle?" In response, a student demonstrated, "I had it like this (using gestures)" to which Ms. Verhey responded: "Okay, that might explain some things" [765-770]. In another excerpt, Ms. Verhey asked the students who were reporting to reenact the investigative set-up: "Okay, this is what they're talking about. This is their screen... hold this up Nicole....Here's the flashlight. Here's the material. This is the light. So, they put the flashlight here on the material and they say, then they'd move it a little bit and if they saw this... the light on there, then they knew it was..." *Ss*: "Reflecting" [1076-1086].

Using Discourse To Advance Investigation and Investigation To Advance Discourse. Finally, recall that our conceptualization of inquiry-based instruction is that it is an iterative process in which students move from investigating to reporting and back to investigating across several cycles. While we have depicted discourse moves relative to the Reporting phase only, ideally, the threads that are introduced in the Reporting phase are carried through and influence what occurs in the next investigation phase. There are several ways in which Ms. Verhey made this process explicit for her class. First, it was included in the way that she began the Reporting when she signaled to the students that: "Some of our claims may end up being 'think-abouts'... let's think about that some more... let's maybe investigate that some more" [262-267] and "I may need to check that out in case we investigate again" [301-302]. Then, near the end of the Reporting phase, she set the stage for the next cycle of investigation by stating: "Do you think that when you go back to investigate you might, I know that Emma's probably, and you too Alan and Bobby – I know you are saying in your head... 'I'm getting that number 9 out (*in reference to a particular material that was observed*) and I'm going to check that out'" [966-971].

Second, she chose to handle certain questions raised by class members during Reporting by suggesting there was a need to investigate. In one case she said, "There's a good question. We'll have to investigate that won't we?" [793-794]. Similarly, when a group appeared stymied in the process of reporting, Ms. Verhey commented, "Sounds to me like they need to investigate" [1099].

CONCLUSION

Schwab, one of the earliest scholars to write extensively about the nature of inquiry-based instruction in science (1962), indicated that part of learning via inquiry was coming to understand science as "a mode of investigation which rests on conceptual innovation, proceeds through uncertainty and failure, and eventuates in knowledge which is contingent, dubitable, and hard to come by." (p. 5) He went on to write that "[The] treatment of science as enquiry is not achieved by talk about science or scientific method apart from the content of science. . . . [It] consists of a treatment of scientific knowledge in terms of its origins in the united activities of the human mind and hand which produce it." (p. 102) These perspectives are consistent with contemporary views of the nature of science as a human enterprise that takes place in particular communities and is enabled and constrained by the nature of the cultural practices of those communities. In this chapter, we have sought to provide information from writings in history, philosophy, and sociology that articulate this view of science, as well as writings from a sociocultural view of psychology that are consistent in representing knowledge production by humans as a cultural and community-bound process. We have also sought to present ideas about what these contemporary views imply for our conceptualization and enactment of science instruction. For those seeking to advance our understanding about inquiry-based science instruction, we argue that these ideas indicate that there are three dimensions that need to be specified in thinking about this most sophisticated of instructional approaches: a) the nature of science as inquiry, b) the nature of learning via inquiry considering the inquiry-based nature of science, and c) the nature of teaching via inquiry considering the inquiry-based nature of learning.

We submit that there is much yet to be understood about learning via inquiry considering the cultural view of human knowledge production addressed briefly in this chapter. Although there is much that we know in some areas about the kinds of ideas that students bring to the study of particular science topics (e.g., Pfundt & Duit, 1994), we know relatively little about the learning process, particularly under instances of inquiry, and in contexts informed by sociocultural views of learning science (cf. Magnusson, Templin, & Boyle, 1997). Furthermore, as some have already pointed out, we know considerably less about teaching via inquiry (Flick, 1995). We hope that the aspects of our work presented in this chapter – a heuristic representing inquiry-based instruction in phases that support thinking about learning in culture and community-based ways, and categories of teacher moves to solicit, facilitate, and promote students' appropriation of scientific discourse – will be informative to others' thinking about teaching science via inquiry. Other aspects that we have written about elsewhere may also be helpful (Palincsar and Magnusson, 2001). Nevertheless, these ideas are but a few of the ways in which we need research and development to support our understanding of inquiry-based instruction toward the advancement of science instruction in our schools.

REFERENCES

American Association for the Advancement of Science (1989). *Science for All Americans: A Project 2061 Report on Literacy Goals in Science, Mathematics, and Technology*. Washington, D.C.: American Association for the Advancement of Science

American Association for the Advancement of Science (1993). *Benchmarks for Scientific Literacy: Project 2061*. New York: Oxford University Press.

Bacon, F. (1620). *Novum Organum*. In J. Spedding, R. L. Ellis, & D. D. Heath (Eds.), The Works of Francis Bacon, Vol. 4. London: Longman, 1860.

Brown, J. S., Collins, A., & Duguid, P. (1989). Situated Cognition and the culture of learning. *Educational Researcher*. **32**, Jan.-Feb. pp. 32-42.

Champagne, A. B., & Bunce, D. M. (1991). Learning-theory-based science teaching. In S. M. Glynn, r. H. Yeany, & B. K. Britton (Eds.), *The Psychology of Learning Science* (pp.21-41). Hillsdale, NJ: Lawrence Erlbaum Associates.

Cobb, P. Yackel, E. (1996). Constructivism, emergent, and sociocultural perspectives in the context of developmental research. *Educational Psychologist*, 31, 175-190.

Descartes, R. (1628) 1985. Rules for the direction of the mind. In J. Cottingham, R. Stoothoff, & D. Murdoch (trans.) *The Philosophical Writings of Descartes*, Vol. 1, (pp. 9-78). Cambridge: Cambridge University Press.

Doyle, W. (1986). Classroom organization and management. (pp. 392-431). In M. C. Wittrock (Ed.). *Handbook of Research on Teaching*. New York: Macmillan Pub. Co.

Driver, R., Asoko, H., Leach, J., Mortimer, E., & Scott, P. (1994). Constructing scientific knowledge in the classroom. *Educational Researcher*, **23**(7), 5-12.

Flick, L. B. (1995, April). *Complex instruction in complex classroom: A synthesis of research on inquiry teaching methods and explicit teaching strategies*. Paper presented at the Annual Meeting of the National Association for Research in Science Teaching. San Francisco, CA.

Forman, E. A. & Larreamendy-Joerns, J. (1998). Making explicit the implicit: Classroom explanations and conversational implications. *Mind, Culture, and Activity, 5*(2), 105-113.

Galilei, G. (1953). *Dialogue Concerning the Two Chief World Systems*, trans. S. Drake. Berkeley, CA: University of California Press.

Gee, J. P. (1996). *Social linguistics and literacies: ideololgy in discourses, 2nd Ed.* London: Falmer Press.

Hooke, R. (1665). *Micrographia*. New York: Dover. 1961.

Karplus, R. & Their, H. D. (1967). *A new look at elementary school science. Science Curriculum Improvement Study*. Chicago: Rand McNally.

Latour, B. (1987). *Science in Action.* Cambridge, MA: Harvard University Press.

Latour, B., & Woolgar, S. (1979). *Laboratory life: The social construction of scientific facts*. London: Sage.

Latour, B., (1987). *Science In Action*. Cambridge, MA: Harvard University Press.

Lave, J. (1991). Chapter 4: Situating learning in communities of practice. (pp. 63-82). In L. B. Resnick, J. M. Levine, & S. D. Teasley (Eds.) *Perspectives on Socially Shared Cognition.* Washington, DC: American Psychological Association.

Lemke, J. (1990). *Talking science*. Norwood, NJ: Ablex.

Magnusson, S. J., & Palincsar, A. S. (1995). Learning environments as a site of science education reform. *Theory into Practice,* 34(1), 1-8.

Magnusson, S. J., & Palincsar, A. S. (in press). Teaching and learning inquiry-based science in the elementary school. In J. Bransford & S. Donovan (Eds.), *Visions of teaching subject matter guided by the principles of how people learn*. National Academy Press.

Magnusson, S. J., & Templin, M. (1995). *Scientific practice and science learning: Individual and community aspects of scientific literacy and learning*. Proceedings of the Third International History, Philosophy, and Science Teaching Conference.

Magnusson, S. J., Templin, M., & Boyle, R. A. (1997). Dynamic science assessment: A new approach for investigating conceptual change *Journal of the Learning Sciences,* 6(1), 91-142.

National Research Council (NRC) (1996). *National Science Education Standards.* Washington, D.C.: National Academy Press.

Newton, P. (1999). The place of argumentation in the pedagogy of school science. *International Journal of Science Education, Vol. 21,* No. 5, 553-576.

O'Connor, M. C. & Michaels, S. (1993). Aligning academic task and participation status through revoicing; Analysis of a classroom discourse strategy. *Anthropology and Education Quarterly*, 24, 318-335.

Osborne, R. J., & Freyberg, P. (1985). *Learning in science: The implications of children's science.* Portsmouth, NH: Heinemann.

Palincsar, A. S. & Magnusson, S. J. (2001). The interplay of first-hand and text-based investigations to model and support the development of scientific knowledge and reasoning. In S. Carver & D. Klahr (Eds.), *Cognition and Instruction: Twenty-five years of progress* (pp. 151-193). Mahwah, NJ: Lawrence Erlbaum Associates.

Palincsar, A. S. (1986). The role of dialogue in scaffolded instruction. *Educational Psychologist*, 21, 71-98.

Palincsar, A. S., Magnusson, S. J., & Hapgood, S. (April, 2001). *Trafficking ideas through the rotaries of science instruction: Teachers' discourse moves and their relationships to children's learning*. Paper in a symposium entitled: Learning through Conversation: Discourse that Advances Student Understanding in Academically Diverse Classrooms. Presented at the annual meeting of the American Educational Research Association, Seattle, WA.

Palincsar, A. S., Magnusson, S. J., Ford, D. J., Marano, N, & Brown, N. (1998). Design principles informing and emerging from the GIsML community: A community of practice concerned with guided inquiry science teaching. *Teaching and Teacher Education*, 14(1), 5-19.

Pera, M. (1994). *The Discourses of Science*. Chicago: University of Chicago Press.

Pfundt, H. & Duit, R. (1994). *Bibliography: Students' alternative frameworks and science education*, 4th Edition. Kiel, Ger: University of Kiel Institute for Science Education (Institut für die Padagogik der Naturwissenschaften).

Rogoff, B. (1994). Developing understanding of the idea of communities of learners. *Mind, Culture, and Activity.* 1(4), 209-229.

Scheiner, C. (1612). De maculis solaribus et stellis circa Iovem errantibus accuratior disquistio. In Galilei, *Le Opere di Galileo Galilei*, vol. 5 (pp. 39-70).
Schwab, J. J. (1962). The teaching of science as enquiry. In J. Schwab & P. Brandwein (Eds.), *The teaching of science* (pp. 1-103). Cambridge, MA: Harvard University Press.
Tharp, R. G. & Gallimore, R. (1988). *Rousing minds to life: teaching, learning, and schooling in school context.* Cambridge, MA: Cambridge University Press.
von Glasersfeld, E. (1989). Cognition, construction of knowledge, and teaching. *Synthese*, 80(1), 121-140.
Weiss, I. R., Banilower, E. R., McMahon, K. C., & Smith, P. S. (2001). Report of the 2000 National Survey of Science and Mathematics Education. Horizon Research, Inc. www.horizon-research.com.
Wertsch, J. V. (1991a). Chapter 5: A sociocultural approach to socially shared cognition. (pp. 85-100). In L. B. Resnick, J. M. Levine, & S. D. Teasley (Eds.) *Perspectives on Socially Shared Cognition.* Washington, DC: American Psychological Association.
Wertsch, J. V. (1991b). *Voices of the mind : a sociocultural approach to mediated action.* Cambridge, MA: Harvard University Press.
Woolgar, S. (1988). *Science: The very idea.* New York: Routledge.

NOTES

[1] As in "the language peculiar to an occupational group" as well as "a method of argument or exposition that systematically weighs contradictory facts or ideas with a view to the resolution of their real or apparent contradictions." (American Heritage, p. 515). Pera distinguishes rhetoric from dialectics by defining rhetoric as the "*practice* of persuasive argumentation" whereas dialectics refers to "the *logic* of such a practice or act" (1994, p. viii).

[2] Note the similarity of this statement to a radical constructivist position such as that articulated by von Glasersfeld (1989): it does not matter whether or not there is an objective reality because we cannot come to know it; we can only come to know our construction of it.

[3] Elsewhere, we have presented a theoretical framework featuring the types of community described in this paper, as well as two other types (Magnusson & Templin, 1995), which are more extensively described, and then discussed in terms of how they simultaneously influence scientific practice. We believe this typology of community is useful to describing and explaining human activity in many social settings.

[4] Scientific practice occurs in many communities. However, to facilitate being clear in our expression of the implication of our knowledge of the nature of scientific practice in communities, which is argued to have common features (AAAS, 1989), we will hereafter *the* scientific community.

[5] Most instances of the Reporting phase take two class sessions.

[6] These numbers refer to lines of a transcript of classroom from which the data were taken. All of the quoted material came from the same day of instruction; hence, the same transcript, so we have eliminated more specific information in the reference.

[7] Numbers were used to distinguish the 20+ to materials that were observed during the investigation, as a shorthand for indicating about which items the data or claims pertained.

CHAPTER 8

LAWRENCE B. FLICK

DEVELOPING UNDERSTANDING OF SCIENTIFIC INQUIRY IN SECONDARY STUDENTS

This chapter is about what secondary teachers and students do when scientific inquiry is the focus of instruction. Selected historical antecedents in science education, educational psychology, and direct experiences in classroom teaching guide this analysis. This framework is similar to the one Siegler (2001) used to reflect on recent research in cognition and instruction. He envisioned a tetrahedral framework with learner characteristics, instruction, and target domain at the base and learning at the apex. My model begins with historical elements of a science-specific pedagogy, selected characteristics of learner cognition, and teachers discussion of instructional experience in teaching science as inquiry. Like Siegler (2001), an analysis of a pedagogy of science, learner cognition, and teaching experience matters little if it does not affect learning at the apex of the framework.

The educational challenge posed by making inquiry a focus of instruction is to strike a balance between science as a productive body of knowledge and science as the creative insight fashioned from skepticism and conjecture. Unbalanced, either aspect of science by itself leaves the subject matter superficial. Poincaré (1913) captured the point: "To doubt everything and to believe everything are two equally convenient solutions; each saves us from thinking" (p. 27). Contemporary science educators and scientists take the view that to learn about the knowledge of science without appreciating the rational, empirical, fallible, human basis for this knowledge is to miss the most salient point about the value of this knowledge (NRC, 1996; AAAS, 1993).

Viewing science as human inquiry takes a reflective frame of mind, an awareness of one's own processes of making meaning, and a critical perspective on the evidence used to support knowledge claims. This chapter will consider how students exercise newly acquired cognitive faculties and how teachers work to provide opportunities to exercise those cognitive skills in the context of inquiry.

In the four parts to this chapter, I, first, briefly review the history of understanding inquiry as an expected outcome of the secondary classroom. Next, I consider the adolescent as a learner in the process of developing cognitive skills relevant to engaging in inquiry tasks. The final two parts are concerned with teaching practice (a) from the perspective of educational research on classrooms then (b) from teacher discussion on what they see as their role and the role of students.

L.B. Flick and N.G. Lederman (eds.) Scientific Inquiry and Nature of Science, 157-172.

INQUIRY IN THE SECONDARY SCHOOL

Valuing student understanding of science as inquiry is deep in the history of science education in the US. As early as 1893 the Committee of Ten, appointed at a meeting of the National Education Association, emphasized that science should be taught through direct contact with the natural world making laboratory work an absolute necessity. Knowledge of science from a text only "is of little, if any, value" (United States Bureau of Education, 1893, p. 119). Implications for instruction and for student learning goals were a matter of teacher discretion. Instruction focused on doing laboratories that involved inquiry skills but did not explicitly teach students about the nature of scientific inquiry. Sophisticated inquiry tasks were generally included in courses for students who were expected to pursue higher education.

Contemporary standards (NRC, 1996) take the position that important principles of science including the concept of science as inquiry should be taught to all students and not to just those that persist in taking more advance classes. This position also has historical antecedents dating from The Committee of Ten who recommended "there should be no difference in the treatment of (science) for those going to college or scientific school, and those going to neither" (United States Bureau of Education, 1893, p. 118). After World War I, the Commission on the Reorganization of Secondary Education (United States Bureau of Education, 1918) established the comprehensive high school as a prototype of American Democracy. By this time science had become a standard part of the secondary curriculum, and schools were challenged to provide an enriched curriculum similar to what more affluent families were providing their children outside the public schools (Hurd, 1997).

An "enriched" science education suggested that teachers providing students with higher quality instruction that added value to what students were learning. Historically, an underlying message in these ongoing reforms was to emphasize science as a creative and disciplined process of inquiry. Under the influence of proponents of progressive education and the subsequent movement for "relevance" in the 1970's, secondary schools revamped science curricula to include examples from everyday life. To emphasize the value of science to the citizen the Carnegie Institute recommended that the National Science Foundation set up programs where scientists would talk directly to students (Hurd, 1997). Scientists could convey the work of real science as posing questions about the world, devising creative approaches to gathering evidence to answer those questions, and communicating results to other scientists.

But hearing about real science is not the same thing as teaching students to think like scientists. High schools were criticized for teaching laboratories as a "rhetoric of conclusions" (Schwab, 1962) meaning that students were not asked to understand the motivating scientific questions nor the reasoning behind how scientific procedures were designed but only to see that the data led to predetermined outcomes. Helping students think like scientists was still a tangential goal of doing labs as scientists would do them or hearing how scientists work.

In the 1960's, the American Association for the Advancement of Science ushered in the era of "hands-on" science. The "basic" and "integrated" science process skills presented teachers with new content (AAAS, 1967). The term "hands on science" is ensconced in the vernacular of science teaching and has taken on a wide variety of meanings including becoming a euphemism for inquiry (Flick, 1993). Science processes have been a part of the curriculum ever since and has also added new targets of and procedures for instruction. Students were taught how to make graphs, tables, observations, and inferences. But analyzing the overall skill of inquiry into a set of sub skills did not lead directly to student understanding of the larger purpose of inquiry. The parts did not sum to the whole. In fact, the concern that "hands-on" instruction was being interpreted as just activity without challenging student cognition led to adding the phrase "minds on science" to the professional vernacular (Duckworth, et al., 1990). Science educators felt it necessary to specifically emphasize the cognitive aspects of laboratory work.

The current mandate that all teachers will teach all students the processes and significance of scientific inquiry makes clearer what was implied by reforms earlier this century. Teaching science should develop student understanding that all knowledge claims in science are rooted in scientific inquiry. Current reforms do not come with curricula, therefore, teachers must modify curricula to create the experiences necessary to teach science as inquiry. This leads teachers to ask, what does science inquiry in the classroom look like and what should be expected from students with inquiry activities demanding more reflection and critical thinking?

STUDENT'S SENSE OF INQUIRY IN SCIENCE

Teachers who engage students in the complex instruction of treating science as inquiry develop a deep understanding of the nature of their students. Teacher knowledge of the physical and psychological characteristics of students is a fundamental component of the professional knowledge base of teaching (Shulman, 1986). This view is echoed in national reports focusing on teaching adolescents (Carnegie Council on Adolescent Development, 1989; 1995). This knowledge is a prerequisite to the instructional scaffolding necessary to support sophisticated and subtle forms of thinking that are generally beyond the unaided capabilities of students.

Understanding science as inquiry requires a broad range of cognitive, social, and physical skills that adolescent students are in the process of developing during high school. For the teacher, it requires creating appropriate experiences that not only communicate the nature of science and its grounding in inquiry but also sensing how this content interacts with the developmental characteristics of adolescents. The National Science Education Standards (NRC, 1996) propose that students in high school "design and conduct scientific investigations." To do this students must "formulate and revise scientific explanations and models using logic and evidence… Recognize and analyze alternative explanations and models…

Communicate and defend scientific argument" (p. 173-174). The cognitive demands of these activities include understanding both the form and content of scientific positions. Thus, the National Science Education Standards expect high school students will construct a principled argument about a scientific problem based on data generated from procedures consistent with the problem that is itself based on relevant scientific concepts.

Educators often rely on Piagetian theory to make the case that the cognitive skills of an adolescent are capable of formal thought to support doing and understanding scientific inquiry. Under this view adolescents are able to manipulate propositions, construct hypothetical situations, and form the logical implications necessary for carrying out experimental controls (Inhelder & Piaget, 1958). This implies that high school teachers should expect students to be able to handle scientific problems based on the science content being studied. However, making a connection from science content to reasoning with this content has proved problematic. Reif and Larkin (1991) have shown how students tend to confuse goals of everyday thinking with goals necessary for carrying out and understanding scientific inquiry. The discipline of scientific inquiry for instance, requires that some goals be made explicit that in everyday thinking are left implicit. The requirements for achieving these goals are more stringent in science than everyday problem solving. For instance, it is valuable to anticipate the behavior of the parking brakes when a car is parked on a hill. However, the scientific study of similar phenomena is more deliberate, conscious, and focused on extending knowledge rather than simply confirming current understanding that results in a practical decision. Even students schooled in scientific inquiry construct inappropriate arguments (Reif & Larkin, 1991).

Competent researchers disagree on the role Piagetian theory should play in the design of instruction. This is especially important in issues of instructional design that makes significant cognitive demands on students in the context of science inquiry. Elementary aged children are credited with being inquisitive and using strategic skills to make generalizations from observations. In a synthesis of the literature, Kuhn (1997) summarizes observed capabilities of young children as including elements of Piagetian formal thought, such as drawing appropriate inferences from contrary-to-fact propositions (p. 145). However, her review makes clear that research also documents developmental change. Children's strategic thinking about scientific problems is limited to simple problems that do not challenge existing beliefs. Older children present improved control over the theory-evidence relationship (Kuhn, 1997).

Metz (1995, 1997) has challenged a view that uses Piagetian theory to constrain science instruction for young children within the bounds of concrete operations. She argues that a competency model is more effective in supporting instructional design than a deficiency model. With appropriate scaffolding, children can differentiate theory from evidence and make data-based inferences. Her recommendations for developing children's scientific reasoning are well worth considering for adolescent learners. They are (a) begin with problems in an engineering frame and proceed to a scientific frame, (b) exploit the power of socially distributed expertise, and (c) develop and support a community of discourse as the norm not exception (Metz, 1995).

Keating (1990) makes an important observation relevant to the debate over cognitive competencies. Where the competencies of children may be underrated, those of older children and adolescents may be overrated especially in complex context of classroom learning (Keating, 1990). Children and adolescents employ strategic skills in thinking about scientific problems but older children show more frequent use of better strategies and less frequent use of inferior strategies (Brown, Bransford, Ferrara, & Campione, 1983). Adolescents, however, do not use strategies in a consistent manner implying the need to support metacognitive capabilities by appropriate instructional methods (Pressley & McCormick, 1995). Targeted instruction is necessary to help students coordinate knowledge for evaluating hypotheses in terms of relevance, testability, and compatibility with other knowledge in science. Instruction must also support development of logical competence to evaluate the predictive or explanatory power of hypotheses.

High school age students have trouble using logical competence in scientific reasoning. Examining ninth graders through adults, Kuhn's (1992) results show broad problems in argumentation skills. These problems include confusing co-occurrence of events with cause and effect, preference for confirming rather than disconfirming evidence, and failure to consider potentially important factors by judging them irrelevant. A critique of this work by Koslowski and Maqueda (1993) suggested that Kuhn's evaluation may be overly restrictive. However, Koslowski and Maqueda emphasized that these capabilities require purposeful practice involving reflection on the relationships between theory and evidence and how they mutually constrain possible conclusions. In their review of these issues, Driver, Newton, and Osborne (2000) emphasize the significance of the teacher in supporting and developing practice in thinking through various interpretations of evidence. The message is that relevant cognitive skills are not developed ready for use in classrooms or daily experience, but must be prompted, exercised, and coached.

Capabilities observed under clinical conditions may overstate what it is possible for the adolescent to produce in classroom settings. Classroom contexts often lack the supports, cues, time, and focus afforded by controlled clinical environments (Keating, 1990). While each of these key clinical parameters have implications for how high school teachers might like to structure their learning environment, this observation implies what was made explicit by Driver et al. (2000) that the teacher has a central role in developing cognitive capabilities through learning environments that support using and understanding scientific inquiry.

From the perspective of the psychological capabilities of the adolescent, the role of the teacher is to create opportunities for and scaffolding to support the exercise of cognitive skills important to understanding and using scientific inquiry. Teachers should be prudent when interpreting research claims of adolescent skills and measure those claims against normal classroom conditions that may not afford the kinds of support needed to apply appropriate cognitive faculties. This same research, however, may have important implications for designing instruction that supports inquiry in the classroom.

TEACHERS AND INSTRUCTION

The role of the secondary teacher is to translate their own understanding of discipline content and understanding of science as inquiry into curriculum and instruction for students. This is a tall order. To accomplish this takes a unique synthesis of knowledge that includes discipline content, pedagogy, curriculum, students, and community. The goal is to lead students through a process of thinking *with* science so that they can think *about* science. Schwab (1962) characterized science education as presenting inquiry as "stable" and unquestioning, where established principles determine experiments. He argued that stable inquiry is critical to science, but it fails to capture the essential, dynamic side of science where contradictory data challenge old truths and stimulate new approaches. Schwab characterized the stable enquirer as "technically creative" and the fluid enquirer as "conceptually creative" by seeing principles as problems (Schwab, 1962).

How do you get a classroom of adolescents, possessing a full developmental range of behavior and skills, to see scientific principles as problems? The science lab is a common strategy. Hands-on laboratory work is one of the most "distinctive feature of science instruction" (Shulman & Tamir, 1973, p. 1118), but there are many activities clustered around the actions of manipulating materials during labs. These include making charts, graphs, answering questions, writing reports or descriptions, and scientific discourse. These and many other aspects of inquiry must be facilitated and managed by the teacher to scaffold emerging cognitive skills, prompt use of appropriate knowledge, and stimulate curiosity that motivates investigative thinking.

Hofstein and Lunetta (1983) reviewed research on laboratory work and found it lacking information that would guide teacher practice in the classroom-lab. Researchers often worked with comparatively small groups of students of limited diversity that ignored significant subsets of students such as those with low socio-economic status, those who were less able, or those with traditionally low motivation. They often used standardized tests that were not designed to measure effects produced by laboratory instruction. Most did not examine teacher behavior, classroom learning environment, or teacher-student interactions. The materials themselves were often not clearly described leaving the content and nature of the instructions and information presented to students in doubt. In short, the reader did not get a clear picture of what was actually going on in the classroom.

More recent work has given attention to some details of instruction but has tended to focus on the upper range of student ability (Roth, 1994; Cavallo & Schafer, 1994; Sanford, 1987) and average to small classrooms (< 26) (Roth & Roychoudhury, 1993; Keys, 1994). Under ideal conditions small, well supported classrooms with motivated students, teachers with sufficient science and teaching backgrounds can achieve significant results with some students by being a guide or advisor and keeping a low profile (Roth & Bowen, 1994; Roth and Roychoudhury, 1993). However, these studies apply to a relatively small percentage of U.S. classrooms and do not offer clear guidance for teachers with more typical classrooms.

Research on typical classrooms offers some insight. Conditions important to inquiry may be lost where teachers provide a variety of safety nets that lower cognitive demands in an attempt to engage more students and maintain interest and motivation. Sanford (1987) observed four high school teachers, two in general science and two in honors biology, and made a qualitative analysis of their management of classroom tasks. Teachers often felt compelled to reduce the grade value of a task when students had difficulty or offer a series of hints that led students to the desired answers. This had the effect of leading students to expect more instruction at the beginning of the project, when they should have been thinking over the problem on their own, rather than later.

How teachers handle classroom tasks that promote science inquiry are telling with respect to how cognitive demands are communicated, supported, and maintained. The following three studies make comparisons between more and less capable teachers. The first pair of studies present qualitative analyses of science classrooms. The third study quantitatively examines instructional tasks in science across two dimensions.

Gallagher and Tobin (1987) observed 15 teachers from elementary to high school over a 14 week period to assess teacher and student engagement. The students were considered to be at least average and there was adequate support for teaching science. Teachers were classified as either average or exemplary. For average teachers, labs tended to be handled more informally than lecture-discussion sessions with more student socializing allowed. The teacher seemed to maintain a relaxed atmosphere by focusing on procedural matters and lower the inquiry-orientation of the tasks. There was a general sense by both teachers and students that the job was to finish assigned tasks. These attitudes and behaviors contrasted with teachers identified as exemplary. Tobin and Fraser (1990) made eight observations each of 20 exemplary teachers. Their practices also involved a relaxed laboratory atmosphere but instructional goals were made clear and activities for overt student engagement were maintained throughout. Teachers actively monitored students and reinforced the goal of learning for understanding through targeted questions. They provided time and opportunity for students to elaborate, clarify, summarize, and to react to other students. Teachers gave clear feedback on incorrect answers, but made the atmosphere safe for making mistakes and stating a point of view.

Blumenfeld (1992) conducted an in-depth study of instructional tasks and how teachers maintained an attitude of thoughtfulness in science. Sixty tasks were observed in 10 classes taught by five teachers at the 5th and 6th grade levels. Tasks were rated across two dimensions: product (cognitive level, form of product, length) and social organization (small or large group). Teachers were selected from those who did hands-on activities and who had superior math and science backgrounds by elementary and middle school standards. Researchers administered measures of motivation and active learning after each of the 60 tasks. The active learning measures were higher for two of the five teachers. Comparison of teacher behavior between the high and low active-learning teachers revealed several marked differences. Teachers who fostered more active learning minimized the number of terms and facts to be remembered and modified published worksheets to include higher level cognitive items. These teachers also consistently and repeatedly maintained a focus on the main point by relating facts to concepts and pointing out relationships. The whole class was engaged in responses through frequent questioning that produced high levels of success. They modeled thinking procedures and provided explicit directions for group work. Risk-taking and mistakes were valued and used in creating new learning episodes.

Teachers who engendered less active learning from their students did many of the same things as the more successful teachers but often clouded the main points. For instance, they would ask synthesis questions that were too hard for students and spent insufficient time relating new information to student knowledge. They allowed the text to carry too much of the discussion and told students what they were to learn from assigned tasks. Like the Sanford (1987) and Gallagher and Tobin (1987) studies, these teachers repeatedly reduced cognitive demand by changing deadlines and de-emphasizing grades. Assessment was public and competitive and emphasized correctness. Student responses were often perfunctory.

Classroom research on instruction in scientific inquiry presents a picture consistent with psychological studies of adolescents. Students engage with instruction in scientific inquiry where (a) instructional supports are present, (b) students are disposed to work with those supports to apply reflective and critical thinking, and (c) teachers have the requisite knowledge about science and the nature of science. In those instances where teachers offer fewer cognitive supports, do not press for critical thinking, and lower cognitive demands, then students focus less on inquiry task meaning and more on task completion. But even where the teacher has designed instruction that supports development of skills, the student must exercise responsibility for engagement and persistence in using those supports.

TEACHERS DISCUSS TEACHING

I was the evaluator for a professional development program conducted by a state department of education. The program was motivated by the implementation of state assessments that in part required student work samples in scientific inquiry. The purpose was to foster teaching of science as inquiry as presented in the Benchmarks for Scientific Literacy (AAAS, 1993) and the state Science Content Standards. A Teacher-Leader Cadre (Cadre) of 30 science teachers was selected through a statewide application process. Following a program of "developing professional developers" (Loucks-Horsley, et al. 1998), the Cadre designed and presented a series of four workshops at locations around the state. The Cadre was a primary mechanism by which the state department promoted the implementation of teaching and assessment strategies that would lead to the creation of work samples in scientific inquiry. Because of their pivotal role in the professional development program, the Cadre teachers themselves were the focus of the most intense professional development.

The members of the Cadre were interviewed after completing 18 days of professional development over 12 months. During this time they also presented four workshops to groups of teachers through out the state. The interviews were part of a larger evaluation of the program involving written surveys and direct observation of Cadre activities. They were asked to respond to three questions: (1) What does it mean for students to engage in inquiry in science at the grade level(s) you teach? (2) What are responsibilities of the teacher? and (3) What are responsibilities of the student? Over all, teachers expressed an emerging understanding of their role in teaching and supporting the development of student thinking skills used in inquiry tasks.

Teacher discussion of what student engagement with inquiry looks like involved three themes (a) learning goals, (b) discourse, and (c) student thinking. Teaching inquiry highlighted goals for learning. These teachers used the term "goal" to imply that under expectations for inquiry the learning environment had changed for students.

Instead of a rote lab with expected outcomes, that there is a goal—a learning goal—that is put up in front of the kids and that within that goal each student uses their prior knowledge to engage in a dimension of that goal...

Inquiry raises the level of expectations between teacher and student. Inquiry becomes a vehicle for letting students know that while some lessons or labs offer interesting things to see and do, there is more to it. The teacher sets the context, provides the tools, and the student is expected to provide a direction.

> ...if you can sort of imbed that idea of the gee-whiz with a "there's an actual reason for this," that makes it much more powerful for the student to be able to put the two together. So as far as inquiry, I think on my level it means that what I do... is meaningful to them, or it has some meaning, it's not just on a whim.

These teachers conveyed the view that inquiry had in fact transformed their view of the entire science curriculum. Inquiry was a new message they could deliver to students that science is meaningful in the immediate sense of investigating problems in the classroom. In doing so, students and teacher were engaged in a higher level of discourse. Classroom discourse was often described in the context of small group discussion.

> (Working) as a group helps contribute to achieving a greater depth of knowledge as a group than as individuals doing the same thing... share their information with the group, and because each person is doing it, looking at the same goal but from a slightly different angle, the overall depth of learning in the class is greater.

This perspective is consistent with the situated view of cognition that views the classroom as a community of science learners (Greeno, Collins, & Resnick, 1996). Through discourse on inquiry students learn what is valued by this community by interacting with members of the community and thereby learn to coordinate what they do with others. The implicit message to students is that knowledge about the world and knowledge about investigating the world is distributed. Through group discourse, students learn new ways of thinking within the group and also individually.

> ...I think inquiry on a student's level is to me just a way to think. I'm finding more and more, definitely with my situation at least, what problem-solving requires of yourself,...

> I've never thought of it as a big deal that you're able to inquire or think or whatever, it's just kind of is natural, but it isn't natural, it isn't always, I mean most people probably possess those type of skills, but it just seems to be less common… for any walk of life.

> So we need to talk about synthesizing thinking at some level I suppose… In one sense it can be just a question in which you ask people to pose some sort of thinking synthesis on that, or it can be some sort of full-blown activity that we normally think of as a lab.

Reflective of the field of cognitive psychology, these teachers implicitly contrasted a situated, or learning community, perspective with an individual, cognitive perspective of learning (Greeno, et al., 1996). Cognition is individual as well as corporate. Further, problem solving in science, as the quote above implies, is not as natural as one might expect. This observation from the classroom reflects Keating's (1990) point made earlier that adolescents present a potential for critical thought that can be more potential than real without instructional scaffolding. Hence, the teacher's role is to create opportunities for students to employ these emerging cognitive skills.

> …Make sure that you're giving the opportunities, and that the students are able to engage from whatever place on the board they are, rather than expecting them to all be at the starting line and to hit the same finish line.

The teacher has the responsibility for setting up the guidelines within which students function with goal of inquiry. This is a more complex view of instruction than existed without inquiry as part of the curriculum. Instruction develops three kinds of competencies (a) background in science principles and concepts, (b) skills with lab procedures, and (c) cognitive skills for engaging with and understand scientific inquiry. The guidelines for this complex instruction are established through the structures of a state assessment, which posed major challenges for these teachers.

> …Making sure that you can assess that. I think the biggest responsibility of the teacher is learning to assess that and in a way that validates the student work as well as meets the needs of communication within the state or the community or whatever. But that's the greatest challenge.

> (Inquiry) has made it more difficult for me—and this is one of those positive, more difficult things—it's made it more difficult for me to learn how to assess the kids… What helps is a scoring guide. That gives me a general bar, did they cross over that bar in these four areas. But as far as each individual it becomes more difficult to make sure that they understand what they learned and that I understand what they learned.

The professional development program focused on use of the state scoring guide in the design and assessment of work samples. Yet a guide with criteria for four dimensions of inquiry, (a) framing the investigation, (b) designing the investigation, (c) collecting data, and (d) interpreting results, does not translate directly into daily instruction and formative assessments. As teacher knowledge of the nature of inquiry deepened and as their awareness of the challenges of implementation in the classroom grew, the complexities of assessment became more evident.

A second theme within the topic "responsibility of the teacher" was the teacher's major role as instructional leader. Along with assessment comes a significant responsibility of how to lead students in the achievement of expectations. The teachers perceived a conflict between the need to structure student work in anticipation of specific assessments and yet leave the atmosphere of the classroom open for inquisitive thinking. These two teachers express two common problems (a) breaking out of a mold of teacher as *the* source of information and (b) keeping a central focus while allowing a measured amount of divergence.

> Just being open-minded, I think. Really open-minded. Like getting out of the idea that I know I grew up with was the teacher knows everything... Whereas the teacher is a resource or maybe even you don't know the answer but you can find out. Or help to find out or even just say I don't know the answer.

> ...I like structure, but then, I think you have to be very open-ended ... it's really hard I think to write curriculum, you know, a script of what you're going to say, what you're going to say next, and so forth, but you kind of have to have an idea but maybe one question or answer that students have will lead you in another direction that would maybe get you to the point. But they have to be structured but not structured I guess, they have to be able to see, I think, the big picture at times, keep that in mind.

Skilled teachers learn to navigate complexities of the classroom and still keep the focus on learning goals. It used to be that the implicit strategy was that the teacher defined the content and therefore to do well students had to listen to the teacher. When students are expected to take a more cognitively active role, the dynamics among students and with the teacher change. The way the teacher maintains the academic focus is less authoritarian and based more on a new kind of authority.

> The teacher needs to have a very clear outcome in mind to start with. You've got to know where you're going with it. And you've got to know what level of inquiry you're looking for, how many components of the activity the kids are going to be involved in. What is the final outcome they really are looking for? ...So decide ahead of time. And it's all about just deciding. But you have to play with it for a while to get to that level.

After a year of professional development in teaching and assessing inquiry, the Cadre constructed a new basis for authority in the classroom. To be sure, most of these teachers were already allowing student thinking to direct some instructional activity, but the goal of channeling cognitive effort toward specific understandings of a deeper order raised an old dilemma of control and focus vis-à-vis allowing appropriate student input. With greater responsibility for cognitive involvement, the role of the student is brought into high relief. Two major themes emerged (a) student engagement in learning tasks and (b) student skills. Tobin et al. (1994) said that the inquiry classroom "begins with motivated students" (p. 47) but implicit in the following statements is that a classroom does not being with this premise. This premise must be established.

> ...responsibilities you hope they take, that is to be curious and to wonder, and I mean, that's the minimum. But to really do it they have to think and be willing to ask questions and listen to other people and try things.

> ...they must be engaged. They must interact with their peers, specifically on the task stuff. So it sounds like a question, or it sounds like asking for assistance, or it sounds like what page, whatever, explain this to me, or why did you do it that way...

Students may be commanded into attention in class but they can not be commanded into thinking. It is the student's responsibility to engage in behaviors that suggest they are pursuing information that applies to the investigative purpose of the class. In broader discussions across the year, the teachers described the challenges of getting all students to the level of meaningful participation. Participation takes many forms and requires a set of skills that must not only be performed but performed at the right times to achieve the intended goal.

> The responsibility of the student is to pursue their inquiry, to challenge their own thinking, to gather information about their ideas, and look and see if those are valid. Evaluate whether they were right or wrong, and why. Learn from that experience and present that experience and be willing to communicate...

> I guess the frustration I see in inquiry or in any type of high-school level of science is the unwillingness of students to repeat things, and to change things, and go through a cycle... Try to make them understand that it's a spiral that you keep coming back to different parts of that and trying over again, and I think that's something that the students really have to be willing to do rather than "Ok, I did my inquiry. It's done."

Teacher comments are consistent with Blumenfeld (1992) who reviewed studies of teaching showing a disconnect between cognitive demand of tasks and cognitive output. Students operated on the basis of getting the task done and skirted opportunities to take the time necessary to understand it. She observed that this cognitive behavior can result from long tasks where students become fatigued or sidetracked or high-level tasks that require large amounts of self-discipline in reflecting, use of resources, or cooperation. The list of social and intellectual skills required of students is long and require teacher support.

> In the ideal world, they must actually do some processing of this (instructional material) away from the classroom. ...they need to know where they're going, they need to start out by knowing that, so they need to make sure that they're clarifying what's going on... So it's just not doing the activity, but it's also saying I'm gonna take and have an outcome that everybody can see.

> The student has to become responsible for their own learning. They have to learn how to do that at the middle school level. They don't know how to do that. They're learning what it means to be responsible as a human being let alone be responsible as a learner. And so it's a struggle...

After a year of concentrated work, these teachers developed broad concepts of classroom implementations of inquiry. Descriptions of classroom inquiry reflected attention to learning goals that went beyond task completion. Improved classroom discourse was directly influenced by increased cognitive engagement of students such that teachers developed skills for listening and interpreting student thinking. Expressing ideas became important for students as they engaged in group work that leveraged individual thinking with contributions from a classroom community of inquirers. At the same time, students depended on classroom structures and the teacher, as instructional leader, to create opportunities to generate, express, and act on their own thinking. This scaffolding was a critical support for students whose cognitive abilities were newly formed. Metacognitive support was needed to consistently apply emerging abilities in timely and effective ways. These emerging cognitive skills were being applied to challenging tasks that teachers were just now learning how to assess. Broad guidelines offered by national and state standards were descriptive in nature and had to be operationalized by teachers skilled in the operation of classrooms and knowledgeable in science and of science as inquiry. This challenging instructional environment operated only if students understood the broad goals of instruction and could maintain a larger framework of intended outcomes that were focused on understanding the human-generated, inquiry nature of science. This understanding came with the integration of a wide range of social and cognitive skills.

FINAL THOUGHTS AND FUTURE DIRECTIONS

Maintaining high levels of cognitive engagement required in inquiry-oriented science instruction involves a significant interplay of student and teacher responsibilities. In direct, transmission-type instruction with minimal modeling and practice, the role of teacher and student are nearly independent. The teacher and student might as well be separated by video transmission. However, the interplay of student and teacher roles becomes critical as expectations of the quality and content of student responses increase and those responses are used to assess the effectiveness of complex instruction.

Effective instruction provides student support for seeing scientific problems where once students saw only science demonstrations. Students must recognize instructional supports for their intended purpose, to prompt questions requiring reflection or that provide feedback on behaviors critical to furthering an inquiry. The teacher must make high level goals clear and students must recognize these goals as meaningful relative to learning science. Cognitive skills critical to understanding scientific inquiry should be seen as integral to learning science and become the objective of instruction.

A set of essential features of classroom inquiry have been stated by the National Research Council in their recent guide for teaching and learning inquiry (NRC, 2000). An adaptation of these outcomes are presented here in Figure 1. Each of these statements imply instructional objectives. The figure is marked to indicate how one might read these statements to identify instructional objectives to support inquiry in the classroom. Using the first line as an example: Learners *are engaged* by **scientifically oriented questions**. The portion in bold indicates an aspect of inquiry that students must be taught. Students may be able to generate questions about a phenomena but instruction, modeling, and practice is needed to express a scientific problem that can be shaped into a testable question. Sometimes teachers use a technique of forming a hypothesis as an If..., then... statement. From here they help students form questions that afford empirical investigation. More broadly stated in the first line, is the portion in italics that implies the cognitive skill of *engagement*. What does it mean for a high school student to be engaged? Answering this question may entail teaching the italicized elements in the rest of the list. For example, teaching engagement means teaching students how to *give priority* to scientific evidence over other possible influences on directing the investigation. *Giving priority* could mean teaching students that even when your hunch agrees with your friend's, if the hunch is inconsistent with the evidence, the evidence must be carefully checked before going further.

- Learners *are engaged* by **scientifically oriented questions**.
- Learners *give priority* to **scientific evidence**.
- Learners *formulate explanations* from **scientific evidence** to address **scientifically oriented questions**.
- Learners *evaluate their explanations in light of alternative explanations*, particularly those **explanations that reflect scientific understanding**.
- Learners *communicate and justify their proposed explanations* based on **scientific criteria such as quality of evidence, soundness of theory, and clearness of reason**.

Learner actions that will be in part scaffolded by instruction.
Content elements of classroom inquiry that will be explicitly taught.

Figure 1. Essential Features of Classroom Inquiry

**Modified from Table 2-5, National Research Council (2000). Inquiry and the national science education standards: A guide for teaching and learning. Washington DC: National Academy Press*

The National Science Education Standards (NRC, 1996) discuss a range of activities that are involved with student engagement in classroom inquiry. These activities are skills needing instructional support and emphasis. "When students are engaged in inquiry they are observing, collecting data, reflecting on their work, analyzing events or objects, collaborating with teacher and peers, formulating questions, devising procedures, deciding how to organize and represent data, and testing the reliability of knowledge they have generated" (p. 33). Incorporating these skills in classroom activities carries significant implications for instructional support for developing and applying those skills. Learning science as inquiry puts the secondary science teacher in a new relationship with students offering direct support for teaching skills important to critical thinking, communicating effectively in groups, and to reflect on one's own understanding of science content.

REFERENCES

American Association for the Advancement of Science. (1993) *Benchmarks for science literacy*. New York: Oxford University Press.

American Association for the Advancement of Science Commission on Science Education (1967). *Science - A Process Approach*. Washington DC: Author.

Blumenfeld, P. C. (1992). The task and the teacher: Enhancing student thoughtfulness in science. In J. Brophy, (Ed.). *Advances in research on teaching Vol. 3: Planning and managing learning tasks and activities*. Greenwich, CN: JAI Press, Inc.

Brown, A. L., Bransford, J. D., Ferrara, R. A. & Campione, J. C. (1983). Learning, remembering, and understanding. In Flavell, J. H. & E. M. Markman. *Cognitive Development, Vol. 3* of P. H. Mussen (Ed.). *Handbook of child psychology, 4th Edition*. New York, NY: John Wiley & Sons.

Carnegie Council on Adolescent Development (1985). *Great expectations: Preparing adolescents for a new century*. Washington, D.C.: The Author.

Carnegie Council on Adolescent Development (1989). *Turning points: Preparing American youth for the 21st century*. Washington, D.C.: The Author.

Cavallo, A. M. L. & Schafer, L. E. (1994). Relationships between students' meaningful learning orientation and their understanding of genetics topics. *Journal of Research in Science Teaching, 31*, 393-418.

Driver, R., Newton, P., Osborne, J. (2000). Establishing the norms of scientific argumentation in classrooms. *Science Education, 84*. 287-312.

Duckworth, E., Easley, J. A., Hawkins, D., & Henriques, A. (1990). *Science Education: A Minds-on Approach for the Elementary Years*, Hillsdale, NJ: Lawrence Erlbaum Associates, Inc.

Flick, L. B. (1993). The Meanings of Hands-On Science. *Journal of Science Teacher Education, 4*, 1-8. Also reprinted in Rezba, R. (1994). *Readings for teaching science in elementary and middle schools*. Dubuque, IA: Kendall/Hunt Publishing Company.

Gallagher, J. J. & Tobin, K. (1987). Teacher management and student engagement in high school science. *Science Education, 71*, 535-555.

Greeno, J. G., Collins, A. M., & Resnick, L. B. (1996). Cognition and Learning. In D. C. Berliner & R. C. Calfee (Eds.). *Handbook of Educational Psychology*. New York, NY: Simon & Schuster Macmillan.

Hofstein, A. & Lunetta, V. N. (1982). The role of the laboratory in science teaching: Neglected aspects of research. *Review of Educational Research, 52*, 210-217.

Hurd, P. D. (1997). *Inventing science education for the new millennium*. New York, NY: Teachers College Press.

Inhelder, B., & Piaget, J. (1958). *The growth of logical thinking from childhood to adolescence* (A. Parsons, & 5. Milgram trans.). New York: Basic Books,.

Keating, Daniel P. (1990). Adolescent thinking. In S. S. Feldman & G. R. Elliott (Eds.). *At the threshold: The developing adolescent*. Cambridge, MA: Harvard University Press.

Keys, C. W. (1994). The development of scientific reasoning skills in conjunction with collaborative writing assignments: An interpretive study of six ninth-grade students . *Journal of Research in Science Teaching, 31*, 1003-1022.

Koslowski, B. & Maqueda, M. (1993). What Is Confirmation Bias and When Do People Actually Have It? *Merrill-Palmer Quarterly, 39*, 104-30.

Kuhn, D. (1992). Thinking as argument. *Harvard Educational Review, 62*, 155.

Kuhn, D. (1997). Constraints or guideposts? Developmental psychology and science education. *Review of Educational Research, 67*. 141-150.

Loucks-Horsley, S., Hewson, P. W., Love, N., & Stiles, K. I. (1998). *Designing Professional Development for Teachers of Science and Mathematics*. Thousand Oaks, CA: Corwin Press. The book was produced by a grant from the National Science Foundation to the National Institute for Science Education.

Metz, K. E. (1995). Reassessment of developmental constraints on children's science instruction. *Review of Educational Research, 65*. 93-127.

Metz, K. E. (1997). On the complex relation between cognitive developmental research and children's science curricula. *Review of Educational Research, 67*. 151-163.

National Research Council (1996). *National science education standards*. Washington, D.C.: National Academy Press.

Poincaré, H. (1913). Science and hypothesis. In G. B. Halsted (Trans.)_*The foundations of science* (pp. 1-196). New York, NY: The Science Press.
Pressley, M. & McCormick, C. (1995). *Cognition, teaching, and assessment.* New York, NY: HarperCollins College Press.
Reif, F. & Larkin, J. H. (1991). Cognition in scientific and everyday domains: Comparison and learning implications. *Journal of Research in Science Teaching, 28*, 733-760.
Roth, W-M. (1994). Experimenting in a constructivist high school physics laboratory. *Journal of Research in Science Teaching, 31*, 197-223.
Roth, W-M. & Bowen, G. M. (1994). Mathematization of experience in a grade 8 open-inquiry environment: An introduction to the representational practices of science. *Journal of Research in Science Teaching, 31*, 293-318.
Roth, W-M. & Roychoudhury, A. (1993). The development of science process skills in authentic contexts. *Journal of Research in Science Teaching, 30*, 127-152.
Sanford, J. P. (1987). Management of science classroom tasks and effects on students' learning opportunities. *Journal of Research in Science Teaching, 24*, 249-265.
Schwab, J. J. (1962). The teaching of science as enquiry. In J. J. Schwab & P. F. Brandwein, *The teaching of science*, Cambridge, MA: Harvard University Press.
Siegler, R. (2001). Cognition, instruction, and the quest for meaning. In
Shulman, L. S. (February 1986). Those who understand: Knowledge growth in teaching. *Educational Researcher, 15*, 4-14.
Shulman, L. S. (1987). Knowledge and teaching: Foundations of the new reform. *Harvard Educational Review*, 57, 1-22.
Shulman, L. S. & Tamir, P. (1973). Research on teaching in the natural sciences (pp. 1098-1148). In R. M. W. Travers (Ed.), *Second handbook of research on teaching.* Chicago: Rand McNally.
Tobin, K. & Fraser, B. J. (1990). What does it mean to be an exemplary science teacher? *Journal of Research in Science Teaching, 27*, 3-25.
Tobin, K., Tippins, D. J., & Gallard, A. J., (1994). Research on instructional strategies for teaching science. In D. L. Gabel (Ed.). *Handbook of research on science teaching and learning.* New York: Macmillan Publishing Company.
United States Bureau of Education (1893). *Report of the committee on secondary school studies.* Washington, D.C.: Government Printing Office.
United States Bureau of Education (1918). *Cardinal principles of secondary education.* Washington, D.C.: Government Printing Office.

CHAPTER 9

SANDRA K. ABELL, DEBORAH C. SMITH, & MARK J. VOLKMANN

INQUIRY IN SCIENCE TEACHER EDUCATION

The term "inquiry" is used throughout the science education literature to describe goals for science learners as well as approaches to science teaching (e.g., National Research Council (NRC), 1996; 2000). Inquiry can also be viewed as an *orientation* to science teaching (Anderson and Smith, 1987; Magnussen, Krajcik, and Borko, 1999)--a set of knowledge and beliefs that guide the teaching of science. When applied to the teaching of science teachers, such a perspective has been called a "reflection orientation" (Abell, 1996; Abell and Bryan, 1997), which values teachers' ability to inquire. We might also view inquiry in science teacher education as a *stance*, following Cochran-Smith and Lytle (1999), a way to frame our thinking about teaching teachers and analyze knowledge and practice. By adopting such a stance, we view teacher educators, teachers, and students as Learners, not merely Knowers; we adjust our teacher education practices to convey the value of learning in addition to knowing; and we judge our effectiveness by the extent to which we and our students and their students adopt an identity as learners and an inquiry stance in their lives, not only by the degree to which they have come to know science or teaching[*].

In our teaching of future and practicing elementary teachers, we have enacted our inquiry stance in several contexts—within undergraduate science courses, and undergraduate and graduate methods courses, while inquiring into science or inquiring into teaching and learning. We view ourselves as Learners along this path, not as Knowers of the truths of teaching teachers for inquiry. In this chapter, we offer our experiences with inquiry in elementary science teacher education and examine what these experiences tell us about our teaching and our students' learning. We also discuss implications for the role of teacher education in achieving the science education reform vision. First, we present a theoretical framework for teacher learning in which our discussion of practice is grounded.

ON TEACHER LEARNING AND INQUIRY

We take a sociocultural perspective on teacher learning (Cobb, 1994; Lave and Wenger, 1991; Wenger, 1998). We are interested in and believe it is important to consider not only students' ideas--about scientific explanations and scientific work, about science teaching and learning--but also their roles and identities as learners, teachers, football players, sorority members, and parents, and how those shape interactions with us and with other students (e.g., Cobb, 1994; Richmond and Kurth, 1999). We need to know what our students' previous participation in learning

L.B. Flick and N.G. Lederman (eds.) Scientific Inquiry and Nature of Science, 173-199.

cultures, in and out of school, has taught them about who they are as science learners and science teachers, and how the participatory roles they have learned may complement, or be obstacles to, the roles and participation strategies we ask them to practice in our inquiry-oriented classrooms (Heath, 1994; Smith, 2001; Wenger, 1998). It is important for us to pay attention to and make part of the curriculum issues of gender, race, class, and ethnicity (Fradd and Lee, 1999), and their influences on status, discourse, and socio-emotional connections in our classrooms (Cobb and Bauersfeld, 1995; Lemke, 1990; Tannen, 1996). We also recognize that preservice and practicing teachers enter our classrooms with multiple purposes, not always directed at learning with understanding. We, as their teachers, need to design intentionally, not only the sequence of academic content, but also the scaffolding of their entry into communities of inquiry (Collins, Brown and Newman, 1989; Hogan and Pressley, 1997).

We also take a sociocultural perspective on the practices of inquiry, both in scientific work (Latour and Woolgar, 1986; Traweek, 1988) and in teacher education (Bentley and Fleury, 2000; Hammer, 2000; van Zee, 2000). Our work draws on sociocultural studies of scientific work (Bazerman, 1988), and of classroom learning communities (Abell, Anderson, and Chezem, 2000; Anderson, Holland, and Palincsar, 1994; Crawford, Krajcik, and Marx, 1999; Lehrer, Carpenter, Schauble, and Putz, 2000; Warren, Rosebery, and Conant, 1989), as well as views of inquiry in national standards (American Association for the Advancement of Science (AAAS), 1993; NRC, 1996) and reform documents (Bybee, 1997; NRC, 2000; Rutherford and Ahlgren, 1993). We view inquiry as theory-laden, community-validated ways of generating explanations supported by evidence and persuasive argument, and held by the community as tentative and open to further development (Smith and Anderson, 1999).

This sociocultural perspective has implications for how we work with future and practicing teachers (see Putnam and Borko, 2000) as well as how we interpret and write about our practices as teacher educators. Our students may be fairly recent "newcomers" to the community of elementary science teaching, but they are "oldtimers" as members of the school science culture (Lave and Wenger, 1991; Lortie, 1975). We are interested in their views about becoming participating members of a community of inquiry, whether the inquiry is into science or science teaching. For example, how do preservice teachers' views of science influence the roles that they take on in doing scientific work in our classrooms? How do their views of appropriate roles for science teachers influence their willingness and ability to assume the role of investigator into their own teaching and their students' learning? How can we help them see that teachers can be, not only users of knowledge, but makers of knowledge about science, teaching, and learning?

In the three cases that follow, we use this sociocultural perspective to select, describe, and interpret events from our classrooms; and examine and highlight issues that arise from our teaching. The first case takes place within a science course for preservice elementary teachers and illustrates a common dilemma in teaching science as inquiry. The second is situated within an elementary science methods course, and focuses on student roles and identities as they engage in science inquiry. The final case uncovers the tensions that practicing teachers experience when asked to become inquirers in their own classrooms. Following the cases, we offer an analysis of the issues that emerge.

CASES OF INQUIRY IN SCIENCE TEACHER EDUCATION

When is it Okay to Tell?

Recently I (Volkmann) implemented a new physics course, specifically designed for elementary education students: PHYS 290E: Physics for Elementary Education. I used the Powerful Ideas in the Physical Sciences (American Association of Physics Teachers, 1996) curriculum. The syllabus described the course purpose as:

> The search for and use of evidence to provide a basis for developing scientific conclusions. Students are encouraged to verbalize not only what they know, but just as importantly, describe "how they know." The end result is that students achieve an operative, rather than simply a declarative, knowledge of physical science subject matter. (p. 9)

Student performance as both Knowers and Learners were goals of the curriculum. Students were to develop knowledge about a few powerful ideas in physical science as they learned everyday phenomena about light, electricity, and heat. Another goal was for students to use their experiences in learning physics as a model for future science teaching at the elementary level.

In scientific inquiry, no one provides scientists with the answers to their questions. Scientists base their answers, in part, on the theoretical perspective within which they work and on the on the evidence they gather. In school science inquiry, the situation is different. Although we want students to develop explanations and support their explanations with evidence, accepted answers exist for the questions students pursue. Throughout my teaching of PHYS 290E, I struggled to understand the nature of school science inquiry, and how it might compare with the nature of scientific inquiry. My students struggled too. On several occasions they became frustrated with me because I did not provide them answers. They wanted to be Knowers and were frustrated when I did not accommodate. And, at times, I became frustrated because I found myself wanting to respond to their questions with my explanations. As a result of these conflicts, I found myself asking "Does inquiry mean that I can't give answers?" In this case, I address the question: "When is it okay to tell?"

My class consisted of 24 students (20 white females and 4 white males). We met twice each week for two hours in a university lab science classroom (there was no lecture component). Below, I focus on the first three circuit lessons, taught over three class periods during the unit on electricity, and describe the teaching/learning in terms of one 3-member team: Sue, Melissa, and Max.

Circuit Lesson #1:

I started the study of electricity by challenging students to light a bulb using one battery, one wire, and one bulb. The students responded to the challenge in terms of the following prompts:

1. **What's Your Idea?** -- Students responded individually to the question: *Can you make the bulb light?*
2. **What Are the Group's Ideas?** -- Members of each small group shared and recorded ideas about how they planned to make the bulb light.
3. **Observation** -- When members of the group were relatively confident in their ideas, they assembled the circuit and compared what happened to what they predicted. If their idea was not successful, then students tried other ideas. Within 20 minutes, all the groups had successfully lit the bulb.
4. **Making Sense** -- Finally, students were asked to make sense of what happened. The class discussed a variety of observations pertaining to conditions-to-be-met to make a bulb light. They wrote their ideas on the board and the class reached consensus on the following conditions: 1) two parts of the bulb must touch two parts of the battery; 2) parts must touch in a circular pattern; and 3) batteries are the source of energy and bulbs use this energy. At this point I told the class that this list described the term *circuit.* I had not used this term during my introduction to the electricity unit, or in my directions on how to get started with the first activity. The class appropriated the term *circuit* into their discourse from this point forward. No conflict erupted over my telling this term.

The students liked this activity because everyone was successful and everyone was able to complete the lab in a single lab period. Sue wrote, "Today was really fun because I learned something that I didn't even know I didn't know - how to make a bulb light" (Lab notebook, 4/15/01). Melissa said, "I was frustrated for a little while because my group had no clue. Luckily, my roommate in the next lab group showed me how they got the bulb to light" (Lab notebook, 4/15/01). Max wrote briefly, "I finally figured it out" (Lab notebook, 4/15/01).

Circuit Lesson #2:

Students examined a circuit drawing that consisted of a bulb, battery, switch, and wires. Their task was to explain: *What is happening inside the wires?* I asked students to use arrows to indicate their ideas about the direction and intensity of the electricity inside the wires. Before students constructed this circuit, I demonstrated how to use a compass to measure relative direction and intensity of the electricity. [When a circuit wire is oriented north-south over the compass needle, the needle deflects by an amount characteristic of the strength of the current. The direction of the needle deflection indicates the relative direction of the electricity in the circuit (clockwise or counterclockwise).] I did not introduce the idea of positive and negative charge in this or subsequent lessons. However, students used plus and minus designations that they observed as labels on the battery terminals in their descriptions and explanations.

What are Your Ideas and the Group's Ideas?

Sue's drawing showed electricity flowing from each end of the battery and traveling to the bulb. Sue embraced a clashing consumption model (Driver, Guesne, and Tiberghien, 1985) consisting of two kinds of electricity--one that flowed from the each end of the battery to the bulb.. In her model, the bulb consumed electricity from both ends.

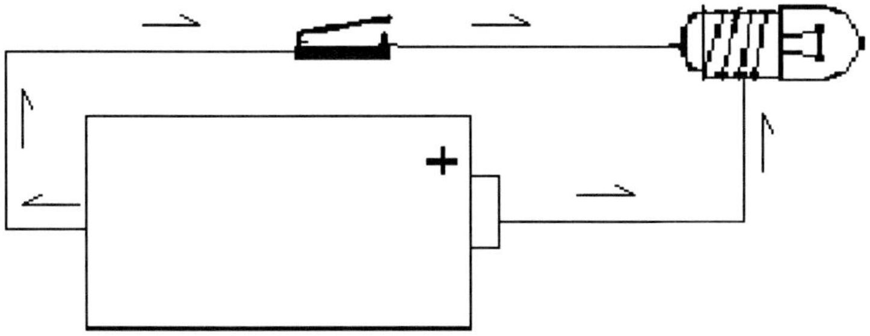

Figure 1.. Sue's Drawing

Melissa's drawing showed electricity flowing undiminished from one end of the battery through the bulb to the other end of the battery. Melissa embraced a constant flow model (Driver, Guesne, and Tiberghien, 1985) of electricity through the circuit.

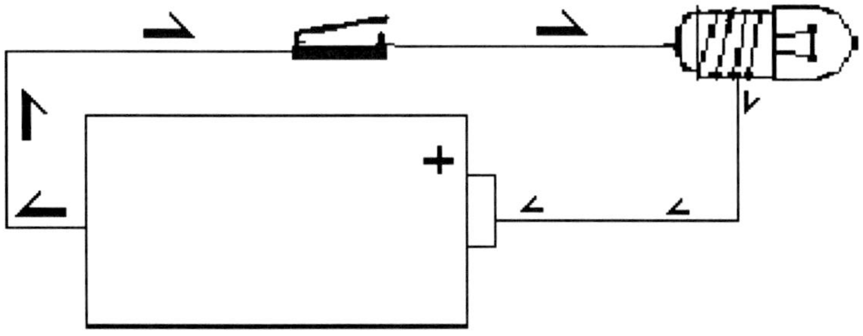

Figure 2. Melissa's Drawing

Max's drawing showed electricity flowing in one direction with diminished flow of electricity after the bulb. Similarly to Sue, Max supported a consumption model (Driver, Guesne, and Tiberghien, 1985) of electricity; similarly to Melissa, he supported electricity flowing in one direction.

After discussing their models, the group members agreed to disagree. Apparently, none of the three students could give any convincing evidence to change anybody else's mind.

Observation and Making Sense

After constructing and testing the circuit, students noted that the bulb was bright and the compass deflected in the same direction and amount at all points of the circuit. Sue and Melissa viewed the evidence as consistent with their models (for different reasons) but the evidence contradicted Max's prediction.

Sue defended her clashing consumption model in terms of the following explanation: "The negative charge caused a clockwise deflection in the compass and the positive charge caused a counterclockwise deflection, but since the positive is moving in the opposite direction to the negative charge, it registered as a clockwise deflection on the compass" (Lab notebook, 4/17/01). Sue understood that the compass needle would respond in the same manner to either negative or positive charges as long as they were moving in opposite directions. Therefore, she had no reason to change her clashing consumption model.

Melissa predicted that the bulb would light and the electricity would flow undiminished through the entire circuit. Since this is what the compass indicated, she had no reason to discard her constant flow model.

Max defended his earlier explanation (consumption model) by questioning the measurements: "None of the deflections were exactly alike, so it was difficult to know if they were all the same" (Lab notebook, 4/17/01).

Unlike the quick rewards experienced during the study of Circuit #1, students experienced increased conflict after completing Circuit #2. No clear explanations emerged from the investigation. Sue, Melissa, and Max continued to question: Is flow constant throughout the circuit? Is something being used up? Is electricity flowing in one direction or two directions?

Circuit Lesson #3:

Circuit #3 (see Figure 3) was similar to #2, except for the addition of a second bulb. Students examined Circuit #3 and continued to address the question: *What is happening inside the wires?*

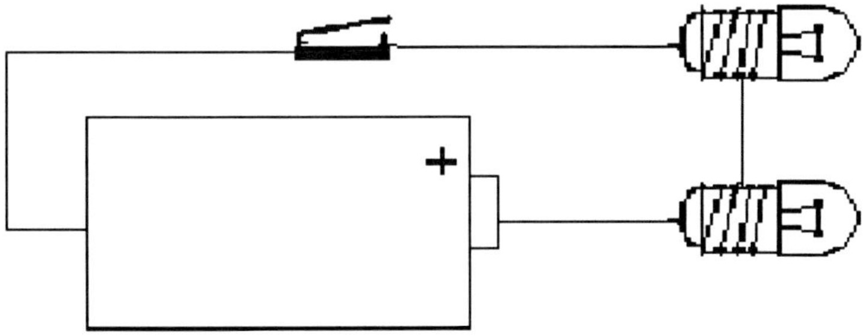

Figure 3. Circuit #3

What are Your Ideas and the Group's Ideas?

Sue held onto her clashing consumption model that the positive electricity would flow from the positive side of the battery to both bulbs and negative electricity would flow from the negative side of the battery to both bulbs. Sue said the compass would deflect in the same direction and same amount at all points in the circuit. I asked Sue to describe how the compass would behave relative to the segment of wire between the two bulbs. She considered my question and responded, "There won't be any deflection in that region because the negative and positive charge will flow against each other and cancel any compass deflections."

Melissa predicted that the electricity would flow from the negative side of the battery through each bulb with the same intensity as the one-bulb circuit. She reasoned that the compass would deflect in the same direction and by the same amount in Circuit #2, and that the two bulbs would be the same brightness as the single bulb.

Max predicted that the electricity would flow in one direction through the bulbs, but the intensity would decrease from bulb #1 to #2. He predicted that bulb #1 would be brighter than bulb #2. He reasoned that, "If bulb #1 is first in line, then it will glow brighter. Bulb #2 would get what's left, and very little would return to the battery. "

Sue and Max explained electricity in terms of something being consumed. Melissa's prediction implied a view of the battery as a constant source of electricity. My own explanation, held back from the class because I wanted the students to focus on the evidence and their own explanations, was that (1) current is not consumed; (2) current flow is reduced, but constant; and (3) current flows in one direction. Thus I predicted that the bulbs would be of equal brightness but dimmer than in Circuit #2.

Observation and Making Sense

After completing the construction of Circuit #3, Sue, Melissa, and Max observed that the bulbs were of equal brightness, but dimmer than the single-bulb circuit. They observed that the compass deflections were smaller in Circuit #3 than #2 and were all in the same direction. I felt good because these results agreed with my unvoiced prediction. I hoped the students would use this evidence to understand that the flow was all in the same direction, that the addition of the second bulb caused a decreased flow, and this decrease resulted in dimmer bulbs.

Sue's prediction of no deflection in the segment between the bulbs proved false. Her level of frustration mounted. Her lab notes said, "I had great faith in the two-direction theory because it was based on what a friend had told me. Josh (a major in Electrical Engineering) said that there are two kinds of charge and that both flow through circuits. I reasoned that our circuit must have two kinds of charge, too. I wonder if I misunderstood Josh?" (Lab notes, 4/19/01). I was frustrated that Sue did not question Josh, but instead questioned whether she had heard him correctly . The evidence we gathered did not support her explanation, but she was still reluctant to let it go. She found it difficult to dispense with her clashing consumption explanation of electricity, having no intelligible alternatives.

Melissa's prediction agreed with her observation that electricity flows in one direction. However, Melissa failed to predict the decreased intensity of flow and the resulting dim bulbs. Her journal indicated that her level of frustration had increased as well: "Either the battery is being used up or it isn't being used up. I don't get it. Why would the flow be smaller with two bulbs? Why would the bulbs be dimmer? Don't two bulbs need more electricity than one bulb--hmmm, unless the electricity is divided equally between the two bulbs" (Lab notes, 4/19/01). Melissa was disappointed and frustrated that her questions remained.

Max's prediction that the electricity would flow in one direction through the bulbs was in agreement with his observation; however, he was disappointed to observe that the intensity of flow did not decrease from bulb #1 to #2. Furthermore, he was disappointed to observe that both bulbs were dimmer. Max began to question his consumption theory, that is, until he heard Melissa's idea that, "Maybe the electricity is divided equally between the two bulbs."

During our large group discussion, several students asked me to help them make sense of the conflicting interpretations. They complained that there were too many competing theories and they were having difficulty keeping them straight. I recognized that the class was feeling frustrated, and attempted to counter their frustration by showing them what they had learned so far about circuits. I asked them if it was okay to be frustrated, and suggested that scientists must become very frustrated sometimes, too. I recommended that they use their frustration as a motivation to figure out the puzzle.

Before taking a short break, I asked the class to think of a test they could do to help clarify the confusion among the competing models. After break, Melissa asked what would happen if the two similar bulbs were replaced with two dissimilar bulbs. That is, replace one with a bulb that glows brightly and the other with a bulb that glows dimly. This was an exciting suggestion because the consumption model predicts that both bulbs light with different intensities. What actually happens is the higher resistance of the dimmer bulb causes a decrease in the current resulting in only the dim bulb lighting. I asked the class to write down what they thought would happen if we followed Melissa's suggestion. I hoped this test would provide the missing evidence that I felt students needed in order to critique the consumption model.

Every student in the class predicted that both bulbs would light--one brightly and the other dimly. I came to the realization that the demonstration might do more harm than good. Although I wanted the students to analyze more deeply, I realized their current level of frustration was a barrier. I wrestled with what to tell and not tell. I decided to offer one more model to consider before I demonstrated Melissa's circuit, and described the resistance model. According to this model electricity flow is constant and in one direction. I told them that bright bulbs have low resistance and high flow while dim bulbs have high resistance and low flow. I shared my prediction in terms of the resistance model that only the dim bulb would light. I performed the demonstration, and only the dim bulb lighted. The students asked me to check various connections to be sure that both bulbs still worked. After showing that they did, I demonstrated that the compass deflection was constant and diminished. Sue asked, "How can I believe the resistance model when it says that the circuit 'knows' what is in the way before it sends out electricity?" She continued, "If current is not being used up, then what is?" This was the question I had hoped for. I responded, "According to the resistance model, there are two aspects of electricity: energy and current flow. The energy in the battery is being used up, not the current. That is why the battery will eventually die."

Reflection

In my teaching journal, I reflected on the model of inquiry I was demonstrating and my students' level of frustration. I wondered what would happen if I told them answers to all their questions. Would my teaching lose all semblance of inquiry? I worried that if I told them answers, I would emphasize science as static, teaching as telling, and learning as listening. At the same time, I feared that my silent response to their pursuit of answers would increase their frustration to the point that they might give up. My silence might also portray science and science learning as a process of discovery, not as a sociocultural process. I valued my students' struggles as Learners; they preferred being Knowers. I wondered, "When is it okay to tell?"

Furthermore, I wanted my students to understand physics, but not at the expense of misunderstanding the nature of science. I wrote about this dilemma in my teaching journal:

> My students believe it is imperative to know the correct answers. If I choose to tell students the correct answers, then my fear is that they will learn nothing about how science works. Unfortunately, this appears to be okay with them. But what happens when I turn this on its head? If my students' feelings about "knowing the correct answers" were replaced with "understanding the process of inquiry," then would it be okay to have wrong answers? How would my students feel about understanding the approach, even if it resulted in misunderstanding the facts?

As I reflected on the circuit lessons, I realized that I had told the students something during each of them. During Circuit Lesson #1, I told them the scientific term, *circuit*. This telling occurred after students had defined *circuit* based on their inquiry and resulted in them using shared vocabulary throughout the electricity unit.

I perceived no negative fallout. During Circuit Lesson #2, I told the class how to use a compass to measure relative direction and intensity of electrical flow. Again, I perceived no negative fallout. In Circuit Lesson #3, I told them my explanation of Melissa's two-bulb circuit. On first appearance, I accomplished very little with this telling (with the possible exception of increased confusion). Later, I realized that my telling had provided the students with an alternative at just the right moment. I learned from reading their journals that they had been dissatisfied with their models but had no viable alternative. My decision to tell gave them a new model at the point when they were the least satisfied with their own. It became clear in later lessons that students appropriated the resistance model to make sense of new evidence. Perhaps knowing when to tell is as important as knowing what to tell.

STRANGERS IN A STRANGE LAND

I (Smith) have been teaching TE 401, "Teaching Subject Matter to Diverse Learners," the elementary science methods course for seniors in our teacher preparation program, for seven years. The course introduces them to current reforms in science education, research on children's thinking in science, developmentally appropriate curriculum design, and guidelines for authentic assessment of children's understanding of scientific ideas. It is not intended to be a science content course, yet time and time again, I have watched students' progress in "methods" founder on the rocks of shallow and inadequate content understanding. Most of the students who come through the methods classroom door have not had the opportunity to pursue scientific inquiry in their own learning of science. Therefore their conceptual understanding of both the science they teach with children and of scientific inquiry may be flawed. For those who have been "successful" in school science, authentic science inquiry is as strange a land as it is for those who have been "unsuccessful." Few of them have built the habits of mind, discourses, or practices to engage in authentic scientific inquiry. The authors of the National Research Council document, *Inquiry and the National Science Education Standards* (2000), point out, "Most teachers have not had opportunities to learn science through inquiry or to conduct scientific inquiries themselves" (p. 87).

One of my solutions to this dilemma is to create opportunities in TE 401 for students to pursue their own inquiries into scientific questions, and to spend several lessons in the beginning of the semester modeling and supporting scientific inquiry. By inquiry, I refer to both the " abilities students should develop to be able to design and conduct scientific investigations and to the understanding they should gain about the nature of scientific inquiry;" as well as "the teaching and learning strategies that enable scientific concepts to be mastered through investigations" (NRC, 2000, p. xv).

I want my students to have the opportunity to raise questions about natural phenomena, like, "Why does your hand shadow get bigger when you bring your hand closer to the light source?" I want them to articulate their current ideas, speculations, predictions, and explanations; to propose models and theories that could account for the change in the size of the hand shadow; to challenge each others' theories and speculations by bringing in their everyday experiences with shadows; to devise and carry out investigations that test their ideas; to make observations and measurements, and find ways to record what happens; to represent their findings for others and look for patterns in their data; and then to use their evidence to make clearly articulated claims about a model of light that accounts for why the shadow gets bigger. In what follows, I provide examples from our classroom discussions, and students' journals and interviews, that highlight the dilemmas, struggles, and small victories that we encounter together.

From my students' pre-course interviews and their in-class skits on teaching and learning science, I knew that most of them (especially the women) had hated science (especially in high school and college), had never had the opportunity to pursue their own questions and ideas in science, and thought the way to make science easier to understand was to make it more fun, more exciting, more visual, more comfortable and friendly, and more personal. I also knew that only a few had mentioned the roles of discourse, scientific authority, evidence, theories, or the larger scientific community in their descriptions of how scientific knowledge was generated and validated. And I knew that, on the pre-course interview, no one had drawn or explained adequately what would happen with shadows on a wall, if a small child stood close to a light source, in front of a larger child who was much further away from the light source and closer to the wall. Instead, even the most confident had explained that the larger child's shadow would appear on the wall (even though a simple ray diagram would have disproved that idea).

Encountering Phenomena, Raising Questions, Proposing Ideas

I began by reading Ezra Jack Keats' book, *Dreams*, in which a tiny mouse floats down away from a building, creating a bigger and bigger shadow as it falls, and scaring away a dog. Then, I posed the question, "Could a shadow really get bigger and bigger -- even bigger than the mouse itself?" Students immediately jumped in with questions and ideas. Norman asked, "Where could the light could be coming from, for the shadow to be really small at the top and really big at the bottom (there was no light pictured in the story)?" Mary speculated that the light was at the bottom of the picture, because as the mouse passed it, the shadow started to get bigger. Linda noticed that the mouse was closer to the building at the start, then further away, and as it got further away, it looked like the light was beneath him. Other students speculated on whether there was a lamppost across the street and how far the mouse had floated when he made the biggest shadow.

In this short discussion, students were clarifying the events in the story, raising questions about possibilities, and starting to speculate on how the shadows were being made. However, there was little discussion of how or why the shadow had gotten bigger, so I asked each student to draw a picture of what they thought could happen to make a shadow bigger, and then write about why they thought that would happen. Norman, for example, wrote:

> Of course shadows can get bigger with a change in distance from the light source. When a shadow is cast on an object that is tight against a wall, it will be just about the same

> size, because all the light can get around it. When you pull the object out from the wall, then less light can get around and the greater size of the image. So the further from the wall and closer to the light source, the bigger the potential shadow, because more light is blocked and there will be more darkness.

Norman's drawing accurately showed the rays of light diverging from the light source, a smaller shadow when the mouse was closest to the wall, and a larger shadow when the mouse was closer to the light, although he had not been successful on the 2-child pre-interview task.

In a similar vein, Jeri wrote:

> I think that as an object gets closer to a light source it actually blocks more light and therefore creates a bigger shadow. .... I think the light was either a street light or a porch light. So, as the mouse fell he got closer to the light and therefore created a bigger shadow.

However, her drawing showed the mouse coming down the side of the house and making a smaller shadow when it was closest to the light on the side of the building, and a larger shadow when it had fallen below and further away from the light. Despite her writing, which sounded like she had a model of diverging light and of shadows created by blocking some of that light, her drawing showed a contradictory representation.

Another student, Jane, drew a picture of a streetlight facing the building, and showed in succeeding pictures how, as the mouse fell, it moved away from the light source. She drew rays of light in a tunnel and the shadow the same size as the mouse in the first picture; in subsequent pictures, the rays diverged more, and in each picture the mouse's shadow grew bigger. In her journal, she wrote:

> As it (the mouse) left the light source, the area touched by the light increased. Therefore the shadow could get larger. It may also be that the mouse is getting further from the wall, allowing a larger stretched out shadow.

Nowhere in her explanation did the amount of light blocked, relative to the light available, figure into her thinking.

In contrast, Karen drew a person's shadow getting longer and skinnier, but not bigger overall as the mouse had done in the book. Her account offered an accurate description of patterns--"shadows get longer in the late afternoon"--but no explanation for why that happened, and her drawing showed only the person and the shadow, omitting the light source.

All of these students had some valuable prior knowledge which they brought into their consideration of the question, "Can the shadow really get bigger, and how could that happen?" However, their writing and drawings showed some confusions and contradictions, and either the lack of a model of divergent light, or the lack of ways to use such a model to explain the increase in the mouse's shadow size.

Using Prior Knowledge to Develop and Represent Testable Theories

The next day, I asked several students to draw their ideas on the board, and talk about their ways of thinking about the question. Bob came up to draw a picture of himself in a hammock outside his house, where, the summer before, he had made a big shadow of his hand on the side of the house.

> When light comes out of the flashlight, it's diffusing, not staying -- the beam -- the size of the flashlight, so it [the light] keeps getting bigger like this, I guess (drawing the light coming out in straight lines from the sides of the flashlight), so it takes the light that's being blocked from it, it [the light] makes it [the shadow] as big as the light that goes out.

One might expect that Bob's description and drawing of the light as diverging from the source would have been sufficient for others in the class to understand how shadows bigger than the object making them could occur. But Linda responded:

> I think that if a person were standing like right in the middle of the house and Bob -- can you picture? -- and Bob stayed in the same spot, I think their (the person in the middle) shadow would be bigger on the house, but I can't explain why. Because they're closer to the house, their shadow would be bigger.
>
> DS: Why?
>
> Linda: There's more of a distance between the light source and the object blocking it. If the distance were closer, it would be smaller on the, whatever the shadow projected to. As the distance were larger, actually the image would be larger.
>
> DS: Could you come up and draw that for us?

Linda came to the board and drew a person in between Bob and the house, and a larger shadow on the side of the house. Kara then came to the board to question Linda's drawing:

> When the light rays are coming from the flashlight right here (pointing to the area around the person), how would they be blocked (pointing to where Linda has drawn the big shadow on the house), if there's nothing blocking them right there (pointing to the light going around the person)?

At this point, I suggested that we think of our talking and thinking about light and shadows as "rough draft" thinking and talking, to take the burden off the person who was presenting. We would make it okay to think things through in front of other people, and not worry about whether or not we had it quite together yet. Linda responded, "Kara made a lot of sense, when she said if the light is still going past that object, there's a contradiction there. I was thinking that Kara made sense when she said that."

Other students then shared their experiences in elementary school, seeing children make big "bunny ears" shadows in front of the film projector, as well as experiences the previous night, when they tried this at home. Yolanda raised the question of whether a person standing in the middle of the yard would have a shadow at all, since Bob's hand shadow is so big, and no light might be reaching that person. Norman suggested an analogy for understanding the path of the light:

> If you had a can of paint, and a person standing next to the wall, and you threw the paint, the person would have a distinct outline [on the wall]. If they were standing in the middle of the room and you threw the paint, they would block more paint, so the paint

> getting around them would be the paint they didn't block. There is a scatter pattern, the light doesn't stay together. I'm using paint as an analogy for light.

Students continued the discussion by speculating about what would happen to the shadow if Bob knelt down and angled the light up towards the top of the house, using their experiences of seeing long shadows at sunset and small shadows at noon. Then Aaron came to the board to draw his idea, a cone of light extending into space.

> The way I think about it, if there's a <u>cone</u> <u>emanating</u> from <u>rays</u> from a single <u>point,</u> then you take a <u>cross section</u> of the <u>cone</u> which is going to be a circle, then you put your hand in here (draws a person holding his hand in front of the cone of light), then take the <u>rays</u> <u>emanating</u> from this point, and it's going a little bit like this (drawing light going by the hand) – see the <u>proportions</u> (of the hand) of the circle (of light), to the <u>proportion</u> here (on the house)? Then if there's a smaller circle (of light, closer to the source) here, then you put your hand here (close to the light) it will take a bigger part of the circle (of light), it (the light stopped) will be a bigger <u>proportion</u>. [emphasis added]

As Aaron returned to his seat, Jeri cheered, "Yeah, Aaron!" Others clapped and said things like, "Aaron's really smart."

At this point, I was worried that others in the class would stop trying to make personal sense for themselves, because Aaron had spoken so authoritatively and used technical scientific language (as underlined above). In fact, the three white male students in the class had already decided that they "knew" all about this, and were impatient with others (mostly women and minorities) who were struggling to make sense of their own ideas and those of others. The culture of scientific inquiry they had previously experienced did not include an emphasis on personal sense making or community validation, just the usual authoritative sources and memorization. Thus once Aaron had drawn and explained it, everyone else should "have it"—that's the way things had always worked in science courses. So, in order to raise the key role of evidence in making personal sense and establishing claims in science (in contrast to simply accepting an authority's claims), I asked:

> How do we know that what Aaron told us really works? … This is the question I raised earlier about the person who gives you an idea, it sounds, I heard people say, 'Aaron's really smart.' How do we know that he's got an idea that works?

Immediately, Kara suggested, "We could test it, we could, um, set it up so we could take the proportions of our shadows, find out if there's a pattern." I agreed, as did others in the class, and recommended that we test these ideas the next day.

Designing and Interpreting Data Collection to Address the Questions

In the next class, students worked with materials in small groups to test out the ideas they had discussed. They quickly and delightedly found that Aaron's cone model worked to predict and explain changes in shadow sizes. They traced around shadows made by objects at different distances from the light, then measured the distances and corresponding sizes of the shadows, and looked at the distance/size patterns. Students made charts of their data, and some groups graphed the data, to show how the shadow size grew as the distance between the light and the object diminished. They also tried angling the light, to make shadows longer and shorter, and discussed how the light was stopped or got past the objects they were using.

The small group work gave me the opportunity to push those in the class who viewed themselves as Knowers, by posing challenges they could not explain readily. For example, Marvin could barely conceal his impatience and distaste for the length of time we spent talking about ideas and drawing them on the board. Yet during small group work, he revealed his idea that light from the sun lost its "strength" at some point in its journey to Pluto and just stopped. I asked him how the light from Alpha Centauri got to us on Earth, so that we could see the star. If that light did not stop, why would the light from our sun stop? He was taken aback by this seeming paradox and I left him to think about it, more uncertain about his "knowing" than he had been before.

Unfortunately, because of the limited number of class sessions available, we had to move on to a consideration of children's ideas about light and shadows, and to preparation for teaching children in a public elementary school, and missed the opportunity to have students present their newly constructed understandings to each other.

Participating in Inquiry as a Learner and as a Future Teacher

After we had finished the inquiry work on light and shadows, I asked students to write about what each had learned as a learner of science and as a teacher of science. Jeri's writing and post-interview responses serve as an example of what most of the students learned, thought, and felt about the inquiry into light and shadows. Jeri wrote:

> I learned that a shadow gets bigger when it [the object] gets closer to a light source and smaller when it [the object] is further away. This occurs because the object is blocking less light when it is further away. First of all, I learned because I saw a picture of the shadow in the book, Dreams. This helped me to visualize the concept that we were talking about in class. Visualizations are very helpful to me as a learner. Secondly, I heard many students' ideas of how the shadow worked. This helped me to broaden and challenge some of the ideas I had about shadows. Next, I was given time to think about how shadows work and was able to draw my own conclusions. Finally, I was able to actually try these ideas out by playing with shadows. I think that the actual doing really reinforced the concepts for me.

And in her post-course interview, she reported:

> The shadow thing was something that, I thought, "Oh, I know shadows," but there was a lot about it that I hadn't really thought that in depthly about.... And so, that might be with science, I think that's something that's really helpful, because you might come up with questions or misconceptions that you might have had, or you can work with those with the kids as well.

When asked if her own work with light and shadows was scientific work, she said:

> A little--because that's the first step, investigating what you think about it, and trying to draw a hypothesis, um and then we talked about it, and had other people present their

> ideas, so we kind of did, in that way … so I guess it was in that way, but I would have liked to maybe go a little further.

In the post-unit interviews, we also asked students to evaluate several different scenarios which could be interpreted as scientific work or not. Jeri's criteria for evaluating the scenarios included whether scientists were talking to people about their work, questioning their own work, being open to other people's ideas, going out and testing things, collecting data, documenting the evidence, evaluating other scientists' work, and arguing about claims and evidence with colleagues. For me, her consideration of the roles of discourse, ideas and models, evidence, and the larger community of scientists were important indicators of the progress she had made in constructing a more sophisticated understanding of scientific work.

In the methods course, the knowledge and dispositions students had for doing inquiry with children were as important as students' own learning about scientific inquiry. In her final assignment for the course, Jeri wrote about her teaching and how she created opportunities for children to do authentic scientific work. She wrote:

> To work on conveying an authentic view of science, we developed our lessons so that they reflected much of what real scientists do when they are trying to figure things out. We started out our lesson with questions and encouraged students to develop a hypothesis with the information that they already had. Then we tested out the hypothesis, by actually experimenting with flashlights and objects, such as the doll. After experimenting we had a group discussion to talk about what the children had found and to pull all of their ideas and evidence together. We also had the students write and draw pictures to express their ideas, in their journals. Finally, we had the students prepare a final presentation, that they presented to other scientists (the children in the class), at the science meeting.

As Jeri's instructor, I was pleased with the progress she made over 15 weeks--in understanding and participating in scientific inquiry, in constructing her own understanding of light and shadows as a learner, and in planning and teaching inquiry lessons with children. However, I do not wish to convey the impression that three days' inquiry into light and shadows was sufficient for helping my students develop deep understanding of inquiry or of scientific models of light. I have written elsewhere (Smith and Anderson, 1999) about the limitations of even a 15 week course on the physics of light and shadows, in helping students construct sophisticated ways of scientific knowing, working and talking.

My elementary teacher education students and I have struggled mightily with strange new identities, roles, discourses, and feelings, in the strange new land of science. There were days when I felt triumphant about what they had tackled and accomplished and they felt discouraged. There were days when they questioned whether any of this was worthwhile for them as future teachers, and I was discouraged. And there were days when their smiles, writing, and excited talk convinced all of us that we were making good progress in understanding big ideas in science, as well as becoming more familiar and comfortable with the discourses and practices of scientific inquiry.

JUST A TEACHER

I (Abell) had not taught EDCI 605, Teaching Science in the Elementary School, a graduate level methods course, for several years. As I began to plan for a new semester, I saw an opportunity to rethink and restructure, and to apply my latest learning to this course. I wanted to craft the course in such a way to be responsive to the learners, who would be experienced teachers. I did not want the course to be a list of topics about science teaching to be covered, but a series of questions to explore. Thus I stated in the course syllabus:

> Inquiry into Science Teaching and Learning
>
> What is it that you need to learn in order to enhance your science teaching and improve your students' science learning? This is the question that should guide your participation in this course, rather than, "What does this professor want me to learn?" I would like for us to build a workshop atmosphere, where together we explore issues, try out strategies, and solve problems. We bring different kinds of expertise to the course; this expertise should be shared and developed further. My roles as the course instructor include helping to organize the learning community, facilitating communication, sharing resources, and assessing progress. Your roles as student/participant include preparing for course meetings, collecting classroom-based data, reflecting on practice, and synthesizing and communicating knowledge. (Course Syllabus)

My plan to start with the students' needs and expertise became even more relevant when I received my class list and met my students, who had a wide range of science teaching experience. Of the 13 participants, 6 were practicing elementary school teachers (including one first year teacher), 2 were high school teachers (one chemistry, one biology), 4 were graduate students who had had careers as elementary, middle, and/or high school science teachers and were currently teaching science or science methods courses for elementary education majors, and one was a special education faculty member interested in integrating issues about diverse learners into regular education courses. I was hoping that, through structuring my course from a problem based foundation, my students would become Learners in this course and in their own classrooms. Maybe we could achieve Freire's vision (1970), "The teacher is no longer the-one-who-teaches, but one who is himself taught in dialogue with the students, who in turn while being taught also teach" (p. 61).

Setting the Stage for Classroom Inquiry

Of course it is one thing to make a statement in a course syllabus about the kind of atmosphere you want to create, and quite another to put the plan into action. One of my strategies for having students direct the course was to attempt to infuse an inquiry stance throughout. For example, during the third week of the course, I asked students to apply our class discussion and reading about discourse in science teaching (e.g., Gallas, 1995) to their own classrooms:

> Try having a science talk with your students this week. After the talk, write about how it felt as the teacher to conduct a science talk. What did you hear that surprised you?

> What was difficult about conducting the talk? What will you do differently in your next talk? (Course Study Guide)

Students willingly took this assignment into their classrooms, and came back eager to share their detailed reports. Mike's report is one example:

> I asked my fifteen kids in the trail group to go dig in the ground and see what they could discover. About thirty minutes later I had to drag them out of the woods. We sat in a circle and shared our findings. The students placed all their evidence and artifacts in a pile in front of them and then proceeded to share their findings. What a blast!.... The students couldn't wait to share their ideas and co-compare their findings. We looked at bugs, amphibians, reptiles, and various stages of decomposition, soil textures and soil "ingredients." I was truly shocked by how much the students gathered and knew about soil.
>
> Next, I launched the question, "How is soil made?" It was like I opened the floodgates. Every child had a theory. Every child participated. Every child used collected artifacts and evidence to demonstrate their thinking. I applied my usual role of cheerleader and question asker/prober. The students readily advanced their thinking and built upon each other's ideas. The forty-minute discussion period flew by in a flash. When it was over, one girl asked if I could be the trail guide for the rest of camp!
>
> Thinking about this experience gave me the missing link to my science talks. In all three settings that I have previously tried science talks I had met with limited success. Trust and group dynamics played a partial role. What was missing was the common experience. The soil dig provided every child with the opportunity to be an "expert." Every child had collected evidence that they could use during their discussion. Every child was validated because we were in a neutral setting that none of them had previously experienced. (Mike, 11/20/99)

Mike's story of his outdoor science talk illustrates the power of his inquiry. Through making sense of this experience, he opened himself to learning from his students. He was able to compare what he learned with issues he had faced in previous science talks, and to create an implicit plan for next steps in his science teaching. By stepping out of the Knower role into the Learner role, Mike had an "aha!" experience of his own.

Another course assignment, interviewing children about their science ideas, required a more systematic inquiry. I asked the teachers to devise, administer, and analyze an interview with science learners on a particular science concept. In the written analysis, they were to:

- Discuss the scientifically appropriate responses to interview questions.
- Compare and contrast the learners' ideas with those of scientists. (Use learner quotes and drawings to support your ideas.)
- Compare and contrast the learner's ideas with those in the research.
- Discuss the implications to teaching and learning, i.e., how this information might affect the design and teaching of a unit. (Interview Assignment)

Paula, a former elementary and middle level teacher, now teaching a science methods course, decided to study children's and preservice teachers' ideas about the seasons and compare her findings with the extant literature on the topic. She found that both a fifth grader and a preservice teacher employed scientific vocabulary when discussing the seasons, but could not accurately apply their ideas about seasons to various geographical locations.

> The fifth grade subject contradicts his theory in order to explain why seasons are different in the Northern and Southern hemispheres, while the preservice teacher applies her ideas and completely contradicts the scientific conception of seasonal change... It is apparent that [the fifth grader] knows that the seasons are opposite for Northern and Southern hemispheres, yet he continues to hold onto the alternative conception that the seasonal change is based on the rotation of the Earth and affects the Eastern and Western hemispheres...When asked about the present season in South America compared to Indiana, [the preservice teacher] responds, "South America? Hmmmm. It's either, they're either going to be a little ahead of us or a little behind us depending on the tilt of the Earth, but they're on, South America is on the same side as we are, so I'd say it's either mid-summer or mid-fall...just because the tilt would be different." It is interesting to note that she now introduces the idea of the tilt of the Earth having something to do with seasons, but she manages to fit this information into the conception that she has created to explain seasons [that it is the Earth's rotation and revolution that makes the seasons]. (Paula, Interview Paper)

Paula synthesized her findings from the interview with ideas about science talks from earlier in the semester to produce instructional implications for teaching the seasons:

> It may be necessary to allow students to discuss prior knowledge and beliefs rather than infusing students with scientific knowledge. For example, a teacher might hold a discussion encouraging each student to discuss what they believe causes the seasons to change. In this discussion, the teacher has the opportunity to evaluate the various conceptions that exist in the classroom, as well as observe how students accommodate new information and ideas. (Paula, Interview Paper)

Upping the Ante for Inquiry

These assignments were attempts to facilitate teachers acting as inquirers in their own classrooms. Both Paula's and Mike's responses demonstrate their willingness to see themselves as Learners, not Knowers. However, up to this point in the course, the classroom inquiries were based on my questions, not theirs. The culminating project in the course asked teachers to raise their own questions and develop their own methods for inquiring into their classrooms. For the first time, I labeled the activity of inquiry, "research," and asked students to complete a "Teacher as Researcher" assignment:

> By its very nature, teaching is an inquiry process. Teachers constantly collect data about their teaching and student learning, analyze it, and use their findings to make teaching decisions. However, there is not often time in a teacher's day to be systematic about this inquiry process. The purpose of this assignment is for you to raise questions about science teaching and learning, systematically examine your practice, and use what you find to inform your practice. (Course Syllabus)

The reaction of the teachers to this assignment was very different from their reactions to the other classroom inquiry assignments. They seemed anxious, unsure, and uncomfortable. They asked many questions and needed constant reassurance about their projects. The initial discomfort emanated from their views of research, which held an exalted position in their thinking: "Research is something that researchers do. I am just a teacher." Furthermore, in their minds, research meant doing experiments, controlling variables, and running statistical analyses. "What I've been doing can't be research. I've just been listening to my students." Perhaps most importantly, research for them as yet had no purpose connected to their classrooms or their practice. "Research is published for researchers in research journals. I get my ideas from teacher magazines."

In order to scaffold their work and ease their discomfort, I established checkpoints during the project. The first checkpoint required teachers to commit to a classroom problem/research question. Remember, this was the first time in the course where the topic of the inquiry was in their hands. I felt a wave of panic seize the class. I wondered what caused the panic. Could they not define issues in their science teaching or their students' learning that would be interesting to investigate? Were they having trouble thinking about what and how much data to collect? How could I help them? I conversed with many of the teachers after class and via email to better understand the situation. I found out they were not blocked about what they wanted to learn. Instead, they were worried about being right and pleasing me, their teacher. "Is this what you want?" was a common question. One student, who had developed a fascinating question about how other teachers view science teaching remarked, "My concern is that this is too close to the interview assignment. What do you think?" Despite my best intentions to center the course around their needs and expertise, I would still evaluate their efforts. Could inquiry be so natural when it was required?

By the first checkpoint, the teachers had developed viable questions to guide their classroom inquiries. Some of them wanted to extend their inquiries about science talks from earlier in the semester. Their questions included:

- Who participates more in science discussions, boys or girls?
- What is my role as teacher in whole class and small group discussions?
- How does a science talk work in my homeroom compared to 2 afternoon classes?

Others wanted to continue to inquire into their students' science ideas, including their views of science and scientists. Another set of teachers became fascinated by notions of inquiry and wanted to know:

- How do my teaching colleagues view inquiry?
- How do preservice teachers view inquiry?

The next checkpoint in the project came when I asked the teachers to "develop a research plan." Their stilted views of educational research again constrained their thinking. They wondered if they needed to measure things and run statistical tests on the results. They wanted to know how big of a sample to select to get "good results." Somehow my goal for the assignment, that teachers become Learners in their classroom, had faded compared to their need to satisfy what they thought were my views of educational research. Once they trusted me that the methods we had used earlier in the semester—observation, interview, reflection—would work for this assignment, they were able to craft their research plans.

Shelly, an experienced teacher, wanted to know how other elementary teachers taught science and defined inquiry, since she was immersed in reflecting on these issues in her own teaching. After several worried conferences with me, she decided to approach the issue inductively:

> I plan on interviewing several teachers (possibly ranging from veteran—beginning—preservice teachers) about their best science teaching experience. Then based on their answers, and using the National Science Education Standards as my guide, I'll determine what types of science (inquiry?) is happening in different classrooms. (Shelly, Checkpoint #2)

Lisha, who was a substitute teacher at the time of the course, also expressed uncertainty about her research plan. She wanted to know, "What are students' ideas about (some science topic) and how are they affected by a unit of instruction?" Her written plan used the conditional "would," demonstrating the tentative nature of her thinking about inquiry:

> This project would be somewhat of a collaborative project with the [regular] classroom teacher. I would design the surveys, conduct the talks, observe students, etc. The regular teacher would do the actual instructing of the unit. (Lisha, checkpoint #2).

The teachers' discomfort was further illustrated by what happened in class when I asked them to help construct a scoring rubric for this assignment. My intent was to ease their concern by developing shared expectations for the project. What I discovered, however, was a new sort of discomfort, this time stemming from a lack of knowledge. What should a paper based on classroom research look like? What parts of the project should be described? Would my paper look like the articles in research journals I've read? And if so, can I write in that genre? I realized that I was asking students to do something so new to them that they could not envision the final product. Yet I also wondered how they would be able to guide students through an inquiry project if the process was so foreign to them. The final scoring rubric contained standards for outstanding, average, and inadequate performance as related to the purpose of the research, data collection methods, data interpretations, and implications for practice.

Teachers as Researchers

With the help of this scoring tool and the encouragement of their instructor and classmates, the teachers did complete their inquiry projects. During a dinner party at my house, they shared their products. Greg, a fifth grade teacher, extended his inquiry into science talks by conducting two talks with three different fifth grade classes. He wondered how the "level of community" that had been developed in each group would relate to the success of a science talk. After analyzing videotapes of six science talks, Greg found some surprising results. He had expected that a "well-behaved" class with a higher level of community would have more fruitful talks than a "not well-behaved" class.

> Class 2 that is well-behaved, higher academically, and appreciates order was not as willing to share ideas that may be wrong.... Class 3 that is lower academically and appears to have very little community soared with this type of activity. The students

> who are usually the focal points of disruptions were entirely involved and shared their knowledge that they may not have been able to do in some other form. It seemed that the science talks promoted community for this class. (Greg, Teacher as Research Paper)

Greg learned that, rather than being a prerequisite for successful science talks, establishing "classroom community" might actually be a learning outcome. Greg also learned some things about his students and himself through this inquiry.

> Students definitely love science talks and are excited about having them. When questions arise in class, students now suggest having a science talk to find out what everyone believes. The fact that they are in charge and get to talk, debate, ask questions, and argue ignites their curiosity and participation.
>
> I was able to learn more about my students' thinking with this method more than any other I've tried. Little did I know the level of understanding of my students on each topic. The debates that occurred in each of the talks allowed me to use them throughout the unit. Since this form of pre-assessment involved the entire class together, it can then be built upon involving the whole class as inquiry continues. (Greg, Teacher as Research Paper)

William studied his teaching of preservice teachers in a course called, "Exploring Teaching as a Career," a discussion-based course. He wanted to understand his role in discussions while considering how to improve discussions in the course. He recorded five hours of classroom discussion and looked for patterns in the interactions. His conclusions demonstrate his learning:

> I seem to be willing to let the students guide the direction of the large group discussions within reason. By rewording responses and introducing student experiences, I also provide the students with appropriate background and reduce misunderstanding. Directly related to these ideas are the control issues. Most of the students seem comfortable taking control of the discussions when allowed and tend to talk to the other students as well as to me. The environment encourages students to speak, and the students themselves tend to deal with their peers that go off to far on tangents or try to monopolize the discussion.
>
> Since 25% of the class did not participate in the recorded discussions, at least based on the data sources I have, I need to improve the classroom environment. Or in terms presented by the literature, I need to work on making all of the students feel more relaxed. ... I need to work on becoming part of the small groups. Hopefully by becoming more accepted by the students in the small groups, I can get those that participate there to do so in the large group. I can also monitor the pace and direction of the small groups more closely.
>
> The ... data indicate that not all students are comfortable talking to the entire class and still look to me for permission to speak and for acceptance of their responses. I need to find ways to get all of the students to participate in discussions without taking control of the discussion. ...Too many of the students are not participating on a regular basis. (William, Teacher as Researcher Paper)

During the dinner party, as students shared their inquiries, our sense of community grew. Our interactions throughout the semester had coalesced, and our trust had grown. By the end of the evening, it was clear that many of us had come to accept that teaching and inquiry might be different sides of the same coin. Some teachers even expressed an interest in continuing their inquiries. I believed that my goals for the course were starting to be realized. Unfortunately, our official time as a learning community had ended with the close of the semester.

DECONSTRUCTING THE CASES

Each of these classroom cases demonstrates students at work, involved in the essential features of inquiry as outlined by the NRC (2000): they were engaged in raising testable questions and in planning and conducting investigations; they used appropriate tools and technologies to make observations and gather data; they thought critically about and gave priority to evidence in order to formulate viable explanations; and they communicated, justified, and evaluated their explanations. While certainly not sophisticated practitioners of science, these students were on the pathway toward a more inquiry-oriented stance in their scientific work, including the appropriation of scientific discourses and practices.

The three cases also demonstrate different types of and purposes for inquiry (see also Minstrell and van Zee, 2000). In the case of the physics course, Volkmann used his ideas about scientific inquiry to model classroom inquiry. In the preservice methods course, Smith used classroom inquiry as both a model of scientific discourses and practices, and as a way to examine and question beliefs about science learning and teaching. And in the graduate course, Abell used inquiry as a means to develop an inquiry orientation to science teaching. Thus inquiry in teacher education, although adopting the same features as classroom inquiry, can serve different purposes.

Our cases also illustrate that the dilemmas of science teacher educators are similar to those faced by all science teachers in teaching science as inquiry (e.g., Hammer, 2000; Krajcik, Blumenfeld, Marx, and Soloway, 2000; McDermott and DeWater, 2000). In the physics course, Volkmann struggled with when and what to tell students. He worried about accurately representing scientific inquiry, but realized that other learning goals factored into his choice of teaching strategies, goals that might at times compromise his nature of science beliefs. He recognized that some kinds of telling (e.g., term introduction, equipment usage) are substantively different from other kinds of telling (e.g., scientific explanations), and demanded different pathways for him and his students. He also found that the timing of the telling is critical, and dependent on each learner's readiness to make sense of the telling. Such dilemmas of telling are typical of inquiry-based teaching.

As a teacher educator in a science methods course, Smith faced another dilemma of inquiry, the dilemma of who holds the authority, particularly in terms of prior knowledge, prior experience as a science learner, and gender and race. Instead of being concerned about her own telling in the course, she made decisions about how to handle some of her male students' roles as tellers. She struggled with ways to provoke more uncertainty in the knowers and tellers, more reliance on evidence, and more confidence in personal explanations in those initially willing to accept the authority of the tellers (see also Minstrell, 2000). Based on student writing and interviews after several inquiry-oriented lessons, Smith found evidence that students had constructed a working model of light diverging and being blocked. She also found that even a small immersion in inquiry provided students with an image of what scientific work could be like and what might be possible with children. Moreover, Smith found that the inquiry provided all students (not only the "supersmart" white men) with personal evidence that they could pursue an inquiry in science that would provide them with deeper understanding and satisfaction as learners. Her decision to allocate time for student inquiry, when there were so many issues to be addressed in the methods course, turned out to have positive consequences for students' views of inquiry in elementary classrooms.

In the graduate course, Abell faced a dilemma common to all teachers of inquiry, the dilemma of supporting student inquiry without inquiring for them. Abell found that her students were comfortable with inquiry when the questions were laid out and the procedures relatively clear. However, their comfort zone diminished as the inquiry required more student ownership. It became Abell's job to understand why the inquiry was difficult for the students and create ways to support their progress. A second dilemma, related to the dilemma of how to support students, arose when evaluation came into the picture. Could inquiry really be open and directed by students when the teacher mandated the assignments and gave the final grade? Abell involved the students in developing an evaluation tool for the inquiry, yet hers was still the final word when it came to grades.

Our sociocultural perspective on teacher learning and scientific inquiry provided a lens for examining the roles of teacher educators and their students. All three cases demonstrate the various roles a teacher educator assumes in inquiry: one who pushes students to develop evidence-based arguments; one who tells at appropriate times; one who scaffolds instruction to facilitate student ownership of inquiry; one who assesses and evaluates student learning. These roles create tensions for us as we wonder when and what to tell and who does the telling, how to help our students question the tellers, and how to assess students yet maintain mutuality in our classes. These roles create tensions for our students who become frustrated, uncomfortable, and yet more confident in their knowledge across the space of an inquiry. Together with our students, we negotiate our roles throughout a course, hoping that 15 weeks is enough time to develop the trust and communication needed for success. Because the shift from Knowers to Learners, and the integration of those roles, is difficult for all of us raised in authority-driven science learning environments, this negotiation process can be painful. However, we believe that for us and for our students, it is worth it.

CONCLUSION: IMPLICATIONS FOR SCIENCE TEACHER EDUCATION

After experiencing and reflecting on our teaching of teachers, we are left with questions that are similar to those with which we entered: How can we support teachers in their learning of science through inquiry? and How can we support them in developing a stance and a practice of science teaching as inquiry? Our cases illustrate that these questions are difficult to separate in practice. Their answers have the potential to be mutually reinforcing. Engaging teachers in inquiry, whether into science or into science teaching and learning, provides opportunities for them to explore a culture of personal sense-making within a community of learners, and to establish the language and norms for what is accepted and validated as knowledge.

Because we are teacher educators, our inquiry dilemmas, although similar to those faced by all science teachers, create some unique challenges. Engaging students of teaching in inquiry is only a first step. Making our inquiry goals and struggles explicit to our students is critical for their development of an inquiry stance. For example, Volkmann's case of physics instruction informs us that we should consider how we help students of teaching recognize and address dilemmas of telling in their own science teaching. Smith's case implies that teacher educators should articulate and highlight questions about authority, evidence, and validated knowledge in inquiry. Finally, Abell's case demonstrates the need to be explicit about the levels of support we provide for inquiry so that teachers can build a repertoire of support strategies for their own inquiry-based instruction.

The cases also demand that we better understand students of teaching, and how they come to make sense of science and science education. How do our views of teaching and learning assist and constrain our ability to support student learning through inquiry? How do future teachers experience the frustrations and excitement of sense making in science and in the process recognize that they are capable of learning science? How do teacher views of research (that it must be experimental, statistical, and far removed from their lives) limit their thinking about inquiry? How does teacher learning of science in an inquiry setting influence the development of their pedagogical beliefs? How do teacher identities as Learners, not merely Knowers, develop throughout a teaching career? The cases in this chapter only begin to uncover some of these questions. The answers demand more research.

We view our work as examples of inquiry into our teaching practice and our students' learning. Like other teacher researchers (Ball, 1993; Cochran-Smith and Lytle, 1993; Gallas, 1995; Roberts, 2000), we hope to contribute to a scholarship of teaching teachers that generates knowledge for the community to consider, critique, and build upon (e.g., Hatch and Austin, n.d.; Hutchings, 2000; Hutchings and Shulman, 1999; Shulman, 1993). For now we are content that our inquiry into practice has moved us further along our own Learning paths and raised new questions for continuing inquiry.

*For another argument on learning over knowing, see Frank Smith, 2001.

REFERENCES

Abell, S. K. (1996, April). *Building a pedagogical content knowledge base for elementary science teacher education: Reflection as an orientation to teaching teachers of science.* Paper presented at the Annual Meeting of the National Association for Research in Science Teaching, St. Louis, MO.

Abell, S., Anderson, G., and Chezem, J. (2000). Science as argument and explanation: Inquiring into concepts of sound in third grade. In J. Minstrell and E. van Zee (Eds.), *Inquiring into inquiry learning and teaching in science* (pp. 65-79). Washington, DC: American Association for the Advancement of Science.

Abell, S. K., and Bryan, L. A. (1997). Reconceptualizing the elementary science methods course using a reflection orientation. *Journal of Science Teacher Education*, 8(3), 153-166.

American Association for the Advancement of Science. (1993). *Benchmarks for science literacy*. New York: Oxford University Press.

American Association of Physics Teachers. (1996). *Powerful ideas in the physical sciences*. College Park, MD: Author.

Anderson, C., Holland, J.D., and Palincsar, A. (1994). Canonical and sociocultural approaches to research and reform in science education: The story of Juan and his group. *The Elementary School Journal, 97*, 357-381.

Anderson, C. W., and Smith, E. L. (1987). Teaching science. In V. Richardson-Koehler (Ed.), *Educators' handbook: A research perspective* (p. 84-111). New York: Longman.

Ball, D. (1993). With an eye on the mathematical horizon: Dilemmas of teaching elementary school mathematics. *The Elementary School Journal, 93*, 373-397.

Bazerman, C. (1988). *Shaping written knowledge: The genre and activity of the experimental article in science*. Madison, WI: The University of Wisconsin Press.

Bentley, M., and Fleury, S. (2000). Of starting points and destinations: Teacher education and the nature of science. In W. McComas (Ed.), *The nature of science in science education: Rationales and strategies* (277-293). Boston: Kluwer Academic Publishers.

Bybee, R. (1997). *Achieving science literacy*. Portsmouth, NH: Heinemann.

Cobb, P. (1994). Where is the mind? Constructivist and sociocultural perspectives on mathematical development. *Educational Researcher, 23*(7), 13-20.

Cobb, P., and Bauersfeld, H. (1995). *The emergence of mathematical meaning: Interaction in classroom cultures*. Hillsdale, NJ: Lawrence Erlbaum Associates, Publishers.

Cochran-Smith, M., and Lytle, S. (1993). *Inside/outside: Teacher research and knowledge*. New York: Teachers College Press.

Cochran-Smith, M., and Lytle, S. L. (1999). The teacher research movement: A decade later. *Educational Researcher, 28* (7), 15-25.

Collins, A., Brown, J.S., and Newman, S. E. (1989). Cognitive apprenticeship: Teaching the craft of reading, writing and mathematics. In L.B. Resnick (Ed.), *Knowing and learning: Essays in honor of Robert Glaser* (pp. 453-494). Hillsdale, NJ: Erlbaum.

Crawford, B., Krajcik, J., and Marx, R. (1999). Elements of a community of learners in a middle school science classroom. *Journal of Research in Science Teaching*, 37(9), 916-937.

Driver, R., Guesne, E., and Tiberghien, A. (Eds.). (1985). *Children's ideas in science*. London: Milton Keynes.

Fradd, S., and Lee, O. (1999). Teachers' roles in promoting science inquiry with students from diverse language backgrounds. *Educational Researcher, 28* (6), 14-20.

Freire, P. (1970). *Pedagogy of the oppressed*. New York: Continuum Publishing Company.

Gallas, K. (1995). *Talking their way into science: Hearing children's questions and theories, responding with curriculum*. New York: Teachers College Press.

Hammer, D. (2000). Teacher inquiry. In J. Minstrell and E. van Zee (Eds.), *Inquiring into inquiry learning and teaching in science* (pp. 184-215). Washington, DC: American Association for the Advancement of Science.

Hatch, T., and Austin, K. (n.d.). *Toward the scholarship of teaching*. Unpublished manuscript. Menlo Park, CA: Carnegie Foundation for the Advancement of Teaching.

Heath, S. B. (1994). *Ways with words*. New York: Cambridge University Press.

Hogan, K., and Pressley, M. (1997). Scaffolding scientific competencies within classroom communities of inquiry. In K. Hogan and M. Pressley (Eds.), *Scaffolding student learning: Instructional approaches and issues* (74-107). Cambridge, MA: Brookline Books.

Hutchings, P. (2000). *Opening lines: Approaches to the scholarship of teaching and learning*. Menlo Park: Carnegie Foundation for the Advancement of Teaching.

Hutchings, P., and Shulman, L. (1999). The scholarship of teaching: New elaborations, new developments. *Change, 31*(5), 11-15.

Krajcik, J., Blumenfeld, P., Marx, R., and Soloway, E. (2000). Instructional, curricular, and technological supports for inquiry in science classrooms. In J. Minstrell and E. van Zee (Eds.), *Inquiring into inquiry learning and teaching in science* (pp. 283-315). Washington, DC: American Association for the Advancement of Science.

Latour, B., and Woolgar, S. (1986). *Laboratory life: The construction of scientific facts*. Princeton, NJ: Princeton University Press.

Lave, J., and Wenger, E. (1991). *Situated learning: Legitimate peripheral participation* . Cambridge, UK: Cambridge University Press.

Lehrer, R., Carpenter, S., Schauble, L., and Putz, A. (2000). Designing classrooms that support inquiry. In J. Minstrell and E. van Zee (Eds.), *Inquiring into inquiry learning and teaching in science* (pp. 80-138). Washington, DC: American Association for the Advancement of Science.

Lemke, J. (1990). *Talking science: Language, learning and values*. Norwood, NJ: Ablex Publishing Corporation.

Lortie, D. C. (1975). *Schoolteacher: A sociological study*. Chicago: University of Chicago Press.

Magnusson, S., Krajcik, J., and Borko, H. (1999). Nature, sources, and development of pedagogical content knowledge for science teaching. In J. Gess-Newsome and N. G. Lederman (Eds.), *Examining pedagogical content knowledge: The_construct and its implications for science education* (pp. 95-132). Dordrecht, The Netherlands: Kluwer.

McDermott, L. C., and DeWater, L. S. (2000). The need for special science courses for teachers: Two perspectives. In J. Minstrell and E. van Zee (Eds.), *Inquiring into inquiry learning and teaching in science* (pp. 241-257). Washington, DC: American Association for the Advancement of Science.

Minstrell, J. (2000). Implications for teaching and learning inquiry: A summary. In J. Minstrell and E. van Zee (Eds.), *Inquiring into inquiry learning and teaching in science* (pp. 471-496). Washington, DC: American Association for the Advancement of Science.

Minstrell, J., and van Zee, E. (2000). Introduction. In *Inquiring into inquiry learning and teaching in science* (xi-xx). Washington, DC: American Association for the Advancement of Science.

National Research Council. (1996). *National science education standards*. Washington, DC: National Academy Press.

National Research Council (2000). *Inquiry and the national science education standards*. Washington, DC: National Academy Press.

Putnam, R., and Borko, H. (2000). What do new views of knowledge and thinking have to say about research on teacher learning? *Educational Researcher, 29*(1), 4-15.

Richmond, G., and Kurth, L. (1999). Moving from outside to inside: High school students' use of apprenticeships as vehicles for entering the culture and practice of science. *Journal of Research in Science Teaching, 36*, 677-697.

Roberts, D. (2000). Learning to teach science through inquiry: A new teacher's story. In J. Minstrell and E. van Zee (Eds.), *Inquiring into inquiry learning and teaching in science* (pp. 120-129). Washington, DC: American Association for the Advancement of Science.

Rutherford, J., and Ahlgren, A. (1993). *Science for all Americans*. New York: Oxford University Press.

Shulman, L. (1993). Teaching as community property. *Change, 25*(6), 6-7.

Smith, D. (2001). Content and pedagogical content knowledge for elementary science teacher educators: Knowing our students. *Journal of Science Teacher Education, 11*(1), 27-46.

Smith, D., and Anderson, C. (1999). Appropriating science discourses and practices with future elementary teachers. *Journal of Research in Science Teaching, 36*, 755-776.

Smith, F. (2001). Just a matter of time. *Phi Delta Kappan, 82*, 572-576.

Tannen, D. (1996). *Gender and discourse*. New York: Oxford University Press.

Traweek, S. (1988). *Beamtimes and lifetimes: The world of high energy physicists*. Cambridge, MA: Harvard University Press.

van Zee, E. (2000). Ways of fostering teachers' inquiries into science learning and teaching. In J. Minstrell and E. van Zee (Eds.), *Inquiring into inquiry learning and teaching in science* (pp. 100-119). Washington, DC: American Association for the Advancement of Science.

Warren, B., Rosebery, A., and Conant, F. (1989). *Cheche Konen: Science and literacy in language minority classrooms*. Newton, MA: Bolt, Beranek, & Newman.

Wenger, E. (1998). *Communities of practice*. New York: Cambridge University Press.

CHAPTER 10

WILLIAM G. HOLLIDAY

A BALANCED APPROACH TO SCIENCE INQUIRY TEACHING

Good science inquiry teaching in the classroom often means many things to practitioners including less teacher intervention, less expository teaching, less explicit instruction, less direct teaching, or fewer teacher explanations. Researchers and other scholars commenting to teachers don't mean to leave this false, unbalanced impression: Less explicit teaching is better inquiry teaching, by definition (Harris, & Graham, 2000). But reading portions of the National Research Council's (1996a) *National Science Education Standards* and numerous science method textbooks probably suggest to too many teachers that science inquiry teaching normally occurs:

- When teachers say very little about the meanings of concepts,
- When teachers use implicit (indirect) rather than explicit (direct) approaches,
- When students discover a large proportion of school science on their own; and
- When students figure out for themselves how to grapple with problems and construct knowledge while engaging in, for example, project-based activities.

LIMITATIONS IN OVERDOING IMPLICIT AND/OR DISCOVERY TEACHING

Teachers occasionally need to intervene and provide explicit teaching during science inquiry-oriented lessons for reasons of instructional efficiency, reduced chances of errant learning, and increased productive interactions between teachers and students (Holliday, 2001a). Overdoing implicit and/or discovery teaching deprives students of explicit explanations, teacher modeling and teacher scaffolding, which are all teaching strategies supported by established research (Pressley, and McCormick, 1995). Such explicit instruction, in turn, efficiently provides students with needed background knowledge and other information on how, why, and when to use learning strategies leading to learner independence (Zimmerman, 1998), and productive dispositions toward achievement motivation (Alderman, 1999).

We need to present viewpoints to teachers about implicit and discovery teaching not presented by most authors commenting on science inquiry teaching, but views that are characteristic of scholarly treatments such as influential works produced by Lee Shulman about forty years ago and Michael Pressley in the 1990s and their respective colleagues. Teachers and others may profit from these balanced

L.B. Flick and N.G. Lederman (eds.) Scientific Inquiry and Nature of Science, 201-217.

discussions when planning how to teach science content and associated learning strategies. The following remarks may constitute a beginning in this regard.

Let's define inquiry teaching for this part of our discussion as having some common characteristics with discovery teaching and including some implicit teaching approaches where teachers provide limited amounts of instructional support in their guidance of students learning science knowledge. Science inquiry teaching obviously includes many forms or approaches to teaching but rarely encourages explicit, direct explanations instructionally articulated by teachers. Discovery and inquiry learning often emphasize the use of implicit instruction, where students are encouraged to acquire learning strategies and other knowledge by exploring ideas and physical phenomena on their own, with little teacher intervention.

Terms "discovery teaching" and "discovery learning" perhaps were more often used decades ago than the terms "inquiry teaching" and "inquiry learning." Today, the term "discovery" is seldom used by science educators writing for teachers but still often used by learning researchers publishing in curricular areas other than science education (see Tuovinen, & Sweller, 1999). One problem is that these terms seem to change in popularity, making a comparative analysis of these terms increasingly difficult over time. Research linked to teacher education and related analyses suggests that forms of discovery teaching, according to comparative reviews (Pressley & McCormick, 1995; Shulman & Keislar, 1966; Tuovinen, & Sweller, 1999), are limited in their effectiveness when critically assessed by many learning researchers.

Obviously, discovery and inquiry teaching are not identical in meaning and use, but today's science educators seemingly decline from differentiating between these two approaches perhaps because their meanings change over time as a result of ever-changing conventions. Thus teaching of science is much more complicated than perhaps is suggested by simple recommendations from some science educators who suggest high doses of inquiry teaching where teachers rely heavily on forms of discovery teaching which implicitly suggest minimal teacher interventions (Holliday, 2001a). Placing a modifier such as "guided" in front terms like "discovery," "inquiry" or "teaching" fails to ameliorate a misconception held by some science teachers that good inquiry science teachers should try *not* to intervene whenever possible or practical. Again, the research simply fails to support this simple notion of minimal teacher intervention.

In the next three sections of this chapter, three problems for students identified by some learning researchers are described as these issues may relate to excessive discovery or implicit teaching often associated with science inquiry teaching.

Errant Learning and Misconceptions

First, discovery and/or implicit teaching is considered an approach producing uncertain outcomes including learning that can be errant (Harris, & Graham, 1994). When students inquire and learn a wrong or inefficient way of solving a problem, for example, it may be difficult to correct the student's misconception within a reasonable time. A study (van Lehn, 1990, as cited by Harris & Graham 1994, and others) and others reported more than a decade ago substantiated the suspicion that students learning how to subtract can discover at least 100 ways of doing it incorrectly, unnecessarily borrowing from columns, subtracting from just a few columns and writing answers on paper by leaving out the digit zero (e.g., completing the exercise by writing the number 35 rather than the correct number, 305). Some students, even after comprehending how to solve problems, failed to understand completely how to make such calculations with some students unpredictably solving problems using a mixture of reasonable and unreasonable procedures. Students, of course, need to practice on their own and need to have opportunities to make mistakes. But allowing students to make mistakes repeatedly fails to make much common sense.

Struggling Students may Struggle More with Discovery Teaching

Second, discovery teaching also may be an ineffective approach when teaching students especially those students with weak academic backgrounds. Placing a heavy learning load on students unfamiliar with a school subject like science may result in these less-fortunate students acquiring very little new knowledge after a discovery teaching experience compared to equivalent students with reasonable background knowledge (Tuovinen & Sweller, 1999). Indeed, experimental evidence suggests that weaker students perhaps are differentially more susceptible to being distracted by irrelevant information mixed with relevant information when problems are described to them. Such students may need added teacher-directed help or scaffolding to support their learning (Tuovinen & Sweller, 1999) of science knowledge (i.e., content, information, conceptual and strategic knowledge). Discovery learning, like any other teaching approach, can be taken to an extreme, so teachers need to watch each student's progress closely to reduce the chance of less-fortunate students "permanently" learning incorrect ways of solving problems or adding to their deficiency of background knowledge (Pressley, Harris, & Marks, 1992).

Time Allocation

Third, discovery teaching requires an inordinate amount of time with little solid evidence of comparatively improved student construction of meaning and understanding. Discovery teaching may represent a superior approach, but the reported research evidence fails to support this assumption in a material, competent fashions. The notion that implicit teaching is a more successful approach to increasing ordinary achievement scores, when controlling for instructional time, is an idea without reasonable empirical support. Nevertheless, implicit teaching offers other important benefits leading to scientific literacy that may be more difficult to measure but sometimes of greater importance. Such approaches in science generally require the elimination of considerable amount of large chucks of curricular content (Pressley, & McCormick, 1995). No one argues that students need to be exposed to unmanageable amounts of science content, as is often the case. But, realistically, teachers are going to consider the administrative and collegial pressures placed on them to produce students who will likely score high on local, state and other assessment instruments, and be somewhat familiar with large amounts of background knowledge in preparation for subsequent science courses. Other goals in science teaching such as those published in national (for example, NRC, 1996a) and state curricular guidelines, of course, exist and warrant serious consideration. So, extensively engaging students in discovery learning experiences may not be the best use of valuable class time (Harris, & Graham, 1994). Simply explaining some scientific notions to students and having them read or listen to explanations, especially students beyond early elementary school, may be more rational than spending extraordinarily large amounts of time sometimes hoping and praying that students understand and construct the many complexities of science (Pressley, & McCormick, 1995; Shulman, & Keislar, 1966).

Balancing Explicit with Implicit Teaching

No one, including teachers, social science researchers and practicing physical scientists can, of course, provide fixed, rigid, cookie-cutter, one-size-fits-all, precise, rule-like algorithms for when teachers should favor more explicit or more implicit science teaching approaches. Like other teaching and learning strategies, what works best is a mixture of back-and-forth, non-linear combinations of implicit and explicit teaching depending on teachers' professional judgments. In addition, teachers also need to make clear explanations about the science content being learned and explicitly coach students' studying and learning of new material through teacher modeling of studying and learning strategies.

This chapter explores one complex and somewhat artificial continuum with regards to science inquiry teaching, recommending a difficult-to-define mixture or a balanced use of implicit and explicit science teaching. Teachers who believe that science inquiry teaching means "the less the teacher intervenes the better the teaching" are incorrect, and are overdosing their students on an off-balanced brand of inquiry teaching, an approach that lacks reasonable empirical and theoretical support. Balanced in this instructional sense does not mean equal parts of both, or that one approach is better compared to other approaches.

No one familiar with contemporary learning research believes that always "telling students about science" (using explicit teaching only) at any level of schooling is a good way to teach science and other school subjects. Examples of teachers using four regrettable applications of explicit teaching include: 1) talking in front of bored students about unfamiliar, seemingly irrelevant ideas, 2) incessantly engaging students in drill-and-kill activities, 3) providing scaffolding support to students by just asking large numbers of questions and barking orders, and 4) using highly prescriptive hands-on activities integrated with other components of science class. But, being explicit and directly explaining to students selected concepts and problem-solving strategies is a reasonable teaching approach if students have ample opportunities to discover through implicit or indirect instruction some other concepts and strategies. This prudent or balanced, integrated approach is in contrast to allowing students to explore and discover almost everything on their own, using excessive forms of implicit instruction. Material and relevant contradictory data arguing against reasonable uses of explicit teaching has yet to be published in highly reputable research outlets (Duffy, 2002).

The Good and Bad News, and the Problem

The good news is that teachers cannot overdose students on science inquiry teaching, providing they teach using a balanced teaching approach where there are opportunities for students to learn on their own through implicit teaching strategies, mixed with opportunities for students to receive explicit teaching.

The bad news is that hundreds of professional documents commenting about science teaching approaches and some read by science teachers may unknowingly suggest to them that good science teachers strongly emphasize laissez-faire, minimal interventionist instruction, which automatically results in increased students' inquiry-based abilities and further develops their inquiry habits of mind. These documents are quite helpful but are seldom complete in one particular sense. They lack an integrated balanced approach, and instead, suggest or strongly emphasize implicit (i.e., indirect, incidental, immersed, informal, natural, exploratory) or low-teacher intervention approaches and pay little attention to explicit (i.e., direct, formal, informational, expository) or high-teacher interventions. Some documents call for guided inquiry approaches, but too often without reasonably worded, operational descriptions with accompanying detailed examples.

These documents mentioning science inquiry teaching, such as commentaries appearing in curriculum guidelines, pre-service teaching methods books, and professional readings discussing science inquiry, seldom even mention the values of explicit teaching and the liabilities of "too much" implicit teaching where students are encouraged to learn on their own, "naturally," discovering for themselves with very limited assistance from teachers.

The problem is not the presentation of invalid advice. Instead, the problem is that such documents referencing science inquiry teaching do not go far enough to present the larger picture, a more prudent or balanced teaching approach in this regard. That is, teachers at times need to be explicit, direct and specific when explaining certain science content and strategies useful in understanding how to tackle problems. Again, an integrated balanced approach in this chapter does not imply that teachers should use equivalent amounts of implicit and explicit approaches to science inquiry teaching. First, it is impossible to bifurcate teaching approaches in reality even though we often do so for discussion purposes. Second, there is no convincing research that suggests what is a proper proportion or mixture of teaching approaches under specific classroom conditions. Instead, teachers engaging in science inquiry teaching should seriously consider both explicit and implicit approaches along with other teaching approaches when planning and implementing science lessons.

Science educators authoring documents recommending approaches to teaching science using an inquiry approach generally call for teachers to help students strongly emphasizing implicit teaching approaches. Many of these recommendations make research-based and common sense. Examples of such documents and, more particularly the cited chapters within these widely read documents, read by some teachers include:

Benchmarks for Science Literacy. (American Association for the Advancement of Science, 1993).

- *Inquiring into Inquiry Learning and Teaching in Science*. (Abell, Anderson, & Chezem, 2000; Bybee, 2000).
- *Inquiry and the national science education standards*. (National Research Council, 2000).
- *Invitations to Science Inquiry*. (Liem, 1987).
- *National Science Education Standards*. (National Research Council, 1996b).
- *Science for All Americans*. (Rutherford, & Ahlgren, 1989).
- *Teaching Children science: A Project-based Approach*. (Krajcik, Czerniak, & Berger, 1999a).
- *Teaching Science as Inquiry*. (Carin, & Bass, 1997).
- *Teaching Science to Children: An Inquiry Approach*. (Friedl, & Koontz, 2001).
- *Teaching Secondary School Science*. (Trowbridge, Bybee, & Powell, 2000).

None of these cited chapters contained within these widely distributed works directly attack the virtues of explicit teaching but none spend much page space discussing the advantages of science teachers being explicit or mentioning disadvantages of too often engaging students in implicit teaching experiences. Most of these authors surely would claim that their work was not meant to cover the entire teaching-science-by-inquiry waterfront. Such claims would be fair but more of a balanced emphasis, perhaps, would leave teachers with a more valid view of what research suggests about effective teaching. Maybe some authors, understandably, were fearful that a developed discussion about explicit teaching would detract from their important and meritorious theme of encouraging teachers to encourage their students to work independently and in groups by inquiry, tackling and understanding problems in science with reduced teacher intervention. Notably, some authors (e.g., Krajcik, Czerniak, & Berger, 1999b; Metz 2000) did make this point clear by mentioning to science teachers the importance of explicit teaching by describing the value of explaining, modeling and scaffolding and related explicit teaching strategies. So advising teachers to use a balanced approach appears in some documents discussing science inquiry teaching.

Teachers need to know this whole story about providing integrated mixtures or balanced blends of good implicit and explicit teaching. Some learning researchers in fields outside of science education fortunately have provided general descriptions of effective explicit and implicit teaching backed by solid, theory-based empirical studies, results of which are discussed later in this chapter.

Teachers need to see the need for a balance in instruction between explicit and implicit approaches. Interestingly, this balanced or bigger picture in science education is reminiscent and somewhat analogous to serious discussion in reading education, where some phonics/direct comprehension-strategy-instruction enthusiasts have over stressed explicit teaching approaches while some whole-language enthusiasts have over stressed implicit approaches. (Pressley, 2002)

In the remaining portion of this chapter, I provide illustrations linked to balancing implicit-explicit teaching, a "balancing" analogy to reading education, and some concluding remarks. Sometimes teachers need to emphasize explicit instruction, as illustrated in the upcoming section, which discusses individual student studying and learning. In contrast, teachers sometimes need to de-emphasize explicit instruction by reducing teacher intervention, as described in the subsequent section on peer discussions. This discussion is followed by a brief description illustrating how reading researchers and teachers handle an analogous problem in their field, whole language (often seen as emphasizing implicit teaching) verses phonics/comprehension strategy instruction (often seen as emphasizing explicit teaching).

EMPHASIZING EXPLICIT TEACHING: MONITORING, FOSTERING, MOTIVATING AND INFORMING

Helping Students to Monitor and Foster their Learning

Explicit intervention by some teaching agents is typically more profitable than just leaving students basically alone to monitor and foster their own thinking (Harris, & Graham, 1994; Holliday, 2001a). Unfortunately, students of all ages, too often, are left to their own devices under the guise of "student-center" approaches with too little direct teacher intervention (Pressley, Harris, & Marks, 1992).

Some science educators including classroom teachers seem to assume that students learned how to monitor and foster their own learning somewhere in elementary school, that students are automatically learning these strategies efficiently as they proceed through school, or that students can learn these strategies with little need for direct teacher assistance (Baker, 1991; Pressley, & McCormick, 1995). In contrast, too much direct or explicit instruction is unwarranted because it reduces students' chances of becoming independent flexible learners (Zimmerman, 1998). But, without question, implicit teaching alone is not enough to equip students to develop and implement needed thinking strategies in today's demanding, ever-changing world (Graham, & Harris, 2000). Such student-thinking strategies include being aware of what the learner knows, what is not known, and what needs to be known in order to comprehend complex instructional materials. Teachers need to explain what learners should do when the learners' understanding in science class breaks down. Teachers specifically need to explain and model repair and bolstering strategies such as convincing students to put energy into focusing attention on important science information, rereading difficult texts and visuals, and asking teachers for clarification (Baker, 1991). For example, if understanding fails, students need to "fix" the problem by engaging in strategies such as rereading a confusing visual or seeking assistance from their teacher.

Less-experienced students are too often lost during their attempts to think and fail to monitor and foster their thinking in science when, for instance:

- Paying serious attention and comprehending safety instructions;
- Transferring or applying learned knowledge to a slightly different context;
- Monitoring studying habits;
- Coping with unfamiliar complex learning assignments;
- Working with others on problems;
- Completing projects containing many components;
- Planning before beginning to embarking on a task;
- Listening to each other's explanations;
- Refining or revising drafts or first attempts at solving problems;
- Retrieving and searching for information needed to tackle problems; and
- Reading from documents and responding in writing to questions.

Reasons why students need such direct teacher help in learning have been described in an excellent article authored by Ruth Garner (1990), and supported by subsequent research work dealing with students' self-regulation during learning episodes (Zimmerman, 1998; Zimmerman, & Schunk, 2001). Teachers need to teach students how they can improve their learning skills about how they learn, and how to become self-reliant learners. This monitoring and controlling of students' own learning, that is, their "knowledge and cognition and regulation of cognition," is called metacognition by cognitive psychologists (Baker, 1991, p. 2) Experimental evidence suggests that some students compared to others in a no-instruction-control group can capitalize on explicit instruction from teachers illustrating how, when and where learning is failing and what might be done to increase understanding (Baker, & Zimlin, 1989). Garner (1990) focuses on metacognition, which is one important aspect of self-regulation (Zeidner, Boekaerts, Pintrich, 2000), a more encompassing concept or theory about motivation and learning and learners' self-reliance (Zimmerman, 2000). She argues that students, especially in science, need to learn how to learn and how to monitor their learning as they study science and other subjects because too many students:

- Do a poor job of keeping track of their learning while studying;
- Too often produce academic products that are poor at best;
- Don't have reasonable background knowledge needed to use strategies;
- Believe that they cannot learn strategies because they are mentally deficient;
- Just want to engage in easy assignments and tests rather than put forth effort; and
- Cannot apply what they learned beyond highly restricted contexts.

Teachers know, for example, that some science students submit homework and project work of very low quality. Yet, these students often don't even realize that their ineffective use of thinking strategies perhaps: 1) does more achievement damage than good, 2) is more than a waste of time, and 3) seldom results in explicit or implicit gratification (Cooper, 2001; Holliday, 2001b).

It is the teachers' job to help ensure that students don't adopt such faulty thinking strategies. Such misguided metacognitive strategies are easy to adopt when they go unattended by teachers. So teachers need to assist students by providing some explicit teaching to individual students so that students can monitor their own learning and are not fooling themselves by engaging in learning and study habits that produce academic products of sub-standard quality (Alderman, 1999). Students with weak metacognitive knowledge are less likely to view efficient learning as: 1) constructing meaning, 2) requiring specific strategies linked to unique tasks, and 3) implementing strategies without continuous assistance provided by teachers (Baker, 1991). Teaching students science process skills, for example, can help them better discover through implicit instruction how they learn science and self-regulate their own learning (Baker, 1991) logically requiring much less explicit metacognitive instruction during subsequent science learning episodes.

Motivating Students to Achieve

So, what can science teachers directly do to motivate students to help themselves study and learn? Teachers need to explain that learning often requires hard work, considerable effort, and that "sufficient intelligence exist" in most students. Teachers need to remind students explicitly, for example, that it is easy to fool yourself into believing that science can be easily understood after simply listening to explanations, experiencing some hands-on activities or devoting time to "reading" documents (Holliday, Yore, & Alvermann, 1994). Many students at all levels of schooling believe that they work hard putting in large amounts of effort, when, in fact, little real effective effort often is expended (Alderman, 1999). Unsuccessful students are less sophisticated in their reflective analyses, and are less academically motivated. Some less-successful science students don't seem to understand what hard work and real effort mean compared to their more-successful peers (Holliday, 2000b). Such struggling students are in need of explicit teaching to help move them into a higher state of achievement motivation (Alderman, 1999).

What adequate effort means in the minds of teachers and students often differ, with some students believing that a few minutes of deep thought on a complex problem or a quick read of a textbook chapter will do the academic job (Garner 1990; Zimmerman, 1998). Such students need explicit instruction because they fail to realize the extensive efforts successful students and adults devote to learning important ideas and completing projects (Alderman, 1999). Some students just give up if the science learning activity takes too much time or cannot be viewed as fun or exhilarating (Holliday, 2000a). So, teachers need to prompt students by directly and systematically:

- Asking them discrete questions about how their learning is going;
- Providing them with a considerable feedback on their progress;
- Providing explicit modeling about exactly what effort it takes to complete a task;

- Assessing student's learning beyond asking them easy, non-thinking questions; and
- Refusing to accept incomplete and shoddy work.

In such situations, indirect teaching approaches alone, such as leaving the learning totally in the students' hands, seldom works. This is an important point that is not typically emphasized when science educators focus on improving science teaching by using a science inquiry approach (Pressley & McCormick, 1995). On the other hand, teachers need to balance such approaches by gradually withdrawing support so each student learns to stand on his or her own two academic feet (Holliday, 2000a). Too much continual explicit instructional support also may prevent students from "learning how to learn" on their own.

Informing Students about Background Science

Teachers also need to check to see if students just plain don't know enough science content to engage in productive thinking strategies needed to comprehend specific content and solve problems. Prior knowledge is a powerful predictor of achievement often forgotten when teaching new material to naïve students (Garner, 1990). For example, students asked to make inferences based on collected data from a local stream may not understand the effects of water erosions, which is capable of carrying natural-occurring products downstream. Students may wrongly conclude that acidity and salt readings indicating poor water quality collected from water samplings are solely the result of nearby industrial or residential wastes. Another example is that with limited prior knowledge in science may be restricted in their use of thinking strategies compared to experts who think differently, as illustrated by the way they view plants and animals. Novices are more likely to use superficial attributes when categorizing living things. Experts using their knowledge of evolution and cell biology engage in much more sophisticated thinking when contemplating the relationships among a wide range of organisms. So explicit teaching is an appropriate way of providing naïve students with sufficient background knowledge needed to engage in more sophisticated academic tasks.

EMPHASIZING IMPLICIT INSTRUCTION: SCAFFOLDING AND PEER DISCUSSIONS

Scaffolding toward Student Independence

Scaffolding in one sense is when teachers use an explicit teaching approach that provides a large amount of instructional support. Then the teacher decreases the amounts of support to students until the students independently handle much more of their own learning as the teachers use other approaches more closely linked to implicit teaching. (Hogan, & Pressley, 1997) Examples of scaffolding outside of science education include training wheels on bicycles for novices, and external metal support structures called braces used at bridge construction sites that hold in place curing concrete reinforced with steel rods (Holliday, 2001c). Instructional scaffolding is reduced by teachers over time and is theoretically removed so that students can become stand-alone, self-regulated learners in less need of explicit instruction (Zimmerman, 1998).

Developing Student Independence in Peer Discussions

A goal in using scaffolding is to provide students opportunities that lead them to scientific literacy while working toward developing independence and self-reliance, as individuals and groups (Zimmerman, 2000; Zimmerman, & Schunk, 2001). This is a delicate balancing act for science teachers (Hogan & Pressley, 1997), especially when classroom management and science inquiry teaching objectives seem to clash (Holliday, 2001a). Scaffolding peer discussions, through a variety of small group strategies and cooperative learning, means supporting student learning by guiding peer discussions. According to research, the teacher must try *not* to interfere with their students' growing learner independence (Evans, 2002), by providing the students with too much support in the form of explicit instruction (Almasi, O'Flahavan, & Arya, 2001). Failing to use adequate implicit instruction under some circumstances may result in students maintaining their dependence on their teacher's support system, or worse, becoming increasingly dependent on the external guidance of their teacher's discourse which has the effect of reducing cognitive demand (Holliday2001c). If we want students to inquire into the scientific universe, we must provide them with opportunities negotiated on their own.

Yet, teachers, who must exercise extensive control over students in peer discussion groups because they feel threaten or intimidated by students and are situated in an unruly school condition, may have good reason to provide more explicit managerial and explicit instructional structure. They may not be able to relinquish their controlling hand, so to speak, because doing so may result into a chaotic classroom (Alderman, 1999; Holliday, 2000a), or because of an overbearing "bossy" group member (Evans, 2002, p. 47). Such approaches call for increasing implicit teaching, over time. Students need to understand that the teachers, like their parents, are not going to follow them around for the rest of their lives making recommendations on how to work with others and solve problems, and interrupting them at every turn by telling them the "right" answer to every question. Yet, many students at all levels of schooling seem to expect such teacher support (Zimmerman, & Schunk, 2001).

Too much support or lack of withdrawing scaffolding while students work in groups perhaps is illustrated when teachers talk too much or interrupt students' attempt to work things out for themselves. Teachers should avoid, whenever practical, saying such statements as: "You'll never resolve this problem if everyone keeps talking at once." "Listen to what other students are contributing to the discussion." Teachers also should avoid interrupting students by continually describing to students how their discussion is proceeding, by interjecting commentary throwing students off track and by increasing student reliance on teacher's support and reducing students attempts to learn on their own (Almasi, O'Flahavan, & Arya, 2001). Once some explicit instruction and modeling is provided, teachers should try *not* to use explicit instruction by consistently running interference and intervening, unless practical reasons such as classroom management problems require teacher intervention. Students, whenever practical, need opportunities to negotiate conventions among themselves leading to productive group discussions.

RELATED RESEARCH IN A BALANCED APPROACH TO TEACHING READING

Reading educators have been dealing with a different yet analogous "balance" problem for a few decades: How much implicit and explicit instruction is helpful? Lately, leading researchers are recommending a balanced approach to teaching youngsters how to read, with one popular approach (let's call it, whole language) emphasizing greater use of more implicit strategies compared to another popular approach (let's call it, phonics coupled with direct comprehension strategy instruction) emphasizing more explicit strategies. This analysis makes a pedagogical analogy between teaching reading comprehension as a higher-order skill in reading and teaching science as inquiry, as a higher-order skill in learning science. Scholarly discussions by reading educators about both reading approaches and their many variations may help shed some clarifying light on why balanced approaches in science teaching, at times, makes sense and deserves our reflective attention.

Many whole language enthusiasts advocate engaging students in reading large qualities of interesting materials, and unsystematically teaching them how to decode word sounds (that is learning decoding on an as-needed basis). A whole language approach focuses on immersing students into authentic, realistic literature including science texts and writing short documents reflecting on such reading materials often during group discussions. Students instructed in this implicit approach also check the meaning of decoded yet unfamiliar words by comparing their inferred meanings of words with syntactic and semantic cues located in the context of the reading materials. Aspects of this approach can be seen in three integrated science and reading programs designed for upper elementary school youngsters (Alao, S. & Guthrie, J.T. 1999; McMahon, O'Hara, & Holliday, 2000; McMahon, O'Hara, McCormick, Gibson, & Holliday, 1999; Romance, & Vitale, 2001).

In contrast, teaching phonics coupled with direct teaching of comprehension strategies and other essential components of reading instruction represents an explicit approach to developing reading fluency in students (Pressley, & Block, 2002). The phonics component of this second general approach consists of instruction whereby students decode words by sounding them out and blending sounds together, or decode words using analogies such as word families or the Benchmark School Word ID program (Pressley, 2002). Such instruction coaches students in checking decoded words for accuracy by using contextual cues contained in documents, in a fashion similar to the whole language method. The second component of the phonics with comprehension-strategy instructional approach focuses on direct teaching of explicit comprehension strategies such as questioning, summarizing, inferring, predicting, visualizing, and seeking clarification.

These two approaches to reading and their variations are similar in two respects to the explicit-implicit continuum within science inquiry teaching. First, the phonics/comprehension-strategy-instruction approach is more often associated with added emphasis on systematic explicit teaching and reduced emphasis on naturally flowing implicit teaching with the whole language approach paying more attention to implicit instruction. Students who receive a "pure" strand of only whole language instruction are fast readers, but can be poor comprehenders, weak spellers and frustrated oral readers (Pressley, 2002). Second, many prominent researchers (in reading) over time are increasingly citing the value of both approaches using an integrated, balanced reading approach in which teachers use their professional judgments and borrow the best from each teaching approach. In one widely respected survey, at least 75% of the polled leaders in reading education favored a balanced approach (Cassidy, & Cassidy, 2001/2002). Indeed, most teachers reportedly are providing such a balanced approach (Cassidy, & Cassidy, 2001/2002). In a somewhat parallel fashion, science educators, perhaps, need to approach science inquiry teaching applying an analogous balanced approach, considering the combination of the virtues of explicit as well as implicit teaching approaches.

Reading educators recommend elements of whole language because immersing students in reading, providing them with non-trivial choices, linking reading and writing together and facilitating small group discussions and self-expressions promote a range of abilities needed to be academically successful in reading. Meanwhile, directly and systematically teaching phonics while introducing comprehension strategy instruction, and other components of good reading instruction such as vocabulary-building instruction, likewise represents a general approach (National Reading Panel, 2000). Reading and writing researchers increasingly are investigating the potential need for added explicit-related teaching, a research direction apparently receiving much less attention in science education. An integrated balanced approach with regard to an explicit-implicit instructional continuum also makes empirical sense in writing (Holliday, 2000b), handwriting (Graham, Harris, & Fink, 2000) and spelling instruction (Graham, 2000). In these school subjects, these researchers are finding new value in added explicit teaching in comparison to contrary views held by many practicing language educators. Of course, a balanced approach to teaching reading and other language arts instruction varies uniquely with individual teachers and ever-changing classroom conditions.

In our field, science educators do not condemn explicit systematic science instruction including explaining ideas and providing discrete, yet direct, intervention by teachers. But, that lack of attention to the explicit side of the explicit-implicit teaching continuum may be a problem (in science education) that is in need of additional attention more than it is now receiving.

SOME OTHER PROMISING RESEARCH

Researchers in science teaching are spending considerable energy exploring other related issues in science inquiry teaching beyond the specific singular explicit-implicit issue discussed in this chapter. Lederman and his associates (see Abd-El-Khalick, Bell, Lederman, 1998; Akerson, Abd-El-Khalick, Lederman, 2000; Bell, Lederman, & Abd-El-Khalick, 2000) in a series of empirical studies, for example, urge teacher educators instructing pre- and in-service science teachers and school students to consider seriously explicit approaches in addition to other considerations when teaching specifically the nature of science. This recommendation is made in "contrast to an implicit approach, which relies solely on the engagement of learners in activities, inquiries, and reiterations of historical case studies without attention to specific (nature of science) issues" (Schwartz, & Lederman, 2000, p. 207). This chapter's delimited examination of science inquiry teaching explores ground tangential to nature of science studies. But, the focus in this chapter was aimed at modifying our advice to science teachers, encouraging them to balance their use of implicit and explicit approaches when teaching students about the nature of science and the rest of their science curriculum. Too often, as stated by Lederman and his associates, explicit approaches to science teaching are not adequately emphasized and described to practitioners. These studies are examples of additional efforts to investigate related issues linked to science teaching inquiry, each of which are distinctive and reside outside of this chapter's focus.

CONCLUDING REMARKS

Research of good quality does not indicate which conceptual or problem-solving strategies should be explicitly or implicitly taught using a specific method to a particular student under a given condition. Such a theoretical or concrete standard makes no research sense. However, teaching and learning research does support the notion of generally providing students with more direct help than is currently suggested by some authors writing specifically about science inquiry teaching (Holliday, 2001a; Pressley, & McCormick, 1995; Pressley, Harris, & Marks, 1992).

On the other hand, science classes where students are never encouraged to inquire, explore, induce, or discover logically represent an example of poor science teaching by almost any standard. Students, instead, need to construct and apply their understanding of our universe by spending some significant time engaging in explicit as well as implicit learning experiences. The exact composition of these integrated balanced experiences should be determined by informed, experienced science teachers using their professional judgments. A goal of this chapter was to provide fodder for improving such judgments by arguing for an increased use of explicit teaching compared to the advice emphasized in many other otherwise productive documents designed for science teachers. Of course, overuse of explicit science inquiry teaching in any classroom setting simply makes no sense.

There is little reason in one sense to dichotomize instruction or teaching approaches into these two separate categories, implicit and explicit. Yet, that is essentially what educators do and are likely do for some time, so we may have to live with these terms. Viewing these teaching approaches on a continuum perhaps makes more sense than viewing them as partitioned ideas, according to Harris, & Graham (1994).

This chapter did not attempt to cover the research-based waterfront of inquiry teaching but, instead, sought to stimulate a provocative discussion about how some science teachers arguably are overdosing on a traditional focus of science inquiry teaching, this is using as little explicit teaching approaches as possible. The principle, "the-less-teacher-intervention-the-better" must be tempered by our data-based knowledge on the effectiveness of explicit teacher help.

There are no magic ways of getting students to learn on their own and work together. But, it's worth the effort to help students manage their own learning, providing explicit teaching followed by reducing the teacher's role of manager and source of needed information including less direct and explicit teacher talk, whenever reasonable and practical. On the other hand, providing students with high doses of implicit teaching, as suggested in some documents commenting about science teaching inquiry, is effectively overdosing students on a brand of instructional inquiry leading to unnecessary amounts of errant and inefficient science learning, especially among struggling students. Research, instead, suggests that to be an effective science teacher one must integrate both formats of instruction so that efficient teaching more likely occurs. In short, balancing implicit with explicit teaching approaches to science inquiry teaching instruction needs added emphasizes in science education because of the research-based arguments cited in this chapter.

REFERENCES

Abd-El-Khalick, F., Bell, R. L., Lederman, N. G. (1998). The nature of science and instructional practice: Making the unnatural natural. *Science Education, 82*, 417-436.

Abell, S. K., Anderson G., & Chezem, J. (2000). Science as Argument and Explanation: Inquiring into Concepts. In J. Minstrell, & E. H. van Zee, (Eds.) *Inquiring into inquiry learning and teaching in science* (pp.65-79). Washington, D.C.: American Association for the Advancement of Science.

Akerson, V. L., Abd-El-Khalick, F., & Lederman, N. G. (2000). Influence of a reflective explicit activity-based approach on elementary teachers conception of nature of science. *Journal of Research in Science Teaching, 37*, 295-317.

Alao, S. & Guthrie, J.T. (1999). Predicting conceptual understanding with cognitive and motivational variables. *Journal of Educational Research, 92*, 243–254.

Alderman, M.K. (1999). *Motivation for achievement.* Mahwah, N.J.: Lawrence Erlbaum Associates.

Almasi, J. F., O'Flahavan, J. F., & Arya, R. (2001). A comparative analysis of student and teacher development in more and less proficient discussions of literature. *Reading Research Quarterly, 36*, 96-120.

American Association for the Advancement of Science (1993). Habits of Mind, In *Benchmarks for science literacy* (pp. 281-300). Washington, DC: American Association for the Advancement of Science

Baker, L. (1991). Metacognition, reading, and science education. In C. M. Santa & D. E. Alvermann (Eds.) Science learning: Processes and applications (pp. 2-13). Newark (DE): International Reading Association.

Baker, L., & Zimlin, L. (1989). Instructional effects on children's use of two levels of standards of revaluation their comprehension. *Journal of Educational Psychology, 81*, 340-346.

Bell, R. L., Lederman, N. G., & Abd-El-Khalick, F. (2000). Developing and acting upon one's conception of the nature of science: A follow-up study. *Journal of Research in Science Teaching, 37*, 563-581.

Bianchini, J. A., & Colburn, A. (2000). Teaching the nature of science through inquiry to prospective elementary teachers: A tale of two researchers. *Journal of Research in Science Teaching, 37*, 177-209.

Bybee, R. W. (2000). Teaching Science as Inquiry. In J. Minstrell, & E. H. van Zee (Eds.), *Inquiring into inquiry learning and teaching in science* (pp.20-46). Washington, D.C.: American Association for the Advancement of Science.

Carin, A. A., & Bass, J. E. (1997). Methods for teaching science as inquiry. In *Teaching science as inquiry* (pp. 143-173). Upper Saddle River (NJ): Merrill/Prentice Hall.

Cassidy, J. & Cassidy, D. (2001, December / 2002, January). What's hot, what's not for 2002. *Reading Today, 19*(3), 1, 18.

Cooper, H., & J.C. Valentine. (2001). Using research to answer practical questions about homework. *Educational Psychologist, 36*, 143-153.

Duffy, G. (2002). In the case for direct explanation of strategies. In C. C. Block, & M. Pressley, (Eds.) *Comprehension instruction: Research-based best practices* (pp. 28-41). New York: Guilford Press.

Evans, K. S. (2002). Fifth-grade students' perceptions of how they experience literature discussion groups. *Reading Research Quarterly, 37*, 46-69.

Friedl, A. E., & Koontz, T. Y. (2001). Methods, In *Teaching science to children: An inquiry approach.* (pp. 1-11) Boston: McGraw-Hill/College.

Garner, R. (1990). When children and adults do not use learning strategies: Toward a theory of settings. *Review of Educational Research, 60,* 517-530.

Graham, S. (2000). Should the natural learning approach replace spelling instruction? *Journal of Educational Psychology, 92*, 235-247.

Graham, S., & Harris, K. R. (1994). Implications of constructivism for teaching writing to students with special needs. *Journal of Special Education, 28*, 275-289.

Graham, S., & Harris, K. R. (2000). The role of self-regulation and transcription skills in writing and writing development *Educational Psychologist, 35*, 3-12.

Graham, S., Harris, K. R., & Fink, B. (2000). Is handwriting causally related to learning to write? Treatment of handwriting problems in beginning writers. *Journal of Educational Psychology, 92*, 620-633.

Harris, K. R., & Graham, S. (1994). Constructivism: Principles, paradigms, and integration. *Journal of Special Education, 28*, 233-247.

Hogan, K., and Pressley, M. (1997). *Scaffolding student learning: instructional approaches and issues.* Cambridge (MA): Brookline Books.

Holliday, W. G. (2000a) Getting students motivated. *Science Scope, 23*(4), 50-52.

Holliday, W. G. (2000b). Integrating writing with science. *Science Scope 24*(1), 72-74.

Holliday, W. G. (2001a). Critically considering science inquiry. *Science Scope 24*(7), 54-57.

Holliday, W. G. (2001b). Homework in science. *Science Scope 25* (3), 58-62.

Holliday, W. G. (2001c). Scaffolding in science. *Science Scope 25* (1), 68–71.

Holliday, W. G., Yore, L., & Alvermann, D. E. (1994). The reading-science learning-writing connection: Breakthroughs, barriers, and promises. *Journal of Research in Science Teaching, 31*, 877-894.

Keys, C. W., & Bryan, L. A. (2001). Co-constructing inquiry-based science with teachers: Essential research for lasting reform. *Journal of Research in Science Teaching, 38*, 631-645.

Krajcik, J., Czerniak, C., & Berger, C. (1999a). Why and How Should I teach Science to Children? In *Teaching children science: A project-based approach.* Boston: McGraw-Hill/College, (pp. 2-25).

Krajcik, J., Czerniak, C., & Berger, C. (1999b). How do children construct understanding in science? In *Teaching children science: A project-based approach* (pp. 27-61). Boston: McGraw-Hill/College.

Liem, T. L. (1987). Introduction. In *Invitations to science inquiry* (pp. XIX-XXXX). Lexington (MA): Ginn Press.

McMahon, M. M., O'Hara, S. P., Holliday, W. G., McCormack, B. B., Gibson, E. M. (2000). Curriculum with a common thread. *Science and Children, 37* (7), 30–35, 57.

McMahon, M. M., O'Hara, S. P., McCormack, B. B., Gibson, E. M., Holliday, W. G., & Kelly, C.K. January (1999). QUINTO: A fifth grade teacher team's electronic professional development Tool, *Meridian: A Middle School Computer Technologies Journal.* <http://www.ncsu.edu/meridian/jan99/quinto/index.html>.

Metz, K. E. (2000). Young children's inquiry in biology: Building the knowledge bases to empower independent inquiry. In J. Minstrell, & E. H. van Zee, (Eds.) *Inquiring into inquiry learning and teaching in science* (pp.371-404). Washington, D.C.: American Association for the Advancement of Science.

National Reading Panel. (2000, April). *Teaching children to read.* Washington, DC: National Institute of Child Health and Human Development.

National Research Council, (1996b). Principles and definitions. *National science education standards* (pp.27-53). Washington, DC: National Academy Press.

National Research Council. (1996a). *National science education standards.* Washington, DC: National Academy Press.

National Research Council. (2000). Inquiry in science and in classrooms. *Inquiry and the national science education standards* (pp. 1-11). Washington, DC: National Academy Press.

Pressley, M. & McCormick. C. (1995). *Cognition, teaching and assessment.* New York: HarperCollins College Publishers.

Pressley, M. 2002. *Reading instruction that works: The case for balanced teaching.* New York: Guildford Press.

Pressley, M., & Block, C. C. (2002). Summing up: What comprehension instruction could be. In M. Pressley, & C. C. Block, (Eds.), *Comprehension Instruction: Research-Based Best Practices.* (pp. 83-92) New York: Guilford Press.

Pressley, M., & Harris, K. R. (1998). Constructivism and instruction. *Issues in Education. 3*, 245-255.

Pressley, M., Harris, K. R., & Marks, M. B. (1992). But, good strategy instructors are constructivists! *Educational Psychology Review, 4*, 3-31.

Romance, N. R., & Vitale, M. R. (2001). Implementing an in-depth expanded science model in elementary schools: Multi-year findings, research issues, and policy implications. *International Journal of Science Education, 23*, 373-404.

Rutherford, J., & Ahlgren, A. (1989). Effective learning and teaching. *Science for all Americans* (pp.145-151). Washington, DC: American Association for the Advancement of Science.

Schwartz, R. S., & Lederman, N. G. (2002). 'It's the nature of the beast': The influence of knowledge and intentions on learning and teaching nature of science. *Journal of Research in Science Teaching, 39*, 205-236.

Shulman, L. S., & Keislar, E. R., (Eds.) (1966). *Learning by discovery: A critical Appraisal.* Chicago: Rand McNally & Company.

Trowbridge, L. W., Bybee, R. W., & Powell, J. C. (2000). Models of Effective Science Teaching, *Teaching Secondary School Science* (pp. 232-251). Upper Saddle River (NJ): Merrill/Prentice Hall.

Tuovinen, J. E., & Sweller, J. (1999). A comparison of cognitive lead associated with discovery learning and worked examples. *Journal of Educational Psychology, 91*, 334-341.

Van Lehn, K. (1990). Mind bugs: The origins of procedural misconceptions. Cambridge (MA): MIT Press. Cited in M. Pressley, & McCormick. (1995). *Cognition, teaching and assessment.* New York: HarperCollins College Publishers.

Zeidner, M., Pintrich, P. R., & Boekaerts, M. (2000). Self-regulation: Directions and challengers for future research. In M. Boekaerts, P. R. Pintrich, & M. Zeidner, (Eds.), *Handbook of Self-Regulation.* (pp. 749-768) San Diego: Academic Press.

Zimmerman, B.J. (1998). Academic studying and the development of personal skill: A self-regulatory perspective. *Educational Psychologist, 3*(2/3): 73-86.

Zimmerman, B. J. (2000). Attaining self-regulation: A social cognitive perspective. In M. Boekaerts, P. R. Pintrich, & M. Zeidner, (Eds.) *Handbook of Self-Regulation* (pp.13-39). San Diego: Academic Press.

Zimmerman, B. J., & Schunk, D. H. (Eds.) (2001). *Self-regulated learning and academic achievement: Theoretical perspectives.* Mahwah, NJ: Erlbaum.

PART III: CURRICULUM AND ASSESSMENT

CHAPTER 11

BRUCE SHERIN, DANIEL EDELSON, & MATTHEW BROWN

ON THE CONTENT OF TASK-STRUCTURED SCIENCE CURRICULA

INTRODUCTION

Many of the recent innovations in K-12 science instruction share the common goal of embedding classroom learning within rich contexts that students find both intellectually and socially meaningful. Though curricular models vary greatly, many efforts to achieve this goal share an important feature: Rather than being organized around a traditional disciplinary structure, these curricula are organized around a task. This approach has been given many names and has taken many forms; these include varieties of "anchored" or "problem-based" instruction (Barron et al. 1998; Williams 1992; Barrows and Tamblyn 1980; Duschl and Gitomer 1997), "project-based" instruction (Krajcik, Czerniak, and Berger 1999), "learning by design" (Harel 1991; Kolodner et al. 1998), and "goal-based scenarios" (Schank et al. 1993/1994).

In all of these approaches, the curriculum is not organized as a systematic progression through a list of traditional content topics. Instead, the structure of the curriculum is provided by problems, goals, or issues that cut across multiple traditional areas of content. We call curricula that are organized around problems, goals, and issues *task-structured*, as opposed to traditional *content-structured,* curricula.

In abandoning the content-structured approach, we should be aware that we are proposing a dramatic change, and setting out on uncertain ground. In this paper, we will argue that task-structured curricula address a fundamentally different slice of content. By their nature, individual task-structured curricula tend to address content across a range of traditional disciplines. Furthermore, because of these fundamental differences in the content that is addressed, there may be implications for the learning processes that are associated with these curricula. There may be substantial differences, for example, in the range of prior knowledge that is relevant for learning, in the nature of the changes to this knowledge that must occur, and in the form of the final understanding that is produced.

These observations suggest the need for a new program of research, one that takes these differences seriously, and seeks to understand their implications. This program is a large one, and our work is at an early stage. Thus, our purpose in this chapter is simply to point the way; we will endeavor to give a sense for the range of issues that need to be addressed, and the analytical methods that we believe should be employed.

L.B. Flick and N.G. Lederman (eds.) Scientific Inquiry and Nature of Science, 221-248.

Furthermore, our discussion here is limited to issues pertaining to science *content,* where "content" is narrowly construed. There are dramatic differences between task-structured and content-structured curricula along a number of dimensions. For example, there are likely substantial differences in the extent to which each type of curriculum addresses epistemic issues, such as beliefs about the nature of science. However, our focus will be on science content, understood in a more traditional sense. We will elaborate on what this means as the chapter proceeds.

In the remainder of this chapter, we begin with a more extended discussion of the differences between task-structured and content-structured curricula. Then, we will illustrate these differences using a task-structured curriculum that we have developed, the Global Warming Project. Finally, we will describe some of our initial attempts to study the learning processes associated with the Global Warming Project, with particular emphasis on an issue that we refer to as the *bootstrapping problem*

HOW TASK-STRUCTURED CURRICULA ARE DIFFERENT

Arguments for task-structured curricula are typically based on two central premises: (1) learning should occur in contexts in which new knowledge is *useful* for students, and (2) students should engage in practices that, in some manner, mimic the practice of scientists. Because of their belief in these premises, designers of task-structured curricula assign a high-priority to certain properties in their designs. First, they attempt to design activities in which learners are given some overarching goals. These goals drive the need for learning, and provide the activities with some degree of connectedness. Second, they attempt to engage learners in something like authentic scientific activity.

The various task structures mentioned above help to endow task-structured curricula with these properties. They provide overarching structure and, it is hoped, they give students experience with fundamental aspects of the scientific endeavor, such as reasoning from data, and reasoning about complex problems. In the literature, the overarching structure takes a number of forms. For example, as summarized in (Kolodner et al. 1998). science is sometimes learned in the service of a design challenge, such as the design of a subway system (Kolodner et al. 1998). Alternatively, in some instances, students learn by investigating a "driving question" that has relevance to their lives (Krajcik, Czerniak, and Berger 1999).

Table1. Examples of task-structured approaches

Name of approach	*Representative sources*	*Organizing structure*	*Example*
Anchored or problem-based instruction	(Barron et al. 1998; Williams 1992; Duschl and Gitomer 1997; Barrows and Tamblyn 1980)	problem	Medical students learn both basic science and diagnosis by studying the records of an actual patient. (Williams 1992; Barrows and Tamblyn 1980)
Project-based science	(Marx et al. 1994; Krajcik, Czerniak, and Berger 1999)	driving question	Students learn about chemistry and watershed ecology by working on the question "What is the water like in our river?" (Krajcik, Czerniak, and Berger 1999)
Learning by design	Harel, 1991; Kolodner, Crismond, Gray, Holbrook, & Puntambekar, 1998	design challenge	Students learn about rocks and rock formations in the context of designing an imaginary subway system.

The increased interest in task-structured curricula is part of a larger trend in the science education community; namely, researchers have increasingly come to believe that the teaching and assessment of scientific content must be integrated with the teaching and assessment of process (Gitomer and Duschl 1998).This trend is reflected in the National Science Education Standards (National Research Council 1996). The standards state: "Students at all grade levels and in every domain of science should have the opportunity to use scientific inquiry and develop the ability to think and act in ways associated with inquiry" (p. 105). Furthermore, the NSES standards are clear that they intend their conception of inquiry to include more than the teaching of isolated process skills, and more than the teaching of a step-by-step sequence that constitutes the "scientific method."

> In the vision presented by the *Standards*, inquiry is a step beyond "science as a process," in which students learn skills, such as observation, inference, and experimentation. The new vision includes the "processes of science" and requires that students combine processes and scientific knowledge as they use scientific reasoning and critical thinking to develop their understanding of science. (p. 105)

Task-structured curricular approaches are certainly consistent with this broader movement. However, in some respects, they may be seen as departing from the default conception of scientific inquiry that is advocated in the standards. The central premise of the task-structured approach is that learning should occur in contexts in which new knowledge is *useful* for students. Students develop scientific knowledge in order to serve a specific overarching goal, such as solving a problem or building a device. In contrast, in the default conception, scientific inquiry may be geared simply toward understanding phenomena in the world; students are asked to describe phenomena, and understand how and why they occur. In this conception of inquiry, the new knowledge acquired by students is useful, but only in the broader sense of providing an explanation of scientific phenomena.

The premises that underlie task-structured curricula are compelling and plausible, and they are consistent with current beliefs about the nature of learning (e.g., Bransford et al., 1999). However, we believe that there is substantial work to do in elaborating what we believe is learned in task-structured, and to establish whether these beliefs are borne out. As we will argue, articulating what is learned in task-structured curricula poses some particular challenges, and there are real reasons to worry that task-structured curricula may not succeed in achieving the content-learning goals that are intended by their designers.

In order to elaborate on the contrast between task-structured and content-structured curricula, we engage in analyses from two perspectives: (1) an analysis of the *content addressed,* and (2) an analysis of the *learning processes* that occur in each style of curriculum. In the remainder of this section, we introduce these two perspectives, and we discuss some of the issues raised by each type of analysis.

Chapter 4: Newton's Laws

4-1 Force and Mass

4-2 The Force Due to Gravity: Weight

4-3 Units of Force and Mass

4-4 Newton's Third Law and Conservation of momentum

Figure 1. A portion of a table of contents (Tipler, 1976)

Content in task-structured curricula

When we refer to the "content addressed" we have a very particular perspective in mind, and will be giving a very particular type of account. Our accounts of the "content" are in terms of the shared, public language that is employed within the discipline that is the subject of the curriculum. Such an account would employ, for example, the terminology used in a textbook. Indeed, the table of contents of a science textbook is, in a sense, an analysis of the content addressed by the textbook; it divides the subject matter to be covered, and it defines a pathway through that subject matter. For illustration, consider the portion of a table of contents, shown in Figure 1, that is taken from a physics textbook. This table provides a particular perspective on the content, the topics that will be covered and how they will be sequenced.

When viewed from this content perspective, it is clear that there is a profound dissimilarity in the driving logic of the two styles of curriculum. When setting out to design a content-structured curriculum we ask: What are the key concepts in this discipline and how can we build them up in a logical manner? In contrast, when setting out to design a task-structured curriculum we ask: What would make for an engaging task and would also allow useful content to be addressed? How can we scaffold students as they move from start to finish on this task? Because of this profound difference in driving logic, we can draw two important conclusions concerning the nature of content in a task-structured curriculum:

(1) The content required to engage in a meaningful task tends to cut across multiple traditional areas of content. Thus, the slice of content associated with a particular task-structured curriculum unit may well differ from the slice that would be associated with any content-structured curricula. This does not necessarily mean that, over the course of a year or multiple years, different content will be addressed. Nor does it mean that the content in a task-structured curriculum is without logic or integrity. But it does mean that, at the least, this content will be organized into units and sequenced in a very different manner.

(2) In a task-structured curriculum, the task not only dictates what will be taught, it also dictates the manner—or "depth"—to which the various portions of disciplinary content must be understood by students. For this reason, some issues will be covered in great depth (measured against our traditional conceptions of a discipline). In other places, students will learn just enough to "get by." In contrast, in a content-structure curriculum, we often attempt to build up content in a manner that we believe reflects the a priori structure of a discipline.

For illustration, consider problem-based medical instruction (Barrows and Tamblyn 1980). In this innovative approach, the learning takes place in the context of medical problems. Students are given the medical history of a patient, including a list of symptoms, and they investigate the potential causes of the patient's complaints. In the course of these investigations, they typically conduct and share research that cuts across several traditional content areas. For example, they might have to learn a little anatomy, some biochemistry, etc. Thus we see that the content covered within any unit of instruction will tend to cut across disciplines.

Furthermore, the issue of depth is also evident in this type of instruction. Suppose, for example that a particular medical problem requires students to understand a case in which a patient appears to have suffered significant blood loss. For this case, students might need to know about some aspects of the cardiovascular system in great detail, such as hemodynamics and the properties of the heart as a pump. In contrast, they might only need a cursory understanding of other related topics, such as how, exactly, the muscles in the heart allow it to exhibit its particular pump-like properties. Where depth is required is thus dictated by the problem at hand.

The need for an account of learning in task-structured curricula

An analysis of the content addressed, as we defined it in the previous section, does not necessarily tell us how this content is understood or learned by individuals.

Consider, once again, the portion of a table of contents that is shown in Figure 1. This table of contents tells us that Newton's Third Law will be part of the content that is addressed. However, it does not tell us what it means for an individual to understand Newton's Third Law — it does not describe the collection of knowledge that underpins this understanding. Furthermore, it also does not describe important aspects of the learning process; it does not tell us, in explicitly cognitive terms, what relevant knowledge students are assumed to possess upon entry, or how this knowledge will change as the students progress toward understanding.

This suggests that another type of analysis is needed, one that provides a more explicitly cognitive account of the learning that occurs. Some curricula, as part of their specification, come equipped with a learning theory, an explicit account of the cognitive changes that are to be engendered. Other curricula do not come with such an account.

One type of science curriculum that generally is accompanied by a learning theory are what we will refer to as *prior-conceptions-driven curricula*. Beginning in the late 1970s and early 1980s, research concerning students' prior conceptions became prominent (Wandersee, Mintzes, and Novak 1994; Driver 1994; Smith, diSessa, and Roschelle 1993; Eylon and Linn 1988; Pfundt and Duit 2000). In response to these observations, researchers were led to advocate for curricula that were designed with the explicit intention of addressing prior conceptions. For example, Smith and colleagues (Smith et al. 1997) developed a successful curriculum for teaching 8th-graders about matter and density. In their design efforts, this research team assumed, very explicitly, that it was necessary to address prior conceptions. They thus began with a systematic investigation of students' understanding of density and the nature of matter, and then articulated precisely what sort of changes were required for students to move to more appropriate understandings. In this regard, an important observation was that, prior to instruction, many of the students seemed to have an undifferentiated notion of weight and density. Thus, one goal of instruction, Smith and colleagues surmised, must be to help students acquire differentiated notions of weight and density.

The central point here is that prior-conceptions-driven curricula were developed *with content-specific learning theories in mind.* Note, for example, that what Smith and colleagues developed was not just a curricular intervention; it was curricular intervention *and* an account of student learning, all rolled into one. They mapped out student conceptions and the relevant conceptual territory, and then carefully moved students through this territory. This is entirely typical of prior-conceptions-driven curricula; part-and-parcel of these curricula is an account, from beginning to end, of the conceptual change that the curriculum is intended to engender.

The situation in task-structured curricula is quite different. The central image is one of students moving through a task, not of students moving through a conceptual space. This is not to say that the designers of task-structured curricula do not worry about drawing out student conceptions and making sure that students engage with those conceptions. But, unlike curricula that are driven primarily by a desire to address prior conceptions, task-structured curricula typically do not come equipped with a single broad account of the conceptual change that is to be engendered.

Thus, an attention to the learning perspective indicates one important issue: task-structured curricula, when developed, tend not to come equipped with detailed accounts of learning. This suggests an obvious goal, we can set out to develop detailed accounts of this sort. However, there are reasons to believe that producing accounts of conceptual change for task-structured curricula will be particularly difficult — more difficult, for example, then describing the learning that will occur in a curriculum that is focused tightly around matter and density. Because task-structured curricula often address content across a range of traditional disciplines, these curricula may produce learning outcomes that are difficult to encapsulate and characterize.

Thus, with respect to learning, we have just made two points. First, we pointed out that, unlike prior-conceptions-driven curricula, task-driven curricula tend not to come equipped with detailed accounts of the accompanying learning process. Second, we stated that there are reasons to believe that such accounts will be difficult to produce — it is simply more difficult to describe the learning that occurs in a task-structured curriculum. However, neither of these two points addresses a third issue: the question of whether task-structured science curricula can be successful in achieving their specific content learning goals. Clearly, this is one of the central questions that our empirical and analytic work must address.

Indeed, there are some very serious reasons to worry whether any particular task-structured curriculum can achieve its learning goals, and lead to rigorous content understanding. In this paper, we want to draw out and emphasize one particular reason for worry, revolving around what we call the *bootstrapping problem*: How can we expect students to work on problems and issues that cut across multiple disciplines if we have not already provided them with a solid foundation in these disciplines? This is just one question, but we believe it is a central and critical one, and we give it special attention in the remainder of paper.

A TASK-STRUCTURED CURRICULUM: THE GLOBAL WARMING PROJECT

In order to provide some grounding for the above discussion, we will now illustrate some of our points in the context of one particular task-structured curriculum, the Global Warming Project (Edelson, Gordin, & Pea, 1999). First, we will discuss the curriculum itself, with issues of content structure in mind. Then we will discuss some empirical work with students in order to further our discussion of learning and, in particular, of the bootstrapping problem.

Overview of the GWP

The Global Warming Project (GWP) is an 8-10 week middle school science unit created by the Center for Learning Technologies in Urban Schools at Northwestern University in collaboration with Chicago Public Schools. The learning goals of the GWP designers were wide-ranging. Included in these goals were content objectives, in the narrow sense. As we will discuss, the curriculum touches on such topics as the Earth's energy balance and the carbon cycle. But the goals also went beyond these narrow content objectives. As is typical of task-structured curricula, the designers had the intention of engaging students in authentic scientific processes (and thus helping them to develop some of the related skills, as well as a better understanding of the nature of these processes). Furthermore, the curriculum was intended to educate a population of students about a potentially serious problem faced by our society, global warming, and, more broadly, to address issues of the relationship between science and public policy.

In the GWP curriculum, the students adopt the role of scientific advisors to heads of state for various countries, and they are given the task of preparing briefings for the leaders of their respective countries. More specifically, the curriculum is organized around three briefings that students must prepare, each of which pertains to some aspect of global climate change:

(1) How could we tell if the Earth were getting warmer?
(2) What might be causing global warming?
(3) What are the predicted implications of global warming for individual countries and what responses should they pursue?

First briefing: How could we tell if the Earth were getting warmer

To answer the first question, students investigate some of the challenges that underlie any attempt to determine whether the Earth is indeed getting warmer. They begin by measuring the temperature in various parts of their school, looking for variation across space and over time. This provides students with an opportunity to think about such issues as: Given measurements of temperature across locations, how should we compute an average value? Given variations over time, how can we draw conclusions about patterns and trends?

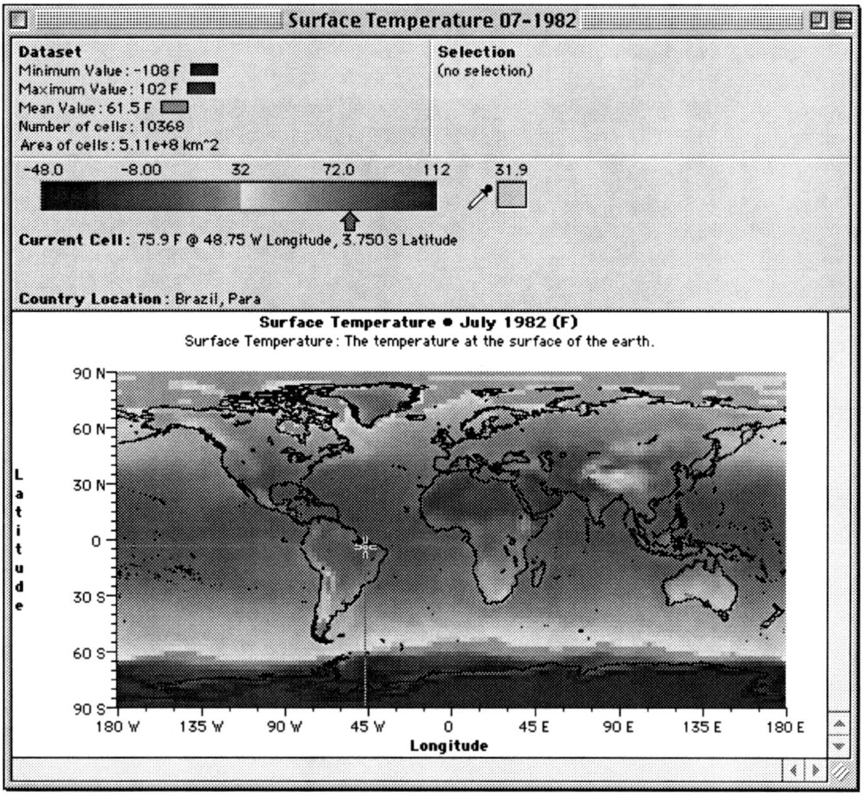

Figure 2. A scientific visualization that shows the variation in temperature over the surface of the Earth. In this figure, the color image has been converted to grayscale.

In addition to these investigations of temperature variation in their schools, the students look at temperature variation over a variety of time-scales given historical temperature data from their city. They also investigate spatial temperature variation by constructing and examining scientific visualizations that show temperature over the Earth's surface. They view these visualizations using software called *WorldWatcher*. (An example is shown in Figure 2.)

Second briefing: What might be causing global warming?

In some respects, the work associated with the second briefing is the heart of the GWP. It is in this part of the curriculum that the students begin to learn about the Earth's energy balance. The diagram in Figure 3, which is taken from the WorldWatcher software, gives a schematic overview of what happens to the solar energy that is incident upon the Earth. As is made clear by the diagram, the process is relatively complex. Some of the incident solar energy is reflected by the Earth, while some is absorbed and later re-emitted. Furthermore, some of this re-emitted energy escapes to space, but some is trapped by the atmosphere due to a process known as the Greenhouse Effect.

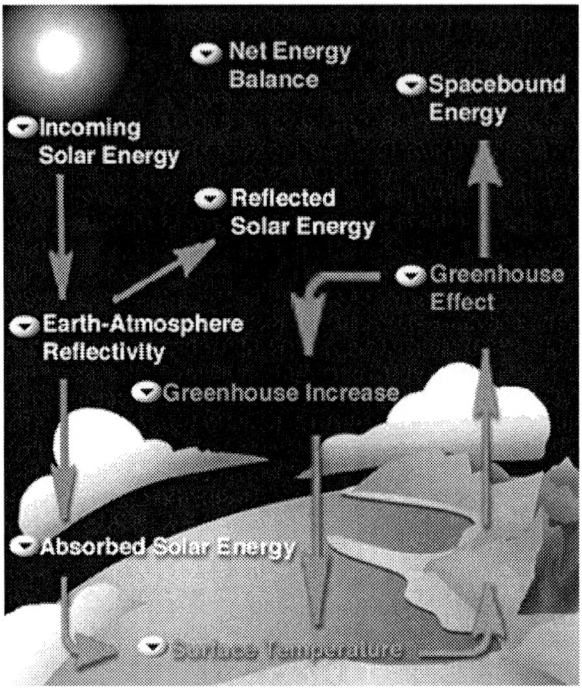

Figure 3. Diagram of the Earth's energy balance. In this figure, the color image has been converted to grayscale

The GWP curriculum takes the time to unpack and elaborate some of the subcomponents of the process shown in Figure 3. To do this, the curriculum employs lab experiments focused on particular properties of light, as well as computer-based work with scientific visualizations. For example, in one laboratory experiment, students shine a penlight on a sheet of paper, while varying the angle of the light. This lab is intended to help students develop an understanding of how intensity varies with the angle of incidence of light. In addition, students perform an experiment that is designed to help them understand how reflectivity of a surface depends on its color. They make envelopes out of sheets of paper of various colors, and they put thermometers inside these envelopes. Then they shine a light on the envelopes in order to see how the color of the paper affects the readings seen on the thermometers.

Third briefing: What are the predicted implications of global warming for individual countries and what solution strategies should they pursue?

The third part of the GWP contains the culminating activities. In this part of the curriculum, the students obtain and study additional background information. For example, they study datasets concerning aspects of human impact on the environment, such as data concerning carbon dioxide emissions. In addition, they look at the predictions of some models for temperature increase in the future years, given assumptions concerning the increase in carbon dioxide. Using this new data, as well as the understanding they have developed in the previous parts of the curriculum, the students make predictions and propose solutions for their individual countries.

Content in the GWP

It is clear that, in order to prepare the briefings specified in the GWP, students need to understand content that is usually taught separately in a number of distinct disciplines, such as biology, chemistry, physics, and the earth sciences. To further convey the range of content that is addressed, Figure 4 shows the energy balance diagram from Figure 3. Figure 3 annotated with some of the topics that are covered. The first briefing is primarily concerned with the bottommost oval, particularly spatial and temporal variation in surface temperature. The second briefing is concerned with the left part of the diagram. Students learn about incoming solar energy, and how this affects temperature on the Earth. They also learn how the temperature at particular locations on the Earth depends on the angle of incident sunlight, and on surface reflectivity.

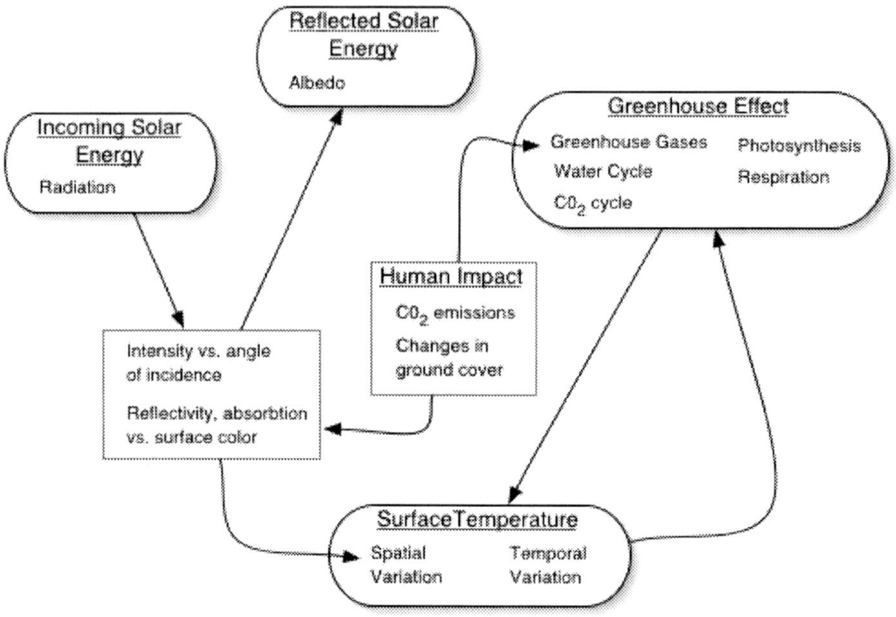

Figure 4. Energy balance diagram with annotations that indicate some of the content in the GWP

Finally, in the activities leading up to the third briefing, the students address topics associated with the right side of the diagram in Figure 4. They learn how the greenhouse affect is tied to the composition of our atmosphere. This, in turn, necessitates some discussion of many sub-topics, including respiration, photosynthesis, the CO_2 cycle, and the hydrological cycle. Finally, they learn about how the activities of humans have had an impact on factors that determine the temperature at the Earth's surface.

In order to understand the potential causes and mitigation strategies for global warming, it is absolutely necessary that aspects of all of these topics be included in the curriculum, at least in some manner. Students must understand the transfer of energy as it passes into, through, and out of the Earth-atmosphere system. They must understand the factors that determine how much solar energy is reflected back into space and how much is absorbed by the Earth's surface and atmosphere. They must also understand the role that atmospheric "greenhouse" gases play in trapping the resultant heat within the Earth-atmosphere system. Finally, they must understand the natural and anthropogenic processes that cause greenhouse gases to be emitted into and removed from the atmosphere. It is easy to see that this selection of content around the threat of global warming leads to a very different slice of content than would be found in any traditional disciplinary curriculum.

The second property of task-structured curricula, the issue of "depth," follows quickly from the first property. Clearly, it would not be possible — or necessary —

for the Global Warming curriculum to address all of the above topics to a high level of scientific detail. As with all task-structured curricula, the task not only dictates what will be taught, it also dictates the manner in which the various portions of disciplinary content must be understood by students. Within the context of the GWP, extensive attention to the processes underlying photosynthesis, for example, would probably not be sensible.

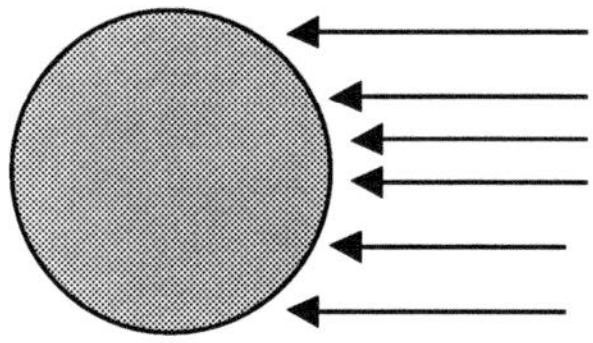

Figure 5. Incoming solar energy striking the earth.

Notice that the arrows near the equator are perpendicular to the surface. In contrast, the arrows nearer the poles strike the Earth at an angle that deviates more greatly from the perpendicular.

Because the issues here are very important, we want to illustrate them further by working through one part of the Global Warming curriculum in some detail. As described above, in the GWP, students must learn what happens to the solar energy that is incident upon the Earth. Some of the relevant information was shown in the diagram in Figure 3, which illustrates the Earth's energy balance. In the teacher's manual that accompanies the Global Warming curriculum, the teachers are given the following description:

> Sun's rays reach the earth's atmosphere.
>
> Some sunlight is reflected by the earth's atmosphere and surface.
>
> Some sunlight is absorbed by the earth's atmosphere and surface.
>
> The energy that is absorbed contributes to the warmth we feel. ...

Part of what students are intended to learn in the GWP is the story that is encapsulated in these few lines. However, in order for students to understand the climate phenomena that underlie global warming, some aspects of this story must be significantly elaborated; that is, more time must be devoted to unpacking some aspects of the relevant phenomena. For example, one place that the curriculum elaborates is around the factors that influence earth-atmosphere reflectivity. Students learn that the fraction of light that is reflected by the Earth's surface and atmosphere varies over the Earth. Roughly speaking, where the surface and atmosphere are lighter in color, more sunlight bounces off, and the Earth is cooler. In contrast, where the surface is darker, less sunlight bounces off, and the Earth is warmer. This content is important primarily because the enhancement of reflectivity is one type of mitigation strategy for global warming.

Another place that the curriculum elaborates is in how incoming solar energy varies over the surface of the earth. The students learn that, because sunlight strikes the Earth more directly near the equator, the amount of incoming solar energy is greater nearer the equator than near the poles (refer to Figure 5). The GWP devotes a significant amount of time to this issue. For example, as mentioned above, the students perform a lab in which a flashlight is shone on a piece of paper, with the flashlight inclined at various angles. In addition, they engage in computer-based analyses of scientific visualizations that show how incoming solar energy varies over the surface of the Earth (refer to Figure 6).

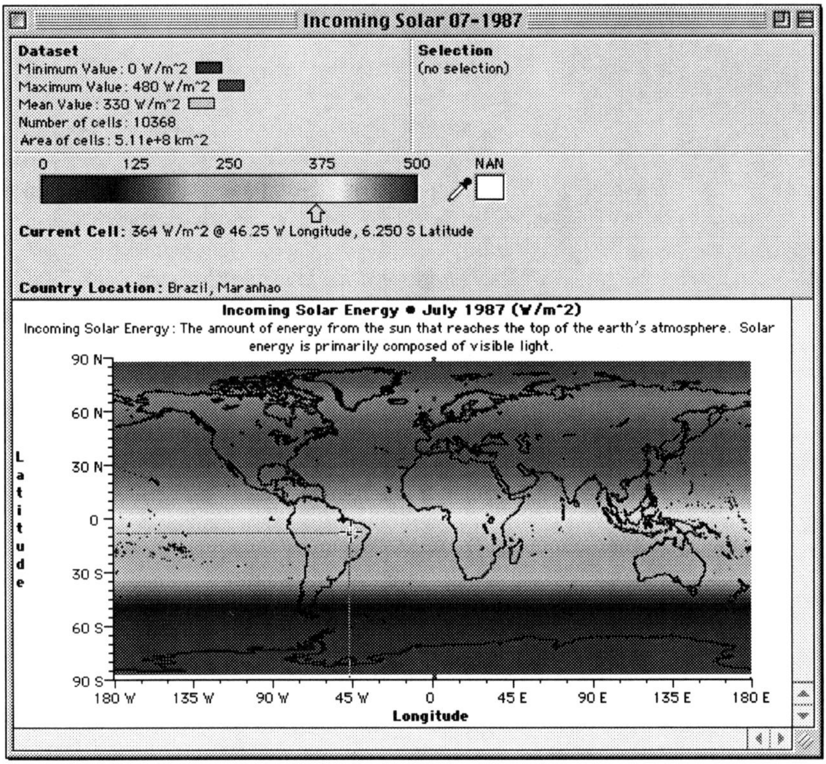

Figure 6. A scientific visualization that shows the variation in incoming solar energy over the surface of the Earth

Once again, the point here is that where the curriculum elaborates depends in a sensitive manner on the requirements of the task. This is made somewhat clear when, as above, we note where the curriculum chooses to focus attention. However, this sensitive dependence on task can be made even more striking if we look closely at what is *not* addressed by the curriculum. For illustration, consider again issues surrounding reflectivity and incoming solar energy. We can ask: How much do students really need to know to understand the relevant issues? For example, do they need to know *why* lighter colors reflect more than darker colors, or simply that they do? More fundamentally, do they need to know what light is? Do they need to know what energy is?

Once again, it would not be possible for the curriculum to answer all of these questions to a high level of scientific precision. The GWP does work hard to teach students the simple fact that lighter colors reflect more than darker colors, but it does not attempt to teach them *why*. More profoundly, the curriculum does not ever directly attempt to help students learn about the basic nature of light. Instead, the GWP chooses to rely on the intuitive understandings of light that students possess prior to the curriculum. This is important, and we will say more in a moment.

Actually, the issues here are a little more subtle than we have painted them. Really, in every curriculum, we must rely on the intuitive understandings of students. For example, although traditional physics instruction seeks to build up content in a systematic manner, virtually all physics curricula assume that students know what a *physical object* is, and that they have some idea of what a *force* is. Strictly speaking, the notion of force is defined by the role it plays in physical laws. Nonetheless, it is traditional for physics instruction to define force as simply "a push or a pull." This is similar, in some respects, to our reliance, in the GWP, on what students know about light. Still, we believe that there are fundamental differences in how task-structured curricula rely on prior understandings; there are differences in the *extent and diversity* of prior knowledge resources that are built upon in a task-structured curriculum.

In summary, in this section we have looked at the nature of content in the GWP curriculum unit, and we have seen how the GWP exemplifies the properties that we ascribed to task-structured curricula: The content covered within this single unit cuts across multiple traditional topics and disciplines, and the manner in which specific issues are addressed depends in a sensitive manner on the task.

LEARNING IN THE GWP: THE BOOTSTRAPPING PROBLEM

Given the above discussion of content in the GWP — and in task-structured curricula generally — there are real reasons to worry about what students may be learning. A potential problem is encapsulated in what we earlier referred to as the bootstrapping problem: How can we expect students to work on problems and issues that cut across multiple disciplines if we have not already provided them with a solid foundation in these disciplines?

In the preceding sections, we hinted at one possible solution to the bootstrapping problem: We bootstrap by, in some cases, "making do" with intuitive understandings. This means that, for any task-structured curriculum, we make critical assumptions about where we can make do with students existing understandings. This, then, is something that we can attempt to validate empirically. We can ask: Are our assumptions valid? What happens in individual cases where students do not have what we assumed?

In this last part of the paper, we turn to our empirical studies of learning in the GWP, with an emphasis on the bootstrapping problem. Our purpose here is not to describe an extensive evaluation of the GWP. Rather, we will only attempt to provide the reader with a sense for the type of empirical work and analysis that we believe is required in order to understand issues of learning in task-structured curricula. Where we use data, we use it as a background for raising issues, rather than as the basis of a systematic analysis.

Data collection

In order to investigate issues of learning in the GWP, we designed clinical interviews to be conducted with students before and after the GWP. Our primary data corpus for this work was videotapes of these interviews.

The interview protocols have gone through several revisions, and have been employed in a number of contexts. Initially, we pilot-tested the interviews with 17 middle-school students in two schools that would provide the context for later study. These pilot interviews gave us a first read on the assumptions embedded in the GWP, and allowed refinement of the interviews.

During the following summer, we conducted pre- and post-interviews in the context of a summer workshop held at Northwestern University. For this summer workshop, we developed a shortened version of the GWP, designed to emphasize portions of the curriculum that were the focus of our interview (see below). In this context, we conducted pre- and post-interviews with 10 middle-school students (graduating eighth graders). These interviews were supplemented with observations of the workshop sessions.

In the final version of our data collection, we conducted interviews in the context of two full classroom enactments of the GWP, in 8th-grade classrooms at two different schools. Pre- and post-interviews were conducted with 9 students, 5 from one school, 4 from the other. A member of the research team also observed classroom sessions on a daily basis. The first school was a relatively high-achieving school, located in a middle class neighborhood on the outskirts of Chicago. The second school was a moderately achieving urban middle school. This latter school has a high percentage of first-generation immigrants, mainly from Southeast Asia.

The interview design

The emphasis in the clinical interviews was on exploring some of the bootstrapping assumptions built into the GWP. In this paper, we will discuss two of these assumptions. First, as we have already mentioned, the GW curriculum assumes that students have a suitable understanding of the nature of light when they enter the curriculum. The GWP curriculum was designed with the implicit assumption that students understood light as radiating out from a central source, covering a larger area with less intensity the greater the distance from the light source. We wanted to explore this assumption.

Another part of the presumed background that we thought might be problematic has to do with the physical structure of the Earth-Sun system. In the curriculum, we essentially presume that students know that the Earth and Sun are both large spheres, and these spheres are separated by some distance in space. If students do not have at least a rudimentary understanding of this sort, much of the curriculum may be nonsensical.

Based on our reading of the literature on prior conceptions, we had some expectations concerning these assumptions. For the case of the physical structure of the Earth-Sun system, we felt we had reason to be optimistic. Although the literature indicates limits to student understanding in this area, it nonetheless suggests that the students in our target population are likely to know that the Earth is a sphere in space (Vosniadou and Brewer 1992).

> *The first thing I want you to do is imagine that we're in a really big room with the lights out. There are no windows in the room, so the room is really dark. If you want, you can imagine that you're in a big room in this school, like your gym. Does your gym have windows?* (Possible little discussion here.)
>
> *Now imagine we've got this lamp in the dark room, and I turn it on.* (Show student the lamp.)
>
> *Imagine that we're pretty close to one of the walls in the gym, about this close, so that the light is on one of the walls.*
>
> *Okay, now I start walking away from the wall, carrying the lamp, and I stop when pretty far away*

Figure 7. An excerpt from the portion of the interview protocol intended to get at students understanding of the nature of light

In contrast, for the case of light, the literature suggests that there is real reason for concern. As summarized in Driver (1994), children seem to have problematic (for our case) conceptions that persist to middle school and beyond. Of particular worry for us was the observation that students in our target population might not even think of light as something that *travels.*

We set out to investigate both of these assumption by interviewing students before and after they participated in the GWP curriculum. In our interviews, we asked a variety of types of questions. Some of these questions were pointedly directed at the assumptions we were investigating. For example, in order to investigate students' understanding of the nature of light, we asked a set of questions based around a situation in which there is someone holding a lamp in an otherwise dark, large room. In addition, we asked questions that were intended to cut across both sets of issues. For example, we asked some questions about climate phenomena:

- Why is it warmer in Florida and colder in Alaska?
- Why is it warmer in the summer and colder in the winter?

When answering these questions, it was hoped that students would reveal some of their understanding of the nature of light, as well as the structure of the Earth-Sun system.

Dedra's understanding of the nature of light

We begin by looking at our interviews with one student, Dedra, who was a student in the summer workshop conducted by researchers at Northwestern University. Here, we will focus on questions pertaining to the nature of light.

As stated above, one set of questions asked students to imagine a simple situation in which there is someone holding a light bulb in an otherwise dark, large room. The students were asked to imagine, first, that the person holding the bulb is standing relatively close to a wall and that the light shines on the wall. Then they were told to imagine that the person gradually walks away from the wall. Finally, the interviewer would ask: "How does what you see on the wall change as I walk backwards from the wall?"

Figure 8. Dedra's drawing from the pre-interview

As mentioned above, the GWP curriculum was designed with the implicit assumption that students understood light as radiating out from a central source, covering a larger area with less intensity the greater the distance from the light source. In fact, most students, when asked the moving light bulb question, responded in a manner that is consistent with this model. However, in response to this question, Dedra, said that there would be an illuminated area on the wall, and that this would get smaller as the light bulb moved away from the wall. Here is an excerpt from her response (I = interviewer, D = Dedra):

> I: How does what you see on the wall change, as I walk backwards from the wall?
>
> D:Like, the reflection gets smaller?
>
> I: Uh-huh
>
> ....
>
> D: It gets smaller when you move back.
>
> ....

I: There's like a circle on the wall that you see?

D: Yeah.

I: And how does that circle change, it gets-

D: It gets smaller till it's gone.

When asked why the circle on the wall gets smaller Dedra responded:

D: Because you're moving further away from the wall.

I: Uh-huh.

D: And the light only shines like in an amount of space.

Like some other students, Dedra is answering these questions as if she is applying what we call the "sphere of illumination" model. This is a version of the problematic conception that we mentioned above. In this model, there is no sense in which the light travels from the light bulb to the wall. Instead, when the light is turned on, it instantaneously creates an illuminated area of fixed size around it; there is a sphere of light around the bulb. Understanding the model in this way can help us to understand, for example, why Dedra says that "light only shines like in an amount of space." It can also help us to understand why she says that the illuminated area on the wall will get smaller. As the bulb moves away from the wall, the intersection of the sphere with the wall is a smaller and smaller circle. Figure 9 shows a drawing that Dedra made to illustrate this situation.

Furthermore, Dedra was consistent across the range of probes we employed in the first interview. In this regard, one of Dedra's later answers is also very telling. As part of the interview, the students were asked what happens at the instant the lamp is turned on. In particular, they were asked if the wall is illuminated immediately or if there would be a short delay. The majority of students responded that there would be a very short delay corresponding to the time it takes for the light to travel from the light bulb to the wall. But students reasoning from a sphere of illumination model said that the wall would be illuminated immediately.

I: At the instant that the bulb comes on does the light appear on the wall right away? Or is there like a little delay from when the bulb lights up till when the light's on the wall?

D: Like, does the bulb come right on or does it like take a while for it to get all lit up?

I: It comes all right on so it's completely bright right away.

D: So, it comes directly on the wall.

I: So it should be right at that instant.

D: Hm-mm.

...

D: It would come right away- the reflection would come right on the wall.

Furthermore, Dedra — and other like-minded students — clung tightly to the assertion that the wall would be lit up instantaneously. In particular, they were clear that there was not even an extremely short delay, too short to be observed.

Given the design of the GWP curriculum, answers of this sort are reason for us to be concerned. What happens to students, such as Dedra, that seem to possess an understanding of the nature of light that is very different from the accepted scientific model? Is their understanding of light sufficient to support the learning that must go on in the curriculum? Will they "pick up" a more useful model of light? As we stated above, students may be exposed to descriptions such as the following:

Sun's rays reach the earth's atmosphere.

Some sunlight is reflected by the earth's atmosphere and surface.

Some sunlight is absorbed by the earth's atmosphere and surface.

The energy that is absorbed contributes to the warmth we feel. ...

If Dedra doesn't even think of light as something that travels, what sense would she make of these statements? There is a range of possible answers to this question. At one extreme is the possibility that these statements will simply be meaningless to Dedra; she may not be able to understand them in any useful way. At the other extreme is the possibility that Dedra will "pick up" a more appropriate way of thinking about light, and she will understand these statements precisely as they are intended.

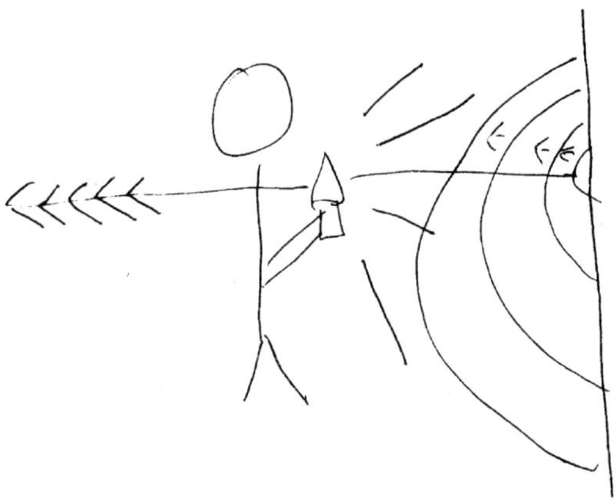

Figure 9. Dedra's drawing made during the follow-up interview

We cannot, of course, give a fully general answer to these questions. However, we can present some excerpts from an interview we conducted with Dedra following her participation in the summer workshop. In this follow-up interview, we asked a series of questions that were similar to those in the earlier interview. When asked what would happen when the bulb is moved away from the wall, Dedra responded that the illuminated area would get bigger rather than smaller (refer to Figure 9), and that the intensity would decrease:

> D: Like the reflection gets bigger. The farther you get away the dimmer the light gets.

Furthermore, throughout the follow-up interview, Dedra answered questions as if she were applying a model in which light travels outward from the source. For example, when asked again about the situation when the light bulb is switched on, she said there would be a very brief delay between when the light comes on and the wall lights up.

> I: Is there a delay between when I turn it on and the wall lights up, or does it light up right away.
>
> D: It lights up right away.
>
> I: ... And is that because if there was a delay it would be too slow to see, or there just isn't a delay, it's instantaneous.
>
> D: There's probably a little delay, but not that the naked eye can see.
>
> I: Okay. Too fast.
>
> D: Hmm-mm.

Some care is needed in interpreting observations of this sort. It is not necessarily correct to say that, before the summer course, Dedra had one model and, after the course, she had the other. The range of possibilities is much broader. For example, both models may, in some sense, have been constructed during the respective interviews. Furthermore, it is possible that both ways of reasoning were accessible to her prior to the course. It may have been the case that, even before the course, Dedra could answer questions about light in a variety of ways, as if she were applying a range of different models of light.

Since this is an important point, we will elaborate just briefly. If we want to be conservative, then we must adopt a general stance: We must assume that Dedra's answers to these questions are generated by a complex ensemble of knowledge. We cannot assume that Dedra's answers can be attributed to any underlying mental model of light; rather, in the more general case, it is attributable to a collection of cognitive bits and pieces. If we view Dedra's learning about light from this perspective, there is the possibility that we are only seeing a shift in how Dedra *tends* to assemble these bits and pieces to generate an answer, at least in the particular situations that we happened to ask about in the interview.

Subtleties of this sort have significant implications for how we must understand the learning that occurs in the GWP, and in task-structured curricula generally. If it turns out that what we are seeing, in Dedra's case, is the tuning of a complex ensemble of knowledge, then the difficulties may not be as bad as one might have thought. It means, first, that Dedra really possesses many of the resources that are necessary for participation in the GWP. Second, it makes it more plausible that Dedra's understanding of the nature of light can evolve, in a useful manner, within the context of the GWP. If it were the case that Dedra possessed a single, monolithic, model of light, then replacing this model might well require explicit attention by the curriculum. In contrast, if there must only be a shift in how Dedra applies some of the bits and pieces of knowledge that she already possesses, then it is more plausible that this learning could happen in the background, without explicitly being addressed by the curriculum.

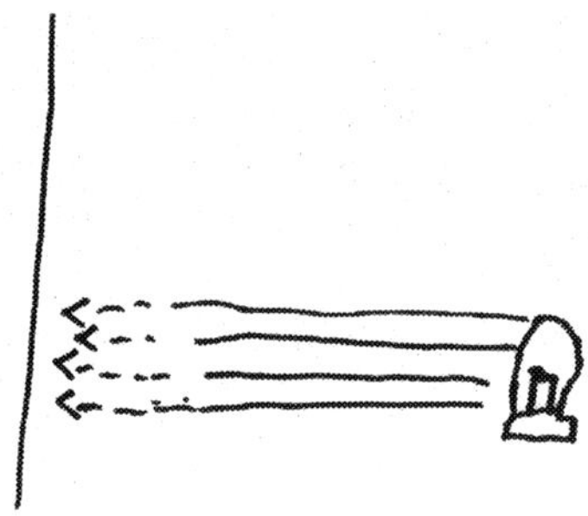

Figure 10. A drawing by a student that shows rays of light weakening as they move farther form the source

For the sake of completeness, and in order to situate our interview with Dedra, we want to say just a little about how the other students we interviewed answered questions about the nature of light. First, during the pre-interviews, all but a few of the 36 students interviewed answered the questions in a manner consistent with a model of light as something that radiates outward from a source. When asked, for example, what would happen when the lamp was switched on, these students would say there would be a brief delay before the wall is illuminated, during which time the light travels from the lamp to the wall.

Nonetheless, there was some interesting variety within these traveling models. For example, students varied somewhat in how they accounted for the fact that the illuminated area on the wall would become dimmer as the lamp was moved away from the wall. Some students, for instance, attributed this dimming to the fact that the light spreads out as it moves away from the source. (This, roughly speaking, is the answer given by scientists.) In contrast, other students maintained that individual rays of light would become weaker as they moved away from the source (refer to Figure 11).

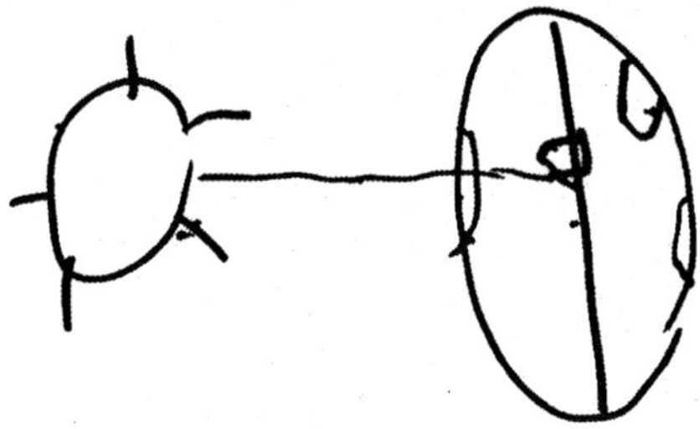

Figure 11. Kelly's drawing showing light striking the equator and then spreading out

In still other cases, even though students answered questions in a manner consistent with a radiating model, their models differed in dramatic respects from the accepted scientific model. For example, one student, Kelly, sometimes spoke of light "spreading out" after striking a surface. When she was asked, for instance, why Florida is typically warmer than Alaska, she said that this is because sunlight strikes directly on the equator, and then spreads out from this point, weakening along the way (refer to Figure 10).

> K: When the light rays travel to the equator it spreads out. (She makes the little circle around the point of contact.) And maybe it would take, um, the light rays; maybe it would take longer to get up here (points near the pole).

Again, our purpose here has not been, principally, to give taxonomy of student models of light. Rather, our purpose has been to illustrate some issues of learning in task-structured curricula, and to demonstrate how we might begin to explore these issues empirically. In particular, we looked at the background assumptions of the curriculum concerning the nature of light. In this regard, the results of our modest presentation are equivocal. It is certainly the case that many students do not possess an understanding of light that is entirely consistent with the accepted scientific model. However, it is not clear whether this will prevent the curriculum from achieving its learning goals. We saw, in Dedra's case, that it is at least possible for a student to "pick up" a more appropriate model of light, even though the nature of light was not explicitly addressed.

Furthermore, even if a student persists in giving non-canonical accounts of light-related phenomena, it is not clear whether this will prevent them from engaging in the task that is at the heart of the curriculum. The ultimate goal is for students to make recommendations as advisors to their chosen country. In order to make these recommendations, it may be enough, for example, for students to know that the amount of incoming solar energy varies with latitude, without knowing why this variation occurs. We will say more about this below.

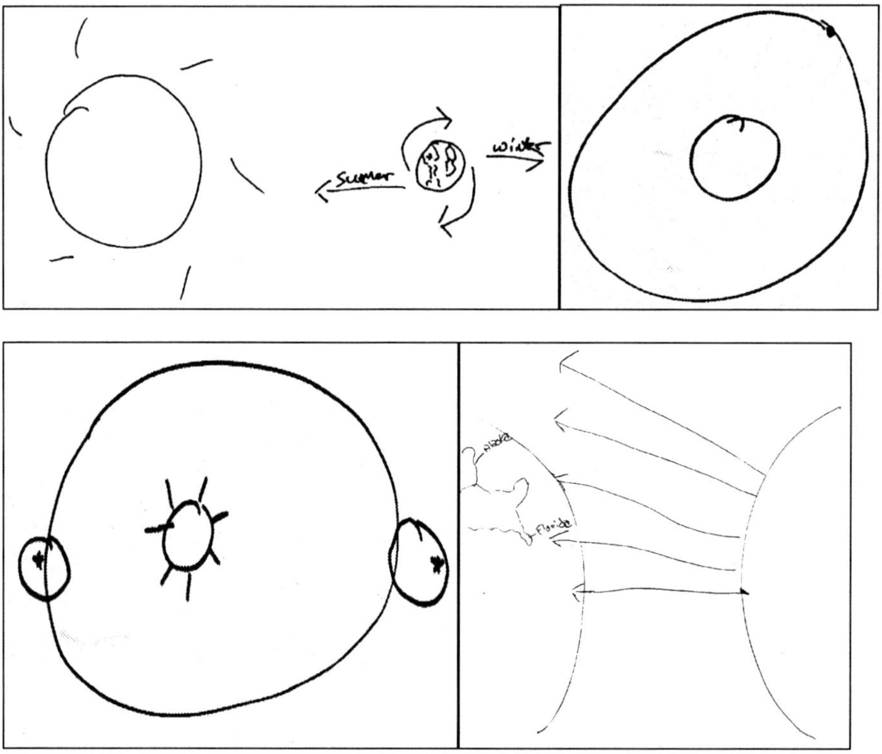

Figure 12. Student drawings made to explain seasonal temperature variation

These drawings illustrate students' understanding of the structure of the Earth-Sun system.

The structure of the Earth-Sun system

We were also concerned with what students knew about the structure of the Earth-Sun system. Did the students entering our curriculum have a sufficient model of the Earth-Sun system? Did they know that the Earth, which is a sphere, travels around the Sun, which is also a sphere? Our observations here are, in some respects, easy to report; this assumption proved to be largely unproblematic. Within the context of the questions that we asked, all of the students we interviewed seemed to be answering as if they had a model in which the Earth and Sun were spheres. This result is consistent with the research mentioned above; for example, prior research in this area suggests that, by this age, students know that the Earth is a Sphere (Vosniadou and Brewer 1992). For illustration, Figure 12 shows a number of drawings made by students when they were asked to discuss seasonal temperature variation.

However, there were some important limitations to students' understanding of the structure of the Earth-Sun system. First, not all of the students seemed to have a clear understanding of the Earth's motion — that it rotates on its axis and revolves around the sun. In addition, there were problems of *scale* that have important implications for the understanding of climate phenomena. If the Earth-Sun system is drawn to scale, it appears roughly as shown in refer to Figure 12 (the Earth is essentially invisible at this scale). Because the Earth is small in comparison to the distance between the Earth and Sun, all locations on the Earth can be treated as if they are the same distance from the sun. Furthermore, it is because the Earth and Sun are small in comparison to their relative separation that we can treat all incident radiation as consisting of parallel rays (refer to Figure 13).

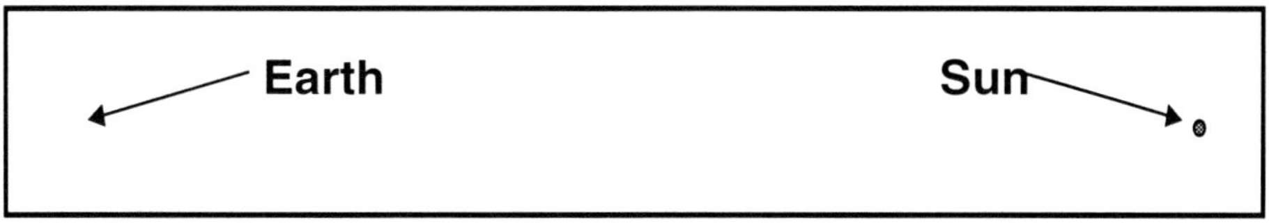

Figure 13. The Earth-Sun system drawn to scale

Clearly, the drawings reproduced in Figure 13 are not consistent with the true scale of the Earth-Sun system. However, we must be careful when interpreting what these drawings indicate about students' understanding. These drawings were made to support an explanation of the seasons, not necessarily to accurately represent scale. Thus, if they were explicitly asked to produce a drawing that was true-to-scale, many of these students might make a very different drawing. Nonetheless, there is a problem here. At the least, the students do not understand that the relative differences in lengths involved are dramatic enough that they undermine their explanations.

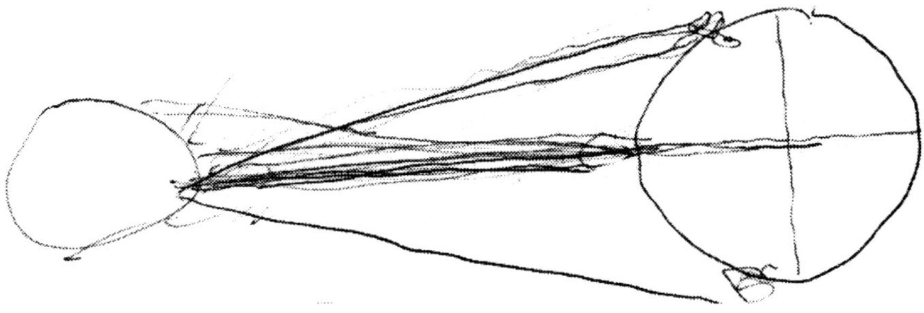

Figure 14. Mitchell's drawing that is intended to illustrate why it is warmer near the equator and colder near the poles. The Earth is on the right

Here, we will just give one brief example to illustration how student explanations were undermined. During the follow-up interview, Mitchell drew the diagram in

Figure 14, while explaining why the Earth is warmer near the equator and colder near the poles. Almost certainly, he did not intend to draw this diagram to scale. Nonetheless, notice that Mitchell's explanation relies on the assumption that the distance from the Sun to the Earth's equator is significantly smaller than from the Sun to the Earth's poles.

> M: Like the Earth's equator is the closest part to the sun. That's why it's hot and humid around there. But like, the top, Antarctica, that's why the North Pole and South Pole are so cold because it takes time for light to travel all the way up there and you'll get less sun light.

As with our discussion of the nature of light, our purpose here is not to give a systematic accounting of all student models of the Earth-Sun system. Rather, our purpose has been to illustrate issues of learning in task-structured curricula. As with the nature of light, the structure of the Earth-Sun is not explicitly addressed in the GWP curriculum. It is assumed that students have a least a rough understanding of this structure, that they know the Earth and Sun are spheres in space. This assumption seems to be largely borne out, but there are a few caveats. Most notably, students' understanding of issues of scale was problematic.

This discussion has implications for our studies of the learning that occurs in the GWP. Issues of scale are not explicitly addressed, but there are places in the curriculum that students could acquire this understanding, essentially as a side-effect of the curricular activities. At the least, they may learn what types of explanations of climate variation are acceptable (i.e., explanations based on angle of incidence, rather than distance of travel). But a question we must ask ourselves in these cases is: What if students never get it? What if, like Mitchell, some students in the GWP never understand why temperature varies with latitude? The curriculum is intended to help students understand why incoming solar energy and temperature vary with latitude; it is one of the explicitly stated learning goals. In that respect, if students never understand the reason for this variation, the GWP has failed. However, it is less clear whether such a failure would undermine other learning goals or the completion of the task. Students do need to know that incoming solar energy varies with latitude. In order to discuss climactic variation and, in particular, to discuss climate issues for the country whose policies they are trying to inform, they need to know that this variation occurs. But, as stated above, it is not clearly necessary for students to know *why* incoming solar energy varies with latitude. When making their proposals, students could essentially treat the amount of incoming solar energy as an input parameter, without further explanation. They would be missing an important part of the curriculum, but they might still be able to complete other aspects of the task.

CONCLUSION

In this paper, we have contrasted two styles of science curricula, task-structured and content-structured curricula. We argued that, by their nature, task-structured curricula embody a different approach to science content. Individual task-structured curricula tend to address content across a range of traditional disciplines. In addition, the manner—or "depth"—to which the various portions of disciplinary content must be addressed is, to a great extent, dictated by the governing task.

Furthermore, we stated that, because task-structured curricula do not build up content in a systematic manner, there is reason to worry whether these curricula can achieve their learning goals. We referred to this issue as the *bootstrapping problem*. In response, we argued that empirical work is required in order to determine the extent to which the bootstrapping problem undermines the success of task-structured curricula. We did not attempt to draw any final conclusions in this regard. Instead, we only attempted to illustrate the sort of empirical work that we believe is required, and to point the direction for future work.

In the future, we believe that both specific and general studies are required. We need to look at the specific curricula that we have designed, articulate the assumptions that underlie these curricula, and look empirically at whether these assumptions are borne out. To a certain extent, this work is already underway in the attempts of individual designers to determine the effectiveness of the learning activities that they create. However, we believe that a more wide-ranging approach is also necessary; we need to look across task-structured curricula with an eye toward developing a broader account of the learning processes that occur within these curricula.

The hope is that such an endeavor can contribute to the formulation of principles for the design of task-structured curricula. Ultimately, we would like better answers to such questions as: How can we design tasks so as to best make connections to interesting content? Where is it necessary to unpack content in more detail and when can we "make do" with the understanding that students bring to the classroom? How can we design partial understandings of more difficult content, so that students can make progress on a task without fully mastering the rigors of an advanced discipline? When we can answer these questions, we will be further along in our attempts to develop a principled practice of science education.

REFERENCES

Barron, Brigid J. S., Daniel L. Schwartz, Nancy J. Vye, Allison Moore, Anthony Petrosino, Linda Zech, John D. Bransford, and The Cognition and Technology Group at Vanderbilt. 1998. Doing with Understanding: Lessons from Research on Problem- and Project-Based Learning. *Journal of the Learning Sciences* 7 (3&4): 271-311.

Barrows, Howard S., and Robyn M. Tamblyn. 1980. *Problem-Based Learning: An Approach to Medical Education*. New York: Springer.

Driver, Rosalind. 1994. *Making Sense of Secondary Science: Research into Children's Ideas*. New York: Routledge.

Duschl, Richard A., and Drew H. Gitomer. 1997. Strategies and Challenges to Changing the Focus of Assessment and Instruction in Science Classrooms. *Educational Assessment* 4 (1): 37-73.

Eylon, Bat-Sheva, and Marcia Linn. 1988. Learning and Instruction: An Examination of Four Research Perspectives in Science Education. *Review of Educational Research* 58 (3): 251-303.

Gitomer, Drew H., and Richard A. Duschl. 1998. Emerging Issues and Practices in Science Education. In *International Handbook of Science Education*, edited by B. J. Fraser and K. G. Tobin. Dordrecht, The Netherlands: Kluwer Academic Publishers.

Harel, Idit. 1991. *Children Designers: Interdisciplinary Constructions for Learning and Knowing Mathematics in a Computer-Rich School*. Norwood, NJ: Ablex.

Kolodner, Janet L., David Crismond, Jackie Gray, Jennifer Holbrook, and Sadhana Puntambekar. 1998. Learning by Design from Theory to Practice. Paper read at International Conference of the Learning Sciences 1998.

Krajcik, Joseph S., Charlene M. Czerniak, and Carl Berger. 1999. *Teaching Children Science: A Project-Based Approach*: McGraw Hill College Press.

Marx, Ronald W., Phyllis C. Blumenfeld, Joseph S. Krajcik, Merrie Blunk, Barbara Crawford, Beverly Kelly, and Karen M. Meyer. 1994. Enacting Project-Based Science: Experiences of Four Middle Grade Teachers. *The Elementary School Journal* 94 (5): 517-538.

National Research Council. 1996. *National Science Education Standards*. Washington, DC: National Academy Press.

Pfundt, Helga, and Reinders Duit. 2000. *Bibliography: Students' Alternative Frameworks and Science Education*. Kiel: Institute for Science Education.

Schank, Roger C., Andrew Fano, Benjamin Bell, and Menachem Jona. 1993/1994. The Design of Goal-Based Scenarios. *The Journal of the Learning Sciences* 3 (4): 305-346.

Smith, Carol, Deborah Maclin, Lorraine Grosslight, and Helen Davis. 1997. Teaching for Understanding: A Study of Students' Preinstruction Theories of Matter and a Comparison of the Effectiveness of Two Approaches to Teaching About Matter and Density. *Cognition and Instruction* 15 (3): 317-393.

Smith, John P., Andrea A. diSessa, and Jeremy Roschelle. 1993. Misconceptions Reconceived: A Constructivist Analysis of Knowledge in Transition. *Journal of the Learning Sciences* 3 (2): 115-163.

Vosniadou, Stella, and William F. Brewer. 1992. Mental Models of the Earth: A Study of Conceptual Change in Childhood. *Cognitive Psychology* 24 (4): 535-585.

Wandersee, James H., Joel J. Mintzes, and Joseph D. Novak. 1994. Research on Alternative Conceptions in Science. In *Handbook of Research on Science Teaching and Learning: A Project of the National Science Teachers Association*, edited by D. L. Gabel. New York: Macmillan.

Williams, Susan M. 1992. Putting Case-Based Instruction into Context: Examples from Legal and Medical Education. *Journal of the learning sciences* 2: 367-427.

CHAPTER 12

SANDRA K. ABELL & JAMES T. MCDONALD

ENVISIONING A CURRICULUM OF INQUIRY IN THE ELEMENTARY SCHOOL

Most authors agree that science is both a collection of knowledge products (i.e., laws and theories), and a set of practices (i.e., observation, experimentation, argument). It would follow that classroom science inquiry should emphasize both science as knowledge products and science as practices. However, our elementary science classrooms have been characterized typically by one of two orientations to science teaching, each of which has emphasized one facet of science to the exclusion of the other. In the didactic orientation (Anderson and Smith, 1987), science instruction emphasizes the products of science, and textbooks dominate. In the 1981 Project Synthesis report, Pratt summarized how elementary teachers depended on textbooks as the authority for science teaching. Recent TIMSS findings (Schmidt, McKnight, and Raizen, 1997) demonstrate that the trend of relying on textbooks and low level facts in elementary science continues. Science is not alone when it comes to an overemphasis on knowledge reproduction in the elementary school. Published curricula such as Saxon mathematics (1992) and the Shurley method of teaching language arts (1992) are being purchased by school districts eager to raise test scores and garner state dollars, without regard for long term learning.

Where the didactic orientation has left off, often activitymania (Moscovici and Nelson, 1998) has taken its place. An activity-driven orientation (Anderson and Smith, 1987) results in students spending much time "doing science," but little time thinking, talking, posing questions, or constructing explanations. In this orientation, science as a practice is emphasized, but developing an understanding of science concepts is neglected. Our preservice elementary teachers often hold this orientation, believing that science should be "active" and "hands-on" (Abell, Bryan, and Anderson, 1998). Their goal is limited to making science fun; the goal of achieving understanding, if it surfaces at all, is only an afterthought.

In our experience as elementary teachers, as student teacher supervisors, and as elementary science methods course instructors, we have witnessed these two orientations reproduced in classroom after classroom. State science education standards often reinforce the product/process dichotomy by placing content standards and inquiry standards in separate sections. Teachers who teach content without process or process without content may deem that standards are being met. We believe this is a limited view of inquiry in elementary science. The two facets of science, the products and the practices, are not separated from each other when scientists do science; they should not be separated in our science classrooms. What we call for is an integrated approach to inquiry (Abell, 1999), where both the

L.B. Flick and N.G. Lederman (eds.) Scientific Inquiry and Nature of Science, 249-261.

products and the practices of science are regarded as critical to developing scientific understanding.

In this chapter we would like to examine the integrated inquiry approach to elementary science. First we present a theoretical framework grounded in a socio-cognitive perspective of learning that supports our vision. Following that, we present two stories of elementary science instruction and analyze them in terms of key features of inquiry in elementary science. We then discuss the constraints that elementary teachers face and the benefits they perceive related to inquiry-oriented science instruction. Finally we claim that, despite such constraints, science is uniquely positioned to lead the way in developing reform-minded, inquiry-based instruction across the elementary curriculum.

A SOCIO-COGNITIVE PERSPECTIVE ON SCIENCE LEARNING

Our view of inquiry is grounded in our beliefs about learning. Learning is mediated by psychological tools such as language, mathematical symbol systems, drawings, works of art, etc. (Vygotsky, 1962, 1987; Wertsch, 1985). Of the psychological tools that mediate thoughts, feelings, and behaviors, language is the most important. Learners internalize language and develop higher mental functions. Language is a tool that enables the emergence of self-awareness and control of actions, yet language is also part of action. "All [symbol] systems are tools embedded in action and give rise to meaning" (Knox and Stevens, 1993, p. 15). If we want students to generate meaning from school science, they must have access to language in action. This is strong support for an integrated approach to inquiry that attends to both the products and practices of science.

Learning is situated. Lave (1988) argued that learning, as it normally occurs, is a function of the activity, context, and culture in which it occurs. This contrasts with much of school science, which involves individual students dealing with knowledge of the abstract outside of an experiential context. Social interaction is a critical component of situated learning—learners take part in a "community of practice" which embodies certain language, beliefs, and behaviors. "A person's intentions to learn are engaged and the meaning of learning is configured through the process of becoming a full participant in a sociocultural practice" (Lave and Wenger, 1991, p. 122). Integrated inquiry involves learners in the collaborative social practice of doing science and communicating about their doing and thinking.

This social view of learning assumes that learning is more apt to occur when students work on authentic tasks that reflect the real world (Jonassen, 1994). Brown, Collins, and Duguid (1989) recommended a cognitive apprenticeship to support learning in a subject domain by enabling students to acquire, develop, and use cognitive tools in authentic domain activity. We view integrated inquiry as a strong example of authentic domain activity. Thus, our view of learning science is built on the assumptions that language is critical, that knowing is social, and that authentic tasks are required. We believe that integrated inquiry in elementary science is an instructional orientation aligned with these assumptions.

STORIES OF INQUIRY IN ELEMENTARY SCIENCE

We generated the following stories from our observations of elementary science classrooms engaged in various types of inquiry. We selected these stories because we worked closely with the classroom teachers, Jessica Smithton and Shirley Morrow, as observers and participants in their classrooms, and found these examples of classroom inquiry to be instructive. Jessica and Shirley, although experienced, had typically taught science from a didactic or activity-driven orientation, and were experimenting with a more integrated approach. We have selected parts of one unit instruction from each classroom to illustrate our views on classroom inquiry. In the first story, Mrs. Smithton's fourth grade is involved with teacher-designed sinking and floating lessons. The second story takes place during Mrs. Morrow's fifth grade erosion unit, where the lessons were based on a published science curriculum. Following each story, we analyze the features of inquiry that were present.

Inquiring into Sinking and Floating in Fourth Grade

"Welcome to the Acme Toy Company. You are members of a research and development (R & D) team assigned to the Water Toys division. Our company operates in 2 main types of settings. One is the scientists meeting, where everyone comes together to discuss their work. For this meeting of the minds, participating by sharing and listening is essential. The second setting is your team work. Think about what will help your team function smoothly. Today's problem is for your teams to try to figure out which kinds of things sink and which kinds of things float. We will use this information later when we design our water toys. Record your team's findings, and be ready to share with other teams at the scientists meeting."

So began the first day of a sink and float unit. The fourth grade textbook chapter on water covered the following: the water cycle; properties of water such as surface tension, dissolving, and sinking/floating; the water supply; weathering; bodies of water; adaptations to underwater life; saltwater habitats; and water and our bodies. Mrs. Smithton believed that such a laundry list of topics would fail to lead to understanding, and decided to delve into sinking and floating for 5 weeks, basically ignoring the rest of the chapter. She set a context for investigation that her students could relate to, the Acme Toy Company. Students would work first as scientists, trying to understand the concept of density. Next they would be engineers, designing sinking and floating toys. Finally the teams would plan and carry out their own investigations related to a new water toy, the Cartesian diver.

During the first week of the unit, students engaged in common sink/float activities—predicting and testing which objects would sink or float, comparing weights of floaters and sinkers, observing discrepant events (like a huge, heavy log floating), and developing explanations for their observations. They redesigned sinkers to float and floaters to sink. At the end of the week, Mrs. Smithton asked students to consider the variety of ideas teams had generated about sinking and floating:

- Things that are heavy sink. Things that are light float.
- Some things that are big can float and some really small things can sink.
- Things that are small and heavy sink. Things that are big and heavy float.
- It depends on the material the object is made of whether it sinks or floats.
- Water pressure keeps things up.

There was certainly no consensus explanation for sinking and floating, but lots of ideas were part of the public discussion in this school science community.

During week two, students explored liquids of different densities by making liquid layers. In week three they explored sinking and floating of various solids in different liquids. Punctuating the explorations were teacher-constructed design problems related to the Acme Toy Company. For example: "Make a prototype submarine toy that will float in the middle of salt or fresh water." Throughout these two weeks, students continued to invent explanations and compare them with each other. Many ideas about sinking and floating from the first week were repeatedly mentioned, sometimes presented with counter evidence from class activities. A new idea also arose: "Things that sink weigh more than water. Things that float weigh less than water." Students wondered how they might test this new theory. Discussion revolved around how much water to use in the test. Several students believed that the test should be fair, that they should use the same "amount" of water as the "amount" of the thing they were testing. Mrs. Smithton realized that by "amount," students meant "volume," and helped them to design an ingenious test. They decided to compare the "weight" of a wooden block that floated to the "weight" of the same volume of water. Rather than teach the students the formula for the volume of the block, Mrs. Smithton asked the students how they would know when the volumes were equal. Students suggested she build a container for the water out of cardboard lined with wax paper that would be the same size and shape as the block. Interest was high as the demonstration proceeded. Mrs. Smithton poured water into the container. She then placed the wooden block on one side of a balance, and poured the water from the cardboard "block" out into the other. The water side went down. One student recommended that they revise their new theory: "If something is heavier than the same amount of water, it will sink. If something is lighter than the same amount of water, it will float."

Of course, not every student was ready to understand or accept this explanation. Most, however, realized that the simple explanation, "Heavy things sink and light things float," was not good enough to account for all of the evidence they had collected. As the students designed their own sinking and floating toys, and later carried out student-generated Cartesian diver investigations, they had the opportunity to test and refine their sinking and floating explanations.

The sinking and floating unit ended with group presentations and a series of written assessments--asking students to explain sinking and floating, to design a new divers investigation, and to assess their own and their classmates' performance in team activities. Mrs. Smithton was somewhat frustrated with the outcomes of the assessments, because not all of the students had arrived at sinking/floating ideas that were consistent with scientific explanations. "I feel we have opened up a lot of ideas and challenges to the students' thinking, but we have not done a very good job of helping them put all of this stuff together." She also wondered if it was reasonable to expect all students to achieve the same outcome given their diverse starting points.

What Does the Sinking/Floating Story Illustrate about Inquiry?

In the sinking and floating unit, many opportunities for inquiry occurred. Not all of them were opportunities for "full inquiry" (National Research Council, 1996), but taken as a whole, the activities reflected an inquiry orientation. "In successful science classrooms, teachers and students collaborate in the pursuit of ideas, and students quite often initiate new activities related to an inquiry" (National Research Council, 1996, p. 33). This focus on the pursuit of ideas was evident throughout the sinking and floating unit. Whether the students were carrying out teacher-designed sink/float activities, creating as a class a way to test the water and weight theory, or planning and carrying out a divers investigation in teams, they were searching for understanding. They were engaged in the processes of science--observing, designing, planning, collecting data and the like--but all of their activity was directed toward building understanding of the science products. We agree with Mrs. Smithton that it may have been unreasonable to expect that everyone would understand sinking and floating in the same way by the end of the unit. A more reasonable expectation might be for students to move beyond their initial science ideas as they articulate their ideas, support their explanations with evidence, question their ideas, and the like.

We believe that the sinking and floating unit represents an integrated inquiry approach grounded in a socio-cognitive perspective. Although firsthand experience was a major component of the instruction, the priority of language is evident throughout the unit. Students were not only asked to do, they were asked to think, talk, and write about their ideas. The instruction alternated between team activities and large group scientists meetings. In each setting, the emphasis was on the social—how knowledge could be built together—vs. on an individual knowledge reception model. And finally in terms of authenticity, although some lessons came from a somewhat artificial school science repertoire (e.g., making liquid layers in drinking straws), students were often engaged in authentic activities of science using materials in contexts that were familiar to them. The teacher acted as a more capable adult who scaffolded activities and monitored learners throughout the unit. By the end of the unit, Mrs. Smithton expected teams to be more independent in asking questions, designing tests, and reporting their findings.

Civil Engineering Design Teams In Fifth Grade

It was a cool crisp autumn afternoon, and the fifth graders were busy with a task proposed by the curriculum: design a dam in a stream table that would protect a fictitious town prone to flash flooding from the Gaveo River. The design teams, three to five students each, had been working together for two months, studying erosion using stream tables. For this design task, the teams had to consider several factors: the location of the town and dam, the materials they would use to construct the dam, and other modifications that might be useful. According to Mrs. Morrow, they would make a presentation on the dam design to their peers and the Gaveo Town Council in two days. At this presentation they would be asked to share their solution to the problem and to demonstrate the effectiveness of their design. The challenge for the first part of the lesson would be to plan and run a test on the structural design of their dam. They could make any modifications that they felt were necessary after the test.

What prior knowledge would the students apply to the design of their dam? In earlier lessons, the *Land and Water* curriculum (National Science Resources Center, 1997) had addressed several aspects of erosion: slope; rate of flow or speed of the water; soil components; the water cycle; rain; human impact on the land in the form of dams and other structures; and how vegetation affects erosion. The dam-building lesson would allow the students to apply their observations, knowledge and experiences with the stream table to a real world problem. This lesson would also let the students work as a team to negotiate a viable solution. In these ways, the students would be emulating the work of scientists and engineers. The students would present their ideas to a group of parents, the Gaveo Town Council, who would rank the proposals and provide feedback to each engineering team.

Students met with their design teams to deliberate the location of the dam, the materials to use in its construction, and how to protect the town from flash floods. Once they had chosen materials, team members built structures in their stream tables. Interaction among the students was high, and they exchanged many ideas. In one of the teams, the following conversation ensued. "I have an idea. Why don't we use craft sticks to construct two walls that are close to one another? Then we can fill in the space in between the walls with gravel and sand." Another group member offered, "Yeah and then we can pack dirt up against the dam to reinforce it." A third "engineer" entered the discussion with, "We can also build walls down channel from the dam to reinforce the area around the town and the houses."

Teams, once they had completed their plans, conducted a test run with a liter of water. When Mrs. Morrow asked what considerations they had made when planning their dam, team members mentioned the speed of the water, how much sediment the water was carrying, and the slope of the stream channel leading up to the dam. The test run confirmed that one team's design was basically sound, but clarified a need to reinforce their structure. They used toothpicks to hold up the walls of the craft stick dam. Other modifications included reinforcing the retaining walls in and near the town, and lining the channel with gravel to slow down the speed of the water.

The engineering teams finished the first day of the lesson by writing down their observations, drawings their structures, and preparing their presentations for the Gaveo Town Council. Some teams constructed posters with drawings of their prior stream table research for their presentations. All of the teams decided that using the stream table itself would be a good way to demonstrate their solutions to the problem of flash flooding. They recorded their ideas in science journals, on planning sheets, and with presentation materials (posters, index cards, reports). Later Mrs. Morrow used these individual and group artifacts as assessments to gauge student understanding.

Two days later, students acted as experts as they presented their proposals to the Gaveo Town Council. Five parents came to class to witness each team's solution to the flash flooding issue. After each presentation, members of the class and the Council gathered around the team's stream table to watch as 2 liters of water streamed through a cup with a hole in the bottom and drained into the channel of the Gaveo River. Only about half of the teams constructed a dam that held the water back and away from the town. However, the discussion that followed among the teams and the Council was rich. Because there were differences in the designs and their effectiveness, students were required to analyze the factors that affected erosion in the stream table.

Student design teams dominated the discussion, which took on the form of a debate where students broke down the problem and considered different points of view. The first part of the discussion centered on the choice of materials used in construction of the dams. The students determined that using several earth materials (gravel, sand, clay and humus) enhanced the design of the successful structures. Sand could fill in the spaces in between pieces of gravel, and clay could pack the entire structure into a solid wall. Another factor was how the earth materials were supported. Students decided that craft sticks worked the best when placed on both sides of the dam.

The discussion then shifted to where the dams were located in relation to the fictional town downriver. When the dam was placed farther away from the town, the rate of water flow slowed before it reached the houses, represented by small cubes. Slower water did not threaten the town; the river was allowed to flow slowly past the town through a channel. The class also determined that reinforcing the walls of the channel prevented landslides and soil slippage.

During the discussion, students mentioned most of the aspects of erosion that they had been investigating in prior work with the stream table: rate of flow, soil components, slope, and human impact on erosion. The adults merely listened to the discussion while students asked the questions and provided the answers. The lesson, in the opinion of Mrs. Morrow and the parents present, far exceeded their expectations.

What Does the Dam Building Story Illustrate about Inquiry?

How did the dam building lesson use inquiry to help students understand something about erosion? The lesson contained many essential features of classroom inquiry. According to the National Research Council (2000), inquiry occurs when:

- Learners are engaged by scientifically oriented questions.
- Learners give priority to evidence, which allows them to develop and evaluate explanations that address scientifically oriented questions.

- Learners formulate explanations from evidence to address scientifically oriented questions.
- Learners evaluate their explanations in light of alternative explanations, particularly those reflecting scientific understanding.
- Learners communicate and justify their proposed explanations. (p. 28)

The fifth grade dam builders addressed a real world question and prepared a solution to address that question. Students used evidence from earlier stream table activities to develop and evaluate their solutions to the problem of flash flooding. The team presentations allowed the class to consider alternative explanations and compare them to the causes of erosion that they had observed in earlier lessons. Students used the presentation forum to communicate and justify their explanations. Thus the dam building lesson demonstrates all five of NRC's "essential features of classroom inquiry" (p. 28). Although not all teams arrived at perfect solutions to the flash flood problem, these students viewed science in a new light. Their new impressions of the nature of science included that scientists and engineers can benefit from their mistakes. Here we see inquiry taking students beyond the learning of concepts and processes, to exploring the broader issues of the nature of the scientific enterprise.

We believe that the dam construction activity allowed these elementary students the opportunity to experience science based upon our socio-cognitive assumptions. Language was critical to this investigation when the fifth graders developed their construction plans, tested their solutions, created products, and presented results to the Gaveo Town Council. Knowing was social as demonstrated by the exchange among members of one group about building materials, and by the whole class conversation during the Town Council meeting. Learning was authentic in that students used the stream table and a real life scenario of dam building as protection from flash floods. The stream table was familiar to students from the prior lessons in the erosion unit, and they used the context of the stream table to help them figure out a solution to the Gaveo problem. Even though this activity was from a published curriculum, for this lesson, students directed their own learning, manipulated materials to control erosion based upon their prior experience, talked and worked together, and figured out and presented their solutions to a life-like problem.

Summary: Features of Integrated Inquiry

Mention "inquiry" to any group of elementary teachers, and they will most likely have visions of full blown student-designed investigations taking place over long periods of time. After all, isn't that what scientific inquiry looks like? Our view of inquiry is much broader—it is an orientation to science teaching, not a mere teaching method. Inquiry-based science classes encompass the kinds of doing, thinking, and communicating of scientists, whether in short lessons or more extended investigations. In the two classroom stories we told, science lessons included different types of science experiences:

1. students engaging in hands-on activities provided by the teacher;

2. students testing their science theories;
3. students planning and carrying out investigations based on their own question; and
4. students applying their ideas to solve design problems, both on paper and with models.

These stories include some commonalities—firsthand experiences with phenomena; students working together in teams; and opportunities for students to express their ideas orally, in writing, and in drawing. In each case it was difficult to tell when students were attending to science processes vs. science products; the two were integrated. Sometimes the teacher initiated the questions, sometimes she helped students initiate the questions. But always questions, evidence, and explanation were given high priority.

Clearly there is no one-size fits all approach to elementary science classroom inquiry. By shifting our thinking from inquiry as a specific method to inquiry as an orientation, we place the emphasis on learning goals instead of instructional strategies. By adopting an integrated view of inquiry, we recognize that the learning goals must include attention to both the products and the processes of science.

CONSTRAINTS AND BENEFITS TO INQUIRY IN ELEMENTARY SCIENCE

According to the National Science Education Standards (National Research Council, 1996), "All teachers of science have implicit and explicit beliefs about science, learning, and teaching. Teachers can be effective guides for students learning science only if they have the opportunity to examine their own beliefs" (p. 28). One way that teachers examine their beliefs about inquiry is by listening to the stories of other teachers.

The vignettes in National Science Education Standards (National Research Council, 1996; 2000) and the two episodes in this chapter present success stories. They describe inquiry oriented lessons that worked, where teachers were confident and students motivated. Do these success stories mirror the situation most elementary teachers face when attempting to implement inquiry-based science instruction? Do these success stories tell the whole story?

Our conversations with Jessica Smithton and Shirley Morrow provided a sense of authenticity about the pitfalls associated with inquiry, and helped us understand why the inquiry vision is not yet a reality in elementary schools. Jessica and Shirley were veterans, who had typically taught science from the textbook-based curriculum in their schools, infrequently using demonstrations or hands-on activities suggested by the textbook. Both teachers were excited by their students' learning in the units, yet both felt constrained in using inquiry more often in their science teaching.

Constraints to Inquiry

One of the biggest constraints faced by elementary teachers of science is materials. Elementary schools lack science equipment, the funds to purchase them, and places to store them. Even when the materials are readily available, as was the case in erosion unit, there are difficulties with time for materials preparation, management, and clean up when teaching science using an inquiry-based approach. Both the sinking and floating and erosion units were messy—they presented they ultimate challenges in preparing and managing materials. Jessica and Shirley were thrilled to have us in their classrooms—we brought equipment to support their teaching and helped with materials management. We are to wonder what will happen to science in these classrooms in the absence of such support.

Another frequently mentioned constraint is time—time to plan and time to enact inquiry. Jessica and Shirley made time for science in these units, partly because they felt an obligation in their partnership with us. We also provided some support with time as we assisted with lesson development, materials management, and assessment. We do not know to what extent they continue to devote such time to science.

In the present climate of high stakes testing, elementary teachers also have serious concerns about how they will assess student learning. The sinking and floating unit relied on teacher made assessments and scoring rubrics. This required additional time and expertise on Jessica's part. Would another teacher, less confident in her own conceptual understanding, be willing to undertake such efforts? The assessments provided by the publisher of the erosion curriculum (National Science Resources Center, 1997) were different than textbook assessments. These assessments did not ask students questions about facts and definitions, but allowed them the opportunity to look at photographs of different landforms and interpret what they noticed in the pictures. However, the curriculum did not provide much guidance on how to score student products or provide grades, something that was required of Shirley in the fifth grade. Shirley was concerned about her lack of content knowledge, and how that would place out when teaching the unit and when assessing student progress. Finally, both Jessica and Shirley were concerned how their students, who had covered less ground in the district science curriculum, would perform on standardized assessments of student achievement in comparison to other fourth and fifth graders.

Benefits of Inquiry

Both Jessica and Shirley felt overwhelmed by the amount of material "covered" in the textbook-based curriculum. The inquiry-based units allowed them to focus on a subset of concepts, delving into their topic in greater depth. They believed that one of the benefits of inquiry was addressing fewer ideas in a more connected way, even as they worried about test score results.

Like Shirley, elementary teachers, new to teaching inquiry-based science, are often anxious about what students will ask and how they, as teachers, will perform. Yet elementary teachers also believe that, through inquiry, they will be able to learn alongside their students, becoming more confident teachers in the process. Our stories occur when Jessica and Shirley were teaching their sinking and floating or erosion units for the first time. Both teachers reported that they learned new science concepts in the process of teaching their units. They also learned which lessons worked well, which did not, and what they would leave out or add the next time they taught the unit. Inquiry-based instruction led to their reflection on both their science and their pedagogical knowledge. By the end of these units, neither of the teachers was completely comfortable with inquiry teaching, nor did they want to totally give up teacher-directed science instruction, teacher demonstrations, or memorizing terminology. However, they recognized the value of lessons where students could struggle with ideas and apply their knowledge to real world problems.

Jessica and Shirley strongly believed that their students benefited from inquiry. They appreciated the level of engagement, both physically and mentally, of their students. They recognized the connections that students made across concepts in the unit, and between science and the real world. These were the positive aspects of inquiry-based instruction that balanced out the frustrating parts of getting inquiry-based science underway. According Shirley, it was what the students received from this type of instruction that would make her continue to improve her inquiry-based teaching skills.

CONCLUSION: THE UNIQUE POTENTIAL OF ELEMENTARY SCIENCE

Despite the constraints that elementary teachers perceive related to time, materials, and assessment, they often value the less is more approach of inquiry-based science and recognize the benefits for student learning that accrue. Implementing inquiry-based elementary science takes time, involves frustrations, and requires a positive attitude about the benefits. It also requires a degree of expertise and confidence on the part of elementary teachers.

Elementary school classrooms are experiencing a paradox in curriculum and instruction. On the one hand, teachers are driven by high-stakes testing to enact a curriculum of reproduction, where the focus in on low level facts, algorithms, and right answers. This curriculum of reproduction is in effect across the disciplines (e.g., Saxon, 1992; Shurley, 1992). What is worse, in our work we have seen the curriculum of reproduction most strongly supported in underachieving urban and rural schools where the hope of raising test scores becomes the primary objective.

On the other hand, in this era of standards-based reform, we find the vision to be that of a curriculum of inquiry. The vision for a curriculum of inquiry is likewise in effect across the disciplines (e.g., National Council for the Social Studies, 1994; National Council of Teachers of Mathematics, 2000). This vision of inquiry is a reality in a few schools where students are valued as active producers of knowledge, where understanding is the aim, and where teachers act to facilitate doing, thinking, talking, and writing about ideas.

We know that science has never held an exalted place in elementary schools. "Science has never been an important part of the elementary curriculum. Teachers have a great deal of experience in the teaching of writing, reading, and mathematics, but not science" (Hall, 1998, p. 27). Thus, claims Hall, teaching science in elementary schools is difficult. We would like to posit a different claim. Since the curriculum of inquiry is a vision for literacy, and for mathematics, and for social studies, then science no longer has to be apart and different in elementary school.

Science could be the model for establishing an inquiry orientation across the disciplines. Science could reinforce efforts in the other disciplines, instead of competing with them for class time. Science could be the core of interdisciplinary inquiry-based instruction. Thus, inquiry-based science instruction has a unique potential for success in elementary schools.

The implications of this vision are far-ranging. How we prepare future elementary teachers must emphasize not only the distinctions among the disciplines, but also the connections across them. This implies that elementary science teacher educators have a unique responsibility to work with teacher educators from other disciplines to present a coherent view of inquiry across the curriculum. When we work with practicing teachers, we must help them see how an inquiry-based science orientation is reinforced in other curricular areas and likewise reinforces those areas. What we seek is an orientation that supports a curriculum of inquiry in science and in other disciplines. And in science, we seek an integrated view of inquiry that engages students in doing and thinking and talking and writing about science. This we hope, will lead students to understand the scientific enterprise more accurately, and to make sense of the big ideas in science.

REFERENCES

Abell, S. K. (1999). What's inquiry? Stories from the field. *Australian Science Teachers Journal, 45*(1), 33-40.

Abell, S. K., Bryan, L. A., and Anderson, M. A. (1998). Investigating preservice elementary science teacher reflective thinking using integrated media case-based instruction in elementary science teacher preparation. *Science Education, 82*, 491-510.

Anderson, C. W., and Smith, E. L. (1987). Teaching science. In V. Richardson-Koehler (Ed.), *Educators' handbook: A research perspective* (p. 84-111). New York: Longman.

Brown, J.S., Collins, A., and Duguid, P. (1989). Situated cognition and the culture of learning. *Educational Researcher, 18* (1), 32-42.

Hall, J. S. (1998). *Organizing wonder: Making inquiry science work in the elementary school.* Portsmouth, NH: Heinemann.

Jonassen, D. (1994). *Situated learning and social constructivism.* University Park, PA: Penn State University Press.

Knox, J.E., and Stevens, C. (1993). Vygotsky and Soviet Russian defectology. An introduction to Vygotsky. In R.W. Rieber and A.S. Carton (Eds.), *The collected works of L.S. Vygotsky. Vol.2. Problems of abnormal psychology and learning disabilities* (pp.1-25). New York: Plenum.

Lave, J. (1988). *Cognition in practice: Mind, mathematics, and culture in everyday life.* New York: Cambridge University Press.

Lave, J. and Wenger, E. (1991). *Situated learning: Legitimate peripheral participation.* Cambridge, MA: Cambridge University Press.

Moscovici, H., and Nelson, T. H. (1998). Shifting from activitymania to inquiry. *Science and Children, 35*(4), 14-17.

National Council for the Social Studies. (1994). *Expectations of excellence: Curriculum standards for social studies.* Washington, DC: Author.

National Council of Teachers of Mathematics. (2000). *Principles and standards for school mathematics.* Reston, VA: Author.

National Research Council. (1996). *National science education standards.* Washington, DC: National Academy Press.

National Research Council. (2000). *Inquiry and the national science education standards.* Washington, DC: National Academy Press.

National Science Resources Center. (1997). *Land and water.* Washington, DC: National Academy of Sciences.

Pratt, H. (1981). Science education in the elementary school. In N. C. Harms and R. E. Yager (Eds.), *What research says to the science teacher* (Vol. 3) (pp. 73-93). Washington, DC: NSTA.

Saxon, J.H. (1992). *Math65: An incremental development*. Norman, OK: Saxon Publishers.

Schmidt, W. H., McKnight, C. C., and Raizen, S. A. (Eds.). (1997). *A splintered vision: An investigation of U.S. science and mathematics education.* Dordrecht, The Netherlands: Kluwer.

Shurley, B. (1992). *The Shurley method.* Cabot, AR: Shurley Instructional Materials.

Vygotsky, L.S. (1962). *Thought and language*. E. Hanfmann & G. Bakar (Translators). Cambridge, MA: MIT Press.

Vygotsky, L.S. (1987). Thinking and speech. In R.W. Rieber and A.S. Carton (Eds.), *The collected works of L.S. Vygotsky. Vol.1 Problems of general psychology* (pp. 121-166). (N. Minick, translator). New York: Plenum (originally published in 1934).

Wertsch, J.V. (1985). *Vygotsky and the social formation of mind.* Cambridge, MA: Harvard University Press.

CHAPTER 13

EDITH GUMMER & AUDREY CHAMPAGNE

CLASSROOM ASSESSMENT OF OPPORTUNITY TO LEARN SCIENCE THROUGH INQUIRY

INTRODUCTION

Assessment has a profound impact on the learning that occurs in classrooms and standardized student achievement testing has considerable impact on what goes on in schools (Stake and Theo bold, 1991). Wilson & Corbett (1991) report that increasing the consequences of assessments at the state level results in refocusing educational efforts away from improving curriculum and instruction to improving test scores by emphasizing basic content and skills. According to Ferrara, Willihoft, Seburn, Slaughter and Stevenson (1991), another effect of state assessment systems is the development of local assessment systems that are parallel to that of the state. They describe several benefits and drawbacks to these local assessment systems that use the state assessment frameworks as organizing principles. Benefits include teachers' and administrators' perceived belief in assistance with targeting instruction for particular students and groups, in improving classroom assessment practices and in decreasing anxiety about the targets of the tests. However, these benefits come at some cost including a decreased focus on learning objectives outside those of basic skills, loss of instructional time to administration of more assessments, and inappropriate classroom teaching and testing practices. The effect of assessment as testing on the practices of classroom teachers is difficult to measure though O'Sullivan (1991) provides evidence that teachers identify increasingly negative effects as the stakes of the testing increase.

This paper examines the image of scientific inquiry that New York State teachers receive through the tasks included in state standards and supporting documents. Etienne Wegner (1999) uses the term 'reification' to refer to the ways in which the objects that are produced during practice make concrete the abstract processes that are invoked.

> With the term reification I mean to cover a wide range of processes that include making, designing, representing, naming, encoding and describing, as well as perceiving, interpreting, using, reusing, decoding, and recasting. Reification occupies much of our collective energy: from entries in a journal to historical records, from poems to encyclopedias, from names to classification systems, from dolmens to space probes, from the Constitution to a signature on a credit card slip, from gourmet recipes to medical procedures, from flashy advertisements to census data, from single concepts to entire theories, from the evening news to national archives, from lesson plans to the compilation of textbooks, from private address lists to sophisticated credit reporting

L.B. Flick and N.G. Lederman (eds.) Scientific Inquiry and Nature of Science, 263—297.

> databases, from tortuous political speeches to the yellow pages. In all these cases, aspects of human experience and practice are congealed into fixed forms and given the status of object. (p. 59)

The documents we analyze are objects of reification that describe what students might know and do as a result of engaging in standards-based learning experiences, taking state-level mandated tests, and developing examples of scientific inquiry work samples. We explore the challenges faced by classroom teachers as they struggle to provide their students the opportunity to meet state-mandated standards. The focus is on the development of understanding science principles at the middle level (Grades 5-8) and the use of inquiry as teaching method. We illustrate our paper using standards from New York from the physical and life sciences. We trace the path teachers must follow that begins with the standards as statements about what students should understand, travels through classroom practices that provide all students the opportunity to meet the standards, and ends with evidence of what students have learned. The path is not well marked and teachers have little time or adequate preparation to reach its end. Our intention here is not to be critical of teachers, but to illuminate the challenges they face. The central question is what image of scientific inquiry supports a teacher as she begins with a standard and provides students adequate opportunity to achieve the standard? The products of our analysis are criteria that may be used for classroom assessment of opportunity to develop science understanding using inquiry.

For a teacher who intends to be consistent with state and national standards three elements must be coordinated as the teacher embarks on the development of inquiry teaching that results in an inquiry assessment task. The first includes the curriculum materials provided by the local district, which varies by locale. The second includes state standards, and the third state mandated assessments. The descriptions of inquiry in each of the three often are quite different making teachers' attempts to implement inquiry in the classroom difficult. The language used in the standards to describe inquiry and the scientific principles students are expected learn compounds the challenge. We have chosen to focus on inquiry as teaching method rather than on student understanding of scientific inquiry because states generally place more emphasis on students' developing understanding of science principles than on developing inquiry abilities. Consequently the development of understanding through inquiry is the focus of teachers' attention, even though researchers in the field have given greater attention to students' development of inquiry skills [citation?]. We have focused on the state standards and assessments as they are common to all teachers in the state while instructional materials vary from district to district. Finally, the state standards and the state assessment tasks represent the focus of accountability as targets to which the teachers and students are held.

The two objectives, developing understanding of science principles through inquiry and learning to inquire, are two sides of the same coin. Learning to inquire in science requires the concurrent development of students' understandings of scientific principles and concepts in order to avoid allowing the inquiry to become the algorithmic application of some subset of inquiry processes. Therefore, in terms of a contribution to the thinking of the field, we believe that focusing on teachers' use of inquiry to facilitate the learning of science content may have more influence on both desired outcomes of science education than a focus on learning to inquire which may not give sufficient attention to students' coming to understand scientific principles

The nation is in the midst of standards-based reform. While state standards are arguably the most influential on classroom practice, the *National Science Education Standards* [*NSES*] (NRC, 1996) and *Benchmarks for Science Literacy* (AAAS, 1994) have had considerable influence on most state's science standards. Science inquiry is perceived as the centerpiece of the *NSES* and is a component of many state standards and state mandated tests. The *NSES* identify two facets of science inquiry that all students should attain, the ability to inquire and the understanding about inquiry. These standards also promote science inquiry as a teaching method to achieve understanding. Not all states have incorporated all facets of inquiry in their standards, but most state science standards promote inquiry as the preferred teaching strategy.

The primary source of the image of scientific inquiry teaching that has supported this analysis is found in the National Resource Council text, *Inquiry and the National Science Education Standards: A Guide for Teaching and Learning* (Olson & Loucks-Horsley, 2000). In that text, the elements of instruction that define scientific inquiry teaching include having the students address questions that address scientific phenomena, construct systematic investigations to answer those questions, develop and defend evidence-based explanations of the natural phenomena under question, and communicate and defend those explanations in light of alternative explanations. These are the criteria of teaching scientific concepts and principles through inquiry that we are searching for in the state standards and supporting documents. We are not examining the claim that teaching science content through inquiry automatically results in student development of an understanding about inquiry. Rather, we are making the case that without explicit attention to the details of inquiry in instructional practice, students do not have the opportunity to develop a deep conceptual understanding about either science principles or about scientific inquiry.

In addition to science standards, states are developing instructional frameworks and resource guides to provide additional information for teachers to assist them in teaching with and about scientific inquiry. These instructional and curriculum resources describe how the states suggest that teachers might make standards reachable by all students and provide examples of how teachers are implementing standards-based teaching and learning. They offer concrete examples of materials that inform local curriculum development. Teachers seeking to provide their students with the opportunity to meet the New York State Science Standards must wend their way through a number of documents that are provided by the New York State Education Department. We have coordinated the analysis of the opportunity to learn science by examining the alignment of 4 documents produced by New York State (Figure 1). We have chosen these as examples of what NYS identifies as standards for learning, alignment documents, and assessments. It is extremely cumbersome to identify documents in an analysis such as this. Therefore, we have named documents in italics and also used italics to identify key standards. We have used bold lettering to identify the major subcomponents of the standards.

Document Name	Available at
New York State Learning Standards for Mathematics, Science and Technology	http://www.emsc.nysed.gov/ciai/mst/pub/mststa1&2.pdf
Intermediate Level Science Curriculum Core	http://www.emsc.nysed.gov/ciai/mst/pub/intersci.pdf
Item Map for 8th Grade Test	http://www.emsc.nysed.gov/ciai/testing/sciei/itemapgr8sci.htm 8/19/02
Intermediate Level Science Test ♦ Item Map 2002 ♦ Performance Tasks from Sampler	Contact New York State Department of Education for copy of the *Intermediate Level Science Test* http://www.emsc.nysed.gov/ciai/testing/sciei/itemapgr8sci.pdf http://www.emsc.nysed.gov/ciai/mst/pub/1interscisam.pdf http://www.emsc.nysed.gov/ciai/mst/pub/2interscisam.pdf

Figure 1. New York State Education Documents Analyzed

The New York State image of science inquiry is contained in standards, *New York State Learning Standards for Mathematics, Science and Technology Standards* (*NYS Science Standards*) (NYSEd, 1996), state developed documents that interpret the standards, and state mandated assessments. We examine the image that is presented in statements describing what students are expected to learn, what evidence might be collected to measure achievement, and what science teaching might be expected to produce high achievement. The *NYS Science Standards* are organized into three levels, Elementary (Grades K-4), Intermediate Level (Grades 5-8) and Commencement (Grades 9-12). Our case study focuses on the standards for the Intermediate Level, grades 5-8. A second standards document, the *Intermediate Level Science Core Curriculum* (NYSEd, 1999) repeats and expands the specificity of the more general *NYS Science Standards*. We use that document to identify the ways in which the general statements contained in the *NYS Science Standards* are made more concrete and specific in a text designed for teachers and curriculum developers. In addition, examples of tasks in which students might engage to attain the standards are included in the *Intermediate Level Science Core Curriculum*. We present and analyze the nature of the tasks that are included as exemplars for the intermediate level.

The third set of documents that we examine contains inquiry assessment tasks that are part of the *June 2002 Intermediate Level Science Test* (NYSEd, 2002) that assesses students' achievement of the standards at Grade 8 in an on-demand format. The *NYS Item Maps* (NYSEd, 2002) define the alignment between the standards explicated in the *Intermediate Level Science Core Curriculum* and the state assessment, the *Intermediate Level Science Test*. We examine several examples of assessment items using various formats from the *Intermediate Level Science Test*. We also examine one task from the task sampler that contains exemplars of performance tasks that are included in the station tasks in the on-demand portion of the *Intermediate Level Science Test*. The task chosen for analysis in this paper deals with life science and examines the measurement of cell area.

The next section describes the structure of standards for scientific inquiry and content that makes up the *NYS Science Standards* and includes the examples of tasks that teachers might use to teach about a scientific principle using inquiry. Our focus on inquiry as content in instruction resulted in an analysis protocol similar to the ways in which Stern and Ahlgren (2002) analyzed the alignment of assessment tasks in the middle school science curricular materials with the AAAS *Benchmarks* (1994). We started with the *NYS Item Map* that defines the ways in which the test developers identified alignment by assigning a particular aspect of the standard statements from the *NYS Science Standards* to each item. Each author of this chapter determined the extent to which she agreed with that identified alignment, and similarities and differences were reconciled. We judged whether or not the expected student performance matched standard statements selected by the assessment developers. We also compared the expected performance with the other standards included in the *NYS Science Standards*. Our methods of determining the alignment of task and standard are consistent with our perspective that examining how a concept is measured is a valuable way to determine the operational definition of that concept. (Whitehead, 1932). Using this operational definition is especially beneficial in measuring the alignment of assessments to standards.

The assessment tasks that we evaluated did not include student responses. Therefore, we identified what the student is asked to do to determine the degree to which the requirements of the task allowed students to demonstrate the knowledge the task was to measure We are using our judgments of what the student would say or do in response to the prompts given. Content alignment is determined by the extent to which the content in the task matches that identified in the standards statements in terms of being necessary and sufficient content that a student would have to know to be able to respond correctly to the item or task.

We recognize that a more detailed analysis of the content and reasoning skills required by the tasks would result from the analysis of actual student work. As described in Champagne & Kouba (in preparation), we are only examining the gross anatomy of assessment tasks, not the fine anatomy that is available using more cognitively based evaluation processes. We would have a much better picture of the alignment of content and scientific inquiry knowledge and skills with statements in the standards if we could examine which scientific concepts and principles are used by students. If we had actual student responses we could examine the extent to which those concepts and principles are used by students to develop explanations of the embedded scientific phenomena, and we could analyze the reasoning processes that students invoked. However, we argue that examining the actual material that is available to teachers as they think about curriculum and instructional planning before they teach–is a valuable initial contribution to determining ways that standards are interpreted. The more extensive *Mathematics Science and Technology Resource Guide* (NYSEd, 2000) will be the topic of a future paper.

INQUIRY IN STATE STANDARDS

Teachers must first understand the organization of the *NYS Science Standards*. This document contains standards for K-12 mathematics, science and technology. Standard I has three distinct sections *Mathematical Analysis, Scientific Inquiry and Engineering Design*; Standard II addresses *Information Systems*; Standard VI addresses *Interconnectedness and Common Themes*; and Standard VII addresses *Interdisciplinary Problem Solving*. Standard III is exclusively *Mathematics*; Standard IV is exclusively Science; and, Standard V is exclusively *Technology*. Standards I (*Scientific Inquiry)*, II, IV, VI and VII delineate the principles and abilities that all students should attain in the discipline of science in that they define the content and skills that are considered central to the vision of scientific literacy (Figure 2).

*Standard I ***Analysis, Inquiry and Design***
- Mathematical Analysis
- Scientific Inquiry
- Engineering Design

*Standard II ***Information Systems***

Standard III ***Mathematics***

Standard IV ***The Living Environment and The Physical Setting***
- Living Environment
- Physical Setting

Standard V ***Technology***

* Standard VI ***Interconnectedness: Common Themes***
- System Thinking
- Equilibrium and Stability
- Models
- Patterns of Change
- Magnitude and Scale

*Standard VII ***Interdisciplinary Problem Solving***
- Connections
- Strategies

*These Standards together make up the New York State **Expanded Process Skills**

Figure 2. New York State Learning Standards for Mathematics, Science and Technology

Each standard has component parts called **Key Ideas** and **Performance Indicators** as presented in Figure 2 for the statements for *Scientific Inquiry*. The **Key Ideas** are declarative statements that describe constructs such as scientific inquiry, information systems, science concepts and principles, interconnectedness, and interdisciplinary problem solving and their components. Associated with each **Key Idea** are **Performance Indicators** that are statements of action, such as 'construct', 'interpret', or 'design' that describe expectations for student performance.

KEY IDEAS, PERFORMANCE INDICATORS, AND MAJOR UNDERSTANDINGS FOR INTERMEDIATE LEVEL STANDARD 1 – SCIENTIFIC INQUIRY
***Key Idea 1**: The central purpose of scientific inquiry is to develop explanations of natural phenomena in a continuing, creative process.*
Performance Indicators *S1.1* Formulate questions independently with the aid of references appropriate for guiding the search for explanations of everyday observations. S1.1a formulate questions about natural phenomena S1.1b identify appropriate references to investigate a question S1.1c refine and clarify questions so that they are subject to scientific investigation *S1.2* Construct explanations independently for natural phenomena, especially by proposing preliminary visual models of phenomena. S1.2a independently formulate a hypothesis S1.2b propose a model of a natural phenomenon S1.2c differentiate among observations, inferences, predictions, and explanations *S1.3* Represent, present, and defend their proposed explanations of everyday observations so that they can be understood and assessed by others. *S1.4* Seek to clarify, to assess critically, and to reconcile with their own thinking the ideas presented by others, including peers, teachers, authors, and scientists.

Key Idea 2: Beyond the use of reasoning and consensus, scientific inquiry involves the testing of proposed explanations involving the use of conventional techniques and procedures and usually requiring considerable ingenuity.

Performance Indicators

S2.1 Use conventional techniques and those of their own design to make further observations and refine their explanations, guided by a need for more information.

- S2.1a demonstrate appropriate safety techniques
- S2.1b conduct an experiment designed by others
- S2.1c design and conduct an experiment to test a hypothesis
- S2.1d use appropriate tools and conventional techniques to solve problems about the natural world, including measuring, observing, describing, classifying, and sequencing

S2.2 Develop, present, and defend formal research proposals for testing their own explanations of common phenomena, including ways of obtaining needed observations and ways of conducting simple controlled experiments.

- S2.2a include appropriate safety procedures
- S2.2b design scientific investigations (e.g., observing, describing, and comparing; collecting samples; seeking more information, conducting a controlled experiment; discovering new objects or phenomena; making models)
- S2.2c design a simple controlled experiment
- S2.2d identify independent variables (manipulated), dependent variables (responding), and constants in a simple controlled experiment
- S2.2e choose appropriate sample size and number of trials

S2.3 Carry out their research proposals, recording observations and measurements (e.g., lab notes, audiotape, computer disk, videotape) to help assess the explanation.

- S2.3a use appropriate safety procedures
- S2.3b conduct a scientific investigation
- S2.3c collect quantitative and qualitative data

Key Idea 3: The observations made while testing proposed explanations, when analyzed using conventional and invented methods, provide new insights into phenomena.

Performance Indicator

S3.1 Design charts, tables, graphs, and other representations of observations in conventional and creative ways to help them address their research question or hypothesis.

S3.1a organize results, using appropriate graphs, diagrams, data tables, and
other models to show relationships

S3.1b generate and use scales, create legends, and appropriately label axes

S3.2 Interpret the organized data to answer the research question or hypothesis and to gain insight into the problem.

S3.2a accurately describe the procedures used and the data gathered

S3.2b identify sources of error and the limitations of data collected

S3.2c evaluate the original hypothesis in light of the data

S3.2d formulate and defend explanations and conclusions as they relate to
scientific phenomena

S3.2e form and defend a logical argument about cause-and-effect relationships
in an investigation

S3.2f make predictions based on experimental data

S3.2g suggest improvements and recommendations for further studying

S3.2h use and interpret graphs and data tables

S3.3 Modify their personal understanding of phenomena based on evaluation of their hypothesis.

Figure 3. Key Ideas, Performance Indicators, and Major Understandings for NYSLMST Standard 1 Scientific Inquiry (NYSEd, 1996)

Scientific Inquiry in Standard I reflects the manner in which inquiry is presented in the supplement to the *NSES*, the text entitled *Inquiry and the National Science Education Standards* (NRC, 2001). In particular, the inclusion of a significant emphasis on the development of explanations of natural phenomena and the importance of differentiating among observation, inference, predictions, and explanations reflect criteria that address students' understandings **about** scientific inquiry (Key Idea S1). The emphasis on explanations is a theme that continues into the criteria that define what students should know about **how to do** scientific inquiry including reference to a broad spectrum of scientific investigations beyond simple experiments (Key Idea S2). Communication is emphasized in the criteria that describe how students should present and analyze data from scientific investigations that generate explanations of natural phenomena (Key Idea S3).

These standards are repeated in the *Intermediate Level Science Core Curriculum*. In this second document, which the developers clearly identify as a guideline for curriculum development rather than a syllabus, Standards I, II, VI, and VII are organized into **Expanded Process Skills** (Figure 3). These are not well defined but are identified as supporting a discovery learning approach to teaching in a "student-centered, problem solving approach to intermediate science." (*Intermediate-level Science Core Curriculum,* p.4) In addition to the Expanded Process Skills for science, there are **Process Skills Based on Standard IV** (Figure 4) that identify safety, measurement, and discipline-specific (General, Life and Physical) skills. However, these process skills are not included in the original *NYS Science Standards* documents and are found only in the *Intermediate Level Science Core Curriculum*.

The **Process Skills based on Standard IV** reflect an interesting collection of generic and discipline-specific skills (Figure 4). They range from following safety requirements to using particular techniques or instruments appropriately to identifying cause- and effect-relationships. Broken into General, Living and Physical Science Skills, they describe many of the analytic models that are used in the specific disciplines. They include examples such as "design and use a Punnett square or a pedigree chart to predict the probability of certain traits" (Living Environment) or "use a diagram of the rock cycle to determine geological processes that led to the formation of a specific rock type: (Physical Setting). These content-connected process skills include a level of specificity that is lacking from the **Expanded Process Skills**. In many cases, they reflect process and laboratory skills found in earlier versions of the NY State Regent's Syllabi for the various disciplines of biology, earth science, chemistry and physics. The relationship between the **Expanded Process Skills** and the **Process Skills based on Standard IV** are never well articulated, but the latter appear to be a cognitively simpler set of skills that are needed to carry out simple investigations.

There are several direct connections between the components of *Scientific Inquiry* in Standard I and in the **General Process Skills based on Standard IV**. The focus on safety in scientific investigations is emphasized on both sets of statements, as are general cognitive processes such as representing data correctly, recognizing patterns, ordering events, and identifying cause- and effect-relationships. In both cases, the use of models is articulated as relevant to engaging in scientific inquiry. These overlaps support our interpretation of the **Process Skills based on Standard IV** as representing a relatively simpler subset of cognitive skills of those stated in the **Expanded Process Skills** of which *Scientific Inquiry* is only a small part. However, without any indication of the relationships between the two sets of skills, teachers do not have the guidance to adequately use these statements as guidelines for the identification or development of instructional activities for their students. The disconnect between the two sets of process skills and direct explanation of the relationships among mathematical analysis, scientific inquiry, and engineering design leave teachers with a fragmented perspective on the complexity of inquiry in science education.

General Skills

1. follow safety procedures in the classroom and laboratory
2. safely and accurately use the following measurement tools:
    - metric ruler
    - balance
    - stopwatch
    - graduated cylinder
    - thermometer
    - spring scale
    - voltmeter
3. use appropriate units for measured or calculated values
4. recognize and analyze patterns and trends
5. classify objects according to an established scheme and a student-generated scheme
6. develop and use a dichotomous key
7. sequence events
8. identify cause-and-effect relationships
9. use indicators and interpret results

Living Environment Skills

1. manipulate a compound microscope to view microscopic objects
2. determine the size of a microscopic object, using a compound microscope
3. prepare a wet mount slide
4. use appropriate staining techniques
5. design and use a Punnett square or a pedigree chart to predict the probability of certain traits
6. classify living things according to a student-generated scheme and an established scheme
7. interpret and/or illustrate the energy flow in a food chain, energy pyramid, or food web
8. identify pulse points and pulse rates
9. identify structure and function relationships in organisms

Physical Setting Skills

1. given the latitude and longitude of a location, indicate its position on a map and determine the latitude and longitude of a given location on a map
2. using identification tests and a flow chart, identify mineral samples
3. use a diagram of the rock cycle to determine geological processes that led to the formation of a specific rock type
4. plot the location of recent earthquake and volcanic activity on a map and identify patterns of distribution
5. use a magnetic compass to find cardinal directions

6. measure the angular elevation of an object, using appropriate instruments
7. generate and interpret field maps including topographic and weather maps
8. predict the characteristics of an air mass based on the origin of the air mass
9. measure weather variables such as wind speed and direction, relative humidity, barometric pressure, etc.
10. determine the density of liquids, and regular- and irregular-shaped solids
11. determine the volume of a regular- and an irregular-shaped solid, using water displacement
12. using the periodic table, identify an element as a metal, nonmetal, or noble gas
13. determine the identity of an unknown element, using physical and chemical properties
14. using appropriate resources, separate the parts of a mixture
15. determine the electrical conductivity of a material, using a simple circuit
16. determine the speed and acceleration of a moving object

Figure 4. Process Skills for Standard 4

The *Intermediate Level Science Core Curriculum* presents **Major Understandings** identified with each of the **Performance Indicators** that provide more specific detail to further expand the nature of each **Performance Indicator** in terms of what students should know and be able to do. Figure 2 contains the **Key Ideas**, **Performance Indicators,** and **Major Understandings** for Standard 1 *Scientific Inquiry* at the intermediate level. It is obvious that *Scientific Inquiry* is a complex process as indicated by the 3 **Key Ideas** and 10 **Performance Indicators** and 28 **Major Understandings** that make up the standard at the middle-school level. Figure 5 summarizes the number of **Key Ideas and Performance Indicators** for each of the **Expanded Process Skills**. The **Expanded Process Skills** include a total of 18 **Key Ideas** and 59 **Performance Indicators** and 47 **Major Understandings**. In addition there are 34 **Process Skills Based on Standard 4**. The sheer complexity of **Expanded and Standard IV-based Process Skills** is a daunting vision for the classroom teacher who must develop a coherent curriculum that addresses all of them at an appropriate developmental level. In addition, the teacher must embed the learning of inquiry into the scientific concepts and principles of the *Living Environment* and the *Physical Setting*, the two disciplinary structures of Standard IV.

MIDDLE LEVEL/INTERMEDIATE SCIENCE PROCESS SKILLS From *Intermediate-Level Science Core Curriculum*			
Expanded Process Skills	**Key Ideas**	**Performance Indicators**	**Major Understandings**
STANDARD 1 Analysis, Inquiry, and Design			
Mathematical Analysis	**3**	**3**	**6**
Scientific Inquiry	**3**	**10**	**28**
Engineering Design	**1**	**5**	**8**
STANDARD 2 Information Systems	**3**	**11**	**5**
STANDARD 6 Interconnectedness: Common Themes	**6**	**15**	**None articulated**
STANDARD 7 Interdisciplinary Problem Solving	**2**	**5**	**None articulated**
Totals	18	59	47
Skills Based on Standard IV			
General Skills	***9***		
Living Environment Process Skills	***9***		
Physical Setting Process Skills	***16***		

Figure 5. Complexity of the Middle Level/Intermediate Science Process Skills

The inclusion of mathematical analysis and engineering design in the **Expanded Process Skills** reflects the initial intention of the individuals designing the standards that mathematics, science, and engineering (technology) be taught and tested as a integrated whole. That intention is reflected in the content of Standards I, II, VI, and VII, which as we noted above are applicable to mathematics, science, and engineering. The effect of the integration makes it difficult to sort out what the developers meant to communicate about the nature of science inquiry and the relationships among mathematical analysis, scientific inquiry and engineering design. Are analysis, inquiry, and design processes unique to mathematics or science or engineering? Or does science inquiry involve analysis and design? Which of these processes- analysis, inquiry and design- is applied to seeking answers to scientific questions, inquiring into the natural world, or solving problems in science? What is the nature of the problems and questions referenced in the Standards? Standard VII seems to target problems related to science/technology/and society. How do the problems referenced in Standard VII relate to the kinds of questions addressed in scientific inquiry? The Standards related to the **Extended Process Skills** raise more questions about the nature of science inquiry than they answer.

While the individuals drafting the standards might argue that the integration represented in the **Expanded Process Skill** Standards represents the real worlds of mathematics, science, engineering and society, the integration presents challenges to curriculum developers and test designers. With few exceptions the integration has not been realized in the K-12 curriculum or in the New York State mandated tests. Mathematics and science are assessed in separate examinations and taught as separate subjects. And while the NYSDEd-developed alignment of the *Intermediate-level Science Test* with the *NYS Science Standards* suggests that the test assesses Standards I-II, IV, and VI and VII, the alignment is questionable, as is the assumption that if there is alignment across standards that the conclusion of integration is justified. (We elaborate on the questions about alignment and integration when we review the NYSDEd alignment for test items.)

TASKS IN THE NYS STANDARDS DOCUMENTS

A sample instructional task used as an exemplar for Standard I, *Scientific Inquiry,* in the *NYS Science Standards* and the *Intermediate-level Science Core Curriculum* focuses on an investigation of the disparity of solid waste that might be recycled and the amount that is actually reported as recycled (Figure 6). The example is written with the context of solid waste identified as only one of a variety of contexts of scientific content that might be appropriate for such an investigation. The connection between science, technology, and society is evident in the choice of context that addresses solid waste. However, the task is not representative of the science content that is represented in the *Living Environment* and *Physical Setting* of Standard IV. This lack of connection presents inquiry as more appropriate to applied or social contexts than as a way to teach scientific concepts and principles. In addition, there is no connection between how the task represents Standard I, *Scientific Inquiry*, and how it might connect to the other elements of the **Expanded Process Skills or the Process Skills based on Standard IV**.

Standard I *Scientific Inquiry* from *New York State Learning Standards for Mathematics, Science and Technology*	Sample Task Statements from Standard I Scientific Inquiry
Key Idea 1: The central purpose of scientific inquiry is to develop explanations of natural phenomena in a continuing, creative process.	*After being shown the disparity between the amount of solid waste which is recycled and which could be recycled,** students working in small groups are asked to explain why this disparity exists. They develop a set of possible explanations and to select one for intensive study. After their explanation is critiqued by other groups, it is refined and submitted for assessment. The explanation is rated on clarity, plausibility, and appropriateness for

	intensive study using research methods.
Key Idea 2: Beyond the use of reasoning and consensus, scientific inquiry involves the testing of proposed explanations involving the use of conventional techniques and procedures and usually requiring considerable ingenuity.	Develop a research plan for studying the accuracy of their explanation of the *disparity between the amount of solid waste that is recycled and that could be recycled.** After their tentative plan is critiqued, they refine it and submit it for assessment. The research proposal is rated on clarity, feasibility and soundness as a method of studying the explanations' accuracy. They carry out the plan, with teacher suggested modifications. This work is rated by the teacher while it is in progress.
Key Idea 3: The observations made while testing proposed explanations, when analyzed using conventional and invented methods, provide new insights into phenomena.	Carry out their plan making appropriate observations and measurements. They analyze the data, reach conclusions regarding their explanation of the *disparity between the amount of solid waste which is recycled and which could be recycled.**, and prepare a tentative report which is critiqued by other groups, refined, and submitted for assessment. The report is rated on clarity, quality of presentation of data and analyses, and soundness of conclusions.

Figure 6. Exemplar Task Associated with Standard I, Scientific Inquiry, from New York State Learning Standards for Mathematics, Science and Technology NYSEd, 1996)

* italics in original to denote sample content

The most problematic aspect of this example is the extent to which the question addresses a natural phenomenon. This question deals with a social issue in a scientific setting, rather than a strictly scientific one. Basically it asks why people do not recycle. Moreover, though there may be numerous social theories that could be presented as reasons, the students are unlikely to address those social theories. However, the question does require the students to develop an explanation for the phenomenon, rather than just describe or quantify the lack of recycling. There is the potential for the students to have to coordinate multiple lines of evidence as explanations proposed by various groups are critiqued.

The example does require students to develop their own plans for addressing the question. The variables of interest do not appear to be defined by the teacher or text, nor are the directions provided. The phenomenon that the students are asked to address is complex and the variables are multiple and difficult to sort out. It is not clear to what extent the students will be required to address multiple variables or to collect data on multiple measures. The requirements for dealing with the complexity of the design of the investigation are not well characterized, nor are the issues of addressing observer bias. Survey or qualitative research design is suggested by the social nature of the question.

The activity provides the students with the opportunity to transform data from surveys into graphs or drawings; however, such simple representations are predominant. The peer critique may result in students reflecting on potential errors, particularly those of design to move their analysis beyond simple methodological mistakes. The social nature of the question provides the opportunity for the students to reflect on issues of causal versus descriptive explanations. The reasoning links between the variables that the students choose to identify and the explanation they are investigating is not well characterized, nor is the extent to which they are required to generalize beyond the situation they are investigating. It is not clear what forms of argument the students are required to use to reason from the data they collect to the explanation they are investigating. There is no mention of the extent to which the students must address anomalous data, but the requirement that the conclusions be tentative is included. The initial question requires students to reason about empirical regularities, but the social nature of the question and the fact that the students are asked to investigate one explanation of a phenomenon that may have multiple causal mechanisms is problematic. There is no evidence of a requirement that the students must coordinate multiple lines of evidence to support or refute the multiple explanations that might be proposed. Finally, there is no requirement that the students examine other reports about explanations of why people do not recycle to build on previous research or to compare their results to other studies.

This example demonstrates the criteria that we are using to determine the image of science inquiry in both teaching and in instructional and assessment tasks. In order for the teaching or the tasks to address inquiry, they must have students engaged in the following practices:

- Address a scientific phenomenon;
- Design an investigation to develop an explanation of that phenomenon;
- Use evidence from the investigations to explain the phenomenon;
- Consider alternative explanations; and,
- Communicate their explanations.

As described above, these criteria are derived from those explicated in *Inquiry and the National Science Education Standards: A Guide for Teaching and Learning* (Olson & Loucks-Horsley, 2000). These criteria, together with the statements about scientific inquiry that are included in Standard I, *Scientific Inquiry*, are the lenses that are provided by the science education reform literature and New York State to guide the ways in which teachers examine their own practice.

Other images of teaching with and about scientific inquiry are included in the discipline-based standards that are part of the NYS documentation. A variety of tasks are included in Standard IV, Science – *The Physical Setting,* in both the *NYS*

Science Standards and the *Intermediate-level Science Core Curriculum*, that provide the beginning of an image of the ways in which inquiry might be used as a teaching strategy to develop students' conceptual understanding of selected concepts and principles in physical science (Figure 7). These tasks are represented by single statements that only hint at the instructional sequence that would be needed to flesh them out into classroom experiences or into assessment tasks to provide evidence of what students knew and could do as a result of those experiences. They include a number of types of tasks such as the construction of models, descriptive investigations, construction of devices and designs of apparatuses, and interpretations of data sets. As such, they represent the best glimpse for the teacher of the opportunities that students need to experience in order to develop the conceptual understandings reflected in Standard IV. However, they do not connect that understanding of science concepts and principles with any discussion of Standard I, *Scientific Inquiry*. The teacher is still left to make inferences of how these content-based tasks relate to scientific inquiry.

Standard IV Science – The Physical Setting	**Sample Task Statements from Standard IV, Science – The Physical Setting from *LSMST***
Key Idea PS 4.1 The Earth and celestial phenomena can be described by principles of relative motion and perspective.	• Create models, drawings, or demonstrations describing the arrangement, interaction, and movement of the Earth, moon, and sun. • Plan and conduct an investigation of the night sky to describe the arrangement, interaction, and movement of celestial bodies.
Key Idea PS 4.2 Many of the phenomena that we observe on Earth involve interactions among components of air, water, and land.	• Add heat to and subtract heat from water and graph the temperature changes, including the resulting phase changes. • Make a record of reported earthquakes and volcanoes and interpret the patterns formed worldwide.
Key Idea PS 4.3 Matter is made up of particles whose properties determine the observable characteristics of matter and its reactivity.	• Test and compare the properties (hardness, shape, color, etc.) of an array of materials. • Observe an ice cube as it begins to melt at temperature and construct an explanation for what happens, including sketches and written descriptions of their ideas.

Key Idea PS 4.4 Energy exists in many forms, and when these forms change energy is conserved.	• Design and construct devices to transform/transfer energy. • Conduct supervised explorations of chemical reactions (not including ammonia and bleach products) for selected household products, such as hot and cold packs used to treat sport injuries. • Build an electromagnet and investigate the effects of using different types of core materials, varying thicknesses of wire, and different circuit types.
Key Idea PS 4.5 Energy and matter interact through forces that result in changes in motion.	• Investigate physics in everyday life, such as at an amusement park or a playground. • Use simple machines made of pulleys and levers to lift objects and describe how each machine transforms the force applied to it. • Build "Rube Goldberg" type devices and describe the energy transformations evident in them.

Figure 7. Sample Tasks Associated with Standard IV, Science – The Physical Setting New York State Learning Standards for Mathematics, Science and Technology (NYSEd, 1996)

A parallel set of task statements is included in Standard IV, Science – *Living Environment* in both the *NYS Science Standards* and the *Intermediate-level Science Core Curriculum* (Figure 8). These tasks also provide limited information for the intermediate science teacher about the nature of learning experiences he or she might design to help students meet standards or the nature of the products that students might produce in order to demonstrate what they know and can do. The tasks range from the use of models as analogs to explain scientific concepts and principles, to simple investigations, to extended investigations that extend the learning of science well beyond the classroom. However, again they do not connect learning about science concepts and principles to learning with and about inquiry as described in Standard I, *Scientific Inquiry*, the **Extended Process Skills**, or the **Process Skills based on Standard IV**.

Standard IV Science – The Living Environment	**Sample Task Statements**
Key Idea LS4.1: Living things are both similar to and different from each other and from nonliving things.	•Conduct a survey of the school grounds and develop appropriate classification keys to group plants and animals by shared characteristics. •Use spring-type clothespins to investigate muscle fatigue or rulers to determine the effect of amount of sleep on hand-eye coordination
Key Idea LS4.2: Organisms inherit genetic information in a variety of ways that result in continuity of structure and function between parents and offspring.	•Contrast dominance and blending as models for explaining inheritance of traits. •Students trace patterns of inheritance for selected human traits.
Key Idea LS4.3: Individual organisms and species change over time.	•Conduct a long-term investigation of plant or animal communities. •Investigate the acquired effects of industrialization on tree trunk color and those effects on different insect species.
Key Idea LS4.4: The continuity of life is sustained through reproduction and development.	•Apply a model of the genetic code as an analogue for the role of the genetic code in human populations
Key Idea LS4.5: Organisms maintain a dynamic equilibrium that sustains life.	•Record and compare the behaviors of animals in their natural habitats and relate how these behaviors are important to the animals.
Key Idea LS4.6: Plants and animals depend on each other and their physical environment.	•Construct a food web for a community of organisms and explore how elimination of a particular part of a chain affects the rest of the chain and web.
Key Idea LS4.7: Human decisions and activities have had a profound impact on the physical and living environment.	•Conduct an extended investigation of a local environment affected by human actions (e.g., a pond, stream, forest, empty lot).

Figure 8. Sample Tasks Associated with Standard IV, Science – The Living Environment New York State Learning Standards for Mathematics, Science and Technology (NYSEd, 1996)

These descriptive statements are a beginning step in providing examples of what it means to teach with and about inquiry in the middle school classroom. However, the statements alone are not sufficient detail to provide the teacher with an image of what it means to teach about scientific principles through inquiry. Nor do they indicate any connection between teaching with inquiry and student learning about scientific inquiry together with the learning of scientific principles. In addition, as we shall demonstrate in the following section, the image of inquiry that is exemplified in the NYS assessments makes teachers' interpretations of the brief examples in the standards especially problematic.

INQUIRY ON STATE ASSESSMENTS

Those states that include all or some facets of inquiry do not necessarily test all inquiry standards. The major emphasis on state mandated assessments is student understanding of science principles and those features of inquiry that can be assessed using paper and pencil methods, which include such processes as the use of instruments, design of controlled experiments, identification of variables, and analysis of data. Several factors influence the choices made about testing time devoted to understanding scientific concepts and principles and understanding inquiry and the ability to inquire. These include the cost of performance measures, the relative ease of assessing inquiry in standardized formats, and the efficiency of using paper and pencil measures to gather evidence of what students know and can do. However, perhaps the most important issue is the relative emphasis placed on the understanding science concepts and principles versus understanding and doing inquiry. This issue has been emphasized in earlier publications.

> Because the ability to inquire is less valued, because paper and pencil methods are more efficient, and because paper and pencil methods are less costly, measurement of the ability to inquire is de-emphasized on many state assessments. Those states that have hands-on inquiry tasks are limited by time constraints so the inquiries are short and highly scaffolded. (Champagne & Kouba, in preparation, p. 22)

These issues have serious implications for the alignment of the tasks with standards, and alignment remains generally fragmented and weak. However, inquiry tasks on state mandated assessments may have more influence on classroom practice than inquiry as described in the standards. The operational definition of inquiry that is made apparent when inquiry standards are translated into measurable terms is a strong message to teachers. Items and tasks represent actual performances that students must accomplish and represent the criteria by which students' inquiry abilities are evaluated. The following section analyzes the image of scientific inquiry that is presented to the teacher in the state-mandated test to which they and their students are held accountable. Our argument is that this image of scientific inquiry is seriously distorted.

The New York State Intermediate Level Science Test

We next present our evaluation of the image of scientific inquiry that is evidenced in a number of items and tasks that come from the NYS standards described above. We have chosen exemplars of two different assessments including the following:

- Items on the Intermediate Level Science Test Parts B & C from June 2002; and,
- A task included as an example of NYS on-demand laboratory performance tasks (*Intermediate-level Science Test* Part D).

Assessment is referenced repeatedly throughout the *Intermediate-level Science Core Curriculum*, and the developers clearly identify the purpose of the document to address the content and skills that are appropriate to the *Intermediate Level Science Test*. The *Intermediate-level Science Core Curriculum* identifies the purpose of the *Intermediate Level Science Test*

> To assess student achievement of Standards 1, 2, 4, 6, and 7 of the *Learning Standards for Mathematics, Science, and Technology* and, when appropriate, include aspects of the other six mathematics, science, and technology standards including analysis, inquiry, design, information systems, mathematics, technology, common themes, and interdisciplinary problem solving. (NYSEd, 1999, p.29).

A laboratory performance task is given prior to the paper and pencil aspects of the *Intermediate Level Science Test*. This hands-on, station type assessment involves three laboratory tasks that are identified as addressing Standards I, II, IV, VI, and VII (the **Extended Process Skills**). Scored with a rubric, the laboratory performance tasks make up 15% of the student's score for the *Intermediate Level Science Test*. The bulk of the exam is an on-demand paper and pencil task that includes a variety of multiple choices, short answer, and extended response formats. Part A makes up 25-35% of the exam and uses a multiple choice format to primarily assess the students' knowledge of concepts and principles that make up the two disciplinary subsections of Standard IV, *Science*, the *Living Environment* and the *Physical Setting*. Part B makes up another 25-35% of the exam using multiple choice and short constructed response tasks and items to measure students' understandings and abilities related to both the content of Standard IV and content and skills of Standard I, *Scientific Inquiry*. Finally, Part C makes up the final 25-35% of the exam using extended constructed response items to measure students' understandings and abilities relevant to Standard I, *Scientific Inquiry*; Standard IV, *Science*; Standard VI, *Interconnectedness*; and Standard VII, *Problem Solving*. Clearly a message that is being sent to teachers is that assessing inquiry requires assessment formats other than multiple choice exams.

This test has four parts distinguished by item type. Part A is composed of multiple choice items and assesses "...the student's ability to apply; ***analyze, synthesize, and evaluate*** core material primarily from Standard 4" (*Intermediate-level Science Core Curriculum*, p.29). Part B is composed of multiple choice and constructed response items and assesses "....the student's ability to apply, analyze, synthesize, and evaluate material primarily from Standard 4 (content) and Standard

1 (inquiry) " (*Intermediate-level Science Core Curriculum*, p.29). Part C consists of extended response items and assesses

> Material from Standards 1, 4, 6, and 7 (problem solving) assessed by requiring students to apply their knowledge of science concepts and skills to address real. world situations. Real-world situations (approximately three to five) ... from newspaper or magazine articles, scientific journals, or current events, for example. Students [are] to apply scientific concepts, formulate hypotheses, make predictions, or use other scientific inquiry techniques in their responses to the questions posed. (*Intermediate-level Science Core Curriculum*, p.29).

Part D consists of performance items and assesses " ....hands-on laboratory tasks linked to content and skills in Standards l, 2,4,6, and 7. " (*Intermediate-level Science Core Curriculum*, p.29)

Our analysis of these assessment items and tasks follows the process of cognitive demand analysis laid out in Champagne and Kouba (in preparation). Our criteria for alignment are similar to those used by Stern and Ahlgren (2002) in their analysis of the alignment of assessment tasks in middle school science curricular materials with the AAAS *Benchmarks* (1993) in that we focus on the representation of inquiry as content. We first examined the nature of the inquiry as science content assigned to the item or task by the developers. We started with the *NYS Item Map* that defines the ways in which the test developers identified alignment. Each author determined the extent to which she agreed with that identified alignment, and similarities and differences were reconciled. We then examined the cognitive demand of the item or task in terms of what the item or task asked students to know and be able do. We identified congruence between the assignment by the test developers and our own conclusions about what the task required. We judged whether or not the expected student performance matched the Key Idea(s) or Performance Indicator(s) selected by the assessment developers, the New York State Department of Education. We also compared the expected performance with the other Key Idea(s) and Performance Indicators(s) that make up the **Extended Process Skills** in the *NYS Science Standards* and the **Process Skills based on Standard 4** described above.

The assessment tasks that we evaluated did not include student responses. Therefore, we identified what the student is asked to do to determine the expected student performance in terms of those understandings of scientific inquiry they are being given the opportunity to demonstrate. We are using our judgments of what the student would say or do in response to the prompts given. Content alignment is determined by the extent to which the content in the task matches that identified in the standards statements in terms of being necessary and sufficient content that a student would have to know to be able to respond correctly to the item or task.

We recognize that a more detailed analysis of the content and reasoning skills required by the tasks would result from the analysis of actual student work. As described above in our discussion of examples of inquiry teaching, we are only examining the gross anatomy of assessment tasks, not the fine anatomy that is available using more cognitively based evaluation processes. We would have a much better picture of the alignment of content and scientific inquiry knowledge and skills with statements in the standards through the examination of which scientific concepts and principles are used by students, the extent to which those concepts and principles are used by students to develop explanations of the embedded scientific phenomena, and the reasoning processes that students invoked. However, we argue that examining the actual material that is available to teachers as they think about curriculum and instructional planning before they teach is a valuable contribution to determining ways that standards are interpreted.

Figure 9 shows items identified as addressing the *NYS Science Standards* Standard I, *Scientific Inquiry* and the way in which the NYS Department of Education test developers aligned items from the June 2002 *Intermediate Level Science Test.* Alignment is expressed in terms of Key Ideas and Performance Indicators for the expected student performance. Figure 10 indicates the nature of the item, New York State Education Department alignment with the *NYS Science Standards* and our comment on that alignment. A total of 7 items address Standard I, *Scientific Inquiry*. Four of the items (35, 63, 64, and 67) involve mathematical skills and knowledge such as interpreting data trends on a graph, identifying a minimum point or constructing a graph from given data points. While we do not argue that these skills and knowledge are important in scientific inquiry, the tasks do not connect the mathematical interpretation to the scientific concepts and principles within which they are cast. They are assigned to Standard I *Scientific Inquiry* **Key Idea S3** and **Performance Indicator S3.2** which addresses the use of methods of data analysis to inform interpretation of a hypothesis. The items are better aligned with Standard I *Mathematical Analysis*, particularly **Performance Indicator M1.1b** "identify relationships among variables including: direct, indirect, cyclic, constant; identify non-related material" (*Intermediate-level Science Core Curriculum*, 1998, p. 4).

NYS Assessment Item or Task	Source
Items 35, 45, 62, 63, 64, 65, 66, 67	NYS Intermediate Level Science Test June 2002
Cell Size	NYS Intermediate Test Sampler http://www.emsc.nysed.gov/ciai/mst/pub/2interscisam.pdf

Figure 9. NYS Scientific Inquiry Assessment Items and Tasks

Intermediate Level Science test (ILST) Parts B and C NYSEd Alignment with Scientific Inquiry and Expanded Process Skills			
Item Number	**Task Description**	**NYSEd Alignment with Standard I *Scientific Inquiry***	**Comments**
35	Interpret trends of solubility on a graph	**Key Idea 3** The observations made while testing proposed explanations, when	This item measures mathematical as much as scientific understanding

45	Identify dependent and independent variables and factors to be held constant	analyzed using conventional and invented methods, provide new insights into phenomena **3.1** Design charts tables, graphs and other representations of observations in conventional and creative ways to help them address their research question or hypothesis.	This item addresses elements of scientific inquiry
62	Describe trend in data and identify possible explanation		The first part of this item measures mathematical as much as scientific understanding The second part requires the student to develop an explanation which addresses scientific inquiry
63	Identify lowest point on a graph		This item measures mathematical as much as scientific understanding
64	Interpret a trend on a graph		This item measures mathematical as much as scientific understanding

<table>
<tr><td>65</td><td>Identify a hypothesis</td><td rowspan="2">Key Idea S1
The central purpose of scientific inquiry is to develop explanations of natural phenomena in a continuing creative process.
S1.2 Construct explanations independently for natural phenomena, especially by proposing preliminary visual models of phenomena.</td><td>This item addresses elements of scientific inquiry in that it asks students to identify a hypothesis. However, the Performance Indicator, the construction of an explanation, is not warranted as the student is not requested to do so.</td></tr>
<tr><td>66</td><td>Identify one variable
Explain why the variable has to be held constant</td><td>This item addresses elements of scientific inquiry in that it asks students to identify a variable. However, the Performance Indicator, the construction of an explanation, identified by NYSEd is not warranted. The student is requested to develop an explanation of the experimental design rather than an explanation of the phenomena.</td></tr>
<tr><td>67</td><td>Construct a graph</td><td>The observations made while testing proposed explanations, when analyzed using conventional and invented methods, provide new insights into phenomena
3.1 Design charts, tables, graphs and other representations of observations in conventional and creative ways to help them address their research question or hypothesis</td><td>This item measures mathematical as much as scientific understanding.</td></tr>
</table>

Figure 10. Analyses of NYS Intermediate-level Science Test Items

In addition, the **Performance Indicator S3.2**, which invokes interpretation and meaning making as well as pattern recognition in data analysis, is significantly more complex than the item to which it is assigned. The individual item in each case is not connected to any explanation of the scientific concept or principle, for example, solubility in the case of item 35. The item only asks the student to identify a value given the linear representation of the data and a reference point. In question 62 the student is asked to give a summary of a trend of the data and to provide a possible explanation of that trend. This is an example of connecting the mathematical interpretation to a natural phenomenon that requires some explanation.

Students' understandings of scientific inquiry are addressed by the remaining three items, though the extent of the match between the item and the Performance Indicator remains problematic. Item 45 asks the student to identify the dependent and independent variable in a physical science context. The student is also asked to identify two factors that need to be maintained at constant levels during an investigation. The test developers have identified **Key Idea S3** which addresses data analysis rather than **Key Idea S2** which addresses the design of an investigation. The item is more closely associated with **Performance Indicator 2.2.2d** which specifies that the students should "identify independent variables (manipulated), dependent variables (responding), and constants in a simple controlled experiment" (*Intermediate-level Science Core Curriculum,* NYSEd, 1999, p.5). Clearly the issues of experimental design in this simple experiment are the focus of this item.

Item 65 requires the student to construct a hypothesis based on the description of an experimental design. This item is aligned with **Key Idea S1** which addresses the construction of explanations of natural phenomena. This task is better aligned with **Key Idea S2** which addresses the design of investigations, including simple experiments, in order to answer a scientific question. The task requires the student to identify the hypothesis that the experiment is designed to answer. We argue that this item assesses reading comprehension as much as scientific inquiry as the response for which the test designers are looking is to be found in the interpretation of the stem of the question. In addition, the second question associated with the item requires the student to identify one variable that is held constant where all of those variables are discussed in the stem of the question. The explanation that the student is required to give addressing the reason for holding that variable constant is the closest element of the question that invokes scientific inquiry.

Overall the multiple choice and short answer items in Parts B and C of the *Intermediate Level Science Test* are better aligned with mathematical analysis than scientific inquiry. While this matches the philosophy of the integrated aspects of the *NYS Science Standards*, it does not represent the nature of the two assessment systems, which are designed independently to measure mathematic and scientific understandings. In addition, it presents a confusing picture of scientific inquiry to the classroom teacher. Scientific inquiry is identified with variable identification, data analysis, and hypothesis generation rather than the identification of scientific questions that address natural phenomena and the production of evidence-based explanations.

Figure (11) illustrates the New York State Department of Education alignment of a sample performance task (*Intermediate Level Science Test* Part D) with the **Extended Process Skills**. Alignment is expressed in terms of **Key Ideas** and **Performance Indicators** for the expected student performance. Two other performance tasks have been previously described in Champagne and Kouba (in preparation).

CELL SIZE		
Task Step	Directions to the Student	Alignment with MST Standards
1	Pick up Slide A, hold it up to the light and look at the squares.	**I (Scientific Inquiry)** S2.1 Use conventional techniques and those of their own design to make further observations and refine their explanations, guided by a need for more information.
2	Slide A is a prepared slide of a tiny piece of graph paper. The lines of the graph paper are all spaced 1.0mm apart.	**I (Scientific Inquiry)** S2.1 Use conventional techniques and those of their own design to make further observations and refine their explanations, guided by a need for more information.
3	Place Slide A on the microscope stage and bring the graph paper into focus, using the lowest power.	**I (Scientific Inquiry)** S2.1 Use conventional techniques and those of their own design to make further observations and refine their explanations, guided by a need for more information.
4	When you look into the microscope, the whole area you see is called the "field of view." Knowing that the lines of the graph paper are 1.0 mm apart, estimate the diameter of the lowest power's field of view to the nearest 0.25mm.	**I (Scientific Inquiry)** S2.1 Use conventional techniques and those of their own design to make further observations and refine their explanations, guided by a need for more information. **1(Mathematical Analysis)** *Key Idea 3*:Critical thinking skills are used in the solution of mathematical problems. VI Interconnectedness: Common Themes *Key Idea 2*:Models are simplified representations of objects, structures, or systems used in analysis, explanation, interpretation, or design. *Key Idea 3*:The grouping of magnitudes

		of size, time, frequency, and pressures or other units of measurement into a series of relative order provides a useful way to deal with the immense range and the changes in scale that affect the behavior and design of systems **VII (Interdisciplinary Problem Solving)** *Key Idea 1*: The knowledge and skills of mathematics, science, and technology are used together to make informed decisions and solve problems, especially those relating to issues of science/technology/society, consumer decision-making, design, and inquiry into phenomena.
5	Return Slide A.	-
6	Place Slide B on the microscope and bring it into focus under the lowest power. Slide B is a piece of onion skin tissue that has been stained and mounted for viewing. See the diagram below for a sketch of what one cell might look like. The cell length has been labeled.	**I (Scientific Inquiry)** S2.1 Use conventional techniques and those of their own design to make further observations and refine their explanations, guided by a need for more information. VI Interconnectedness: Common Themes *Key Idea 2*: Models are simplified representations of objects, structures, or systems used in analysis, explanation, interpretation, or design. *Key Idea 3*:The grouping of magnitudes of size, time, frequency, and pressures or other units of measurement into a series of relative order provides a useful way to deal with the immense range and the changes in scale that affect the behavior and design of systems

7	Look closely at Slide B under the lowest power. Find one row of cells that goes across the middle of the field of view from one edge of the field of view to the other edge. These cells may go from side to side, from top to bottom, or diagonally across the diameter, In the circle at the right, carefully sketch only one row of cells whose lengths go across the field of view.	**I (Scientific Inquiry)** S1.2 Construct explanations independently for natural phenomena, especially by proposing preliminary visual models of phenomena. S3.1 Design charts, tables, graphs, and other representations of observations in conventional and creative ways to help them address their research question or hypothesis. VII Interdisciplinary Problem Solving *Key Idea 1*:The knowledge and skills of mathematics, science, and technology are used together to make informed decisions and solve problems, especially those relating to issues of science/technology/society, consumer decision-making, design, and inquiry into phenomena. VI Interconnectedness: Common Themes Key Idea 2:Models are simplified representations of objects, structures, or systems used in analysis, explanation, interpretation, or design. *Key Idea 3*:The grouping of magnitudes of size, time, frequency, and pressures or other units of measurement into a series of relative order provides a useful way to deal with the immense range and the changes in scale that affect the behavior and design of systems
8	How many cells did you see under lowest power in the row that you drew above?.	**I (Scientific Inquiry)** S1.3 Represent, present, and defend their proposed explanations of everyday observations so that they can be understood and assessed by others. S2.1 Use conventional techniques and those of their own design to make further observations and refine their explanations, guided by a need for more information. VI Interconnectedness: Common Themes

		Key Idea 2:Models are simplified representations of objects, structures, or systems used in analysis, explanation, interpretation, or design.,3 VII (Interdisciplinary Problem Solving) *Key Idea 1*: The knowledge and skills of mathematics, science, and technology are used together to make informed decisions and solve problems, especially those relating to issues of science/technology/society, consumer decision-making, design, and inquiry into phenomena.
9	In step 4 on the previous page, you estimated the diameter of the lowest power's field of view. Record that value again here:	
10	Based on the values you recorded in Steps 8 and 9, calculate the average length of one onion cell in your diagram to the nearest 0.1 mm.	**I (Scientific Inquiry)** S2.1 Use conventional techniques and those of their own design to make further observations and refine their explanations, guided by a need for more information. S2.3 Carry out their research proposals, recording observations and measurements (e.g., lab notes, audiotape, computer disk, videotape) to help assess the explanation. S3.2 Interpret the organized data to answer the research question or hypothesis and to gain insight into the problem. **1(Mathematical Analysis)** *Key Idea 3*: Critical thinking skills are used in the solution of mathematical problems. VI Interconnectedness: Common Themes *Key Idea 2*: Models are simplified representations of objects, structures, or systems used in analysis, explanation, interpretation, or design. *Key Idea 3*:

		The grouping of magnitudes of size, time, frequency, and pressures or other units of measurement into a series of relative order provides a useful way to deal with the immense range and the changes in scale that affect the behavior and design of systems
11	Return Slide B.	
12	Place Slide C on the microscope stage. Bring Slide C into focus under the lowest power. Now bring the slide into focus under the highest power. In the box below, draw an enlarged view of one typical cell on this slide under the highest power. Your drawing should accurately show the shape and structure of the cell.	**I (Scientific Inquiry)** S1.2 Construct explanations independently for natural phenomena, especially by proposing preliminary visual models of phenomena. S1.3 Represent, present, and defend their proposed explanations of everyday observations so that they can be understood and assessed by others. S2.1 Use conventional techniques and those of their own design to make further observations and refine their explanations, guided by a need for more information. S2.3 Carry out their research proposals, recording observations and measurements (e.g., lab notes, audiotape, computer disk, videotape) to help assess the explanation. VI Interconnectedness: Common Themes *Key Idea 2*:Models are simplified representations of objects, structures, or systems used in analysis, explanation, interpretation, or design. *Key Idea 3*:The grouping of magnitudes of size, time, frequency, and pressures or other units of measurement into a series of relative order provides a useful way to deal with the immense range and the changes in scale that affect the behavior and design of systems VII (Interdisciplinary Problem Solving) *Key Idea 1*: The knowledge and skills of mathematics, science, and technology are used together to make informed decisions and solve problems, especially

		those relating to issues of science/technology/society, consumer decision-making, design, and inquiry into phenomena.
13	When you re finished, put the microscope back to the lowest power. Return all materials to their positions as shown in the Station Diagram.	

Figure 11. Alignment of Cell Size Task form

Task steps 1-3 require the student to observe a prepared slide of graph paper and focus the image under low power on the microscope. The test developers have aligned these steps with **Key Idea S2.1**. We argue that the student need only read and follow direction and know rudimentary knowledge and skills with a microscope in order to be successful with these steps. These skills do entail the use of a conventional technique. However, the steps are not set in the context of conducting an inquiry; rather they are the antecedents to making measurements of the cell size. **Performance Indicator S2.1** invokes the notion of observations that a conducted in order to develop or refine an explanation. The task does not require the students to develop such an explanation.

Task step 4 requires the student to estimate the diameter of the field of view of the microscope on low power. Again, **Performance Indicator S2.1** which is the task developer alignment with this step is more sophisticated than what the student is actually being required to do. The **Key Idea M3** involves the use of critical thinking skills used in the solution of a mathematical problem. While estimation is a critical thinking skill, the measurement of the diameter of the field of view is not a significant mathematical problem given that the student has the measuring scale identified on the microscope slide. Two **Key Ideas** from Standard VI *Interconnectedness: Common Themes* are also aligned with step 4. These include the use of models for analysis, explanation, interpretation or design and grouping of magnitudes of size in relative order to measure scale. In both of these cases, the sophistication of the standard is not reflected in what the student has been asked to do. There is no requirement that the student addresses issues of a model or scale in carrying out the instructions to determine the size of the field of view. One **Performance Indicator** from Standard VII *Interdisciplinary Problem Solving* is also assigned to step 4 by the task developer. **Performance Indicator 7.1** requires that integrated knowledge and skills in mathematics, science and technology are used to make informed decisions and solve problems. The student is presented with the requirement of making measurements throughout this task, not solving a problem, especially one that deals with issues of science/technology/society, decision-making in a consumer context, engineering design, or scientific inquiry into natural phenomena. With both of these standards, the level of sophistication of the standard is not well reflected in the requirements of the task.

Task step 6 and 7 require the student to take a stained slide of onion skin, bring it into focus, and draw a row of cells that extends across the middle of the field of view. Many of the standards assigned by the task developer are similar to those assigned above. One addition is **Performance Indicator S1.2** which calls upon the ability to construct an explanation using a visual model of the phenomenon. The drawing of the row of cells does not adequately represent a model as it is given no explanatory power. Rather the student is making a visual representation of an observation, a preliminary step in the construction of a model. **Performance Indicator S3.1** is also listed, and while the student is constructing a representation, there is no hypothesis or research question identified which the student is trying to address. Again, the levels of sophistication of the standards are not well reflected in the requirements of the task.

Task step 8 requires the student to identify the number of cells observed in the field of view. The student is simply making an observation and is not required to make any inferences, draw conclusions or describe an explanation, all of which are the components of **Performance Indicator S1.3** assigned to the step by the task developer. In task steps 9 and 10 the student is asked to use the value previously estimated for the diameter of the field of view and the number of cells observed to calculate the average length of one cell. The task developer assigned **Performance Indicators** from **Key Idea 2** of Standard I *Scientific Inquiry* that deal with the use of investigatory design and analysis and interpretation of data. These two steps represent the core of scientific inquiry that is embedded in this task and they contain the most cognitively complex requirements for reasoning that the student must display. Assignment of the critical thinking **Key Idea** from *Mathematical Analysis* and the models and scale **Key Ideas** from Standard VI *Interconnectedness: Common Themes* have also been made by the task developers. These key ideas are only superficially represented in these task steps.

Task step 12 requires the student to draw a typical cell from a third slide seen under the highest power of the microscope. **Performance Indicators S1.2, S1.3, S2.1** and **S2.3** are assigned by the task developer demonstrating that components of the construction of explanation through visual models of phenomena, representations of proposed explanations, use of investigatory techniques to make observations and explanations, and implementation of research proposals are all represented in the task. While the drawing of a cell is a representation, no explanation of that representation is required.

Overall, the alignment proposed by the task developer presents an image of much more complex knowledge and skills as the target of the assessment task than are actually required by the task. This overblown estimate of the sophistication of the task does not present a rich well contextualized image of scientific inquiry to the classroom teacher. The task is more closely aligned with the more simplistic skills identified in the **Process Skills based on Standard IV** which address procedural skills that are needed to apply certain techniques or use certain equipment. These skills are not included in the alignment analysis by the task developer.

CONCLUSION

The *NYS Science Standards* include standard statements about scientific inquiry that hold the potential for informing the development of rich and detailed scientific inquiry learning experiences for students. The standards map well onto the elements of classroom inquiry found in both the *National Science Education Standards* and the Olson & Loucks-Horsley (2000) addendum to the standards. The standards for scientific inquiry emphasize the importance of the development of evidence-based explanations of natural phenomena in response to questions developed and investigated by students. However, the image of scientific inquiry that is found in the example instructional tasks and state level assessments are not consistent with inquiry as explanation development and defense. Examples of instructional tasks address science/technology/society issues rather than the scientific concepts and principles that are found in Standard IV, *Science*, and the examples given in that standard are not mapped onto scientific inquiry in a way that assists the teacher in making the connection. The nature of the inter-connectedness of the **Expanded Process Skills** is never well addressed, and the *Intermediate Level Science Test* and the *Intermediate-level Science Core Curriculum* as test and expanded test framework are not well tied together. It is unclear what the relationship is between science literacy and the **Expanded Process Skills**. It is not clear how those skills might be incorporated into an intermediate level science curriculum in an articulated fashion to examine how each contributes to science literacy. The test that serves as a measure of evidence of what students know and can do in scientific inquiry is impoverished when compared to the standards with which the items and tasks are aligned.

New York State is in a common dilemma with other states that must include testing of student achievement in science as part of a state assessment system. We do not trivialize the difficulty of assessing scientific inquiry in paper and pencil tests. New York has a long history of attempting to develop performance-based assessment tasks. What we argue in this paper is the language and certainty of alignment that the state intends. The state needs to convey a better image of the limitations of multiple-choice and structured performance assessments at the same time that they work to develop and use assessment tasks that more authentically address scientific inquiry.

We argue that a much more systematic documentation of alignment of scientific inquiry and the concepts and principles of science are crucial to assist the science teacher in the development of coherent learning experiences for students. The fragmented perspective on scientific inquiry supports teaching inquiry separate from the learning of those important content ideas and further burdens the already overloaded science curriculum. A cognitive analysis of the range of responses to the performance tasks would be a starting place for such an analysis. In addition, analysis of the cognitive demand of instructional tasks and examples of student work in other supporting documents currently under development by New York State Education Department would contribute to such understanding. The cognitive analysis of those tasks found in the supporting documents of the *NYS Mathematics Science and Technology Resource Guides* (NYSEd, 2000) is currently being examined by the authors of this chapter. The third paper in this series will complete the picture of the image of scientific inquiry in New York State documents.

REFERENCES

American Association for the Advancement of Science. (1994). *Benchmarks for science literacy* Washington, D.C.: Author.

Champagne & Kouba (in preparation) Optimizing state and classroom tests: implications of cognitive research for assessments of higher order reasoning in subject-matter domains. Cognitive Mechanisms in Assessment Conference. College Park, MD. August 15-16. 2002.

Ferrara, S., Willihoft, J., Seburn, C., Slaughter, F., & Stevenson, J. (1991) Local assessments designed to parallel statewide minimum competency tests: Benefits and drawbacks. In Stake, R. E. (Ed.) *Using assessment policy to reform education. Advances in Program Evaluation, v. 1. part A.* (pp. 41-74). Greenwich, CT: JAI Press.

National Research Council. (1996) *National science education standards.* Washington, D.C.: National Academy Press.

New York State Education Department (NYSEd.). (1996). *New York State learning standards for mathematics, science and technology.* Albany, NY: The University of the State of New York, The State Education Department, Office of Curriculum, Instruction and Assessment.

New York State Education Department (NYSEd.) (1999). *Intermediate level science core curriculum.* Albany, NY: The University of the State of New York, The State Education Department, Office of Curriculum, Instruction and Assessment.

New York State Education Department (NYSEd.) (Spring, 2000). *Intermediate-level science examination: Test sampler draft.* Albany, NY: The University of the State of New York, The State Education Department, Office of Curriculum, Instruction and Assessment.

New York State Department of Education (NYSEd.). (2000). *Mathematics science and technology resource guide.* Albany, NY: The University of the State of New York, The State Education Department, Office of Curriculum, Instruction and Assessment.

New York State Education Department (NYSEd.) (2002) *June 2002 Intermediate level science test.* Albany, NY: The University of the State of New York, The State Education Department, Office of Curriculum, Instruction and Assessment.

New York State Education Department (NYSEd.) (2002). *New York State item maps for intermediate level science test.* Albany, NY: The University of the State of New York, The State Education Department, Office of Curriculum, Instruction and Assessment.

Olson, S. & Loucks-Horsley, S. (2000) *Inquiry and the National Science Education Standards: A guide for teaching and learning.* Washington, D.C.: National Academy Press.

O'Sullivan, R. (1991). Teachers' perceptions of the effects of testing on classroom practices. In Stake, R. E. (Ed.) *Using assessment policy to reform education. Advances in Program Evaluation, v. 1. part A.* (pp. 145-163). Greenwich, CT: JAI Press.

Stake, R., & Theobold, P. (1991). Teachers' views of testing's impact on classrooms. . In Stake, R. E. (Ed.) Using assessment policy to reform education. *Advances in Program Evaluation, v. 1. part A.* (pp. 189-202) Greenwich, CT: JAI Press.

Stern L. & Ahlgren, A. (2002). Analysis of students' assessments in middle school curriculum materials: Aiming precisely at benchmarks and standards. *Journal of Research in Science Teaching 39, 889-910.*

Wegner, E. (1999). *Communities of practice: Learning, meaning, and identity.* Cambridge University Press: Cambridge, UK.

Whitehead, A. N. (1932). *The aims of education and other essays.* London : Williams & Northgate.

Wilson, B. L. & Corbett, H. D. (1991). Two state minimum competency testing programs and their effects on curriculum and instruction. In Stake, R. E. (Ed.) Using assessment policy to reform education. *Advances in Program Evaluation, v. 1. part A.* (pp. 7-40). Greenwich, CT: JAI Press.

PART IV: TEACHING AND LEARNING ABOUT NATURE OF SCIENCE

CHAPTER 14

NORMAN G. LEDERMAN

SYNTAX OF NATURE OF SCIENCE WITHIN INQUIRY AND SCIENCE INSTRUCTION

INTRODUCTION

Criticisms about the quality of pre-college and undergraduate science education continue and, from my perspective, are beginning to wear thin. I certainly concur with many of the problems cited, but I am becoming impatient with critiques that are not accompanied by any concrete solutions to the problem at hand. Although the reasons for concern about the quality of science instruction differ from nation to nation, the primary rallying point is the perceived level of scientific literacy among each nation's populace. Although the words may differ, the situation is not new. One can easily point to "critical" concerns voiced about science teaching and learning, and associated reforms, for well over a century. In each case, whether the label "scientific literacy" was used, concerns have typically focused on the usefulness and relevancy of the subject matter included in K-12 science curriculum. More specifically, educators have historically been concerned with students' ability to apply their science knowledge to make informed decisions regarding personal and societal problems. The ability to use scientific knowledge to make informed personal and societal decisions is the essence of what contemporary science educators and reform documents define as scientifically literacy. Perhaps the most recent reform visions of note have been the *National Science Education Standards* (NRC, 1996) and *Project 2061* (assorted publications of AAAS) of the U.S. As was the case with most of their predecessors in the U.S. and elsewhere, these reform efforts stress the importance of conceptual understanding of the overarching ideas in science (e.g., cause and effect, equilibrium, structure and function, cycles, scale). Such ideas are believed to transcend the individual disciplines within science and are believed to be superior educational outcomes than the mere memorization of foundational discipline-based subject matter. The phrase "less is more" has often been invoked to communicate the desire that instructional time focus on in-depth understanding of a reduced set of unifying scientific concepts. Ultimately, it is believed, a focus on a fewer number of more global themes will result in a more useful and productive understanding of science. Although the words of various reforms are different, the message remains quite familiar. Just as familiar is the lack of progress toward the all too familiar goals of reform efforts.

Although I have lingering concerns about reform efforts in science education, there is an increased emphasis in current reforms that make their visions significantly different from previous efforts: nature of science and scientific inquiry.

L.B. Flick and N.G. Lederman (eds.) Scientific Inquiry and Nature of Science, 301-317.

Up to this point in this text, there has been more than adequate discussion of the importance of scientific inquiry. It is mentioned again here to emphasize its intimate relationship with nature of science. As mentioned previously, helping students develop adequate conceptions of nature of science (NOS) and scientific inquiry has been a perennial objective in science education (American Association for the Advancement of Science [AAAS], 1990, 1993; Klopfer, 1969; National Research Council [NRC], 1996; National Science Teachers Association [NSTA], 1982). Indeed, this objective has been agreed upon by most scientists and science educators for approximately 100 years (Central Association of Science and Mathematics Teachers, 1907; Kimball, 1967-68; Lederman, 1992). Presently, despite their varying pedagogical or curricular emphases, there is strong agreement among the major reform efforts in science education (AAAS, 1990, 1993; NRC, 1996) about the importance of enhancing students' conceptions of NOS and scientific inquiry. In fact, "the longevity of this educational objective has been surpassed only by the longevity of students' inability to articulate the meaning of the phrase 'nature of science,' and to delineate the associated characteristics of science" (Lederman & Niess, 1997, p. 1) or scientific inquiry. Despite numerous attempts, including the major curricular reform efforts of the 1960s, to improve students' views of the scientific endeavor, students have consistently been shown to possess inadequate understandings of several aspects of NOS and scientific inquiry (e.g., Aikenhead, 1973; Bady, 1979; Broadhurst, 1970; Lederman & O'Malley, 1990; Mackay, 1971; Mead & Metraux, 1957; Rubba & Andersen, 1978; Tamir & Zohar, 1991; Wilson, 1954).

Consequently, it is only natural to ask whether there are reasons to believe that the recent reforms in science education are more likely to impact students' understandings than their predecessors. It is my view that the current reform documents' emphasis on the NOS and scientific inquiry are likely to have as little impact as earlier efforts. Two critical and interrelated omissions that have typified previous efforts are, unfortunately, evident in the more recent reform documents. There is not, and there has not been, a concerted professional development effort to clearly communicate, first, what is meant by "NOS" and scientific inquiry and second, how a functional understanding of these valued aspects of science can be communicated to K-12 students. I am not totally pessimistic about the goals of current reforms, but I do believe it is absolutely necessary that our efforts proceed with proper reference to research on the teaching and learning of NOS and scientific inquiry. Situating NOS instruction within a context of scientific inquiry can provide significant progress. The syntax of NOS within inquiry is critical.

Perhaps the lack of professional development related to NOS and scientific inquiry is a consequence of the misunderstanding that NOS and scientific inquiry fall within the realm of affect and process as opposed to cognitive outcomes of equal, if not greater, importance than "traditional" subject matter. NOS and scientific inquiry are as much an aspect of subject matter as the reactions of photosynthesis, atomic structure, plate tectonics, or pH. In reality, however, it is NOS and scientific inquiry that provide a meaningful context for the subject matter specified in the *Standards* and other reform documents. Furthermore, NOS permeates all areas of the discipline-specific standards and it is a critical component of the standards on "science as inquiry." From the perspective of currently advocated pedagogy (i.e., constructivist approaches), an understanding of NOS and scientific inquiry underlies the essence of the *Teaching and Assessment Standards* specified by the *National Science Education Standards*. It is not at all difficult to argue that a teacher who lacks adequate conceptions of NOS and scientific inquiry, and a functional understanding of how to teach these valued aspects of science cannot orchestrate the types of instructional activities and atmosphere, or assess students' progress, as specified in the various reform efforts in science education. Indeed, a functional understanding of NOS and scientific inquiry by teachers is clearly prerequisite to any hopes of achieving the vision of science teaching and learning specified in the various reform efforts. In the following sections, I will clarify the meaning of the NOS and, briefly, scientific inquiry. These terms are used with little precision and high variability within educational circles and it is necessary to insure that we are all on the same page regarding these important educational outcomes. I will also delineate several misconceptions promoted (or ignored) by reform efforts. It will further be argued that without explicit instructional attention to NOS, students will once again learn science subject matter in a context-free environment. Such an environment does not permit the in-depth conceptual understanding of science subject matter advocated in the various visions of reform and will not help create a populace that can be considered scientifically literate.

WHAT IS NATURE OF SCIENCE?

The phrase "nature of science" typically refers to the epistemology of science, science as a way of knowing, or the values and beliefs inherent to scientific knowledge or the development of scientific knowledge (Abd-El-Khalick & Lederman, 2000; Lederman, 1992). Beyond these general characterizations, no consensus presently exists among philosophers of science, historians of science, scientists, and science educators on a specific definition for NOS. Hence, the reason for not placing the word 'the' in front of NOS. This lack of consensus, however, should neither be disconcerting nor surprising given the multifaceted nature and complexity of the scientific endeavor. Conceptions of NOS have changed throughout the development of science and systematic thinking about science and are reflected in the ways the scientific and science education communities have defined the phrase "nature of science" during the past 100 years (e.g., AAAS, 1990, 1993; Central Association for Science and Mathematics Teachers, 1907; Klopfer & Watson, 1957; NSTA, 1982).

Many of the disagreements about the definition or meaning of the NOS that continue to exist among philosophers, historians, and science educators are irrelevant to K-12 instruction (Smith, Lederman, Bell, McComas, & Clough, 1997). The issue of the existence of an objective reality, for example, as compared to phenomenal realities is a case in point. I argue that there is an acceptable level of generality regarding NOS that is accessible to K-12 students and relevant to their daily lives. Moreover, at this level, little disagreement exists among philosophers, historians, and science educators. In the work pursued by my research group, three criteria are used to determine what aspects of nature of science we choose to focus upon in our research and curriculum development efforts. These criteria are:

1. Is knowledge of the aspect of NOS accessible to students (can they learn and understand)?
2. Is there general consensus about the aspect of NOS?
3. Is it useful for all citizens to understand the aspect of NOS?

Using these criteria, the following seven aspects of nature of science are viewed as important to include in science curriculum and instruction: Scientific knowledge is tentative (subject to change), empirically-based (based on and/or derived at least partially from observations of the natural world), subjective (theory-laden, involves individual or group interpretation), necessarily involves human inference, imagination, and creativity (involves the invention of explanations), and is socially and culturally embedded (influenced by the society/culture in which science is practiced). Two additional important aspects are the distinction between observations and inferences, and the functions of, and relationships between, scientific theories and laws. Many of my colleagues, for example, argue that there are no observations, just inferences. Their point can certainly be argued successfully in a philosophy class. Unfortunately, this would not be a fruitful path to take with a seventh grade student. For middle school students, and I dare say for most of the general public, distinguishing between observations and inferences is concrete and a more productive way to view one's everyday experiences. Often there is utilitarian value in what seems intuitively obvious.

Overall, the important point is that there are exceedingly long laundry lists of aspects of nature of science that can be developed if one's purpose is totally academic with little regard for practicality. In addition to the overwhelming strain such additions would have on current curriculum requirements, these lists necessarily include numerous ideas about science that are contentious and/or inaccessible to students in our pre-college schools. Hence, as science educators, we must have some modicum of sensibility in our recommendations regarding NOS. We must consider the value and developmental appropriateness of what we recommend for ALL students.

What follows is a brief consideration of the aspects of nature of science previously identified as having met the criteria for selection used by my colleagues during the past several decades. It is also important to note that the seven aspects described are consistent with what is recommended by the *National Science Education Standards* and *Benchmarks for Science Literacy* relative to nature of science. This brief primer is also being provided since, regardless of numerous discussions elsewhere, there continues to be confusion about the meaning of the various aspects of NOS.

First, students should be aware of the crucial distinction between observation and inference. Observations are descriptive statements about natural phenomena that are "directly" accessible to the senses (or extensions of the senses) and about which several observers can reach consensus with relative ease. For example, objects released above ground level tend to fall and hit the ground. By contrast, inferences are statements about phenomena that are not "directly" accessible to the senses. For example, objects tend to fall to the ground because of "gravity." The notion of gravity is inferential in the sense that it can *only* be accessed and/or measured through its manifestations or effects. Examples of such effects include the perturbations in predicted planetary orbits due to inter-planetary "attractions," and the bending of light coming from the stars as its rays pass through the sun's "gravitational" field.

Second, closely related to the distinction between observation and inference is the distinction between scientific laws and theories. You may recall the so-called "hierarchy of credibility" found in most science textbooks that presents categories of scientific knowledge/ideas (i.e., observations, hypotheses, theories, laws/principles) in an ascending list of credibility or certainty. Individuals often hold the simplistic, hierarchical view of the relationship between theories and laws presented in such lists. That is, theories become laws as they accumulate supporting evidence over numerous years. It follows from this notion that scientific laws have a higher status than scientific theories. You don't have to be a biology teacher to remember the phrase, "evolution is just a theory." The common notions relating theories and laws are inappropriate because, among other things, theories and laws are different kinds of knowledge and one can not develop or be transformed into the other. Laws are *statements or descriptions of the relationships* among observable phenomena. Boyle's law, which relates the pressure of a gas to its volume at a constant temperature, is a case in point. Theories, by contrast, *are inferred explanations* for observable phenomena. The kinetic molecular theory, which explains Boyle's law, is one example. Mendel's Laws of genetics allow us to make predictions about offspring from parents with certain traits, The explanation for these predictions and what is observed is provided by chromosome or gene theory. Moreover, theories are as legitimate a product of science as laws. Scientists do not usually formulate theories in the hope that one day they will acquire the status of "law." Scientific theories, in their own right, serve important roles, such as guiding investigations and generating new research problems in addition to explaining relatively huge sets of seemingly unrelated observations in more than one field of investigation. For example, the kinetic molecular theory serves to explain phenomena that relate to changes in the physical states of matter, others that relate to the rates of chemical reactions, and still other phenomena that relate to heat and its transfer, to mention just a few.

A third aspect of nature of science is that even though scientific knowledge is, at least partially, based on and/or derived from observations of the natural world (i.e., empirical) it nevertheless involves human imagination and creativity. Science, contrary to common belief, is not a totally lifeless, rational, and orderly activity. Science involves the *invention* of explanations and this requires a great deal of creativity by scientists. The "leap" from atomic spectral lines to Bohr's model of the atom with its elaborate orbits and energy levels is just one example. This aspect of science, coupled with its inferential nature, entails that scientific concepts, such as atoms, black holes, and species, are functional theoretical models rather than faithful copies of reality. Most students are quick to accept the idea that scientists must use their creativity in the design of scientific investigations. However, they are less likely to recognize that analysis of data involves creativity and, in many ways, a scientific theory involves as much creativity as a work of art.

Fourth, scientific knowledge is subjective or theory-laden. Scientists' theoretical commitments, beliefs, previous knowledge, training, experiences, and expectations actually influence their work. All these background factors form a *mind-set* that *affects* the problems scientists investigate and how they conduct their investigations, what they observe (and do not observe), and how they make sense of, or interpret their observations. It is this (sometimes collective) individuality or mind-set that accounts for the role of subjectivity in the production of scientific knowledge. It is noteworthy that, contrary to common belief, science never starts with neutral observations (Chalmers, 1982). Observations (and investigations) are always motivated and guided by, and acquire meaning in reference to questions or problems. These questions or problems, in turn, are derived from within certain theoretical perspectives. These frameworks that guide scientists' work at all levels were initially labeled as "paradigms" by Thomas Kuhn. It is important to note that I am not advocating that scientists be subjective. Indeed, many of the procedural conventions of scientific research are designed to limit the negative impacts of subjectivity. However, it is critical for us to realize that subjectivity is unavoidable and in many cases can lead to productive interpretations.

By way of simple analogy, imagine driving your car into a supermarket parking lot that has a fresh cover of snow. There are no other cars present and you are unable to see the lines that separate the parking spaces. If you have ever experienced this situation, the feeling is an uneasy one and may have even worried about whether you were driving in the appropriate lanes or across parking spaces. In reality, it made no difference but this does not relieve the uneasiness. Eventually, you just park the car and enter the supermarket. Upon leaving the store you probably noticed that other cars arrived and they followed your "lead" with respect to what angle and location was appropriate for parking. In short, our paradigms, expectations, biases, and subjectivities provide reference points by which to gauge the meaning of our research findings. Subjectivity becomes a detriment when it is so strong that we ignore overwhelming evidence to the contrary of our expectations.

A fifth aspect of NOS is that science is a human enterprise practiced in the context of a larger culture and its practitioners (scientists) are the product of that culture. Science, it follows, affects and is affected by the various elements and intellectual spheres of the culture in which it is embedded. These elements include, but are not limited to, social fabric, power structures, politics, socioeconomic factors, philosophy, and religion. An example may help to illustrate how social and cultural factors impact scientific knowledge. Telling the story of the evolution of humans (*Homo sapiens*) over the course of the past seven million years is central to the biosocial sciences. Scientists have formulated several elaborate and differing story lines about this evolution. Until recently, the dominant story was centered about "the man-hunter" and *his* crucial role in the evolution of humans to the form we now know (Lovejoy, 1981). This scenario was consistent with the white-male culture that dominated scientific circles up to the 1960s and early 70s. As the feminist movement grew stronger and women were able to claim recognition in the various scientific disciplines, the story about hominid evolution started to change.

One story that is more consistent with a feminist approach is centered about "the female-gatherer" and *her* central role in the evolution of humans (Hrdy, 1986). It is noteworthy that both story lines are consistent with the available evidence. A more contentious perspective related to the cultural and social embeddedness of science is whether there is one science. That is, some would argue that the science that dominates North American schools is Western Science, but there exist other analogous sciences (e.g., indigenous science) in other parts of the world.

Sixth, it should be clear from the previous discussions about creativity, subjectivity, and inference in science that many possible explanations can be offered to explain a particular phenomenon in the natural world. Nevertheless, it is also true that any explanation is not acceptable and all explanations do not necessarily have equal credibility. The consistency between evidence and conclusions is critical in science. In the end, the "creations" and speculations of scientists must be tested in the empirical world. Are predictions accurate? Does what you predict will happen, actually happen? In short, scientific knowledge must be referenced to the natural world as much as possible. In short, this aspect of nature of science is known as "empirical." Science's necessary reliance on empirical evidence is what distinguishes it as a way of knowing from other disciplines such as philosophy and mathematics. The empirical basis of science is the aspect that is most easily understood by students. Perhaps this is because students have the most concrete exposure during their school years to the idea that evidence and experiments are important in science.

Seventh, it follows from the previous discussions that scientific knowledge is never absolute or certain. This knowledge, including "facts," theories, and laws, is tentative and subject to change. Scientific claims change as new evidence, made possible through advances in *theory* and technology, is brought to bear on existing theories or laws, or as old evidence is reinterpreted in the light of new theoretical advances or shifts in the directions of established research programs. Important to note is that "tentativeness" is a characteristic of the scientific knowledge and it is not, as many misconstrue, the hesitation a student may feel when deciding how to answer a scientific question. It should be emphasized that tentativeness in science does not only arise from the fact that scientific knowledge is inferential, creative, subjective, and socially and culturally embedded. There are also compelling logical arguments that lend credence to the notion of tentativeness in science. Indeed, contrary to common belief, scientific hypotheses, theories, and laws can *never* be absolutely "proven." This holds irrespective of the amount of empirical evidence gathered in the support of one of these ideas or the other (Popper, 1963, 1988). For example, to be "proven," a certain scientific law should account for *every single instance* of the phenomenon it purports to describe *at all times*. It can logically be argued that one such future instance, of which we have no knowledge whatsoever, may behave in a manner contrary to what the law states. As such, the law can never acquire an absolutely "proven" status. This equally holds in the case of hypotheses and theories. Some individuals take exception to the use of the word "tentative" when describing scientific knowledge. It is felt that the word implies that scientific knowledge is less stable and durable than we assume it to be. If you are one of these individuals, perhaps you will find "revisionary" or "subject to change" more palatable descriptors. At least this has been the case in my experience with individuals reacting with disdain to "tentative."

Finally, it is important to note that individuals often conflate the NOS with science processes (which is more consistent with scientific inquiry). Although these aspects of science overlap and interact in important ways, it is nonetheless important to distinguish the two. Scientific processes, as you know from the previous chapters in this text, are activities related to collecting and analyzing data, and drawing conclusions (AAAS, 1990, 1993; NRC, 1996). For example, observing and inferring are scientific processes. On the other hand, the NOS refers to the epistemological underpinnings of the activities of science. As such, realizing that observations are necessarily theory-laden and are constrained by our perceptual apparatus belongs within the realm of the NOS. The fact that multiple approaches to answering a scientific question may be acceptable is relevant to scientific inquiry. The understanding that the conclusions drawn from any investigation are subject to change lies within the realm of nature of science.

Teachers' understandings of nature of science, as recent research shows, Abd-El-Khalick and Lederman, 2000) does not necessarily translate into classroom practice. Consequently, professional development efforts designed for teachers must not conclude, as they have in the past, with the development of adequate teacher understandings. Certainly, teachers must have an in-depth understanding of what they are expected to teach. However, professional development efforts must also emphasize how teachers can successfully facilitate the development of students' understandings of the NOS. How nature of science is situated in the curriculum is referred to here as the "syntax" of NOS within instruction. It will be argued that students' understandings of NOS are best facilitated if situated within a context of inquiry. By "context of inquiry" I am referring to an instructional context that includes all aspects of inquiry as specified in the *National Science Education Standards*: a teaching approach and as student outcomes of doing and knowing about inquiry. Given the importance of "inquiry" to teaching about NOS, it is critical to at least provide a general description of scientific inquiry as it is perceived here. Certainly, more extensive and varying treatments of the topic can be found in the preceding chapters.

WHAT IS SCIENTIFIC INQUIRY

As mentioned before, there is much confusion about the distinction between nature of science and scientific inquiry. The two are intimately related and there is critical overlap. However, it is useful to conceptualize scientific inquiry as the process by which scientific knowledge is developed and, by virtue of the conventions and assumptions of this process, the knowledge produced necessarily has certain unavoidable characteristics (i.e., NOS).

Although closely related to science processes, scientific inquiry extends beyond the mere development of process skills such as observing, inferring, classifying, predicting, measuring, questioning, interpreting and analyzing data. Scientific inquiry includes the traditional science processes, but also refers to the combining of these processes with scientific knowledge, scientific reasoning and critical thinking to develop scientific knowledge. From the perspective of the *National Science Education Standards* (NRC, 1996), students are expected to be able to develop scientific questions and then design and conduct investigations that will yield the data necessary for arriving at conclusions for the stated questions. The *Benchmarks for Science Literacy* (AAAS, 1993) are a bit less ambitious, in terms of the doing of inquiry, as they do not advocate that all students be able to design and conduct investigations in total. Rather, it is expected that all students at least be able to understand the rationale of an investigation and be able to critically analyze the claims made from the data collected. Perhaps the most significant difference between the *National Standards* and the *Benchmarks* is that the former expects students to be able to **DO** inquiry as well as **KNOW ABOUT** inquiry, while the latter just focuses on students' knowledge **ABOUT** inquiry.

Knowledge **ABOUT** inquiry is not given nearly as much attention in the literature as it should and is virtually ignored in instructional materials. What follows is a brief primer of what we would like students to know **ABOUT** inquiry.

Scientific inquiry refers to the systematic approaches used by scientists in an effort to answer their questions of interest. Pre-college students, and the general public for that matter, believe in a distorted view of scientific inquiry that has resulted from schooling, the media, and the format of most scientific reports. This distorted view is called **THE SCIENTIFIC METHOD** That is, a fixed set and sequence of steps that all scientists follow when attempting to answer scientific questions. A more critical description would characterize **THE METHOD** as an algorithm that students are expected to memorize, recite, and follow as a recipe for success. The visions of reform, however, are quick to point out that there is no single fixed set or sequence of steps that all scientific investigations follow. The contemporary view of scientific inquiry advocated is that the questions guide the approach and the approaches vary widely within and across scientific disciplines and fields.

At a general level, scientific inquiry can be seen to take several forms (i.e., descriptive, correlational, and experimental). Descriptive research is the form of research that often characterizes the beginning of a line of research. This is the type of research that derives the variables and factors important to a particular situation of interest. Whether descriptive research gives rise to correlational approaches depends upon the field and topic. For example, much of the research in anatomy and taxonomy are descriptive in nature and do not progress to experimental or correlational types of research. The purpose of research in these areas is very often simply to describe. On the other hand, there are numerous examples in the history of anatomical research that have lead to more than description. The initial research concerning the cardiovascular system by William Harvey was descriptive in nature. However, once the anatomy of blood vessels had been described, questions arouse concerning the circulation of blood through the vessels. Such questions lead to research that correlated anatomical structures with blood flow and experiments based on models of the cardiovascular system. To briefly distinguish correlational from experimental research, the former explicates relationships among variables identified in descriptive research and the latter involves a planned intervention and manipulation of variables related in correlational research in an attempt to derive causal relationships. In some cases, lines of research can been seen to progress from descriptive to correlational to experimental, while in other cases (e.g., descriptive astronomy) such a progression is not necessarily relevant.

The perception that a single scientific method exists owes much to the status of classical experimental design. Experimental designs very often conform to what is presented as **THE SCIENTIFIC METHOD** and the examples of scientific investigations presented in science textbooks most often are experimental in nature. The problem, of course, is not that investigations consistent with "the scientific method" do not exist. The problem is that experimental research is not representative of scientific investigations as a whole. Consequently, a very narrow and distorted view of scientific inquiry is promoted in our K-12 students.

Scientific inquiry has always been ambiguous in its presentation within science education reforms. In particular, inquiry is perceived in three different ways. It can be viewed as a set of skills to be learned by students and combined in the performance of a scientific investigation. It can also be viewed as a cognitive outcome that students are to achieve. In particular, the current visions of reform are very clear (at least in written words) in distinguishing between the performance of inquiry (i.e., what students will be able to do) and what students know about inquiry (i.e., what students should know). For example, it is one thing to have students set up a control group for an experiment, while it is another to expect students to understand the logical necessity for a control within an experimental design. Unfortunately, the subtle difference in wording noted in the reforms (i.e., "know" versus "do") is often missed by everyone except the most careful reader. The third use of "inquiry" in reform documents relates strictly to pedagogy and further muddies the water. In particular, current wisdom advocates that students best learn science through an inquiry-oriented teaching approach. It is believed that students will best learn scientific concepts by doing science. In this sense, "scientific inquiry" is viewed as a teaching approach used to communicate scientific knowledge to students (or allow students to construct their own knowledge) as opposed to an educational outcome that students are expected to learn about and learn how to do. Indeed, it is the pedagogical sense of inquiry and the doing of inquiry that it is strongly communicated to most teachers by science education reform documents, with the knowledge about inquiry lost in the shuffle.

THE SYNTAX OF INSTRUCTION AND DEVELOPMENT OF FUNCTIONAL UNDERSTANDINGS OF NOS

As mentioned earlier in this chapter, science education reforms have not handled the teaching of NOS well presently or in the past. On the one hand, it has been assumed that teachers understand NOS (as well as inquiry for that matter) and little professional development has been planned or provided. On the other hand, little has been provided to teachers regarding the teaching of NOS to students. Knowing about NOS is necessary, but it is not sufficient. The provision of professional development is closely linked to financial resources and the purveyors of reform can hardly be held accountable for what school districts, counties, states, and countries decide to provide. One could argue, however, that reform documents could present a much stronger case for the necessity of professional development.

However, there is a critical flaw in the various approaches reform efforts and curriculum developers have taken with respect to the teaching of nature of science. It is the same critical flaw that has existed since the beginning of the science education community's recognition of the importance of NOS as an important educational outcome. Interestingly, this same flaw is also evident with respect to attempts to teach scientific inquiry.

Two general approaches have been clearly evident over the years in both reform documents and the science education literature when it comes to the enhancement of students' and teachers' understandings of NOS and/or scientific inquiry. The first approach, labeled here as the *implicit* approach, suggests that by "doing science" students will also come to understand NOS and scientific inquiry (Lawson, 1982; Rowe, 1974). This approach was adopted by most of the curricula of the 1960s and 70s that emphasized hands-on, inquiry-based activities and/or process-skills instruction. Research studies have clearly indicated that the implicit approach was not effective in enhancing students' and teachers' understandings of the NOS or scientific inquiry (e.g., Durkee, 1974; Haukoos & Penick, 1985; Riley, 1979; Spears & Zollman, 1977; Trent, 1965; Troxel, 1968). There are two interrelated assumptions that underlie the implicit approach and have compromised its effectiveness. The first depicts attaining an understanding of the NOS as analogous to the attainment of an "affective" (as compared to a cognitive) learning outcomes. That is, students will come to understand NOS by simple repeated exposure to scientific endeavors in a manner similar to how one develops a positive or negative attitude toward a particular subject matter discipline. Indeed, it is not uncommon to find scientific inquiry and NOS collapsed under the label of "scientific attitude." This first assumption entails the second assumption; the assumption that learning about the NOS will result as a by-product of "doing science."

The second approach, the *historical* approach (one that is strongly recommended by the *National Science Education Standards*), suggests that incorporating the history of science (HOS) in science teaching can serve to enhance students' views of NOS. This approach is not new and has been advocated ever since the beginning of systematic research related to the teaching of NOS. The *History of Science Cases for High Schools* (Klopfer & Watson, 1957) and *Harvard Project Physics* (Rutherford, Holton, & Watson, 1970) were two of the most notable curriculum development efforts that included substantial attention to the HOS at the high school level. However, a review of the efforts that aimed to assess the influence of incorporating the HOS in science teaching (Klopfer & Cooley, 1963; Solomon, Duveen, Scot, & McCarthy, 1992; Welch & Walberg, 1972; Yager & Wick, 1966) indicates that evidence concerning the effectiveness of the historical approach is, at best, inconclusive. And, most recently, the work of Abd-El-Khalick and Lederman (2000) has indicated that specific courses in the history and/or philosophy of science have little impact on students' understanding of NOS and scientific inquiry. However, the intuitive appeal of the historical approach has helped it maintain its supporters in spite of the clearly one-sided evidence that runs contrary to its expected effectiveness. The International History, Philosophy, and Science Teaching Group has held biannual meetings since the early 1990s with the clear assumption by attendees that inclusion of history of science in science instruction facilitates students' understandings of nature of science. Furthermore, since its inaugural volume in 1992, the journal *Science and Education* has dedicated itself at least partially to the same assumption.

Do not misconstrue my comments to indicate that I do not favor the use of history of science in science curriculum. I have used history of science, and continue to use history of science cases, to engage students and give them an appreciation for the human side of the scientific endeavor. However, at the printing of this text, the empirical research does not support the effectiveness of using history of science to promote students' understandings of nature of science.

There is a more than promising alternative instructional approach to those just discussed. The approach recognizes that the goal of improving students' views of the scientific endeavor "should be planned for instead of being anticipated as a side effect or secondary product" of varying approaches to science teaching (Akindehin, 1988, p. 73). This approach, labeled here as the *explicit* approach, uses instruction specifically focused on various aspects of NOS to improve learners' views of NOS. In general, relative to the implicit and historical approaches, the explicit approach has been more effective in helping learners achieve enhanced understandings of the NOS and scientific inquiry (e.g., Abd-El-Khalick & Lederman, 2000; Akindehin, 1988; Bell, Blair, Crawford, & Lederman, 2003; Billeh & Hasan, 1975; Carey & Stauss, 1968, 1970; Jones, 1969; Lavach, 1969; Lederman, 1998; Lederman, 1999; Ogunniyi, 1983; Olstad, 1969). It is important to note that history of science could prove to be an effective venue for teaching NOS if instructional implementation included an explicit orientation.

A functional understanding of NOS and/or scientific inquiry is best facilitated through an explicit reflective approach. I cannot over emphasize the importance of taking time, at the conclusion of any activity, to *explicitly* point out to students the aspects of NOS and scientific inquiry that are highlighted. To encourage reflection, teachers must discuss with students the implications such aspects of NOS and scientific inquiry have for the way they view scientists, scientific knowledge, and the practice of science. It is important to note that "explicit" is not synonymous with what is labeled as direct instruction. Explicit attention to NOS simply means that the various aspects of NOS are made "visible" within instruction through reflective discussions with students about the practice of science. As much as possible, these discussions should be student centered with the various characteristics of science and scientific knowledge elicited from students rather than communicated in a didactic manner.

In order to engage students in such reflective discussions, students need to have experiences upon which to reflect, In my opinion, the most efficient way to do so is to develop an instructional syntax focused within an inquiry-oriented environment. In particular, as much as possible, the teacher should attempt to promote student understanding of "traditional" science subject matter using an inquiry-oriented teaching approach, as specified in the *National Science Education Standards*. Such an approach provides students with experiences in "doing" science, an experiential base upon which to reflect about the process and nature of science. Nature of science can be made instructionally explicit by having students reflect on what they did procedurally, why they did it, and what implications this has for the knowledge produced. For example, ask students if there are alternative explanations for the data collected, or why different lab groups arrived at different valid conclusions even though they had the same data. If the laboratory activity is more open-ended, engage students in conversations about how and why they chose their questions of interest and discuss how their individual creativity and backgrounds influenced data collection and interpretation. Closure to such experiences should make it clear that what students have experienced, although not exactly the same as science in the "real world," is a reasonable facsimile of the daily practice of science. This last step is critical because students often believe that the science they do in school is in no way related to the practice of science. There is not an extensive literature of classroom tested activities for the successful teaching of NOS. However, one clear exception to this statement is the extensive collection of activities that have been used in my research group's work with teachers and their students over the past 15 years (Lederman, N. G., & Abd-El-Khalick, F., 1998)

I have stated that the aforementioned instructional scenario is the "most efficient" for one very obvious reason. Nature of science is just one of numerous cognitive goals of science instruction. The instructional syntax described allows the teacher to smoothly integrate attention to nature of science into an instructional format that facilitates students' understandings of other subject matter goals within an engaging climate. Nature of science instruction need not be and should not be an additional topic dumped on top of all other topics teachers are expected to address. It can be integrated into what teachers currently do in a relatively seamless manner. In the best of all worlds, nature of science, and scientific inquiry, can serve as unifying themes that provide a meaningful context for the learning of the more "traditional" science concepts and principles.

In conclusion, I can not emphasize enough that we should no longer assume that students will come to understand NOS or scientific inquiry as a by-product of "doing" science-based or inquiry activities. The empirical research completed since the 1960s is fairly conclusive on this point. Nor should we assume that if teachers understand NOS they will automatically teach in a manner "consistent" with those understandings. Nature of science, and scientific inquiry, should be thought of as a "cognitive" rather than as an "affective" instructional outcomes. If K-12 students are expected to develop more adequate conceptions of NOS and scientific inquiry (knowledge about, not just skill development) then, as any cognitive objective, this outcome should be planned for, explicitly taught, and systematically assessed. All this can be facilitated through concerted and continued professional development efforts designed to promote science teachers' understandings, pedagogical abilities provide to facilitate students' conceptions, and foster teachers' commitment to the idea that NOS and scientific inquiry are instructional objectives of primary importance that permeate all aspects of curriculum and instruction. Naturally, students will only come to value the importance of learning about NOS if it is assessed following instruction instead of just included as part of instruction. A discussion of the assessment NOS outcomes has not been included in this text because of its extensive discussion elsewhere (Lederman, Abd-El-Khalick, Bell, & Schwartz, 2002; Lederman, Wade, & Bell, 1998).

CONCLUDING REMARKS

If you are at all like me, you may be wondering what the title of this chapter has to do with what has been presented. As you may recall, I began this discussion by distinguishing current reform efforts in science education from their predecessors with respect to the heightened interest and emphasis on scientific inquiry and NOS. The primary reason for this heightened, but certainly not new, advocacy is the belief that students need to develop in-depth understandings of how scientific knowledge is generated and the implications this has for the status of the knowledge. Science educators have come to believe that if students understand the source and limits of scientific knowledge they will be better equipped to make informed decisions about personal and societal problems that are scientifically-based. In short, understandings NOS are believed to be critical and essential components of the modern day battle cry of "scientific literacy."

With respect to students' achieving in-depth understanding of subject matter, which is another component of the current reform efforts, it can be argued that such a goal is unachievable unless students understand NOS and scientific inquiry. For example, can it be said that a student truly understands the concept of a gene if he/she does not realize that a "gene" is a construct invented to explain experimental results? Does the student who views genes as possessing physical existence analogous to pearls on a necklace possess an in-depth understanding of the concept? Does the student who is unaware that the atom (as pictured in books) is a scientific model used to explain the behavior of matter and that it has not been directly observed have an in-depth understanding of the atom?

Misconceptions about the scientific validity of biological evolution commonly appear in the media and courts of law. Many of these misconceptions relate to whether evolution is a testable scientific theory. The arguments against the validity of evolution usually proceed to point out that evolution can not be tested using the scientific method. Therefore, evolution can not be a valid scientific theory. Many feel that the problem is at least partially created by the public's misunderstanding of scientific theory. These few examples should make it clear that understanding NOS, as well as scientific inquiry, provides a guiding framework and context for the meaningful understanding of scientific knowledge. Without an understanding of how scientific knowledge is derived and the implications the process of derivation has for the status and limitations of the knowledge, all students can ever hope to achieve is knowledge without context. Context is necessary for students to understand what the knowledge means. In short, lack of context is the equivalent to playing a game of chess without knowing the rules of the game. Unless students can derive meaning for the scientific knowledge they acquire, there is little hope that they can use their knowledge to make informed decisions.

In many ways the state of science education and science education reform is exactly where it was 100 years ago. Today, just as we did at the turn of the century, we are seeking the holy grail of in-depth understanding of scientific concepts for our K-12 students. We have progressed in the realization that students can not meaningfully learn a long laundry list of terms, vocabulary, and factoids. We have recognized the sensibility of attempting to focus our educational efforts on fewer, unifying themes/concepts. However, we continue to fail at providing students with the most important organizing themes of all, NOS and scientific inquiry. Despite volumes of research we continue to believe that students will come to understand NOS and scientific inquiry simply by "doing science." Such an expectation is equivalent to assuming that individuals will come to understand the mechanism of breathing simply by breathing or miraculously come to understand photosynthesis by watching a plant grow. Obviously, this is not the case. Doing science is certainly a start, but students need to reflect on what it is they are doing. They need to be engaged in discussions of why scientific investigations are designed in certain ways. Students need to discuss the assumptions inherent to any scientific investigation and the implications these assumptions have for the results. Furthermore, students need to discuss the fact that science is done by humans and the implications this has for the knowledge that is produced. These questions are but a few and you can certainly develop many more. NOS and scientific inquiry are different, but intimately related. The conventions of how inquiry is practiced has clear implications for the knowledge produced. The syntax of science instructions should include the themes of inquiry with reflections on the experience. My point, however, is quite simple. NOS and scientific inquiry need to be addressed explicitly during science instruction. They need to be given status equal to that of traditional subject matter. Would we ever expect a student to implicitly learn pH? Without such explicit instructional attention, students will continue to learn subject matter without context and the visions of reform in science education will progress no further than they have in the past.

REFERENCES

Abd-El-Khalick, F. (1998). *The influence of history of science courses on students' conceptions of the nature of science*. Unpublished Doctoral Dissertation, Oregon State University.

Abd-El-Khalick, F., & Lederman, N.G. (2000). Improving science teachers' conceptions Of the nature of science: A critical review of the literature. *International Journal of Science Education*, *22*(7), 665-701.

Abd-El-Khalick, F., & Lederman, N.G. (2000). The influence of history of science courses on students' views of nature of science. *Journal of Research in Science Teaching*, *37*(10), 1057-1095.

Aikenhead, G. (1973). The measurement of high school students' knowledge about science and scientists. *Science Education, 57*(4), 359-349.

Akindehin, F. (1988). Effect of an instructional package on preservice science teachers' understanding of the nature of science and acquisition of science-related attitudes. *Science Education, 72*(1), 73-82.

Alvarez, W., & Asaro, F. (1990, Oct.) An extraterrestrial impact. *Scientific American*, 78-84.

American Association for the Advancement of Science. (1990). *Science for all Americans.* New York: Oxford University Press.

American Association for the Advancement of Science. (1993). *Benchmarks for science literacy: A Project 2061 report.* New York: Oxford University Press.

Bady, R. A. (1979). Students' understanding of the logic of hypothesis testing. *Journal of Research in Science Teaching, 16*(1), 61-65.

Bell, R., Blair, L., Lederman, N.G., & Crawford, B. (2003). Just do it? Impact of a science apprenticeship on high school students' understandings of the nature of science and scientific inquiry. *Journal of Research in Science Teaching, 40*(5), 487-509.

Billeh, V. Y., & Hasan, O. E. (1975). Factors influencing teachers' gain in understanding the nature of science. *Journal of Research in Science Teaching, 12*(3), 209-219.

Broadhurst, N. A. (1970). A study of selected learning outcomes of graduating high school students in South Australian schools. *Science Education, 54*(1), 17-21.

Carey, R. L., & Stauss, N. G. (1968). An analysis of the understanding of the nature of science by prospective secondary science teachers. *Science Education, 52*(4), 358-363.

Carey, R. L., & Stauss, N. G. (1970). An analysis of experienced science teachers' understanding of the nature of science. *School Science and Mathematics, 70*(5), 366-376.

Central Association of Science and Mathematics Teachers (1907). A consideration of the principles that should determine the courses in biology in the secondary schools. *School Science and Mathematics, 7,* 241-247.

Chalmers, A. F. (1982). *What is this thing called science?* (2nd ed.). Queensland, Australia: University of Queensland Press.

Courtillot, V. (1990, Oct.) A volcanic eruption. *Scientific American*, 85-92

Durkee, P. (1974). An analysis of the appropriateness and utilization of TOUS with special reference to high-ability students studying physics. *Science Education, 58*(3), 343-356.

Glen, W. (1990). What killed the dinosaurs? *American Scientist, 78*, 354-370

Haukoos, G. D., & Penick, J. E. (1985). The effects of classroom climate on college science students: A replication study. *Journal of Research in Science Teaching, 22*(2), 163-168.

Hrdy, S. B. (1986). Empathy, polyandry, and the myth of the coy female. In R. Bleier (Ed.), *Feminist approaches to science* (pp. 119-146). Perganon Publishers.

Jones, K. M. (1969). The attainment of understandings about the scientific enterprise, scientists, and the aims and methods of science by students in a college physical science course. *Journal of Research in Science Teaching, 6*(1), 47-49.

Kimball, M. E. (1967-68). Understanding the nature of science: A comparison of scientists and science teachers. *Journal of Research in Science Teaching, 5,* 110-120.

Klopfer, L. E. (1969). The teaching of science and the history of science. *Journal of Research for Science Teaching, 6,* 87-95.

Klopfer, L. E., & Cooley, W. W. (1963). The history of science cases for high schools in the development of student understanding of science and scientists. *Journal of Research for Science Teaching, 1*(1), 33-47.

Klopfer, L. E., & Watson, F. G. (1957). Historical materials and high school science teaching. *The Science Teacher, 24*(6), 264-293.

Lavach, J. F. (1969). Organization and evaluation of an inservice program in the history of science. *Journal of Research in Science Teaching, 6,* 166-170.

Lawson, A. E. (1982). The nature of advanced reasoning and science instruction. *Journal of Research in Science Teaching, 19,* 743-760.

Lederman, N. G. (1992). Students' and teachers' conceptions of the nature of science: A review of the research. *Journal of Research in Science Teaching, 29*(4), 331-359.

Lederman, N.G. (1998). The state of science education: Subject matter without context. *Electronic Journal of Science Education* [On-Line], *3*(2), December. Available: http://unr.edu/homepage/jcannon/ejse/ejse.html

Lederman, N.G. (1999). Teachers' understanding of the nature of science and classroom practice: Factors that facilitate or impede the relationship. *Journal of Research in Science Teaching, 36*(8), 916-929.

Lederman, N.G., Abd-El-Khalick, F., Bell, R.L., Schwartz, R.S. (2002). Views of nature of science questionnaire: Toward valid and meaningful assessment of learners' conceptions of nature of science. *Journal of Research in Science Teaching, 39*(6), 497-521.

Lederman, N. G., & Abd-El-Khalick, F. (1998). Avoiding de-natured science: Activities that promote understandings of the nature of science. In W. McComas (Ed.), *The nature of science in science education: Rationales and strategies, pp.83-126.* The Netherlands: Kluwer Academic Publishers.

Lederman, N. G., & Niess, M. (1997). The nature of science: Naturally? *School Science and Mathematics, 97*(1), 1-2.

Lederman, N. G., & O'Malley, M. (1990). Students' perceptions of tentativeness in science: Development, use, and sources of change. *Science Education, 74*(2), 225-239.
Lederman, N.G., Wade, P.D., & Bell, R.L. (1998). Assessing the nature of science: What is the nature of our assessments? *Science and Education*, *7*(6), 595-615.
Lovejoy, C. O. (1981). The origin of man. *Science, 211,* 341-350.
Mackay, L. D. (1971). Development of understanding about the nature of science. *Journal of Research in Science Teaching, 8*(1), 57-66.
Mead, M., & Metraux, R. (1957). Image of the scientist among high school students. *Science, 126,* 384-390.
National Research Council (1996). *National science education standards.* Washington, DC: National Academic Press.
National Science Teachers Association. (1982). *Science-technology-society: Science education for the 1980s.* (An NSTA position statement). Washington, DC: Author.
Ogunniyi, M. B. (1983). Relative effects of a history/philosophy of science course on student teachers' performance on two models of science. *Research in Science & Technological Education, 1*(2), 193-199.
Olstad, R. G. (1969). The effect of science teaching methods on the understanding of science. *Science Education, 53*(1), 9-11.
Popper, K. R. (1963). *Conjectures and refutations: The growth of scientific knowledge.* London: Routledge.
Popper, K. R. (1988). *The open universe: An argument for indeterminism.* London: Routledge.
Raup, D. (1991). *Extinction: Bad genes or bad luck?* New York: W W Norton & Co.
Riley, J. P., II (1979). The influence of hands-on science process training on preservice teachers' acquisition of process skills and attitude toward science and science teaching. *Journal of Research in Science Teaching, 16*(5), 373-384.
Rowe, M. B. (1974). A humanistic intent: The program of preservice elementary education at the University of Florida. *Science Education, 58,* 369-376.
Rubba, P. A., & Andersen, H. (1978). Development of an instrument to assess secondary school students' understanding of the nature of scientific knowledge. *Science Education, 62*(4), 449-459.
Rutherford, F. J., Holton, G., & Watson, F. G. (1970). *The project physics course.* New York: Holt, Rinehart & Winston.
Smith, M.U., Lederman, N.G., Bell, R.L., McComas, W.F., & Clough, M.P. (1997). How great is the disagreement about the nature of science? A response to Alters. *Journal of Research in Science Teaching, 34*(10), 1101-1103.
Solomon, J., Duveen, J., Scot, L., & McCarthy, S. (1992). Teaching about the nature of science through history: Action research in the classroom. *Journal of Research in Science Teaching, 29*(4), 409-421.
Spears, J., & Zollman, D. (1977). The influence of structured versus unstructured laboratory on students' understanding the process of science. *Journal of Research in Science Teaching, 14*(1), 33-38.
Tamir, P., & Zohar, A. (1991). Anthropomorphism and teleology in reasoning about biological phenomena. *Science Education, 75*(1), 57-68.
Trent, J. (1965). The attainment of the concept "understanding science" using contrasting physics courses. *Journal of Research in Science Teaching, 3*(3), 224-229.
Troxel, V. A. (1968). *Analysis of instructional outcomes of students involved with three sources in high school chemistry.* Washington, DC: US Department of Health, Education, and Welfare, Office of Education.
Welch, W. W., & Walberg, H. J. (1972). A national experiment in curriculum evaluation. *American Educational Research Journal, 9*(3), 373-383.
Wilson, L. (1954). A study of opinions related to the nature of science and its purpose in society. *Science Education, 38*(2), 236-242.
Yager, R. E., & Wick, J. W. (1966). Three emphases in teaching biology: A statistical comparison of results. *Journal of Research in Science Teaching, 4,* 16-20.

CHAPTER 15

RICHARD A. DUSCHL

RELATING HISTORY OF SCIENCE TO LEARNING AND TEACHING SCIENCE: USING AND ABUSING

ABSTRACT

The application of history of science to inform the design/curriculum, implementation/instruction and learning/assessment of science education is a process full of choices. What history and whose history to select and for what purposes ultimately defines the models of curriculum, instruction and assessment employed. Three organisational approaches for using history of science are examined: (1) A "How did we come to know and believe . . .? scientific thema approach; (2) A film/video reenactment approach focusing on key conceptual issues and on critical questions; and (3) A critical examination of competing explanations from modern history of science (1850 to present) approach focusing on epistemic reasoning. Respectively, the approaches represent the application of history of science in a university course sequence for non-science majors, as a context for teaching science concepts and images of science in Key Stage 3 (UK) or Middle School (USA), and as a framework for engaging learners in scientific inquiry. The author participated in the development of each of the approaches. A description of each approach precedes a critical review of the use of history of science in science education.

INTRODUCTION

The confluence of history and philosophy of science is often located with the publication of Thomas Kuhn's critically important book *The Structure of Scientific Revolutions* (1962/1970). Here Kuhn argues for a model of the growth of scientific knowledge that challenges the received view of the status of scientific knowledge. Scientific knowledge is not cumulative, there are occasions when modification, adaptation and/or outright abandonment of core theoretical and methodological beliefs occur. Kuhn's thesis, simply stated, is that the accumulation of anomalous data and information, where anomalies represent "a persistent discrepancy between observation and theory" (Brush, 2000), begins to challenge the status of normal science knowledge, beliefs, claims and activities. For Kuhn this set of ideas and commitments constitute the disciplinary matrix of a domain of science. Scientists

L.B. Flick and N.G. Lederman (eds.) Scientific Inquiry and Nature of Science, 319-330.

enter periods of revolutionary science for the purpose of reconciling anomalies and in the process spawns new ideas and commitments that alter the disciplinary matrix.

Given the theme of this conference, I think it is appropriate to point out for those individuals who may not know, that the development of Thomas Kuhn's ideas about paradigms, incommensurability and about science shifting between normal and revolutionary periods of inquiry occurred while he was working on a science education curriculum development project. Namely, *The Harvard Case Studies in Science Education* (Conant, 1957), directed by James Conant, then president of Harvard University. The spirit of science education at Harvard at the time is perhaps best demonstrated by the fact that it was while engaged in the preparation of materials for a case study that Thomas Kuhn began to develop the ideas for this seminal work *The Structure of Scientific Revolutions*.

As a science educator, I find this confluence of events so very fascinating for it explains so very much about the role history of science has come to play in American science education. Again, a bit more context. The Cases were being developed in the 1950s for the purpose of providing an alternative yet meaning science education for the liberal education of undergraduate students who would not be majoring in a science. The names of some of the individuals at Harvard then working on the Cases and on other curriculum development efforts (e.g., *Harvard Project Physics*, see Holton, 1978) include such luminaries as Thomas Kuhn, I.B. Cohen, Gerald Holton and Stephen Brush in the history of science and Fletcher Watson, Leopold Klopfer, James Rutherford, William Cooley, and Jack Easley in science education. Now understand, that this collection of scholars were working on science curriculum reform issues at precisely the time that science curriculum reform became a national and international agenda. Just a mile or two down the road from Harvard, a group of physicists at MIT were setting in motion the development of the first of many National Science Foundation (NSF) funded projects – Physical Science Study Committee (PSSC). – that would adopt a science for scientists approach. The Harvard group had serious concerns with this approach questioning whether it was a significant education objective.

The role of history of science in science education in the USA has been contested and examined ever since the NSF curriculum development period. Klopfer (1969) outlines a strategy for using the case history approach in secondary schools. Russell (1981) asks us to consider what and whose history we choose to teach, and how we choose to teach it. Brush (1974) warns us of the "x-rated" accounts history of science can bring to students' studies of scientific method. Rutherford and Ahlgren (1990) include history of science at the high school level (Key Stage 4) level only in the AAAS Project 2061 reform proposal. Allchin (1995) guides us in "How *Not* to Teach History in Science". Brush (2000) carries out a small examination of physics textbooks to discover that certain important details are historically wrong. He makes a plea that science textbook writers should take into account historical research.

The purpose of this paper is examine some of the ideological and theoretical clashes that occur when history of science is used as a framework for designing curriculum, instruction and assessment models. Within a discipline – science education or didactic of science – that draws upon a diverse set of theoretical perspectives (e.g., philosophy, psychology, pedagogy), trade-offs between such frameworks are inevitable. The early clash between Harvard and MIT foreshadows, I will argue, the ongoing and continued clashes that emerge whenever history of science is invoked as an organising tool. What history and whose history to select and for what purposes ultimately defines the models of curriculum, instruction and assessment employed. The purpose of this paper is to briefly examine 3 organisational approaches for using history of science, ones with which I was personally involved, and to examine the tensions that arise when history meets philosophy, psychology and pedagogy.

My three involvement's were being a member of staff for a full-year university course sequence at Hunter College in New York City, serving on the advisory board for the NSF funded "Mindworks" project, and directing my own curriculum development efforts. Following a brief elaboration on the MIT/Harvard clash, the paper will discuss each of the my three forays into using history of science in science teaching and curriculum design. The paper concludes with summary statements and perspectives about issue of using and abusing history of science in educational contexts.

WHAT PLACE HISTORY OF SCIENCE IN SCIENCE EDUCATION?

Following World War II, the National Science Foundation(NSF) was established. NSF was charged with guaranteeing that the USA's potential in science research and science education would be exemplary. Under the direction of practising scientists, the first curriculum development grant was award to the Physical Science Study Committee in 1956. The position was taken that the summer institutes NSF was offering to science teachers since 1951 would not have any impact if the teachers were using outdated textbooks and curriculum materials. By 1964 NSF was supporting seven elementary/junior level (Key Stages 1,2, and 3) projects and five secondary (Key Stage 4) projects. By 1966, there were 26 project (19 science and 7 maths) receiving NSF funding. Being the first co-ordinated and funded effort the PSSC project established the procedures all other curriculum projects would ultimately follow. Typically, project teams were composed of scientists, teachers, and administrators. It was clear from the very start though that the scientists were in charge (Welch, 1979). Individual projects were directed by prestigious scientists, co-ordinated by advisory boards of prestigious scientists, and written by scientists. Useful overviews of the NSF science education curriculum reform period can be found in DeBore (1991), Duschl (1990), Jones (1977) and Welch (1979).

The role of the teachers and administrators in the development of curriculum was primarily to provide feedback to the scientist-writers. Once a curriculum draft was ready, it would be distributed to teachers in test classrooms across the nation. Based on the trial run of the materials, changes would be made. More often than not, however, teacher's feedback had very little effect on subsequent versions of the curriculum (Jones, 1977; Welch, 1979).

A critical factor, then, in understanding the history of NSF curriculum projects is the dominant and decisive role members of the scientific community played. This role had a significant effect in determining the focus of the curriculum – science for scientists. But as early on as 1958 the history of science group at Harvard raised concerns about whether the type of science education being proposed by PSSC was what public education wanted or needed.

In 1959 the *Harvard Educational Review* dedicated a major portion of one issue (V 29, N1) to reporting the papers delivered at a symposium on the PSSC program. Given the changes in science education taking place at Harvard University, the critical reactions were quite predictable. Initial reactions by educational researchers to PSSC's curriculum were quite critical of the dominant emphasis on scientific method rooted in experimentation and not in educational theory (Easley, 1959). The plea was for an integration of philosophy, logic, statistics, and psychology frameworks that would inform the scientific knowledge and processes being packaged into PSSC and other NSF curriculum efforts. A particularly poignant issue was whether the teaching of the scientific method by getting students to operate as scientists was a significant educational objective. The Harvard team (Holton, 1978) went on to receive their own NSF funding and developed *Harvard Project Physics* for high school students. Other history of science and science teaching efforts include the college level textbook *Introduction to concepts and theories the Physical Science* (Holton & Brush, 1952/1969) and an adaptation of Conant's Harvard Cases *History of Science Cases* (Klopfer & Cooley, 1961) for use in secondary level science programmes.

For a variety of reasons, some of which are similar to those I will raise below, the infusion of history of science into the teaching of physics never caught on. While evaluation data clearly showed positive gains in students attitudes toward science and understanding of the nature of science (Welch, 1973), the gains on physics achievement tests, the benchmark used by elite universities to admit students to physics programmes, were minimal. The purpose of a science education was on the line then and still is today – science for all v science for future scientists.

Hunter College – Foundations of Science

With support from the Andrew W. Mellon Foundation and under the direction of Ezra Shahn (1988), the Foundations of Science course was designed as a 1-year introductory course for non-majors. The goal of the course is to provide students with the background knowledge needed to appreciate "how we know what we know and why we believe what we believe". I was brought to Hunter College to become the science educator on the team which comprised myself, Professor Shahn in biology, a professor of chemistry, a professor of physics and a professor of anthropology. My roll was to evaluate the programme models for curriculum, instruction and assessment as well as evaluate student learning. For the former task I attended all lectures, taught a lab section, graded student essays, and participated in all team meetings. For the later task, I set up a series of interviews over the course of the course with a number of target students.

The interviews revealed quite early on that the students did not know what to learn. That is, there was a confusion about knowing the history and knowing the science. The course is designed around three big ideas that are still a prominent component of our contemporary scientific view of nature. The first was a physics focused question – How did we come to know motion and forces that govern planets and objects on our planet? The second was a chemistry focus – How did we come to know that matter is particulate in nature? The third focus both geological and biological – How did we come to know that the Earth and life on Earth have a history?

The course was designed to have both lectures and labs. Final grades are based on two exams, four or five essays, and laboratory work. The essays were short 4-5 page long assignments with revisions required.The laboratory exercises were frequently replications of historical experiments (e.g., Galileo's inclined plane experiment using a water clock). Many of the readings were from original sources supplemented by either professor's course notes or chapters selected from various history of science anthologies. The inclusion of the essays with revisions in the course speaks to the importance the course designers held for language and reasoning development. The inclusion and design of the lab exercises reflects the designers' commitments to observing and doing science (See Shahn, 1988 for a more complete description of the course and of the designers' guiding frameworks).

From my position of lab instruction and grader of essays, I quickly began to see that students were not using the evidence from the labs and from the history readings and lectures to reason about the "How" path of changing concepts and evolving explanations. An age old problem – students making links between lecture and lab. The problem, I felt, was the decision to base the labs on the "learning cycle" approach. This is a Piagetian model of instruction in which students explore, analyse, and apply knowledge. On theoretical grounds, using the learning cycle also commits one to a hypothetico-deductive (H-D) philosophy of science. The idea was that this discovery mode would replace the didactic mode, particularly in the labs where the lab instructors were informed to provide non-intrusive guidance.

The students were not discovering. The learning of concepts employing the learning cycle did not fit nor keep pace with the evolving historical dynamics of observation and theory changes. Sorting out the history and sorting out the science and how to bring them all together was an enormous challenge for students. Adopting a model of instruction based on a theory of learners abilities (e.g., concrete v formal operational thinkers) and on a context of justification as opposed to instructional models based on theories of learning strategies (e.g., cognitive problem-solving theory) and a context of discovery, was in my opinion a serious error. In my opinion, the models of psychology and philosophy did not cohere with the historically based curriculum that was telling the story of theory restructuring and development.

The student were just inundated with concepts, dates, names, instruments, evidence, etc. and they thought it was all equally important to know and to learn. It wasn't. History of science was the context within which the students were to learn important concepts and how these concepts came into existence. Students needed guidance about what to a pay attention to and strategies for co-ordinating information from the lectures for use in the labs. It was my learned opinion that the historical laboratory approach would need to be adapted to accommodate psychological theories of learning which embraced knowledge restructuring and philosophical views of theory change (e.g., Kuhn, Lakatos, and Laudan).

Southwest Regional Laboratory - Mindworks

Under the direction of Barbara Becker and with support from NSF, Mindworks represents a series of eight physical science instructional modules for use in secondary schools. The goals of the modules are (1) to motivate students with low interest in science, (2) to invoke conceptual change in students understanding of the structure and workings of the physical world, and (3) to enhance understanding of the process and culture of scientific activity.

Building on her commitment to using original historical materials for teaching secondary and college science (Becker, 1992), Dr. Becker sought to bring this approach to the design of curriculum materials for slightly younger children grades 7-9 in the US or Key Stage 3 in the UK. Her position is that "exposing students to the social construction of scientific knowledge through historical episodes that emphasise the intellectual struggle involved in developing key physical science concepts will help them articulate their own theories about the world and recognise a need to change the form and structure of these theories" (Becker, 2000, 270). Not unlike the position proposed by Nancy Nersessian (199x), history of science would be used "as a heuristic device to anticipate and address alternative conceptions in science across the grade levels" (Becker, 2000, 270). Historical examples would be used to integrate content, context, and method.

Careful not to fall into the trap of providing quasi-history, pseudo-history or simplified history, Becker did set out to capture a simplified history of science that would illuminate the subject matter while not abusing the history. Her endeavour is to expose students "to the social nature of scientific knowledge via a mix of observation and discourse" (Becker 2000, 272). The NSF funding was used to create 8 professionally produced short video dramatisations focusing on core concepts in the physical sciences as well as raising social, philosophical and/or political issues that will interest students.

Once the videos were written and produced, teachers and staff worked together to develop the full package of curriculum units, lesson plans and student activities. The teachers were provide with original historical documents and summaries of biographic and historical information and asked to develop activities through writing, debates, and discussions that would immerse the learners in the work of scientists and inventors.

The tension that arose, not unlike the Hunter course, was the kind of science investigation activities being developed for some of the modules. The Mindworks modules had to meet the California State guidelines. This meant, like here in the UK, the inclusion of investigations or practical work in the curriculum units. Teachers were selecting investigations that would quite clearly work against students seeing the social nature of science. Standard already-in-use cook-book style investigations were being proposed rather than the more effective design based investigations like that developed for the module on George W. Ferris' plans for the observation wheel at the 1893 Chicago World's Fair. Here students would design their own Ferris Wheels and test forces acting on their structure. Science lessons with fixed outcomes do not engender debates and discussions about the nature of science. There are also compelling arguments that these sorts of science lessons do not promote conceptual change learning.

The culprit to me seemed to be the pressure of time and the pressure of politics. There wasn't enough time to developed project-based or design-based activities for each of the modules nor did t he existing curriculum framework allow instructional time for such extended inquires. The issue for me was that goals of the project would be compromised. While the motivation and cultural goals , see 1 and 3 above, would be the purview of videos and original historical documents and the writing, debating and discussing activities around these, the goal of concept learning and conceptual change, number 2, would be delegated to the investigations demonstrating concepts. The integration of history of science with science teaching contained tensions. A coherent educational theory that functioned across curriculum design, instructional strategies and assessment frameworks seem to be missing. The evaluation of the project is presently underway. I hope I am proved wrong because the history of science materials Barbara Becker has produced are, in my opinion, excellent.

Causes of Earthquakes or How Not to Teach History of Science

Let me now turn the microscope on myself. At the Third IHPSST conference held in Minneapolis, MN, USA, Douglas Allchrin (1995), presented a paper that uses one of my curriculum design approaches as an example of how <u>not</u> to teach history of science. Let me first outline the approach and then his criticisms.

A major area of my research has been understanding the role of explanation and theories in science education. Duschl and Wright (1989) sets out the problem – teachers do not consider the structure of scientific theories in their planning or teaching. Duschl (1990) sets out some strategies for addressing the problem. The basic proposal is that philosophical models that account for theory change and evaluation can be used as strategies and frameworks to inform the design of curriculum, instruction and assessment models. Over the years, we have come to call these pedagogical models Growth of Knowledge Frameworks (Duschl & Erduran, 1996).

Inspired by Stephen Brush's work on modern history of science (1988), I hit upon the idea that getting teachers to examine with their students contemporary competing explanations for the same phenomenon would be an effective way to teach about the nature of theory change and the dialectic processes between observation and theory. A focus of my research has been developing students epistemic reasoning in general and the evaluation of knowledge claims in particular.

Using both contemporary and historical sources from textbooks and popular science magazines and books, I located 5 competing explanations for the causes of earthquakes proposed over the last 100 years. A complete description of the original activity can be found in Duschl (1987). The five explanations, in chronological order, are (1) gravitational forces of the moon and sun (2) barometric changes, (3) isostatic (floating) movements of mountains, (4) plate tectonic theory, and most recently, (5) rising methane gas. These were provided in original text formats and in summary formats. The first part of the activity was to determine the explanation for earthquakes from each source.

The context for evaluating each of the explanations was the focus of the second activity. Students were guided to ascertain the pattern of earthquakes found in a 10-year data base map produced by the US Geological Survey. The map presented both the epicentre (surface location – position on map), the focus (depth below surface – red shallow, green middle, blue deep), and intensity of earthquakes - those > 7.0 on Richter scale appear as circles rather than dots.

Employing Laudan's general problem solving criteria for evaluating research programs, namely empirical and conceptual problem solving, students were guided to consider the extent to which the evidence supported or refuted each explanation. Thus, for example, the barometic pressure explanation does not account for earthquakes at the bottom of the ocean floor nor does it explain deep focus earthquakes. Five criteria based on the location and patterns of earthquakes were used to evaluate the empirical adequacy of the explanations. These criteria reduced the pool of options to two since both explanations 4 and 5 had equal empirical adequacy. The last part of the activity then turn to conceptual problem-solving tasks. How did the models and theories scientists hold about the Earth help determine the most plausible explanation for earthquakes? The fact that the jury is still out on this one suggests that both explanations must be considered.

The framework of using competing explanations was used by my graduate students as a context for applying HPS to science teaching. The students designed curriculum units on theories of breast cancer treatment, theories about life on Mars, theories about origin of the moon, and theories about the interior of the earth (Kachman & Sutton, 1993). For this collection, the GKF used was Giere's Theory Testing Argument Scheme (see Duschl & Erduran, 1996). In brief, the Giere Framework uses an argumentation structure to frame both the background knowledge and initial conditions used to propose a theoretical hypothesis or model. The framework is there for teachers to use as a planning device and students to use as learning device.

Allchin took issue with the reconfiguring of history. That is, there was never an instance in the history of science when geologists were actually confronted with this mix of explanations, "no one in history sat down to consider these five theories all at once" (Allchin, 1996; 16). While Allchin recognises the value of the exercise in terms of getting students (1) to appreciate that different explanations of the same phenomenon are possible, (2) to consider how theories are rooted in certain assumptions, and (3) to engage in the higher cognitive reasoning skills associated with the evaluation and interpretation of knowledge claims, he takes issue with several things.

> The exercise reconstructs rather than simulates science. The exercise is further reconstructed – or artificially contrived – in its pre-established data set (Part 2). Why this data? Students receive only a small part of evidence is relevant, once they know the multiple explanations; and the data is largely significant only from the current theoretical perspective. (Allchin, 1996; 16).

Douglas goes on to critique the decision to provide students with the data and with the criteria/questions/standards to judge the explanations as an example of not being concerned with the process of history but only its product. He writes:

> The problem with teaching through rational reconstructions is that the history . . . is backwards. The aim is to find the route to (from?) the final answer. We should, instead, be tapping history to model the blind forward-moving context of science. The generation of hypotheses, the search for relevant information, the design and critique of experiments, the elaboration of alternative explanations, the struggle with experimental anomalies – all the elements of scientific discovery – cannot be taken for granted. There is more to science than just justifying the final outcome – or assuming that it is correct (ibid., p 16-17).

There is much that I agree with here. And with a full semester in college classes or 6 to 8 weeks in a high school class it would be possible to design the curriculum model he suggests. But the reality is that typical classrooms do not address theory or explanation evaluation, teachers have not been provided in their training with pedagogical, philosophical and/or psychological models for addressing the evaluation of explanations, and schools do not provide the resources for students to collect certain types of data (e.g., earthquake data). On epistemic and cognitive grounds, the comparison of plausible explanations has merit. On historical grounds, the approach presents problems.

A MODEL-BASED VIEW OF SCIENCE EDUCATION

Shapere (1977) introduced the idea of domains of science and as intellectual spaces where scientific activity and reasoning function across traditional disciplinary boundaries. Giere (1989) further develops these naturalised philosophy views into a model-based view of science. For Giere, theories are composed of sets of models and only the models have a foot in the real world. The connection between the models and the theories he asserts is non-linguistic in nature while the connection between models and the natural world is linguistic in nature and best understood in terms of cognitive processes. This account has the benefit of explaining the revisionary nature of knowledge claims within a modest realism framework.

The amelioration of the tensions between history of science and science teaching can, I think, be decided in terms of a model-based view of science teaching. We work in a domain that is informed by intellectual disciplines of history and philosophy of science, psychology, and pedagogy. Our models of philosophy, of psychology and of pedagogy need to come together and cohere. This union, if you will, will constitute a theory of science education. The adoption of cognitive frameworks to propose a cognitive model theory for articulating the design of science curriculum was first proposed by Merce Izquierdo and colleagues (Izquierdo, Cabello, & Solsona, 1992) and quickly developed into a comprehensive 'didactic theory' (Izquierdo, Sanmarti, Espinet, Garcia, Pujol, in review). Their didactic theory draws heavily from Giere's model-based view.

The adoption of cognitive models and decision making within human judgement, or more precisely cognitive structures and cognitive processes, as a mechanism to explain science shifts, Giere argues, the philosophical goal from the justification of scientific knowledge, processes and methods to an understanding of scientific knowledge, processes and methods. This semantic-realism model-based view(MBV) of science has the advantage of allowing one to talk about models "fitting" the world rather than of being truth of statements about the world.

This global MBV perspective of scientific processes, has the advantage over rival instrumental-based and justification-based philosophies of science in that the MBV can embrace, where the others can not, the inherent variation and complexity of the natural world and cognitive processes that seek to make sense of that world. On these grounds I feel the MBV embraces the psychological human information-processing theory (Newell and Simon, 1972). This theory speaks of individuals developing models of reality to cope with the inherent limitations that exist when the task environment one encounters provides far too much information to process. Classroom teachers, as all professionals, must cope with this dilemma (Duschl & Wright, 1989). MBV can also embraces socio-cultural theories of learning (Wertsch, 1985) since both recognise the importance acquiring cognitive resources have for providing mechanisms which drive the development, evaluation and deployment of scientific thinking and judgements.

> These cognitive resources and interests, combined with various judgmental strategies, provide the *mechanisms* - the analogs of genetic mechanisms in organic evolution - which drive the evolution of scientific fields. And this evolution takes place in an "environment" of cultural and material resources required to support modern, high technology research . . . cognitive models represent some of the *mechanisms* by which various interests influence the evolutional development of scientific fields. The remaining enterprise is to work out the details of this process. (Giere, 1986; p 324., italic in original)

The agenda for aligning history of science with science teaching is one that requires consideration of coherent models of philosophy, psychology and pedagogy. The recent attention in educational research to investigating and understanding the design of learning environments (Bransford, et al; 1999) that support the development of cognitive resources, interests, judgmental strategies and mechanisms for the purpose of developing, in our case here, learner's scientific understanding, is, on my view, an enterprise quite similar to that proposed by Giere. So, too, is the need to "work out the details of the process" a similarity.

Clearly, history of science is a critical element in the model-based view of science teaching precisely because it is a critical element in the philosophical domain of the model. The historians of science and educational researchers at Harvard in the 1950s understood then that a proper model of science education needed to deal with an integration of philosophy, logic, statistics, and psychology frameworks. It needed to have an educational theory. My three personal stories show some of the complexities of the integration process when philosophical, psychological and pedagogical frameworks are at odds with each other. When such tensions exist, important details can drop out in the design and implementation processes. The 3 scenarios certainly mask or hide the diverse and successful ways that history of science has informed the design of science curriculum. However, for two disciplines that are both in the first century of belonging to academia, a great deal of progress has been made already. The historians of science are setting exemplary examples of working with educators to improve science education. We both benefit from recognising that each of our communities have comprehensive guiding conceptions and programmes of research. Let us continue to come together and discuss the problems and potential solutions.

REFERENCES

Allchin, D. (1995). How *not* to teach history of science. In F. Finely, D. Allchin, D. Rhees & S Fifield (eds.), *Proceedings of the Third International Seminar on History and Philosophy of Science and Science Teaching,* Vol. 1, University of Minnesotta: Minneappolis, MN. 13-22.

Becker, B.J. (1992). Incorporating primary source material in secondary and college science curricula. In K. Hill (ed.) *Proceedings of the Third International Seminar on History and Philosophy of Science and Science Teaching,* Vol. 1. The Mathematics, Science, Technology, and Teacher Education Group, Queen's University: Kingtson, ON 69-76.

Becker, B.J. (1999). Mindworks: Making Scientific Concepts Come Alive. *Science & Education, 9,* 269-278.

Bransford, J., Brown, A., & Cocking, R. (eds.) (1999) *How People Learn: Brain, Mind, Experience and School.* Washington, DC: National Academy Press.

Brush, S. (1974) Should the history of science be x-rated? *Science, 183,* 1164-1172.

Brush, S. (1988). *The History of Modern Science: A guide to the second scientific revolution, 1880-1950.* Iowa State University Press: Ames, Iowa

Brush, S. (2000). Thomas Kuhn as a historian of science. *Science & Education, 9,* 39-58.

Conant, J.B., (1957). *Harvard Case Histories in Experimental Science.* Harvard University Press: Cambridge, MA.

DeBoer, G.E., (1991) *A History of Ideas in Science Education: Implications for Practice.* New York: Teachers College Press.

Duschl, R. A. (1990). *Restructuring Science Education: The importance of theories and their development.* Teachers College Press: New York.

Duschl, R. A. (1987). Causes of Earthquakes: An inquiry into the plausibility of competing explanations. *Science Activities, 34,* 8-14.

Duschl, R. A. & Erduran, S. (1996). Modelling the growth of scientific knowledge. In G. Welford, J. Osborne, & J. Leach, (eds.) *Research in Science Education in Europe: Current Issues and Themes.* Falmer Press: London, 153-165.

Duschl, R. A. & Wright, E. (1989). A case study of high school teachers' decision making models for planning and teaching science. *Journal of Research in Science Teaching, 26,* 467-501.

Easley, J. (1959). The Physical Science Study Committee and educational theory. *Harvard Educational Review, 29,* 4-11.

Giere, R. (1986). Cognitive Models in the Philosophy of Science. In *PSA 1986,* V2, 319-328. East Lansing, MI: Philosophy of Science Association.

Giere, R. (1988) *Explaining Science: A cognitive approach.* University of Chicago Press: Chicago.

Holton, G. (1978). On the educational philosophy of the Project Physics Course. *The Scientific Imagination: Case Studies.* Cambridge University Press: Cambridge, MA.

Holton, G & Brush, S. (1952) *Introduction to concepts and theories in Physcial Sciences. 2nd Ed.* Reading, MA: Addison-Wesley.

Izquierdo, M., Cabello, M., & Solsona, N., (1992). Using analogical models for articulating science curriculum, "Science 12-16". (531-543). In S. Hills (Ed.) *History and Philosophy of Science in Science Education.* Kingston, Ontario: Queen's University.

Izquierdo, M., Sanmarti, N., Espinet, M., Garcia, M.P., Pujol, R.M. (in review) Characterization and Foundation of School Science, *Science Education.*

Jones, H. (1977). The past, present, and future of science education before, during and after the year of the golden-fleeced MACOS. In G. Hall (ed.) *Science teacher education: Vantage Point 1976, AETS Yearbook.* ERIC Clearinghouse Ohio State University: Columbus, OH. 189-213.
Kachman, K. & Sutton, C. (EDS.) (1993). *Curriculum reform in college science*. Department of Instruction and Learning, University of Pittsburgh: Pittsburgh, PA.
Kuhn, T. (1962/1970). *The Structure of Scientific Revolutions.* University of Chicago Press: Chicago.
Klopfer, L. & Cooley, W. (1961). *The Use of Case Histories in the Development of Student Understanding of Science and Scientists.* Harvard University Press: Cambridge, MA.
Klofper, L. (1969). The teaching of science and the history of science. *Journal of Research in Science Teaching, 6,* 87-95.
Newell, A. & Simon, H. (1972) Human Problem Solving. Englewood Cliffs, NJ: Prentice-Hall.
Russell, T. (1981). What history of science, how much, and why?. *Science Education, 65*, 51-64.
Rutherford, J. and Ahlgren, A., (1990) *Science for All Americans.* New York: Oxford University Press.
Shahn, E. (1988). On scientific literacy. *Educational Philosophy and Theory, 2*, 42-52.
Shapere, D. (1977) Scientific theories and their domains. In F. Suppe (ed.) *The Structure of Scientific Theories, 2nd ed.* University of Illinois Press: Champagne-Urbana, IL.
Welch, W. (1979). Twenty-five years of science curriculum development. In D. Berliner (ed.) *Review of Research in Education. Vol. 7.* American Educational Research Association: Washington, DC. 282-306.
Welch, W. (1973). Review of the Research and Evaluation Program of Harvard Project Physics. *Journal of Research in Science Teaching, 10*, 365-378.
Wertsch, J. (1985). *Vygotsky and the Social Formation of Mind.* Harvard University Press: London.

CHAPTER 16

RENEÉ S. SCHWARTZ & BARBARA A. CRAWFORD

AUTHENTIC SCIENTIFIC INQUIRY AS CONTEXT FOR TEACHING NATURE OF SCIENCE

Identifying Critical Elements for Success

"Students should develop an understanding of what science is, what science is not, what science can and cannot do, and how science contributes to culture."

(National Research Council [NRC], 1996. p. 21)

This statement encapsulates the most recent reforms that reclaim the importance of introducing students to the culture of science wherein they can develop conceptual understanding of traditional science subject matter, the nature of science (NOS), and scientific inquiry (American Association for the Advancement of Science [AAAS], 1993; NRC, 1996). It is the intercept of these three domains, along with an understanding of the utility of that knowledge to the individual and society that represents the conceptual foundation for a scientifically literate individual. Teaching emphasis has shifted from presenting science as a final body of knowledge to presenting science as a human endeavor that produces a solid (empirically-based and internally consistent), yet fallible, understanding of the natural world (see, for example, Duschl, 1990; Hodson, 1988). Such emphasis on the inclusion of NOS and scientific inquiry in science education extends back nearly a century (e.g. Linville, 1907) and has been described as central to generating an informed citizenry (e.g., Driver, Leach, Millar, & Scott, 1996; McComas, 1998; Schwab, 1962; Smith & Scharmann, 1999). As a pioneer in describing scientific inquiry in the classroom, Joseph Schwab (1962) drew attention to the significance of understanding the source and justification of scientific knowledge. He stated, "The knowledge won through enquiry is not knowledge merely of the facts but of the facts *interpreted*. And this interpretation, too, depends on the conceptual principle of the enquiry" (p. 14) [emphasis added]. Without understanding the qualities and assumptions that are inherent to the knowledge (NOS) and the processes by which the knowledge was created and accepted (scientific inquiry), the learner can do little more than construct an image of science consisting of isolated "facts" void of context that make the knowledge relevant, applicable, and *meaningful* (Lederman, 1998). This means that science is to be understood as an interpretive body of knowledge with foundation in

L.B. Flick and N.G. Lederman (eds.) Scientific Inquiry and Nature of Science, 331-355.

assumptions that influence the processes as well as the products of the enterprise. This means that science and the scientific community are made accessible through personal relevance and real-world context.

If teachers are to present an image of science that is consistent with the practice of scientific inquiry and NOS, it is imperative to consider the educational context. Brown, Collins, and Duguid (1989) argued that the context of the learning situation impacts cognitive achievement, and few would doubt this claim. They assert that knowledge is situated within the context in which it is learned. Based on this view of contextual learning, engaging learners in scientific inquiry for the purpose of teaching them about NOS has intuitive appeal. In this chapter we explore the literature on the effectiveness of various inquiry approaches to teach NOS, consider fundamental differences between school-based and authentic scientific inquiry, and describe examples wherein NOS has been effectively addressed within the context of authentic scientific inquiry experiences. Considering this body of work, we propose critical elements necessary for successful teaching of NOS within authentic scientific inquiry contexts.

THE EMPHASIS ON INQUIRY TO TEACH NATURE OF SCIENCE

The literature demonstrates students and teachers persist in holding positivist views and misconceptions relative to NOS (Duschl, 1990; Gallagher, 1991; Lederman, 1992; Meichtry, 1992; Pomeroy, 1993; Ryan & Aikenhead, 1992, among others). The question of how one comes to understand NOS has led science educators to look toward the creative, cognitive, and contextual source of scientific knowledge itself; that is the scientist and the accepted habits of mind and action within the scientific community that lead to the construction of scientific knowledge. Naïve NOS views and misconceptions has, thusly, been attributed, at least in part, to learners' lack of exposure to authentic scientific contexts or experience in conducting scientific investigations (Gallagher, 1991; Harms & Yager, 1981; Roth, 1995; Schwab, 1962; Welch, Klopfer, Aikenhead, & Robinson, 1981). Reform recommendations that aim to improve learners' conceptions of NOS strongly emphasize including opportunities for learners to engage in scientific inquiry and inquiry-oriented activities within science instruction (AAAS, 1990, 1993; NRC, 1996; Schwab, 1962; Welch et al., 1981).

Such efforts are consistent with the contextual basis of cognitive achievement (Brown et al., 1989). Scientific inquiry may provide a critical context for discussion and reflection within which learners can more fully develop an understanding of NOS (AAAS, 1990, 1993; NRC, 1996; Schwab, 1962; Welch et al., 1981). Four decades ago Schwab (1962) claimed that science should simply be taught *as science*:

> What is required is that in the very near future a substantial segment of our publics become cognizant of science as a product of *fluid enquiry*, understand that it is a mode of investigation which rests on conceptual innovation, proceeds through uncertainty and failure, and eventuates in knowledge which is contingent, dubitable, and hard to come by... No more updating of course content will suffice. (p. 5)

Recent calls are similar:

> Education in science is more than the transmission of factual information: it must provide students with a knowledge base that enables them to educate themselves about the scientific and technological issues of their times; it must provide students with an understanding of the nature of science and its place in society; and it must provide them with an understanding of the methods and processes of scientific inquiry. To achieve these goals, science should be taught as science is practiced at its best (AAAS, 1990, p. xii).

WHAT THE RESEARCH SAYS ABOUT THE USE OF INQUIRY TO TEACH NOS

Inquiry-based instruction is intended to facilitate the learners' development of conceptual understanding of science concepts, skills of scientific inquiry (what students should be able to *do*), understanding *about* scientific inquiry (knowledge about the nature of scientific inquiry), and understanding of NOS (AAAS, 1993; Lederman, 1998; NRC, 1996; Schwab, 1962; Welch et al., 1981). Typically, however, inquiry-based instruction alone has not consistently achieved these desired goals, especially in enhancing learners' understandings about scientific inquiry or NOS (Lederman, 1992; Schwartz, 2002). To avoid making the same mistakes in our current efforts, we can learn from past efforts. A brief review of the literature examining the use of inquiry to teach NOS offers insight into effective practices and serves as the theoretical framework for the inclusion of authentic scientific research experiences as a context for learning NOS.

THE EFFECTIVENESS OF INQUIRY-BASED APPROACHES TO TEACHING NOS: AN HISTORICAL GLIMPSE

In the wake of the Sputnik success of 1957, the push for improved science achievement and advancement in the United States sparked the development of a multitude of science curricula that focused on inquiry. The main emphasis was toward active learning, stressing hands-on laboratory experiences. The curricula were intended to be inquiry approaches to science, encouraging students to make observations, define problems, formulate hypotheses, test hypotheses, interpret data, formulate inferences and make generalizations (Shulman & Tamir, 1973; Sund & Trowbridge, 1973). Included in the stated laboratory objectives of the 1960s secondary science curricula, such as BSCS (Biological Sciences Curriculum Study) biology and PSSC (Physical Science Study Committee) physics, was the objective of enhancing students' understanding of NOS, specifically pertaining to "the scientific enterprise, the scientists and how they work, the existence of multiplicity of scientific methods, the interrelationship between science and technology and among the various disciplines of science" (Shulman & Tamir, 1973, p. 1119).

These curricula were designed and implemented under the assumption that by "doing science" learners would come to understand NOS. Actions of the "hands-on" activities were considered sufficient learning opportunity. This approach to teaching

NOS is "implicit because NOS learning is assumed to be a consequence of the engagement in the inquiry-based activities. As is often the case, this assumption, albeit appealing, was unsupported. In fact, research indicated the curricula were generally unsuccessful in impacting students' understandings of NOS (Lederman, 1992; Ramsey & Howe, 1969; Tamir, 1970). Studies comparing traditional science instruction, which relies heavily on the textbook and "cook book" style laboratories, with inquiry-based instruction revealed NOS conceptions were not significantly enhanced in secondary students (Crumb, 1965; Tamir, 1970, 1972; Trent, 1965) or in middle school students (Meichtry, 1992). In fact, neither approach was particularly effective in enhancing students' views. Students' NOS views remained naïve and unchallenged. So what next? The voice of Schwab and the calls for increased critical thinking, scientific problem solving, and understanding NOS are still strongly heard. But the solution isn't as simple as inquiry-based vs. cook-book style.

Science educators attempted variations in inquiry-oriented teaching approaches. For example, studies compared NOS learning through engagement in structured (guided inquiry) versus unstructured (left open for students to design and conduct) laboratory investigations (Spears & Zollman, 1977). Investigations of their relative success in improving secondary students' (Yager & Wick, 1966; Yager, Engen, & Snider, 1969) and college students' (Haukoos & Penick, 1985; Spears & Zollman, 1977) NOS views indicate that, regardless of approach, there is little or no difference in students' understandings of NOS. Even though these studies employed different inquiry-based approaches (open vs. guided and active vs. demonstration), they were still considered implicit in reference to NOS. That is, none of the approaches offered clear discussion or instruction relating the inquiry activities to specific elements of NOS. Moreover, none of the approaches was considered effective in engendering acceptable conceptions of NOS as measured by the instruments employed in the studies (e.g. TOUS, SPI). Despite engaging in hands-on inquiry-based activities, learners tended to maintain views of science as authoritative and objective, void of creative inference.

Various attributions have been placed towards the failure of the inquiry-based approaches to engender the desired NOS conceptions, including problems from lack of consistent and contemporary philosophy and intentions within the curricula (Hodson, 1988). "...modern science courses have failed to achieve their goals because of inadequacies in the philosophical stance underpinning course design and in the implicit philosophies of science teachers" (Hodson, 1988, p. 35). Other reasons for the continued failures include the curriculum developers, who were primarily scientists rather than science educators (Duschl, 1985), influential differences in teachers and teaching approach (Trent, 1965; Yager, 1966; Yager & Wick, 1966; Yager et al., 1969), the reliance on implicit NOS messages during the prescribed activities (Lederman, 1992; Meichtry, 1992; Yager et al., 1969; Yager & Wick, 1966), the lack of "authentic" inquiry engagement (Herron, 1971; Sund & Trowbridge, 1973), lack of teacher experience conducting their own scientific investigations (Gallagher, 1991; Harms & Yager, 1981; Welch et al., 1981), and the lack of teachers' understanding of NOS (Hodson, 1988; Lederman, 1992). The bottom line is that the learner was not coming to understand science and the scientific endeavor in a manner consistent with the current post-positivist philosophy of science (e.g. Kuhn, 1970) simply through engaging in the classroom activities.

Based on this work, questions are raised concerning the influence of the teacher (knowledge of NOS and experiences with inquiry) and the influence of the actual level of scientific inquiry in which the learners are engaged. Examining the literature on developments in teachers' NOS knowledge through inquiry experiences will lead us into the discussion of "authenticity" and the role of authentic scientific research experiences in NOS learning.

TEACHERS' EXPERIENCES WITH CLASSROOM-BASED SCIENTIFIC INQUIRY AND LEARNING OF NOS

In as much as the teacher must have an adequate understanding of NOS to effectively teach NOS, he/she must also have an understanding of the processes by which scientific knowledge is created to effectively incorporate inquiry-based activities or projects as pedagogical approaches to teaching NOS (Gallagher, 1991; Herron, 1971; Horner & Rubba, 1978; Ramsey & Howe, 1969; Robinson, 1969; Rutherford, 1964; Schwab, 1962). "Prospective teachers have limited knowledge of, and experience with, the processes by which scientific knowledge is generated. This puts serious limitations on their ability to plan and implement lessons that will help the students develop an image of science that goes beyond the familiar 'body of knowledge'" (Gallagher, 1991). Furthermore, Rutherford (1964) recognized the need for teachers to have adequate philosophical views concerning the nature of scientific inquiry in order to teach science as an inquiry process. Teachers "must come to understand just how inquiry is in fact conducted in the sciences. Until science teachers have acquired a rather thorough grounding in the history and philosophy of the sciences they teach, this kind of understanding will elude them, in which event not much progress toward the teaching of science as inquiry can be expected" (p. 84). Again, these arguments are consistent with others who purport the influence of context on cognitive developments (Brown et al., 1989).

In response to the disappointing outcome of the inquiry-based curricula, teacher education programs incorporated scientific inquiry experiences for the purpose of enhancing teachers' views of science processes and NOS. Programs that utilized an implicit approach, however, produced similar results as those previously described for the inquiry-based curricula. Results suggest that the inquiry-oriented activities alone did not enhance teachers' NOS conceptions (Billeh & Hasan, 1975; Riley, 1979). However, programs that involved inquiry-oriented activities along with explicit NOS instruction and reflective components were more successful in improving participants' views of NOS (Billeh & Hasan, 1975; Shapiro, 1996). Shapiro (1996) and Bianchini and Colburn (2000) demonstrate progress in NOS learning even though their participants were engaged in different levels of inquiry. As described in these reports, the preservice teachers in the Shapiro study experienced more open inquiry (independent investigations chosen, designed, and conducted by the students), and those in the Bianchini and Colburn study experienced guided inquiry (following the learning cycle). Both studies reported the inclusion of NOS instruction and guided reflections throughout the time students were engaged in inquiry-based investigations. Both reported developments in NOS views. They suggested that the teacher played a pivotal role in bringing relevant NOS aspects to the attention of the students through discussion and questioning. It should be noted that these relatively successful attempts to enhance participants' NOS views employed *explicit* NOS instruction within the context of the classroom-based inquiry investigations. That is, there appears to be more success in enhancing NOS views when NOS is a valued cognitive learning outcome, and treated as such within the context of these inquiry investigations. We characterize "explicitness" in a later section of this chapter.

It is important to note that an assumption of the inquiry-based curricula and activities was that the learners were engaged in scientific inquiry similar to the activities of scientists. However, examination of some of the actual activities of the popular inquiry-based curricula, such as the BSCS series and PSSC physics, revealed the inadequacies of the activities as "high level" or open scientific inquiry (Herron, 1971; Shulman & Tamir, 1973). Very few investigations enabled students' creative input in investigative designs or questions. They were far more traditional and held closer resemblance to verification exercises than open-ended challenges wherein learners would develop new scientific knowledge. In this respect, learners were not engaged in truly authentic scientific inquiry, where "authentic" inquiry is that which takes place within the scientific community and adheres to the norms and expectations of that community. We clarify this description shortly. Nonetheless, desired changes in NOS conceptions resulted in those cases where NOS was provided instructional priority within the context of the classroom-based inquiries (Bianchini & Colburn, 2000; Billeh & Hasan, 1975; Shapiro, 1996). Similar outcomes have been reported within other contexts such as history of science courses (Abd-El-Khalick & Lederman, 2000), and constructivist-based classrooms (Carey, Evans, Honda, & Unger, 1989; Carey & Smith, 1993; Smith, Maclin, Houghton, & Hennessey, 2000).

AUTHENTIC SCIENTIFIC INQUIRY

All the inquiry activities in the reports thus mentioned took place within the context of a classroom or education setting. We ask the question, "How does scientific inquiry within the classroom context compare with scientific inquiry within the context of the scientific community?" To begin to address this question, we examine the meaning of "authentic scientific inquiry."

What is authentic scientific inquiry? The responses to this question are as varied as the methods of scientific inquiry itself. In its most basic form, authentic scientific inquiry is that which scientists experience in everyday practice (Roth, 1995; 1997). In a much more complex description, Martin, Kass, & Brouwer (1990) present several meanings of "authentic." For example, "Science may be authentic if it is in accord with a commonly held agreement over what constitutes science" (p. 542). They continue to explore different contexts of science, including personal science, private science, public science, historical science, and societal science. They question whose science should be considered authentic. Additionally, they propose a slightly different question related to authenticity: *What is an authentic view of science?* They suggest that a view of science may be authentic if it is a "reliable or trustworthy or genuine representation of what science really is" (Martin et al., 1990, p. 542). The circular arrangement of these two questions and responses introduces a dilemma regarding "authenticity." Which came fist – the science or the view of science?

Rather than dwell on esoteric notions that parallel the great chicken-and-egg debate, we prefer here to describe "authenticity" as it pertains to the practice of scientific inquiry, conducted by scientists, within the community of science. For the purpose of discussion related to the learning of NOS within an authentic context, we believe a distinction must be made between scientific inquiry as conducted within the classroom community (school-based scientific inquiry) and that which is conducted within the scientific community (authentic scientific inquiry).

It is important for the purposes of this chapter to also define "authenticity" in relationship to science classrooms. Given the notion that authentic scientific inquiry is that which occurs within the scientific community by practicing scientists in efforts to gain understanding of the natural world, school-based scientific inquiry cannot be considered authentic in the strictest sense. In general, the world of science as experienced by scientists is not represented in the classroom experience (Brown, et al., 1989; Driver et al., 1996; Roth & Roychoudhury, 1993; Roth, 1991; Ryder, Leach, & Driver, 1999). Although both contexts provide opportunities for social construction of understanding, the school community rarely promotes the complexity of reasoning and negotiation of meaning as it is expressed within the scientific community (e.g. Chinn & Malhotra, 2002). Negotiation is a critical process in the authentic practice of science. Negotiation involves argumentation; communication of findings; the sharing of ideas; identification of where the findings fit within the scientific community; the provision of exemplary support; and eventually earned acceptance among peers that the knowledge is valid within the norms of the scientific enterprise. Such open "negotiation of meaning" is lead by scientists and directed by the evidence and construction of explanations. Final acceptance of knowledge claims is gained through true negotiation. In the classroom, similar elements of negotiation are desirable and achievable (sharing of ideas, identification of relevance, argument and support), but most often the teacher has a pre-planned direction in which to lead the students. Given additional limitations of the classroom environment that also distinguish the school context from the scientific community (e.g. time, equipment, motivation, existing knowledge base, purpose), it is difficult for a teacher to enable students to engage in truly open negotiation to derive meaning from their inquiries. This is not to say students do not, cannot, or should not share their ideas and be challenged to defend them within the classroom in manners *similar* to conventions of scientific practice. These processes are the essence of scientific inquiry. Crawford, Krajcik, & Marx, (1999) proposed that establishing a network for negotiation within a community of learners is part of the learning process. Similarities and parallels between the two communities of practice are evident. It should be noted, however, that processes and practices of reasoning and knowledge construction may differ by level of sophistication.

A second distinction between school-based inquiry and scientists' inquiry deals with the notion of meaningfulness. In addition to relating to what scientists do, authenticity to students depends on the relevancy of the investigations to students' own lives (Crawford et al., 1999; Krajcik, et al. 1994). A scientific investigation that parallels one carried out in a scientific setting may likely be so sophisticated in concepts and experimental design as to be outside the students' own Vygotskyan zone of proximal development. Experts (scientists) and novices differ in such areas as level of scientific reasoning (Niemi, 1997; Samarapungavan, 1992), structure of knowledge base (Chi, Feltovich, & Glaser, 1981), time frame and investment (professional career versus unit in school) (Samarapungavan, 1992), among others. In other words, authentic science may have a different meaning for scientists, science students, and science teachers, depending on each person's frame of reference. As we've stated, we consider authentic inquiry to be that which scientists do within the scientific community. This argument is intended to point out distinctions between the activities and contexts of authentic and school-based scientific inquiry. One of the hurdles so many educators face is understanding and appreciating this distinction. If a teacher feels the expectation is to present the school-based inquiry as authentic scientific inquiry, and the teacher understands the complexities of authentic scientific inquiry in relation to the classroom context, the task at hand may seem, and likely so, as unachievable and overwhelming. Furthermore, if a teacher attempts to portray school-based scientific inquiry as one and the same as that which occurs within the authentic context of the scientific community (Lave & Wenger, 1991) isn't the teacher portraying a skewed image of authentic science? The two communities are different. The two contexts are different. It seems that the task of the teacher is to engage students in school-based scientific inquiry and foster skills and cognitive practices similar or in parallel to those of the scientific community, but appropriate to the context and motivation of the learning situation. The teacher draws parallels between the two contexts and ideally is able to explicitly address NOS and scientific inquiry in order to present a more authentic image of the scientific endeavor *within the limitations and strengths* of the classroom environment and participants.

Learners' experiences within authentic scientific contexts can serve to reinforce the parallels and provide first-hand and often eye-opening and unique learning opportunities (Barab & Hay, 2001; Bell, Blair, Crawford, & Lederman, 2003; Richmond & Kurth, 1999). Although few engage in truly authentic scientific inquiry, teachers and students may also develop and deepen conceptions of NOS through such experiences. An examination of programs that involve authentic science research experiences provides insight into effective instructional components for achieving the goal of enhancing NOS conceptions.

AUTHENTIC RESEARCH EXPERIENCES AND LEARNING NOS

Science apprenticeship programs have emerged to provide teachers experiences with authentic scientific inquiry. These programs offered teachers opportunities to work "at the elbows" of the experts. Lave and Wenger (1991) referred to the role of "legitimate peripheral participation" and movement toward the "core" in one's learning about the scientific community and practice through apprenticeship. Investigations of the outcomes of apprenticeship programs have indicated positive achievements in inquiry and technological skills, science content, and confidence in conducting and teaching inquiry, as well as a better understanding about what scientists do and careers in science (e. g. Gottfried, Brown, Markovits, & Changar, 1993; Kielborn, & Gilmer, 1999; Spiegel, Collins, & Gilmer, 1995; Westerlund, Garcia, Koke, & Taylor, 2000, and references therein). Although these programs provide learners opportunities to engage quite "authentically" in scientific research, they did not focus explicitly on developments of NOS conceptions, nor did they formally assess NOS views.

Using our definition of authenticity, there have been surprisingly few studies investigating the influence of authentic science contexts on students' and teachers' understandings of NOS. The assumption made in many of these studies is that the more authentic the research experience by situating students in scientific settings, the more likely students will learn about NOS and of scientific inquiry. On the surface this appears a reasonable assumption. However, the results of these studies do not fully support this notion. The studies we highlight below include one by Ritchie and Rigano (1996), in which high school students worked with a scientist mentor in a university chemical engineering laboratory, and a study by Barab and Hay (2001) in which the researchers studied middle school students in a summer science camp with "real" scientists. A third study by Richmond and Kurth (1999) investigated the influence of a summer science research apprenticeship on high school students' understandings of the nature of scientists' work and the students' roles in different communities. A fourth study by Bell, Blair, Crawford, and Lederman (2003) investigated immersing high school science students in a summer long science internship and examining the effect of the research experience on their ideas about the nature of science and scientific inquiry.

Finally, we present a case example that focuses on creating an intervention based on findings from previous studies. The case example is drawn from the work of Schwartz, Lederman, and Crawford (2000; in press), who designed a course to integrate an authentic research experience for prospective secondary science teachers with a campus-based, theory-driven seminar, rich in opportunities for discussion and reflection. The design of the course for prospective science teachers reflected results of previous research explored in this chapter.

Inquiry Outside the Walls of the Classroom

Ritchie and Rigano (1996) investigated the influence of situating high school students in a real laboratory setting on their conceptual understandings and abilities to carry out scientific practices. High achieving high school students (grades 11 and 12) were selected to work on small-scale projects during part of their high school day. Their mentor scientists worked in a university chemical engineering laboratory. This apprenticeship involved one afternoon once a week for about 6 months. The purpose of Ritchie and Rigano's study was to describe work done by students in a scientifically authentic environment. Their findings identified three outcomes from this apprenticeship experience: 1) development of lab skills (for example, learning techniques of titration to ensure reliable results; 2) acquisition of conceptual understanding (e.g. chemistry concepts); and 3) gaining understanding of the meaning of scientific integrity (e.g. dealing with problems). The two boys in the study, both elite science students, experienced a cognitive apprenticeship that was, to a certain extent, positioned to enculturate novices into a community of science. The researchers claimed these students became independent researchers. However there is no evidence they learned specific NOS tenets other than the empirical basis of science.

Middle School Students in a Participatory Science Context

Barab and Hay (2001) studied the participation of 24 middle school students in a two-week summer experience, a Science Apprentice Camp, during which students worked in teams of four students mentored by a scientist and a middle school teacher. Barab and Hay defined authenticity as corresponding to the real world of scientists and they differentiated between two models, a simulation model and a participation model. The focus of the camp was on engaging middle school students in a participatory science context, in which students chose one of six apprenticeship settings. Students spent 2 hours of the 6 workdays with the scientist in his or her laboratory. As a component of this summer experience, students used an electronic apprenticeship notebook. Additional time was spent in developing a PowerPoint presentation of the research study, a part of the experience that Barab and Hay connected to a simulation model. Barab and Hay described the experiences and outcomes of this 2-week summer experience and presented evidence of the middle school students developing an appreciation of the "situated nature of science." Contributing to this appreciation was the opportunity to interact directly with scientists. However, the authors reported no formal assessments of participants' NOS views. The researchers characterized this experience as an authentic research experience, and they concluded the middle school students came to understand that unexpected findings can be a product of the process of science itself, instead of only the result of procedural errors. Limitations of the experience included first, the brevity of the time students actually spent with a scientist, and second, the lack of opportunity for students to develop research questions for these investigations. Although the students were afforded opportunity to experience scientific activities in the context of a real scientific laboratory, it is not evident if these middle school students actually learned about specific aspects of NOS from this experience, as there were no pre and post assessments. The authors stated the students participated in *doing* science. However any scientific processes learned from the experience were confined to those performed specific to each study. Barab and Hay advocated that research of these kinds of participatory science learning environments is applicable to other learning contexts, and that a participatory model is applicable to multiple learning contexts. Yet, no mention was made of how these results realistically translate to science classrooms in general.

Use of Apprenticeships to Facilitate Access to the Culture of Science

Richmond and Kurth (1999) studied twenty-seven 10th and 11th grade high school students during their participation in a 7-week residential summer science program. During the internship program students worked on research projects with scientist mentors, graduate students, and technicians, and participated in discussions in a laboratory setting. Richmond and Kurth examined students' views of the nature of scientific work through interviews conducted at three different points in the apprenticeship. Students reflected in journals, during interviews, and group discussions on their views of science. Richmond and Kurth identified two kinds of knowledge gained. Students learned first about the process of building scientific knowledge and the role of evidence in science, and second, about the tentative nature of science. The researchers described participating students as gaining deeper understandings of the seminal role that evidence plays in the construction of scientific understandings. In addition these researchers pointed out that the main impact of the experience was on students' development of a sense of themselves as scientists. As the students worked in the various communities that formed during the summer, the laboratory-centered community, the peer-centered community, and the program-centered community, they took on varied roles. The authors discuss the importance of students moving from the periphery of the community of science to becoming a part of the community. We would suggest that in addition to moving from the outside to within, the students stepped back again to the periphery in order to engage in reflection about their experiences.

When comparing the impacts of these three programs, it is important to note that both the Barab and Hay (2001) and Richmond and Kurth (1999) investigations were situated in apprenticeship programs that included reflective components that *appeared* to explicitly address NOS and scientific inquiry. These components included direct discussions with scientists, reflective journal keeping, and scaffolding provided by teachers. However, it should also be noted that few details of these activities were provided. This leaves open the question of the relative impacts on student learning of the authentic science experiences vs. the explicit instruction that accompanied these experiences. Differentiating the impacts of these factors appears critical when one considers that designers of authentic science experiences in secondary science instruction may not always include the explicit reflective components.

Influence of a Summer Apprenticeship Program on High School Students Understandings of NOS and Scientific Inquiry

Bell et al. (2003) investigated the impact of an 8-week science apprenticeship program on a group of high-ability secondary students' on their understandings of NOS and scientific inquiry. The high school students appeared to be immersed in a summer internship similar to one described in the Richmond and Kurth (1999) study. Data collected by Bell et al. included interviews with the scientists acting as research mentors as well as the high school students. Based on information from the mentor scientists and the interns, students were exposed to a full range of scientific investigation experiences. In particular, scientists described their students as engaged in the development of research methods, data collection, and data interpretation. This goes beyond other descriptive studies (Barab & Hay, 2001) that describe the context, but not what students actually learn. In addition, this authentic science experience spanned a total of eight weeks, representing a substantial investment of time, versus a shorter program as described by Barab and Hay (2001).

Overall, Bell et al. detected few changes in students' initial conceptions of NOS and scientific inquiry as revealed by pre- to post-apprenticeship assessments. With respect to NOS, students believed (as demonstrated in both pre- and post-apprenticeship assessments) that scientific knowledge is tentative, based on empirical evidence, and involves creativity and subjectivity. However, these beliefs tended to be superficial. Students ascribed tentativeness to the lack of information, and they did not exhibit an in-depth understanding that it is possible for different interpretations of the same data to be valid. Furthermore, students maintained the misunderstanding that, with more evidence, scientific theories eventually turn into laws. Finally, there was still some misunderstanding about the role of creativity in the analysis of data. Since the students participating in the program were identified as high ability science students (as was apparently the case in the Richmond and Kurth (1999) study, the lack of major changes in views of NOS clearly calls to question that these authentic experiences, by themselves, effect dramatic changes in students' NOS conceptions.

Enhancement of NOS Views using an Integrated Approach

In light of the previous inconclusive research on the impact of authentic science experiences on developing informed views of NOS, we present two case examples addressing the problem of helping teachers develop informed views of NOS while providing valuable experiences within authentic scientific research contexts.

Case Example #1: A Science Research Internship Course for Preservice Teachers

Drawing on the findings of the previous research, Schwartz, Lederman, and Crawford (2000; in press) designed a course for a Masters in Arts in Teaching Program that integrated authentic research experiences and campus work. The need for this preservice course was identified from research on teachers' knowledge of NOS (e.g. Abd-El-Khalick et al., 1998; Bell et al., 2000; Lederman, 1992; Meichtry, 1992). The course designers provided opportunity for preservice teachers to work as interns on a laboratory or field research project, and then to focus on the nature of scientific work with consideration of how teachers create inquiry-based environments in science classrooms. The three main components of the course included 1) The Research Setting, 2) Journals, and 3) Seminars.

The first component, the Research Setting, involved the placement of all interns with a scientist who was actively conducting scientific research at the University. The rationale for these settings included providing an *authentic* context for scientific inquiry. The interns spent an average of five hours a week in the research setting for the 10 weeks of the term. For the second component, the Journal, the interns made entries in two sections: a research section and a reflection section. This Journal included detailed records of the scientific work (research section) and reflections to focus questions addressing the tenets of the nature of science (reflective section). The focus questions were intended to help interns make connections between the scientific work and NOS. The third course component, the Seminars, consisted of five, 2-hour whole class meetings of the interns during which the instructor conducted discussions of the interns' research experiences and their views of NOS and scientific inquiry. In order to track changes in interns' NOS views, questionnaires, interviews, journal entries, and participant observations were used as data sources. Schwartz et al. assessed the interns' NOS views in a pre/post format with the Views of Nature of Science Questionnaire, form C [VNOS-C] and follow up interviews (Lederman, Abd-E-Khalick, Bell, & Schwartz, 2002). The purpose of the post-internship interview was to determine perceptions of the interns' authentic research experiences, to assess learning outcomes, and to detect changes, if any, and sources of change in their views of NOS

The results revealed that following the course and internship, all participants demonstrated improved understandings of NOS. The Journal writings guided by the focus questions had the most impact on enhancing understandings of NOS. Contributing to these enhanced understandings, the Research Setting had the least direct influence. This is not to say the interns (as in previously cited apprenticeship studies) did not learn specific laboratory techniques and scientific processes. What the research setting did provide, however, was an authentic context for reflection and provided opportunity for situated cognition--a real research setting upon which to reflect and connect ideas about science with the work of scientists. The scientists' presentations and student discussions were viewed as most beneficial parts of the seminar. There is evidence that the combined seminar, journal writings, and research experience effected positive changes in the interns' NOS views. To this end, the reflective journal writings and corresponding discussions appeared critical in challenging interns to formalize their NOS views. The explicit efforts to make connections and the opportunity for substantive active reflection upon an authentic context appeared key elements in fostering change in their views of NOS. Schwartz et al. suggested learner identity impacted NOS learning outcomes. Those interns who maintained identities as scientists were less reflective about NOS issues and showed little or no change in views. The authors suggest explicit/reflective assignments need to facilitate learners' abilities to cognitively step out of the authentic context and "reflect from the outside" in order to challenge and develop NOS conceptions.

Case Example # 2: A Comparison of Internship Models

Further evidence of the importance of the explicit/reflective components of the internship course was provided by Westerlund, Schwartz, Lederman, & Koke (2001). They compared NOS learning outcomes of the internship model just described (Schwartz et al., 2000; in press) to outcomes when components of the model were changed or eliminated. All the participants in several programs involving authentic science research experiences were administered the VNOS-C (Lederman et al., 2002) in a pre/post format. They found somewhat similar gains in NOS achievement when the same 10-week course that was shown to be successful in enhancing participants' NOS views (Schwartz et al., 2000; in press) was shortened to a 4-week course, with 10 hours a week engagement in laboratory settings. Similar outcomes pertained mainly to shifts away from naïve views and increased abilities to describe views in their own terms. Differences did lie in the extent of the gains, with fewer participants demonstrating abilities to support their emerging views with a variety of examples from their and other's laboratory settings. In addition, fewer participants showed a dramatic shift in conceiving NOS as inherent to all of science, a shift determined to be essential to facilitating effective teaching of NOS (Schwartz & Lederman, 2002). Although NOS issues were brought to a level of cognitive awareness for the participants in the 4-week course, as evidenced through weekly journal writings, discussions, and post-assessments, the time to process and reflect on the research experiences may have limited the extent of NOS learning. Nonetheless, developments in NOS learning for these participants exceeded outcomes in comparison to a third internship model. Another 10-week internship program that had the same goals as the first two models had interns spend the same amount of time in research settings, held two group meetings with the interns (designed as check-ins rather than discussion sessions), assigned similar focus questions for journaling, but did not provide feedback for journal writings nor a forum for discussion. In this latter situation, most participants' NOS views remained, as they reported, unchallenged. Participants in this final model expressed frustration with the lack of opportunity to discuss their experiences with their peers and instructors. Additionally, even though the same focus questions were assigned, participants reported that they delayed most journal writings until the end of the term when the entire collection was due. Instead of learning about NOS within their authentic contexts through continuous reflection, they did not think about NOS until the end, at which time they still were not provided opportunity to share with others.

In short, participants did not feel learning about NOS was a priority of the experience.

A comparison of two programs involving practicing teachers in summer science research experiences highlights differential effects of the level of engagement in authentic scientific research and the inclusion of explicit/reflective NOS instruction (Westerlund et al., 2001). The researchers compared participants' views of NOS as expressed on the VNOS-C (pre/post design) and in post-instruction interviews. The first program involved experienced teachers in a three-week summer institute on a University campus as part of a federally funded teacher enhancement project. For the first two weeks of the institute, participants engaged in explicit/reflective, activity-based sessions that aimed to teach about NOS and scientific inquiry. This instructional sequence was similar to those offered in the internship model described by Schwartz et al. (2000; in press). These 10, four-hour sessions were accompanied by a science research internship with practicing scientists on campus. In pairs, teachers spent four hours per day for two weeks with scientist mentors as participant observers and engaged in various aspects of the ongoing investigations in the research settings. It should be noted that these teachers did not assume the role of an "apprentice" aiming to become full members of the contributing scientific teams. Teachers kept daily journals, guided by focus questions to make connections between their experiences in the research settings with what they were learning about NOS in the classroom component of the institute. Whole group discussions were held each day of the two weeks for further reflection, comparison among research experiences, and feedback from other participants and science educators. The final week of the institute took place one month after the two-week internship/NOS instruction experiences. The final week focused on teachers developing lessons and practicing teaching NOS in an explicit manner within the context of inquiry-based science activities. Participants demonstrated major enhancements in understandings of NOS. Eighty-five percent of the teachers showed enhanced views. Overall, 62% showed positive shifts in their conceptions of four or more NOS aspects (e.g. tentativeness, creativity, subjectivity, observation/inference, empirical basis, cultural embeddedness). Similar to the findings by Schwartz et al. (2000; in press), the research component offered some concrete examples to support and develop NOS views. However, the concurrent learning about NOS, learning about the research setting, and learning instructional methods to teach NOS within the three-week institute appeared to limit the applicability of the authentic science context to the teachers' developing NOS views.

These results were compared to the NOS learning that occurred within a summer institute at a different institution (Westerlund et al, 2001). This second program employed did not provide instruction or guided reflection relative to NOS and the research experiences. However, the teachers in this second program had a more traditional science apprenticeship experience. For 8-weeks, these teachers spent to 40 hours per week in their respective research settings, designing and conducting individual research projects. They were immersed in the research setting and culture. Participants kept journals and were guided by questions intending to elicit descriptions of the activities of the research settings. Each week, participants met for a 2-hour group session to debrief their experiences with each other and a science educator, two scientists, and program organizers. During these sessions, participants discussed applications of their research experiences to their classroom. NOS was not a topic of discussion. VNOS-C responses indicated no differences between pre- and post- administrations, with the exception of the theory/law distinction. A discussion about distinctions between scientific theory and law took place at the request of a participant after the VNOS pretest. Most teachers in the traditional apprenticeship program maintained relatively naïve conceptions of the targeted aspects of NOS.

These results support the notion that authentic scientific research experiences can provide valuable *context for reflection* that serves to challenge and, in turn, enhance learners' conceptions of NOS. Optimal time and level of participation within the context is certainly an area for future research. However, engagement in authentic scientific research is insufficient to challenge NOS views. Topics of NOS in relation to authentic experiences need to be brought to a cognitive forefront for change to occur.

The Influence of the Ultimate Authentic Context: Scientists' Views of NOS

It can be argued that the experiences of the students and teachers within the apprenticeship and internship programs just described were still short of "authentic." This argument stems from the aforementioned distinctions between expert and novice inquirers. Clearly the students and teachers within these programs are considered novices in relation to scientists. Although the settings were authentic, the activities and knowledge development of the inquirers remained at the novice level, albeit to varying capacities. Examination of scientists' views and consideration of NOS, however, presents a similar picture to that already described. Even though scientists engage in truly authentic scientific inquiry within the community of science, their NOS views are not necessarily any more informed than those of teachers or students (Behnke, 1961; Bell, 2000; Glasson & Bentley, 2000; Kimball, 1967-68; Pomeroy, 1993). That is, scientists do not *necessarily* hold views of NOS that alignment with currently accepted views advocated by philosophers of science, science educators, and reform documents. Thus, not all of those who "do science" necessarily understand NOS better than all of those who do not "do science."

Bell (2000) compared views of NOS and decision-making processes of scientists, philosophers of science, science educators, and other professionals with equal levels of education. He reported that those subjects who had experiences in philosophy of science held more informed views of NOS, regardless of participation in authentic scientific research.

Moreover, Pomeroy's (1993) findings suggested the teachers with the least experience in science, the elementary teachers, held more nontraditional views of NOS. Included in the suggested reasons related the nontraditional views to lack of exposure to, or indoctrination into, science and scientists.

Glasson and Bentley (2000) suggest that scientists maintain views of NOS and STS that might not be completely congruent with the descriptions currently promoted because they [scientists] are an integral part of the scientific community. As such, scientists do not typically examine NOS or STS issues from the perspective of one *outside* of that community. This argument suggests that membership within the scientific community could affect one's perspective of NOS. This is not to say all scientists hold naïve or empiricist views of NOS. This is to say that the scientific profession is not one wherein the scientist must consider and debate philosophical issues related to NOS. This is not the typical role of the scientist within the community of science. According to Glasson and Bentley (2000):

> The overriding view among practicing scientists is that science is essentially experimental and empirical; however, the important role of theory, the multiplicity and complexity of science methods, and the value-ladenness of science require that scientists examine the assumptions underlying their own research and what goes into the decision-making that affects research design, funding, and public acceptance of results. (Glasson & Bentley 2000, p. 483)

This is a powerful statement that suggests scientists are such an integral part of the scientific endeavor that they do not necessarily take a step back from that role to reflect upon their discipline. They must assume a different, a reflective or even a philosophical, perspective. Until they do, they do not necessarily recognize the values and assumptions that are an inherent part of the scientific discipline itself. The importance of such a shift in perspective from "working within" to "reflection from the outside" in enabling individuals to develop informed views of NOS through engagement in scientific inquiry has been illustrated through examination of the literature on the use of scientific inquiry to teach about NOS (Schwartz et al., in press; Westerlund et al., 2001). How, then, can students or teachers be expected to "discover" the nature of the scientific enterprise through engagement in scientific inquiry, even if the context is authentic? If those who develop the knowledge do not instinctively view their discipline with a philosophical eye, it is unreasonable to assume students or teachers will become intuitive philosophers during the course of an inquiry investigation. How, then, can authentic scientific inquiry experiences be used to teach NOS?

CRITICAL ELEMENTS FOR SUCCESS

The literature and case examples discussed in this chapter indicate authentic scientific inquiry experiences can be effectively used to teach about NOS *provided that* certain critical elements are integrated. It has been shown that experience within the authentic context, by itself, is insufficient. This collection of work suggests there are critical elements for success. We summarize these critical elements as our recommendation for effective teaching of NOS within the context of authentic scientific inquiry.

1. *Treat NOS as science content by bringing aspects of NOS to the forefront of learners' thoughts.* Cognitive engagement needed to generate and enhance one's conceptions of NOS is not an automatic consequence of the authentic experience. Like with school-based scientific inquiry, programs

providing authentic scientific research experiences that rely on implicit messages to teach about NOS fall short of their desired NOS learning achievement. In contrast, advances in NOS conceptions are more likely to occur for those programs designed and implemented in a manner consistent with the perspective that NOS is a cognitive learning outcome, just like understanding of protein synthesis or acid/base reactions lie within the cognitive domain. Importantly, NOS *is* considered science content in current national science standards (AAAS, 1990; NRC, 1996). From this framework, these programs then treat NOS accordingly by employing explicit instruction and reflective opportunities to stimulate learners' thinking toward NOS issues in relation to their research settings and experiences. We have shown that explicit/reflective treatment of NOS within school-based or authentic scientific inquiry leads to the development of NOS conceptions that are more aligned with currently accepted positions. Explicitness in this sense takes clearer form when one considers qualities of effective teaching. Teachers provide demonstrations, explanations, and opportunity for student engagement in activities that are relevant to the students and to the subject matter being taught. Teachers provide clear and intentional opportunities for students to practice and apply their newly acquired knowledge in a variety of situations. Teachers monitor students' progress through formative and summative assessments and provide feedback. In other words, NOS is treated as any other science content held in priority within the curriculum. The teacher purposely introduces and reinforces topics of NOS, providing terminology and examples to demonstrate the concepts. Discussions are then critical to engage the learners in shared dialogue that helps build meaningful knowledge about NOS.

2. *Facilitate reflection: From outside to inside and back out again.* Apprenticeship learning has been described as the movement of the learner from the periphery of the community to the center of the community where the learner is integral to the process and essentially inseparable from the community (Lave & Wenger, 1991; Richmond & Kurth, 1999). Richmond and Kurth (1999) referred to this movement as moving from the "outside to the inside" in the process of science apprenticeship. The suggestion is that to gain awareness of the culture of science and engage in authentic scientific inquiry, one must transition from the periphery to the core through gradual progression toward becoming an independent contributor within the community. This *is* the essence of apprenticeship. However, an active participant within the scientific community does not necessarily need to reflect upon the nature of the community, the knowledge, or the endeavor to be a successful contributing member (Glasson & Bentley, 2000; Longbottom & Butler, 1999; Schwartz et al., 2000; in press). Reflection is a metacognitive activity and a skill that can serve to inform the learner of his/her own developments in concept understanding and lead to deeper understanding (Hodson, 1998; Paris & Ayres, 1994). We suggest that the ability to express one's views of NOS and certainly to modify

one's views through participation within the context of authentic scientific research requires the learner to shift back to the outside, or periphery, in order to assume a reflective stance conducive to challenging and developing one's conceptions of NOS. Without such reflection from the outside, one does not consider the nature of the enterprise in which he/she has become a part. For the purpose of science education, a type of catalyst can serve to facilitate the transition in perspective. This catalyst might be guided discussions or journal focus questions. Figure 1 offers a sample of focus questions used by Schwartz et al. (2000; in press) and Lederman et al. (2002) that guided interns' journal writings.

Interns are to consider their research experiences in response to the following questions:

1. At the start of each daily journal entry, respond to these three questions:

Today I learned....

Today I felt ___________ about my experience in the lab because.....

I think I will be able to use this experience in my science teaching by....

Look for instances where conclusions were drawn and supported by multiple data sources, including the scientific literature. How are contradictions or anomalous data handled? What are the reasons given for the decisions the scientists make regarding contradictory or anomalous data?

All decisions and conclusions made are unavoidably subjective in nature. Try to identify instances where subjectivity was openly recognized in your research setting. Try to identify the assumptions (theories, other biases, etc.) that influence the decisions made in the research.

What is a scientific model? How are models used in the research project you are most familiar with in your research setting?

In what ways does what is happening in your research setting relate to "science is a complex social activity?"

Figure 1. Sample focus questions

3. *Understand that one does not "do NOS."* If the goal is to develop and enhance conceptions of NOS, the scientific inquiry experience must be viewed as a *context* for learning about NOS, not as the final product in and of itself. Understanding the difference between the "doing of science" and the "NOS" is essential in order to explicitly teach the content of NOS within an inquiry context. Conflation of inquiry and NOS can be an unrecognized obstacle for many educators attempting to reach the reform's vision for NOS and inquiry teaching. Educators too often assume the students are "doing NOS" because they are engaged in data collection and analysis. Likewise, experiences within authentic science contexts might lend themselves to such expressions as "The scientists are doing NOS in their research" or "We did NOS when we collected only certain data during the experiment and chose to eliminate other data." Statements like

these need to be attended to during instruction through questioning and discussion, asking for examples and elaboration, to dispel the misconception that NOS is an activity, versus a cognitive concept gained through reflection upon activity and experience. We have shown that regardless of the level of inquiry, one does not "do NOS," and learning *about* NOS does not occur unless learners are challenged to do so. Furthermore, the conflation offers complications to researchers who rely on teachers' self reports of NOS instruction. Such teachers might be highly efficacious regarding their NOS teaching knowledge and practices, but their reports are not necessarily reflective of effective NOS instruction (Lederman, Schwartz, Abd-El-Khalick, & Bell, 2001; Lederman et al., 2002). A clear understanding that NOS is not a skill or activity is essential for being able to bring NOS to the forefront of learners' thoughts and to facilitate reflection.

ALTERNATIVES TO AUTHENTIC CONTEXTS

One major limitation looms in identifying and procuring appropriate authentic science experiences for students and teachers. As reported in studies addressed earlier in the chapter, the *feasibility* of providing authentic science experiences for all teachers and students is a major stumbling block in K-12 science classrooms and teacher preparation programs. In a study of prospective science teachers' NOS learning outcomes during a science research internship (Schwartz et al., 2000; in press), the course instructors encountered two problems. First, the instructors had difficulty in identifying viable scientific internships, and second, the variability of such internship experiences may have compromised the success potential for all students. The problem of limited scientific research opportunities may be amplified in small liberal arts teachers colleges and small, rural schools Realistically, integrating authentic scientific research experiences into the culture of the school or college classroom is challenging.

As a way to ameliorate the challenge of providing authentic science experiences for *all* teachers and students, there are practical alternatives to situating students in actual scientific laboratory or field settings. These alternatives include using 1) community-focused, problem-based learning (PBL) and project-based science (PBS) curricula; 2) place-based experiential learning; and 3) the use of technology-rich environments in science classrooms.

The first of these alternatives involve community-based PBL [e.g., Barrows, 1994; Uyeda, Madden, Brigham, & Luft, 2002] and PBS [Blumenfeld, Soloway, Marx, Krajcik, et al., 1991; Crawford et al., 1999) curricula. Both PBL and PBS approaches involve identifying ill-structured problems that have no simple, known solution and are connected to important science content. In PBL and PBS learners engage in solving problems and taking control of their own learning. The authenticity stems from the real-world science issues and the use of actual scientific data. One example of an ill-structured problem that students might investigate is 'what is the quality of our school's drinking water?'

The second alternative to working in actual scientific laboratories involves place-based experiential learning (Dewey, 1938; Smith, G., 2002). In place-based education the learning is grounded in local phenomena and students' lived experiences. Traced back to John Dewey and the Chicago Lab School, place-based experiential learning affords students the opportunity to investigate issues having social reality. One example includes the Environmental Middle School in Portland, Oregon, in which the curriculum encompasses a living laboratory of the local rivers, mountains, and forests (Smith, G., 2002). In this case teachers become unique curriculum designers within the context of the community.

A third alternative that appears promising is the use of technology-rich learning environments based on data-driven scenarios related to real-world scientific problems. These learning technologies scaffold students in scientific inquiries and create the needed context for reflection on aspects of NOS. Examples of the use of these learning technologies include software and curricular supports developed at Northwestern University and The University of Michigan Hi-Ci Group (Reiser, 2002). These learning technologies have potential to provide 1) authentic experiences for prospective teachers thereby providing a context for explicit/reflective discussions; and 2) opportunity for prospective teachers to envision how to translate their own experiences to their future science classrooms.

An example of a technology-rich experience that provides a context for learning about NOS includes one reported by Crawford and Cullin (2002). The researchers designed a modeling module in a secondary science teaching methods course. During the module prospective teachers designed investigations of real-world phenomena; then built and tested computer models of scientific phenomena using the dynamic modeling software *Model-It* (Jackson, Stratford, Krajcik, & Soloway, 1994). Prior to using the software, the prospective teachers designed and carried out extended inquiries of the relationships among plants, soil, and water using TerrAqua columns (Ingram, 1993). Pairs of prospective teachers built and tested computer models of their plant, soil, and water system. Following the model building activity the instructors used the "black box" activity, Ropes and Tubes (National Academy of Sciences, 1998, pp. 22-25). The purpose of the activity was to explicitly emphasize aspects of NOS, tentativeness and use of evidence. The methods instructors attempted to connect use of models by scientists with the prospective teachers' own model construction. Research findings suggest that prior to the modeling experience the prospective teachers held uninformed views of scientists' construction or use of models. Following the instruction there was evidence that the prospective teachers' began to recognize the importance of models in science and science teaching. However, there is much work to be done to overcome years of didactic instruction and long-ingrained perceptions of what is important to teach. Maintaining the three critical elements for effective teaching of NOS within authentic contexts is essential within the classroom technology-based modeling context as well. The relative impacts of the alternative contexts and authentic contexts on learners' developing conceptions of NOS and inquiry is an area for future research.

REFERENCES

Abd-El-Khalick, F. (2001). Embedding nature of science instruction in preservice elementary science courses: Abandoning scientism. *Journal of Science Teacher Education, 12*(3), 215-233.

Abd-El-Khalick, F., Bell, R. L., & Lederman, N. G. (1998). NOS and instructional practice: Making the unnatural natural. *Science Education, 82*, 417-436.

Abd-El-Khalick, F., & Lederman, N. G. (2000). Improving science teachers' conceptions of nature of science: A critical review of the literature. *International Journal of Science Education, 22*(7), 665-701.

American Association for the Advancement of Science. (1990). *The Liberal Art of Science: Agenda for Action.* Washington D. C.: AAAS Publication.

American Association for the Advancement of Science. (1993). *Benchmarks for science literacy: A Project 2061 report.* New York: Oxford University Press.

Barab, S.A., & Hay, K.E. (2001). Doing science at the elbows of experts: Issues related to the science apprenticeship camp. *Journal of Research in Science Teaching, 38*, 70-102.

Barufaldi, J., Bethel, L., & Lamb, W. (1977). The effect of a science methods course on the philosophical view of science among elementary education majors. *Journal of Research in Science Teaching, 14*(4), 289-294.

Behnke, F. L. (1961). Reactions of scientists and science teachers to statements bearing on certain aspects of science and science teaching. *School Science and Mathematics, 61*, 193-207.

Bell, R. L. (2000). *Understandings of the nature of science and decision making on science and technology based issues.* Unpublished doctoral dissertation, Oregon State University, Oregon.

Bell, R. L., Blair, L., Crawford, B., & Lederman, N. (2003). "Just do it." The impact of a science apprenticeship program on high school students' understandings of the nature of science and scientific inquiry. *Journal of Research in Science Teaching, 40*(5), 487-509.

Bell, R. L., Lederman, N. G., & Abd-El-Khalick, F. (2000). Developing and acting upon one's conceptions of the nature of science: A follow-up study. *Journal of Research in Science Teaching, 37*, 177-209.

Benson, G. D. (1989). The misrepresentation of science by philosophers and teachers of science. *Synthese, 80*, 107-119.

Bianchini, J., & Colburn, A. (2000). Teaching the nature of science through inquiry to prospective elementary teachers: A tale of two researchers. *Journal of Research in Science Teaching,37*(2), 177-209.

Billeh, V. Y., & Hasan O. E. (1975). Factors affecting teachers' gain in understanding NOS. *Journal of Research in Science Teaching, 12*(3), 209-219.

Blumenfeld, P., Soloway, E., Marx, R., Krajcik, J., Guzdial, M., & Palincsar, A. (1991). Motivating project-based learning: Sustaining the doing, supporting the learning. *Educational Psychologist, 26*, 369-398.

Brown, J. S., Collins, A., & Duguid, P. (1989). Situated cognition and the culture of learning. *Educational Researcher, 18*, 32-42.

Carey, S. (1986). Cognitive science and science education. *American Psychologist, 41*(10), 1123-1130.

Carey, S., Evans, R., Honda, M., & Unger, C., (1989). "An experiment is when you try it and see if it works": A study of grade 7 students' understanding of the construction of scientific knowledge. *International Journal of Science Education, 11*, 514-529.

Carey, S., & Smith, D. (1993). On understanding the nature of scientific knowledge. *Educational Psychologist, 28*(3), 235-251.

Carey, R. L., & Stauss, N. G. (1970). An analysis of experienced science teachers' understanding of NOS. *School Science and Mathematics, 72*, 336-376.

Carey, R. L., & Stauss, N. G. (1968). An analysis of the understanding of NOS by prospective secondary science teachers. *Science Education, 58*(4), 358-363.

Chinn, C., & Malhotra, B. (2002). Epistemologically authentic inquiry in schools: A theoretical framework for evaluating inquiry tasks. *Science Education, 86,* 175-218.

Crawford, B. A., Krajcik, J. S., & Marx, R. W. (1999). Elements of a community of learners in a middle school science classroom. *Science Education, 83*, 701-723.

Crawford, B. A. & Cullin, M. (2002, April). *Prospective teachers' use of Model-It: Supporting conceptions of modeling in science.* A poster and paper presented as part of the interactive symposium session for Division C. *Characterizing and evaluating software scaffolds for scientific inquiry. Reiser, B* at the annual meeting of the American Educational Research Association. New Orleans, LA, April 1-5, 2002.

Crumb, G. H. (1965). Understanding of science in high school physics. *Journal of Research in Science Teaching, 3*(3), 246-250.

Dewey, J. (1938). The school and society. In M. Dworkin (Ed.), *Dewey on education.* New York: Teachers College Press.

Driver, R., Leach, J., Millar, R., & Scott, P. (1996). *Young People's Images of Science.* Buckingham: Open University Press.

Duschl, R. A. (1990). *Restructuring science education.* New York: Teachers College Press.

Duschl, R. A., & Wright. E. (1989). A case study of high school teachers' decision making models for planning and teaching science. *Journal of Research in Science Teaching, 26,* 467-501.

Gallagher, J. J. (1991). Prospective and practicing secondary school science teachers' knowledge and beliefs about the philosophy of science. *Science Education, 75*(1), 121-133.

Gilbert, J. K. (1993). *Models and Modelling in Science Education.* Hatfield, UK: Association for Science Education

Glasson, G., & Bentley, M. (2000). Epistemological undercurrents in scientists' reporting of research to teachers. *Journal of Research in Science Teaching,84*(4), 469-485.

Gottfried, S. S., Brown, C. W., Markovits, P. S., & Changar, J. B. (1993). Scientific work experience programs for science teachers: A focus on research-related internships. In P. Rubba (Ed.), *Association for the education of teachers in science (AETS) 1992 yearbook: Exemplary programs in science teacher development.* ERIC Documentation Service.

Harms, N. C., & Yager, R. E. (1981). *What research says to the science teacher.* (vol. 3). Washington DC: NSTA.

Haukoos, G. D., & Penick, J. E. (1985). The effects of classroom climate on college science students: A replication study. *Journal of Research in Science Teaching, 22*(2), 163-168.

Herron, M. D. (1969). Nature of science: Panacea or pandora's box. *Journal of Research in Science Teaching, 6,* 105-107.

Herron, M. D. (1971). The nature of scientific enquiry. *School Review, 79,* 171-212.

Hodson, D. (1988). Toward a philosophically more valid science curriculum. *Science Education, 72*(1), 19-40.

Hodson, D. (1998). *Teaching and Learning Science: Towards a Personalized Approach.* Philadelphia, PA: Open University Press.

Horner, J. K., & Rubba, P. (1978). The myth of absolute truth. *The Science Teacher, 45*(1), 29-30.

Hurd, P. D. (1991). Issues in linking research to science teaching. *Science Education, 75*(6), 723-732.

Ingram, M. (1993). *Bottle Biology.* Dubuque, IA: Kendall Hunt Publishing Company.

Jackson, S. L., Krajcik, J. S., & Soloway, E. (2000). Model-It: A design retrospective. In M. J. Jacobson & R. B. Kozma (Eds.), *Innovations in science and mathematics education : advanced designs for technologies of learning.* Mahwah, N.J.: Erlbaum.

Kielborn, T. L., & Gilmer, P. J. (1999). Meaningful science: Teachers doing inquiry + teaching science. Tallahassee, FL: SouthEastern Regional Vision for Education.

Kimball, M.E. (1967-68). Understanding NOS: A comparison of scientists and science teachers. *Journal of Research in Science Teaching, 2*(1), 110-120.

Lave, J., & Wenger. E. (1991). *Situated learning: Legitimate peripheral participation.* Cambridge: Cambridge University Press.

Lederman, N. G. (1992). Students' and teachers' conceptions about NOS: a review of the research. *Journal of Research in Science Teaching, 29,* 331-359.

Lederman, N. G. (1998). The state of science education: Subject matter without context. *Electronic Journal of Science Education, 3*(2).

Lederman, N. G., & Abd-El-Khalick, F. (1998). Avoiding de-natured science: Activities that promote understandings of NOS. In W.F. McComas (Ed.), *NOS in science education: Rationales and strategies* (pp. 83-126). Dordrecht, The Netherlands: Kluwer Academic Publishers.

Lederman, N. G., Abd-El-Khalick, F., Bell, R. L., & Schwartz, R. S. (2002). Views of Nature of Science Questionnaire (VNOS): Toward Valid and Meaningful Assessment of Learners' Conceptions of Nature of Science. *Journal of Research in Science Teaching, 39*(6), 497-521.

Lederman, N. G., Schwartz, R. S., Abd-El-Khalick, F., & Bell, R. L. (2001). Preservice Teachers' Understanding and Teaching of the Nature of Science: An Intervention Study. *The Canadian Journal of Science, Mathematics, and Technology Education, 1*(2), 135-160.

Lederman, N. G., Schwartz, R. S., Lederman, J., Matthews, L., & Khishfe, R. (2002, April). Project ICAN: A teacher enhancement project to promote teachers' and students' knowledge of scientific inquiry and nature of science. Symposium presented at the annual meeting of the National Association for Research in Science Teaching, New Orleans, LA.

Linville, H. R. (1907). Biology as method and as a science in secondary schools. *School Science and Mathematics Journal. 7(*4), 264-272.

Longbottom, J., & Butler, P. (1999). Why teacher science? Setting rational goals for science education. *Science Education, 83*(473-492).

Martin, B., Kass, H., & Brouwer, W. (1990). Authentic science: A diversity of meanings. *Science Education, 74*(5), 541-554.

McComas, W. F. (Ed.), *The nature of science in science education: Rationales and strategies.* Dordrecht, The Netherlands: Kluwer Academic Press.

Meichtry, Y. J. (1992). Influencing student understanding of NOS: Data from a case of curriculum development. *Journal of Research in Science Teaching, 29*(4), 389-407.

National Research Council. (1996). *National science education standards.* Washington, DC: National Academic Press.

Niemi, D. (1997). Cognitive science, expert-novice research, and performance assessment. *Theory into Practice, 36*(4), 239-246.

Paris, S., & Ayres, L. (1994). *Becoming Reflective Students and Teachers with Portfolios and Authentic Assessment.* Washington, D. C.: American Psychological Association.

Pomeroy, D. (1993). Implications of teachers' beliefs about NOS: Comparison of the beliefs of scientists, secondary science teachers, and elementary teachers. *Science Education, 77*(3), 261-278.

Ramsey, G. A., & Howe, R. W. (1969). An analysis of research on instructional procedures in secondary school science, Part I: Outcomes of instruction. *The Science Teacher, 36*(3), 62-70.

Reiser, B. (2002). Characterizing and evaluating software scaffolds for scientific inquiry. An interactive poster session presented at the annual meeting of the American Educational Research Association, New Orleans, LA.April 1-5, 2002.

Richmond, G., & Kurth, L.A. (1999). Moving from outside to inside: High school students' use of apprenticeships as vehicles for entering the culture and practice of science. *Journal of Research in Science Teaching, 36*, 677-697.

Riley, J. P. (1979). The influence of hands-on science process training on preservice teachers' acquisition of process skills and attitude toward science and science teaching. *Journal of Research in Science Teaching, 16*(5), 373-384.

Ritchie, S. M., & Rigano, D. L. (1996). Laboratory apprenticeship through a student research project. *Journal of Research in Science Teaching, 3*(7), 799-815.

Robinson, J. T. (1969). Philosophy of science: Implications for teacher education. *Journal of Research in Science Teaching, 6*, 99-104.

Roth, W. M., (1991, March). *Students as scientists: Constructivism and the negotiation of meaning in physics teaching.* Paper presented at the annual meeting of the Association for the Education of Teachers in Science, Houston, TX.

Roth, W. M. (1995). *Authentic school science: Knowing and learning in open-inquiry science laboratories.* Dordrecht, The Netherlands: Kluwer Academic.

Roth, W. M., & Roychoudhury, A. (1993). The development of science process skills in authentic contexts. *Journal of Research in Science Teaching, 30*(2), 127-152.

Rutherford, F. J. (1964). The role of inquiry in science teaching. *Journal of Research in Science Teaching, 2*, 80-84.

Ryan, A.G., & Aikenhead, G. S. (1992). Students' preconceptions about the epistemology of science. *Science Education, 76*(6), 559-580.

Ryder, J., Leach, J., & Driver, R. (1999). Undergraduate science students' images of science. *Journal of Research in Science Teaching, 36*(2), 201-220.

Samarapungavan, A. (1992). Children's judgments in theory choice tasks: Scientific rationality in childhood. *Cognition, 45*, 1-32.

Schwab, J. J. (1962). The teaching of science as enquiry. In J. J. Schwab and P. F. Brandwein (Eds.), *The teaching of science* (pp. 1-103). Cambridge, MA: Harvard University.

Schwartz, R. S. (2002). *The use of inquiry to teach the nature of science: The effectiveness of implicit and explicit methods.* Unpublished manuscript, Oregon State University, Corvallis.

Schwartz, R. S., & Lederman, N. G. (2002). "It's the nature of the beast": The influence of knowledge and intentions on learning and teaching nature of science. *Journal of Research in Science Teaching, 39*(3), 205-236.

Schwartz, R., Lederman, N. G., & Crawford, B. A. (2000, April). *Understanding the nature of science through scientific inquiry: An explicit approach to bridging the gap.* Paper presented at the annual meeting of the National Association of Research in Science Teaching, April 28 -May 1, 2000, New Orleans, LA.

Schwartz, R. S., Lederman, N., & Crawford, B. (in press). Developing views of nature of science in an authentic context: An explicit approach to bridging the gap between nature of science and scientific inquiry. *Science Education.*

Shapiro, B.L. (1996). A case study of change in elementary student teacher thinking during an independent investigation in science: Learning about the "face of science that does not yet know." *Science Education, 80*(5), 535-560.

Shulman, L. S., & Tamir, P. (1973). Research on teaching in the natural sciences. In R. M. W. Travers (Ed.), *Second Handbook of Research on Teaching.* Chicago: Rand McNally College Publishing Co.

Smith, C., Maclin, D., Houghton, C., & Hennessey, M. G. (2000). Sixth-grade students' epistemologies of science: The impact of school science experiences on epistemological development, *Cognition and Instruction, 18*(3), 349-422.

Smith, G.A. (April, 2002). Place-based education: Learning to be where we are. *Phi Delta Kappan,* 584-594.

Smith, M. U., & Scharmann, L. C. (1999). Defining versus describing the nature of science: A pragmatic analysis for classroom teachers and science educators. *Science Education, 83,* 493-509.

Spears, J., & Zollman, D. (1977). The influence of structured versus unstructured laboratory on students' understanding the process of science. *Journal of Research in Science Teaching, 14*(1), 33-38.

Spiegel, S. A., Collins, A., & Gilmer, P. J. (1995). Science for early adolescence teachers (Science FEAT): A program for research and learning. *Journal of Science Teacher Education, 6*(4), 165-174.

Sund, R. B., & Trowbridge, L. W. (1973). *Teaching Science by Inquiry in the Secondary School.* Columbus, OH: Charles E. Merrill Publishing Company.

Tamir, P. (1970). Long-term evaluation of BSCS. *The American Biology Teacher, 32,* 354-358.

Tamir, P. (1972). Understanding the process of science by students exposed to different science curricula in Israel. *Journal of Research in Science Teaching, 9*(3), 239-245.

Trent, J. (1965). The attainment of the concept "understanding science" using contrasting physics courses. *Journal of Research in Science Teaching, 3,* 224-229.

Welch, W. W., Klopfer, L., Aikenhead, G., & Robinson, J. (1981). The role of inquiry in science education: Analysis and recommendations. *Science Education, 65*(1), 33-50.

Westerlund, J. F., Garcia, D., Koke, J. R., & Taylor, T. (2000, April). *Teachers as summer scientific researchers: Transformative experiences.* Paper presented at the Annual meeting of the National Association for Research in Science Teaching, New Orleans, LA.

Westerlund, J., Schwartz, R. S., Lederman, N. G., & Koke, J. (2001, March). *Teachers learning about nature of science in authentic science contexts: Models of inquiry and reflection.* Symposium presented at the annual meeting of the National association for Research in Science Teaching, March 25-28, St. Louis, MO.

Yager, R. (1992). Viewpoint: What we did not learn from the 60s about science curriculum reform. *Journal of Research in Science Teaching, 29*(8), 905-910.

Yager, R. E., Engen, H. B., & Snider, B. C. F. (1969). Effects of the laboratory and demonstration methods upon the outcomes of instruction in secondary biology. *Journal of Research in Science Teaching, 6,* 76-86.

Yager, R., & Wick, J. (1966). Three emphases in teaching biology – A statistical comparison of results. *Journal of Research in Science Teaching, 4,* 16-20.

CHAPTER 17

HARRY L. SHIPMAN

INQUIRY LEARNING IN COLLEGE CLASSROOMS

For the times, they are, a changing

PREAMBLE

Readers of a certain age will remember the subtitle of this chapter as the title and beginning phrase of a memorable Bob Dylan song, which was popular in the 1960s (Dylan 1985). This was a decade of major changes. The opening decade of the 21st century is also a decade of major changes in American colleges. Everyone is starting to pay attention to teaching. These changes are nicely illustrated by a little story.

In June of 2000 I hosted a major astronomical conference, the 12th European Workshop on White Dwarf Stars. These conferences are like medieval pilgrimages, events where people from around the world assemble to share their latest findings. Most of the conversation at conferences like these has to do with the astronomical quarter-acre of ground which the participants all share an interest in. For this conference, the topic was white dwarf stars: tiny balls of enormously compacted starstuff, no larger than the earth, which are the final stages in the life cycle of stars like our sun. The printed record of the conference is now published (Provencal, Shipman, MacDonald, and Goodchild 2000).

When the conferees took a trip to a nearby garden in Pennsylvania, I contrived to sit on the bus in a seat next to my old friend Jim Liebert, a senior astronomer whom I have known for over 25 years. Jim is a full professor at the University of Arizona, one of the top departments in the U.S., and shares many other characteristics of a Apundit,@ a leader in the profession. For example, he has played a major role in supervising the management of The Astrophysical Journal, the premier astronomy journal in America. I wanted to talk to Jim about using a forthcoming spacecraft to seek planets and low-mass objects around white dwarf stars. I wanted to have my astronomer's hat on. That's not what happened, and I was pleased that the conversation turned elsewhere.

As we wandered around Longwood Gardens, it turned out that Jim wanted to talk to me about something else: inquiry based teaching. Jim has always been a good teacher, and his teaching, like the teaching of most American professors, has generally been in the lecture mode. I had done a few things at the conference which

L.B. Flick and N.G. Lederman (eds.) Scientific Inquiry and Nature of Science, 357-387.

were designed to suggest that teaching and lecturing were not necessarily the same thing. I was pleasantly surprised that he really wanted to bend my ear for the better part of an hour about teaching, and in particular how to get students to work in collaborative groups in a large class. We did talk about the technical astronomy stuff, but for only a few minutes.

In the 1970s, such a conversation at a research meeting would have been inconceivable. There were no organized sessions on education at astronomy meetings, and while there was an "education" category for papers, I can't remember anyone ever using it. But in the 1990s and 2000's, these conversations are common. Special sessions on education at meetings of the American Astronomical Society and the American Chemical Society are regular events. I have had dozens of conversations similar to the one which I described above. For example, I went to Columbia University to give a technical talk on astronomy and was pleasantly surprised to spend most of dinnertime talking about the University of Delaware's teaching center, which I was then directing. I remember a lunchtime conversation next to the beach in Blanes in Spain's Costa Brava, talking to Maria Teresa Ruiz of the Universidad de Chile, not about the very cool white dwarf stars which she was an expert on, but about teaching. University professors are really starting to think seriously about teaching, and are taking inquiry learning seriously. The times, they are indeed, a changing.

INTRODUCTION: FROM LECTURES TO INQUIRY

Most college classes are usually thought of as lectures. The notion of a college professor as a "sage on the stage" is at least a century and a half old (Boyer 1987, p. 149) and perhaps older. In the nineteenth century at some of America's older universities like Harvard and William and Mary, someone probably stood at the front of a room, in an academic gown, talking for some extended period of time. I wonder if, before the printing press existed, the same script was played out in the old European universities like Bologna, Paris, and Oxford in the middle ages.

Until recently, the same drama was played out in almost every classroom in almost every American university, with a few changes. The academic gown is no longer a professors' standard workday clothing. Centuries ago, professors were always males of European ancestry; by 1985, the professoriate had become a little bit more diverse, though not diverse enough. My principal observation is that for centuries the college professor always stood in the same place: immobile, at the front of the class, and sometimes on an actual stage. I taught my first college class on a stage in the fall of 1971. And I lectured, just like my teachers had always done. I did allow for a few short interruptions for questions from a few courageous students.

In the past fifteen years, a number of university faculty have begun to recognize some of the serious limitations of the lecture style of teaching. We have begun to experiment with some learning techniques which have been used in at least some K-12 classrooms, and in some other disciplines like business, for a longer period of time. In many classrooms, students spend a good deal of time sitting in groups, learning from each other. In other classrooms, student projects have become the focus of part, and sometimes even all, of the curriculum. Technology has begun to make a difference in the way that students and faculty interact.

This chapter will focus on the changing role of inquiry learning in higher education. I will first address the question of why college faculty are changing their practices in the first place. Why are they turning away from the lecture and turning towards other, inquiry-based techniques. I will then struggle with the definition of exactly what inquiry teaching is.. I will then turn to the question of what inquiry teaching looks like in a college classroom, particularly with respect to a teaching sequence in which the nature of science was an important consideration.

REASONS FOR MOVING TOWARDS INQUIRY

College teachers will not change their practices just because some national organization says so. Not even the prestige of the AAAS (one of the oldest scientific societies in America) or the National Academy of Sciences with its limited, elite membership can sway a professor who is comfortable with lecturing, by themselves. The momentum for change has come from two other sources: evidence that lecturing doesn't work and local initiatives.

Relative ineffectiveness of lecturing

Lecturing really doesn't work as well as its aficionados think it does. When I do workshops promoting change on college campuses, I frequently use one of the videos showing articulate Harvard and MIT seniors displaying egregious but well-articulated misconceptions about astronomy, electricity, and biology (see, e.g., Sadler and Schneps 1989). In the astronomy video, 21 out of 23 Harvard seniors, alumni, and faculty were unable to explain why the earth had seasons or why the moon had phases. Subsequent videos of the series show that similar proportions of well-educated students, including at least one electrical engineer, cannot construct an electric circuit from a battery, bulb, and one wire. Students are at a loss to explain where the material comes from to convert a tiny maple seed into a huge tree.

These videos led me and a number of friends and colleagues to investigate the many studies which show the ineffectiveness of lecturing. Some of these papers explicitly demonstrate that inquiry learning works much better. The literature is long and reaches back to John Dewey. One of the most comprehensive studies was done in physics by Hake (1998), comparing the results of inquiry-based instruction to the results of lecturing in a national sample of 6000 students taking courses at dozens of institutions. Some more comprehensive reviews, which have been useful for college faculty, have been provided by Johnson, Johnson, and Smith (1991, 1998).

Why did professors get used to lecturing?

Lecturing comes from an era where sources of knowledge, like books, were scarce. When a young Massachusetts minister left his library and half of his estate to a newly founded academic institution in 1638, the library was thought important enough so that the college adopted the benefactor's name: Harvard College. Only a century ago, William James, preparing his Gifford Lectures for 1901-1902, gave thanks Ato conversations with the lamented Thomas Davidson and to the use of his books, at Glenmore, above Keene Valley@ (James 1935/1902). In the 21st century, such an acknowledgment would be highly unlikely. However, traditions in education, perhaps most particularly in higher education, are rooted in the past when knowledge was scarce and access to it was difficult. In those past times, which cover most of human history, a lecturer who could organize knowledge and deliver it in an inspiring way was extraordinarily valuable.

Information is much more available now. Even by the mid-twentieth century, before the days of the internet, books had become cheap, common, and available -- through interlibrary loan if necessary. Many people find it impossible to read all of the books which they want to, or which are conceivably relevant to their current projects. A few scholars, primarily historians, need to chase off to remote libraries to find rare books. With the internet, the accessibility of information is much greater. The challenge for a learner is not simply obtaining information, but organizing it, making sense of it, and selecting what information to chase down. Using a lecture simply to transmit information is unnecessary if the same information is readily available in student-friendly form in a textbook.

Furthermore, if teaching is just seen as the transmission of information, the explosion of knowledge often leads to information overload in a number of courses. A well known study (Yager 1983) indicated that the number of individual concepts, each of which had an associated vocabulary item, taught in a high school biology course usually exceeded the number of vocabulary items in a foreign language course. A more recent version of Yager's study (Groves 1995) has come up with some less extreme numbers, the curriculum can still be heavily burdened by, for example, the 2,950 terms which Groves found in a leading high school chemistry curriculum. (A typical foreign language course, according to Groves, includes 1250 words in one year of junior high school and 2500 words in high school.) A recent discussion of heavily burdened curriculum is provided by a Project 2061 document (American Association for the Advancement of Science, hereafter AAAS 2001).

Is there still a place for the lectures?

Yes, there is. As I see it there are two places for a lecture: lecturettes, and occasional full-class lectures. Almost everyone who uses inquiry-based teaching uses lecturettes. In a lecturette, an instructor talks nonstop to all of the students in a class for a short period of time. Students need to be guided to the appropriate information before their minds go to work. If a number of groups become stuck on a particular point, it is often helpful for the instructor to intervene and speak for several minutes to try to help these groups over their problem. Once group work is done, the instructor needs to summarize what the groups learned, or should have learned, from the activity. Most importantly, lecturettes can connect the activity to the overall theme of the class.

How long should a lecturette be? Research in my classrooms (see, e.g., Shipman and Duch 2001) confirms some much older work by Verner and Dickinson (1967) which indicates that audience attentiveness tends to fade after 15 minutes. Indeed, my experience with somewhat large classes (70 students and more in the room at the same time) indicates that it is generally best to change what the students are doing, no matter what it is, after 15 minutes. The similarity of this 15-minute time interval to the average time between TV commercials on broadcast shows in the U.S. seems interesting, though I know of no evidence for cause and effect.

Furthermore, there are some classes where it seems most reasonable for an instructor to synthesize a lot of material and simply present it. For example, in my introductory physical science course, I wish to briefly give my students a glimpse into a vision of the future of technology in society. I have not found any books or articles which are sufficiently brief and engaging to do the job. So I lecture for one class. I do provide students who are interested in this topic access to books and articles.

PATHS TO REFORM ON CAMPUS

With all of these reasons to change, what is actually happening on college campuses? Many professors across the nation are changing their practices to include a variety of inquiry-based techniques in their teaching, either supplementing or in some cases nearly replacing the traditional full-class lecture. In this section, I'll begin with a brief description of the kinds of changes that are taking place on the campus I know best B the University of Delaware. The University of Delaware is particularly well known for a particular type of inquiry learning, namely Problem-Based Learning (PBL). I will then fit the changes at this particular institution into the larger national and international picture.

What Goes On In Class

In the fall of 2001, I took a few minutes to make a brief but systematic survey of 18 different classrooms at the University of Delaware. In seven of these classrooms, students were either sitting in groups talking to each other or listening to a student present a brief report on behalf of a group. This picture is quite different from what would have been found a decade earlier, where the vast majority of our classrooms would have contained a professor at the front of the room lecturing and students either listening or not listening to what the professor said..

What are the students doing in these groups? In my classroom, students are asking and answering questions. The starting point of their conversation could be a problem which I pose to them. They could also be reacting to an experiment or demonstration which I did in front of the class or some assignments which they completed before class. Occasionally they might be working on more extended group projects, or on experiments which they do in "lecture" (Shipman 2001). When they are done, all groups in my class will prepare a brief written report which they hand in at the end of class. The results of some group conversations will be shared with the class as a whole when I summarize the activity and connect it to some of the larger curricular questions which we are answering. Later on in this chapter I present some specific examples of activities and some references to books and articles which describe more things that students can do when they work in groups.

Since a number of faculty at our campus actually talk to each other about what we are doing in our classes, I know more than one might expect about what's going on in the other 140 different classrooms on our campus. I work quite closely with about a dozen different faculty from many different disciplines including science education, the other natural sciences, philosophy, English, and physical education. Each of us has developed our own version of inquiry-based pedagogy, often but not always based on problem-based learning, which best fits our discipline, our class level, and class size. We have learned a lot from each other and also fro our colleagues in the schools of education and business who have been doing some of the same kinds of thing for a long time. In all of these classes, students are working with each other in groups, learning from each other under the guidance of a professor.

How widespread are the pedagogies of engagement? A variety of data sources indicate that slightly over half of the faculty on our campus are using some form of inquiry-based learning in at least some of their classes. Institutional surveys of faculty and students (see, e.g., Bauer 2001) confirm the widespread nature of these practices. The data from the surveys is consistent with the informal yet systematic survey which I referred to above (Shipman 2002). Faculty surveys, student surveys, and classroom surveys are all subject to different sorts of limitations and biases. The description of the prevalence of this practice as involving "slightly over half" of the faculty is only an estimate but a reasonably robust one.

We came to this point as a result of a number of individual efforts that began in the early 1990s. Several individual faculty started questioning the effectiveness of the lecture mode. We began meeting for lunch to talk about teaching. A small, internally funded grant from the campus's Center for Teaching Effectiveness sent seven faculty to the University of New Mexico to participate in a workshop on problem-based learning (PBL), led by New Mexico's medical school faculty. Another institution-wide reform grant from the National Science Foundation initiated a campus-wide workshop which combined sessions on PBL and other inquiry-based teaching modes and sessions on technology. By the summer of 2002, the campus-wide workshop became transformed into an international conference with attendees from as far away as England and Singapore. The stories of how this change came about have been described from several viewpoints by Duch, Groh, and Allen (2001), Watson and Groh (2001), and Cavanaugh (2001).

Inquiry Learning on Other Campuses

The voluminous number of papers, publications, and websites devoted to various forms of learning that goes beyond the lecture mode certainly suggests that something is happening in higher education. Try using any of the widely available search engines to look for web pages which include terms like "inquiry learning," "group learning," or "Problem-based learning" and you will find a lot of information about different professors' practices. Well respected organizations like the American Association for the Advancement of Science (AAAS 1989, 1991, 2001), the National Academy of Sciences (1996, 1998), and the National Science Foundation (1996) have used their prestige and resources to push the pedagogies of engagement. The National Science Foundation in particular has emphasized work by interdisciplinary teams of people in their grants programs. Informal national alliances like Project Kaleidoscope (Narum 2001) and disciplinary-based conversations such as the American Astronomical Society's dialogue on the introductory astronomy course are other indicators of change.

How much of this actually reaches down to the classroom level is less clear. An illustrious panel, meeting with the name of the late Ernie Boyer and including major college presidents, national education leaders, and a Nobel Laureate (Kenny et al. 1998), put together a list of institutions which were then seen as examples to emulate. This list includes brand-name colleges like Harvard, Princeton, MIT, and Chicago, as well as less exalted places like the University of Delaware, Rensselaer Polytechnic Institute, Maryland, and Iowa. Surveys Every university that I know of has at least a few faculty who are using inquiry in their teaching.

WHAT INQUIRY-BASED LEARNING IS

In principle, the definition of inquiry learning stems from the use of inquiry as the basic research tool of the scientific enterprise. This enterprise has been one of the most successful in human culture. When you include technology as being closely allied with the scientific enterprise, its importance becomes even greater. So inquiry is seen as being what scientists and engineers do. The notion of scientific "habits of mind" was central to American Association for the Advancement of Science's *Science for All Americans* (AAAS 1989), and is also seen in the National Academy of Sciences' s *National Science Education Standards* (NSES) (National Academy of Sciences 1996). The notion of scientific inquiry has also become part of many state standards, though not all states include it explicitly (Blank and Pechman 1995).

Inquiry in the National Science Education Standards

A careful reading of the NSES document reveals a very careful and important distinction between doing inquiry and teaching inquiry. In the classroom, "doing inquiry" is not simply replicating what scientists do. Rather, it means that students involve their hands and minds in a carefully planned activity which, if carried out thoughtfully, will improve student understanding. As the standards indicate (p. 113), activities which "demonstrate and verify science content" are not particularly well suited to doing inquiry. More suitable are activities in which students themselves seek to answer questions, either questions posed by an instructor or questions which the students develop themselves. The NSES identifies five themes which characterize the scientific process, namely:

- systems, order, and organization
- evidence, models, and explanation
- change, constancy, and measurement
- evolution and equilibrium
- form and function

"Teaching inquiry" represents what teachers do in their classrooms in order to facilitate the doing of inquiry by their students. Most importantly, teachers select and adapt activities for their own classrooms. The adaptation process is crucial; virtually all of the teachers whom I talk to and whom I teach in my in-service classes and workshops find it extremely difficult to simply lift an activity out of some book and use it straight off. Another dimension of teaching inquiry is what a teacher does in an individual class period. A teacher needs to introduce an activity and show how it connects to the overall theme of the class. When students are doing the activity, the teacher needs to monitor what students are doing, react to individual questions, and address the whole class when needed if many groups become stuck on a particular problem. Once the activity is complete, a teacher needs to summarize for the whole class what the students have done and, again, the important connection between the activity and the larger questions that the class addresses.

While the rest of this chapter will focus on teaching inquiry, it is not really possible to describe some individual activities without also describing what students are doing and why they are doing it. Put differently, the distinction between doing inquiry (what students do) and teaching inquiry (what teachers do) is just a little too superficial to be maintained rigidly in practice. What students do is strongly influenced by what teachers do.

The standards documents are often of only limited assistance when it comes to translating inquiry, whether it be teaching inquiry or doing inquiry, into the classroom. NSES does contain a number of delightfully described examples, some of which I've used directly in my own classes. But the examples are too specific for general classroom use. For instance, the NSES contain a really inspiring vignette about teaching the pendulum (NSES, pp. 145-147), which I have used as the basis for an activity which my students do in lab. However, an instructor who cannot simply lift this activity from the NSES book and use it in exactly as presented. To adapt an activity, instructors should have some understanding of what inquiry is about.

Teaching Inquiry and the Nature of Science

A particular concern of the chapters in this book is that the inquiry-based pedagogy incorporate some aspects of the nature of science (NOS) as part of the instruction.

Science is, after all, a significant part of human culture, especially if the "science" curriculum includes the allied area of technology. One of the reasons we teach science to all students is the intellectual motivation for education, that science is an important part of what makes humans special. Another reason we teach science to all students is that the scientific way of thinking and solving problems is something that students should know how to do. It is probably obvious that students who end up in technical occupations will need to know how scientists and engineers solve problems. However, an increasing number of students are in occupations that do not require work-related technical training per se, but do require a technical background. One of my liberal arts students, who is selling cable system installations, neither has nor needs the detailed training that his installers have, but he has commented to me that the technical background he received in science courses like the one he took with me has been very useful in his job. For all of these reasons, we wish to incorporate NOS into our inquiry-based lessons.

But what aspects of the nature of science are the most important in teaching? This is a big question with no universally accepted answer. Over my teaching career I have developed some provisional answers to this question, and these answers will shape my selection of activities to describe in the remainder of this article. But particularly in the past few years, as my understanding of NOS and its relation to science teaching has become deeper, my ideas of what I would like my students to do have grown, sometimes well beyond what can be done in a one-semester course!

The nature of evidence and its relationship to explanation has remained as the most important NOS issue in my courses consistently over a period of thirty years. Virtually all of my teaching has been in courses where most students are not planning to be future scientists. I remain puzzled by the ability of the average American B and the average citizen of the world B to accept appropriately packaged yet dubious claims. The American circus showman P.T. Barnum may have claimed that "there's a sucker born every minute." (Barnum may not actually have said this; see Saxon 1989.) An important job for teachers is to transform their students, who may be suckers or people who readily accept claims, into rational, reasoning citizens with good baloney detection skills. Evaluating evidence may be particularly important in the Internet era where filters like editors and publishers have more limited influence.

An understanding of NOS goes beyond understanding what evidence is and how it is used. My courses often focus on some of the great explanatory ideas that have permitted us human beings, located on a comparatively tiny planet at the edge of a rather ordinary galaxy, to understand a great deal about the entire Universe and how it works. How has life on earth evolved to become the wondrous complexity that it is? How did the Universe evolve, and how did cosmic evolution result in a planet, our Earth, which is capable of supporting life? How do the elements in matter come together to make an enormous variety of materials, including the materials of life, possible?

While these big questions are interesting in their own right, I believe that an important component of a science course is not simply providing the currently accepted answers to these questions but also focusing on why scientists have grown to accept the essential parts of these answers as core scientific knowledge.

Developing student understanding of the paths that link the data that we have to the answers to these big questions has been an important component of almost all of the courses I teach (see, for example, Shipman 1980/1976). A comprehensive study of one of my courses has revealed that I have some success in this area (Brickhouse, Dagher, Letts, and Shipman 2000).

There are of course many different aspects of the nature of science that could be included in a science course. For instance, multidisciplinary courses like the ones I teach naturally lend themselves to some discussion of the different ways in which scientists in different disciplines gather data. Via appropriate selection of topics, courses can deal with the multicultural nature of science. I know of science courses which have dealt with the whole social structure of science, comprising scientific societies, journals, and journal editors, in a fairly explicit way. One of my colleagues teaches a biochemistry course in which students go back to the original literature and analyze scientific papers, paying appropriate attention to the context in which they were written. I team-teach a course on science and religion and include a small bit of this topic in some of my science content courses (Shipman, Brickhouse, Dagher, and Letts 2002).

An instructor needs to keep in mind the limits of a one-semester or even a one-year course that is still supposed to focus on science content. If too many different NOS topics are included in such course, then such rich concepts as "theory," "evidence," "explanation," "model," and "acceptance" can end up being just more foreign words in the course whose incomplete one-sentence definitions are memorized, regurgitated on exams, and then forgotten. Teachers and curriculum developers need to select a few aspects of the NOS for a course and really do them well.

How is inquiry related to teaching the NOS? All of the aspects of the NOS that are mentioned above can be taught very well using inquiry-based pedagogies. The relationship between evidence and explanation, in particular, is a natural topic for inquiry teaching. Many activities ask students to do inquiry by gathering evidence and evaluating explanations themselves. Other activities, including some of the ones given later in this chapter, illustrate the relationship between individual experiments and some of the big, important ideas in science more explicit. By this time I have become so accustomed to inquiry-based teaching that I find it hard to imagine ways to include these ideas in courses that do not involve using the techniques of inquiry. While there undoubtedly are some other ways to teach this material, they may not be as effective.

MANY WAYS TO TEACH: THE PEDAGOGIES OF ENGAGEMENT

Figure 1 is an attempt to illustrate the variety of pedagogies that include inquiry-based teaching along with some other pedagogies that are commonly used in college classrooms. It does require some discussion.

To start with, student learning is placed in the center. A principal characteristic of 21st century classrooms and all of the advice that is currently being given to teachers at every level is that the focus is on what the student learns rather than what the teacher does. In the old days I sometimes asked for, or gave, advice on how to give better lectures. The topic of conversation was what the instructor should do: use more, fewer, or better slides to illustrate the lecture, be more explicit about outlining the lecture, change voice timbre, and so forth. However, now the emphasis has shifted from teacher behavior to student learning. And while student learning depends greatly on what the teacher does, both in structuring a course and in teaching individual classes, the bottom-line question is whether the student learned better rather than whether a lecture was better organized or delivered.

Traditional Transmission Teaching

Transmission teaching is put at the top because it is still the most traditional and it is still widely used. It is described as teaching rather than learning because most evaluations of it refer to the quality of the teaching rather than its effect on students. Can one do inquiry learning using transmission-based teaching techniques? I used to think so B and I was supposed to be a pretty good lecturer; I won an excellence in teaching award as a lecturer. I would pose thoughtful questions that students were supposed to go home and think about. Based on current science education research, I now believe that what was mostly happening was that a few exceptionally well motivated students might have really thought about some of the thoughtful questions that I would ask. Most would ignore them, or possibly use them to try to figure out what questions would be on exams, which were then exclusively made up of multiple choice questions for my classes of 300-400 students. Put differently, I've decided that for me, inquiry learning is not compatible with a course where transmission teaching is the only or even the primary teaching mode that is used. Based on the evidence that we have regarding what actually happens in lecturing, there is good reason to suspect that this conclusion applies to most, if not all, college teachers.

On the right of the picture, skills-based learning is found in a growing area of higher education where students are prepared for particular professions. The mentored student teaching experience is an example of skills-based learning. As more and more students around the world participate in higher education, departments that prepare students for particular professional roles will be using this kind of teaching technique. I have personally experienced skills-based learning in recent years; most applicable to classroom teaching was the course which certified me as a scuba diver. Inquiry learning is not particularly applicable to skills-based learning. In scuba diving, for example, you would not want to spend ten minutes inquiring into what to do if you couldn't get air from your scuba gear. You simply want to know, instinctively and immediately, what the choices are that will permit you to take your next breath. Similarly, I find it hard to believe that inquiry learning would be an appropriate way to teach a student how to use a microscope. If a course were to address how the microscope was invented, that would be another matter.

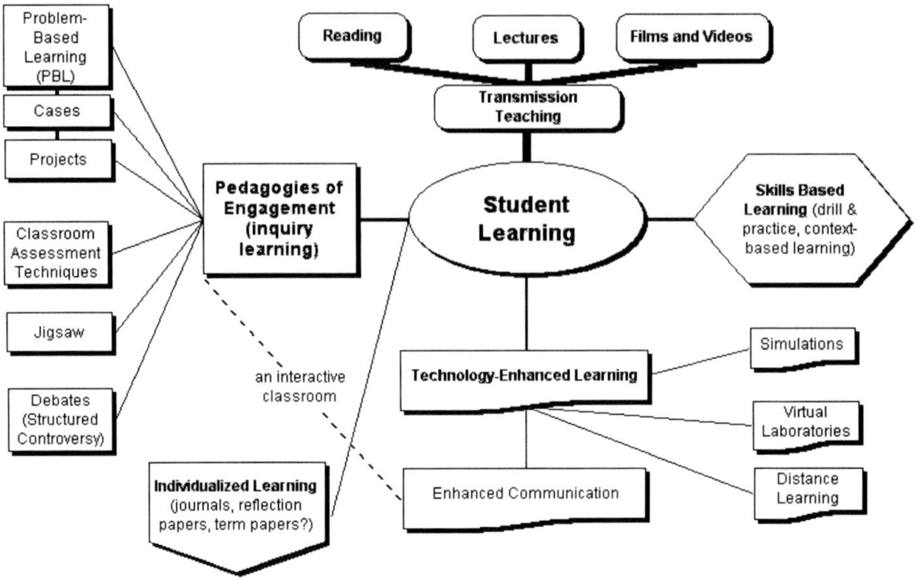

Figure 1. Different kinds of pedagogy in college classrooms. The thickness of the lines connecting the different categories is an indication of how well established these techniques are in current practice. For example, lectures, common practice for a century and a half at least, are firmly connected to student learning; technology-based learning, which is comparatively more recent, is connected with a thinner line. Most of the pedagogies of engagement are found on the left hand side of the picture.

Using Computers and Networks in Instruction

Technology-based learning is sometimes seen as an end in and of itself but really ought to be considered as a way of extending the reach of other forms of learning. Whether it is inquiry-based learning or not depends largely on how it's used. Its position in this picture is a reflection of the dominant way in which it has been used in higher education B as an extension of the lecture mode. Computers and networks have largely been used as a way to deliver lecture classes and other forms of information to people who by virtue of distance or time are not in the lecture room at the same time as the lecturer. Virtual laboratories often take a plain old cookbook lab and simulate it on the computer.

Technology does not have to be confined to the older pedagogical methods. The information superhighway can run from student to student, and from student to instructor, as well as in the traditional direction from the instructor to everyone else in the class. The educational use of this technology for two-way communication has, so far, been largely confined to the exchange of e-mail between professors and students. I've been e-mailing students since 1991, when e-mail was a new technology. Ordinary e-mail seems to be primarily used for rather routine procedural dialogues and less for discussion of meaningful issues. Some of us are using more sophisticated tools like WebCT and Blackboard, which offer the potential to permit students to work in groups, electronically. As an astronomer, I have worked in "virtual collaboratories" where groups of people write proposals and conduct investigations over e-mail for well over a decade. This practice is beginning to make an impact on many parts of the business world. If we can structure classes to persuade students to work in groups that meet electronically, then technology-based learning could become an important part of inquiry learning.

The Pedagogies of Engagement

The principal focus of this chapter are the teaching techniques at the left hand side of the diagram. Readers with even a passing familiarity with the literature may find themselves confused, as I am, by the plethora of descriptors used to describe a variety of closely related pedagogies, which Lee Shulman has called the "pedagogies of engagement" (Shulman 2000). When I started writing grants in this area in the mid-1990s I was careful to use the word "constructivist pedagogies" in all grant proposals since the funding agencies used that phrase in their proposal solicitations. Five years later, the phrase "inquiry-based" has displaced constructivism. The teaching techniques remain the same.

The first three inquiry-based pedagogies all challenge the student to do something. In problem-based learning (PBL), a popular pedagogy at my own institution, students are given a problem to solve and then asked what knowledge they need in order to solve it. The problem is a complex, real-world problem, often without a completely right or wrong answer. For example, "when did cosmology, the study of the evolution of the entire Universe, become a science?" is a question with many defensible answers, no one of which can be called "correct." The purpose of asking this question in an astronomy course is to get students to think about what science is in an applied context, rather than to have them pick the same date as some authority figure. Some problems such as the one I just mentioned are short ones, taking one class period. Other problems, such as asking physics students to analyze a traffic accident, can take up to two weeks for students to work through. (A traffic accident problem is described on the University of Delaware's PBL Clearinghouse, http://www.udel.edu/pblc.)

Cases, projects, and PBL form a set of triplets in the family tree of teaching techniques. They are not identical but are quite closely related. Some "problems," like the one just described, are distinctly problems in that the focus of student work is on the question that's asked and not on the surrounding story. Some cases are distinct in that the story is complex enough that it's not obvious at the onset whether there even is a clear question to answer. In a case, the situation is often a past one from the real world, and the story can be quite complex. The "question" to be answered is often more subtle. A case for advanced geology or biology students is based on what someone should do if he were a graduate student participant in the research group which claimed to find evidence for life on Mars. The case postulates that the conclusions of the research group disagreed with the students' own interpretation of the evidence. (For a real case of this sort, see the on-line library of cases at the University at Buffalo, http://ublib.buffalo.edu/libraries/projects/cases/case.html; search for cases on "Mars," and discover that the case just mentioned is one of three cases which deal with Mars.)

Projects ask the students to investigate something, often of their own choosing. Some very interesting projects ask student groups to design something such as a car that runs off the power of a mousetrap or a device that permits an egg to be dropped from a great height without breaking. Project selection is often a dilemma for student groups and can require a fair amount of consultation between the students and a sympathetic, knowledgeable instructor. The NSES contain several examples of possible projects for grades K-12, such as a class investigation of earthworms (pp. 34-35) and the egg drop design activity (pp. 162-163). I've used the egg drop activity in college classes with considerable success. A nice book which describes project-based science is Polman (2000).

For the practicing teacher, articulating the differences between cases, problems, and projects may be less important than locating ideas for all three types of activities, selecting some promising ones to try in a particular classroom, and gathering evidence to see which ones work in the teacher's class. A write-up of a particular activity can be much more useful if it contains at least some evidence that the activity worked in the author's class. In this chapter I will provide such evidence when I describe three activities in detail below. Defining cases as those activities found on the Buffalo website, problems as those found on the Delaware website, and projects as the kinds of activities described in Polman's book is an oversimplification but may be good enough for the practicing classroom teacher.

Classroom assessment techniques refer to the wonderful collection of short activities assembled by Angelo and Cross (1993). Their advantage, particularly for new practitioners of inquiry-based learning, is that they are very easy to use, and short enough so that mistakes in using them, however unlikely they may be, will have minimal negative impact on one's overall teaching. Many college instructors, including me, started with these techniques and then incorporated them into more conceptually complex activities. The last two entries refer to particular ways or organizing groups to facilitate discussion. In jigsaw activities, students are given different tasks, or different readings to investigate (see Angelo and Cross 1993, p. 65). When groups are assembled, each student has something different to bring to the group. In debates or structured controversies students are assigned particular positions to defend.

Individualized Learning

The pentagonal box at the lower left of Figure 1 lists a number of other things which students do as individuals in a typical college level course. (For American readers, this is the box that is shaped like baseball's home plate.) For many years, I followed the practice of my own professors and regularly assigned long term papers, believing that they would lead students to investigate some topic that was of interest to them in great depth. Even before such papers became widely available over the internet, I gave them up because I found that the vast majority of students treated them as survival exercises rather than as opportunities to investigate something that they cared about. Students who are future academics might well get something out of writing a 15-page paper on something. But the vast majority of our students, who even at elite schools will not become future academics, see these papers as hurdles to be surmounted B increasingly by buying them off the internet. These long papers also do not encourage the compact writing style which is the norm in business.

What my colleagues and I do more of is to give more short writing assignments that are individualized to a particular course and learning situation. When students write journals, they write down their reactions to the learning events in a course and share those reactions with the instructor. Reflection papers are similar to journals yet cover a longer term, sometimes the entire course. Both journals and reflection papers have become institutionalized in the University of Delaware's summer science courses for in-service teachers. My colleagues who teach writing tell me that asking for more short writing assignments is becoming the norm in their courses as well. These short writing assignments can often be well integrated with inquiry learning activities, and as such they are part of inquiry learning.

Any of the techniques described above and shown in Figure 1, even the pedagogies of engagement which lend themselves more naturally to inquiry learning, can be used in ways that either enhance or minimize the inquiry process. For example, a jigsaw activity could ask student groups to respond to lower-level questions which might emphasize recall. To be specific, an instructor could ask students to read the descriptions of cells in three different biology books and ask them to assemble a list of cell structures from each book and then compare the three different lists. If lists were the only thing that were asked for in such an activity, it would be difficult to see how it would really be inquiry learning because it really deals with a rather low level of thinking.

The common element in all of the pedagogies of engagement is that students are taking an active rather than a passive role in their learning. They are inquiring about something. Their minds are working at the same time as they are learning. If it makes sense to have a one-sentence definition of what inquiry learning is, then it is a form of learning which actively engages the student's mind. But a general definition of inquiry learning is only a starting point to helping teachers figure out how to use it, particularly with respect to using inquiry learning to teach about the nature of science. Where can teachers look in order to figure out what to do in tomorrow's class?

SOURCES OF INQUIRY-BASED PEDAGOGY

Curriculum designers and teachers can turn to a variety of sources for ideas as to how to design inquiry into their lessons. These can consist of prior practice, the philosophy of science literature, the cognitive science literature, and the practice of science itself. I will review what's available from each of these sources and will conclude with some observations on how this literature could be made more valuable to those who seek to incorporate the nature of science into their inquiry-based science lessons.

Prior Practice and Textbooks

In all levels of education, particularly in K-12 but also in higher education, what's written in the textbook is a strong determinant of what ends up in the classroom. Unfortunately, the textbooks often don't help much. Publishers seem inordinately reluctant to print and support innovative materials. For example, the classic and misleading drawing of the moon's orbit around the earth (see, for example, Snow 1988) is substantially identical to a similar drawing published in a textbook written a century earlier (Newcomb and Holden 1883). In this picture, the distance between the earth and moon is much too small, leading students to the logical but incorrect conclusion that the phases of the moon come from the shadow of the earth falling on the moon.

One difficulty in developing a reasonable picture of the nature of science from a textbook stems from the warmed-over logical positivism which is the espoused philosophy of science that appears in the beginning of textbooks at all levels, ranging from preschool through university. The positivist outlook sees data as being completely objective and the relationship between explanation and data as being totally rational and value-free. Hypotheses and even major theories come from the data by a process which is totally opaque to the student and is supposed to be driven by pure logic. Even in my much younger days (Shipman 1980/1976, chapter 1) I saw that there was an almost total disconnect between the positivistic picture of textbook science and the research enterprise which I was then somewhat familiar with. There are now much more complete critiques of positivism, stemming from Kuhn's (1970/1962) classic work and continuing through present-day treatments of the philosophy of science (see, e.g. a brief critique in Duschl 1990, pp. 33-37, and , Longino 1998). At its worst, this epistemic outlook presents the scientific method as consisting of five steps which must always be followed blindly. For example, a recent physical science textbook (see, e.g., Tillery, Enger, and Ross 2001, page 12) describes a five-step process as "the scientific method" as though one simple five-step process could describe all of science. The authors and their editors back off a bit with the next sentence, which softens the impact of the five steps by claiming that the exact approach a scientists uses depends on the individual doing the investigation as well as the particular field of science being studied.@ I doubt that many students or instructors will realize that there are at least two huge fields of scholarly inquiry, the history and philosophy of science, are exploring and reframing that last sentence.

A positivistic outlook on what science is can produce some rather strange definitions of inquiry. Some of my physics colleagues, who act like positivists when they are talking about teaching, think of inquiry as whatever students do in lab.

While students can do things in lab which genuinely are inquiry-based, it would be a travesty to think that many of the verification labs which are the core of most beginning science laboratory textbooks really represent inquiry teaching. Students can, for example, measure a well-known physical constant like the acceleration of gravity at the earth's surface. They follow procedures that are carefully laid out, like a cookbook. In the worst cases I have seen, they are then asked to compare their values to well-known ones and describe their "errors." Such a lab is not inquiry. If it's anything, it's a perfect way to use a behavioristic psychological model to convince students that they can't do science, by showing them that all they can up with is bad data.

Some Better Examples

Astute educators in search of inquiry-based materials can find them in the literature of they are quite selective. A few college-level textbooks which incorporate inquiry-based pedagogy, such as McDermott (1996) in introductory physical science. There are considerably more resources at the K-12 level, such as the collection of astronomy exercises edited by Andrew Fraknoi (Fraknoi 1996) and a series of wonderful books by Janice Van Cleave (see, e.g., Van Cleave 1989). A number of college teachers have adapted these exercises to their college classrooms (Shipman 2001, Zeidler 2001). I've already described the wonderful collection of techniques written by Angelo and Cross (1993); while this book is not specifically about science, many of the activities can be used for science.

Not all of these inquiry-based materials address the nature of science explicitly. The starting point of many of these activities is getting students to think and engage their brains, not just memorize. For example, the 340 students who dunked plastic drink bottles into baggies filled with ice, as described in Shipman (2001), only had as their goal convincing themselves that pressure decreased with decreasing temperature. I did not relate this particular curricular activity to Boyle's discovery of the behavior of gases. While I certainly could have, neither my article nor my teaching practice did. Similar comments apply to much of the inquiry-oriented pedagogy that's available. Its purpose is not to teach the nature of science explicitly, but to use constructivist pedagogy to teach science better. Later in this article I'll describe two activities from my own teaching which do explicitly include the nature of science.

The Philosophy of Science Literature

In the latter part of the twentieth century there has been an explosion of publications on the philosophy of science. One of the seminal works was by Kuhn (1970/1962). Following upon Kuhn's suggestion that logic alone cannot explain the way that science advances, a large number of investigators sought to figure out what does produce a changing scientific view of nature, if it's not just logic. Scholars whose works have resonated with my own research experience include Kuhn, Lakatos (Lakatos and Feyerabend 1999/1973), Laudan (1977), and Toulmin (1972). The writings of the senior biologist Ernst Mayr (see, e.g., Mayr 1997) have informed me that the philosophy of science is largely the philosophy of physics, and that a complete view of science has to include all the sciences, most particularly biology. In a similar vein, Sandra Harding (1991) and Nancy Brickhouse (2001), among others, have pointed out the effects of the limitations of gender and ethnicity on what we consider science. While I don't agree with everything he says, Michael Matthews (see, e.g., Matthews 1994) is provocative, appropriately constructively critical, and not hesitant to point out how, in his view, currently fashionable ideas may be leading us in the wrong direction.

However, the philosophy of science literature is addressed to the history and philosophy of science communities, not to practicing teachers and curriculum developers. A much too small literature has been written by a small group of science educators who seek to bridge the disciplines of history of science, philosophy of science, and science education. Examples of this literature are Duschl (1990), Lederman (1992), Mathews (1994), and particularly a book edited by McComas (1998a). Duschl's book contains some teaching sequences which could incorporate the nature of science into science instruction and which have been used in a science course at Hunter College. These activities are quite explicit in the way that they incorporate scientific epistemology into student work.

A particularly valuable set of materials are the 14 activities, some with several variations, described by Lederman and Abd-El Khalick (1998). Two of them were later adapted by a committee of the National Academy of Sciences and incorporated into a booklet which gives teachers of evolution some tools to help students understand that a theory is a really powerful idea rather than just a random guess (National Academy of Sciences 1998). I have used several of these activities in my class and thus have my own data to support the effectiveness of "Tricky Tracks," "Observing and Inferring," and "The Cube Activity (cubes 3 and 4)." Lederman and his colleagues have presented some very thorough evaluations of the effectiveness of a course which uses these activities (see, e.g., Akerson, Abd-El Khalick, and Lederman 2000).

Other activities from this literature, mostly from the McComas book, worth mentioning are one by Cobern and Loving (1998) in which students compare and contrast a number of different general statements about the nature of science (see a review by Lederman 2002). The book also contains an updated version of McComas's article about the "15 myths of science" in which he lists and fully destroys a number of popular misconceptions about how science works (McComas 1998b). McComas has published earlier versions of this argument in several other places which were nice, provocative, and well placed, but did not permit the depth of discussion that is found in the article in his book. Smith and Scharmann (1999) describe an activity where students evaluate a fictitious theory about umbrellas, and show that it is particularly useful in biology classes where using evolution as an example of a theory to be discussed often fails because many students have strongly held, pre-existing attitudes about it. Their exercise is apparently well known in the philosophy community but seems to be new to the science education community.

At least at the university level, one reason that the history and philosophy of science literature has a limited impact is the limited presence of this material in undergraduate science curricula. Even the most seminal thinker, Thomas Kuhn, is barely on the radar screen of many university-level scientists. Kuhn himself remarked on this in a three-day discussion he participated in in 1995 (Baltas, Gavroglu, and Kindi 2000):

> [Structure of Scientific Revolutions] was of course not widely read by scientists. I used to say that if you go through college in science and mathematics you may very well get your bachelor's degree without being exposed to the Structure of Scientific Revolutions. If you go through college in any other field you will read it at least once. That was not altogether what I had wanted. [emphasis in the original]

A few years ago, I surveyed my astronomical colleagues to verify Kuhn's suspicion that scientists did not widely read him (Shipman 2000). He was right B the majority of a randomly selected sample of astronomers did not even recognize his name! An interesting and unanticipated result of my survey was that for those astronomers who had read Kuhn, he had some very interesting and almost always positive influence on their ability to plan their careers. Clearly we have a long way to go before the majority of teachers of university science courses are familiar enough with the original literature in history and philosophy of science to be able to draw from it in any meaningful way.

Science Itself

I have found that reflection on my own experience in the practice of science has proven to be a rich source of activities which incorporate the nature of science. A question I keep asking in these activities is "How do we know what we know?" "What is the connection between evidence and explanation?" "Why is a particular explanation so widely accepted?" The connection between evidence and explanation might seem so basic as to be unnecessary at the college level, but the science education literature, the critical thinking literature, and the experience of many college teachers suggests otherwise.

Teachers who are not working scientists can use a whole variety of autobiographical or biographical books written by scientists. For example, Jonathan Weiner (Weiner 1994) provides an excellent description of the recent work on Darwin's Finches by Rosemary and Peter Grant, including the implications of this research for our understanding of evolution. In my experience, students who read that book can understand why the Grants and their students made the careful observations of the finches that they did, and how those observations show evolution in action, with significant changes in the finch population occurring on time scales of a few years. Similarly useful books that I have found in courses that I have taught or taken recently are by Mather and Boslough (1996) on the microwave background radiation and Trinkaus and Pat Shipman (1993) on the Neandertals.

A challenge that I have not always met successfully is taking books like these and going beyond the simple, basic, and important question of the connection between evidence and explanation. Science as an intellectual enterprise is more than just a collection of explanations of isolated facts. I find myself agreeing with Duschl (1990) in concluding that the larger explanations, which provide a much more comprehensive view of the Universe, are an essential part of science. Darwinian evolution, for example, is one of the key ideas in biology. It unifies the life sciences in a way that goes well beyond any particular explanation of, for example, the appearance of the peppered moth and its changes over time.

Critical Thinking

For nearly twenty years, a small group of aficionados has been conducting summer workshops on critical thinking at Sonoma State University. None of the three workshop principals is a scientist. The founder of these workshops, Richard Paul, is a philosopher. The other principal presenters have backgrounds in educational psychology (Linda Elder) and philosophy (Gerald Nosich). So why am I mentioning their work in an article about science education? Despite the predominance of philosophical language in the workshops and the literature on critical thinking, (see, e.g., Paul and Elder 2001), many of their ideas can help shape lessons in science. The discipline is particularly useful for understanding the interpretation of evidence and the elements of logic which may (or sometimes may not) be present in student explanations. The literature, and the workshops, are quite accessible, since authors are sensitive to the needs to reach people in all disciplines. I know of one college (King's College, Wilkes-Barre, Pennsylvania) where faculty in all disciplines have adopted the critical thinking paradigm as a way of focusing the development of logical habits of mind in all of their courses. These logical habits of mind include the same habits of mind that are found in science standards documents, though they extend beyond even the most flexible definition of what constitutes science. While I suspect that few colleges, universities, or high schools are sufficiently cohesive to make it practical for an entire institution to focus on one particular paradigm to describe logical reasoning, the King's College example can serve as an indication of how widely applicable the critical thinking model can be.

SOME SAMPLE CURRICULAR MATERIALS

What can an individual instructor or curriculum designer make of the wide variety of sources of ideas and organizational structures for designing lessons? What would inquiry learning look like in practice, in a real classroom? In this section of this chapter I describe three teaching sequences which I have successfully used in my own college classrooms. Two of them have been used in my college astronomy course ("Black Holes and Cosmic Evolution", which I'll cal "Black Holes" in what follows). This course is for general audiences, so students in the classroom have no particular background in science and have an interest which is at most avocational. The course has been more extensively described in Brickhouse et al. (2000). The third exercise has been used in the "Physical Science and Technology: the Way The World Works" course, more extensively described in Shipman and Duch (2001). The student audience is generally similar to the audience in "Black Holes," but with an interesting difference. "Physical Science" is a required course for elementary teacher education majors, and depending on the year in which it is taught, the preservice teachers are sometimes the dominant student population in the classroom. "Black Holes" is taken by a few pre-service teachers who have elected to make science their area of concentration. The second and third teaching sequences have also been used in workshops for middle and high school teachers. All three teaching sequences have been used in large lecture rooms with more than 200 students sitting in forward-facing seats. They work much better with smaller classes. (See Shipman and Duch 2001 for a discussion of class size.)

Big Bang and Steady State Theories

The "Black Holes" course does cover the Big Bang Theory as part of cosmic evolution. The usual way of describing cosmology in such a course is to describe the Big Bang Theory, which is currently accepted by the vast majority of the astronomical community. (I believe that the dissenters in the professional community who are still alive can be counted on the fingers of one hand.) This theory explains a wide range of astronomical phenomena by postulating that the Universe we know had a hot beginning approximately 14 billion years ago, when everything that we can see was compressed into a much smaller volume of space and was hot and dense. It has been expanding ever since. One of the most important phenomena that the Big Bang Theory explains is the microwave background, which is seen in all directions in the Universe with almost perfectly equal intensities. When the Universe was hot and dense, it was full of light. The microwave background that we see is that early cosmic light, shifted into the microwave spectral region by the expansion of the Universe.

An alternative explanation, which was quite popular about 40 years ago, is called the "steady state theory." (I wish it was called "hypothesis" or "model" or "proposal" or "idea" instead of theory, and I wonder whether I should follow astronomical tradition or whether I should unilaterally, in the classroom, use a term which will be less confusing to the students.) In this article I'm at least removing the customary capital letters. The steady state theory postulates that matter is being continuously created in intergalactic space, and that an interaction between this new matter and the matter which was there already explains the expansion of the Universe. It is not able to account for some other cosmological phenomena, the most important of which is the microwave background radiation.

In my classroom teaching sequence, I introduce the Big Bang Theory first, and then spend considerable time describing the evidence which supports it. The discovery and confirmation of the microwave background is an interesting story in its own right. In some years I have gone into this story in some detail, including the conflicts between some of the scientists who were responsible for the spacecraft mission that confirmed the nature of the background as a Big Bang fossil (Mather and Boslough 1993). I also mention some other pieces of evidence that support the Big Bang Theory, such as the discovery that all cosmic objects began their life cycles with virtually the same amount of helium, and the fact that the oldest stars and the Universe are approximately the same age.

At this point in the course, rather than go even further into the poorly understood detailed aspects of the Big Bang Theory such as how galaxies formed, I introduce the steady state theory. It takes about 10 minutes of class time at most -- I described its essence above in one sentence. Students in the class then work together in groups and identify pieces of evidence that support the Big Bang or Steady State theories. In some years, I have even challenged the class to put themselves in the position of citizen advisors to NASA and choose a mission which could place our understanding and acceptance of the Big Bang Theory on even more solid ground. (For details on some of these teaching sequences, see the course website, http://www.udel.edu/physics/phys145).

Did it work? The purpose of this teaching sequence is to bring students to the point where they understand the connection between the Big Bang Theory and the evidence that supports it. An extensive study of this course (Brickhouse et al. 2000) concluded that most students in the course did understand the importance of evidence to the Big Bang Theory (see Brickhouse et al. 2001). As far as connecting evidence and theory, the only difficulty which we saw in the student data was that students found that measuring anything in space, where whatever is being measured is hundreds, thousands, or millions of light years beyond earth, is inherently considerably more indirect than they are comfortable with.

However, any teaching sequence can be improved, and thanks to our intensive study of my course, I figured out how. My colleagues and I studied the 1996 offering of my course by collecting 4,000 pages of data from the entire group of 340 students, analyzing some of that data, and interviewing 20 of my students three times during the semester. My students still had a rather incomplete understanding of the general concept of theory (see Dagher et al. 2002). As a result of our investigation, the treatment of theory in this course has been modified. I replaced an intensive lecture on the nature of science with a series of group activities. The data show that students do perform better.

Why Things Fall

To me, one of the most astonishing results from our study of my course was finding out that 19 out of 20 college students had basically a third-grade understanding of gravity (Brickhouse et al. 2001), even after taking a course which described many astronomical phenomena in which gravity plays a major role in explaining them. They could not distinguish gravity as a phenomenon from gravity as a law or a theory. In elementary school, students learn that things fall. Birds and airplanes and orbiting planets, which do not fall, come later. Our study showed that the naive expectation that somewhere between 3rd and 12th grade students would have developed a slightly more sophisticated understanding of gravity was totally unfounded.

Consequently, in the fall of 1999 I developed a new teaching sequence about gravity and added it to the course. I first introduced the naive conception of gravity: “Things fall.” (This delightful two-word sentence was in an early draft of the State of Delaware’s science standards for third grade. It didn’t make it to the final text.) I suggested that there might be two explanations of gravity. One, which I dubbed the "Earth, Wind, Fire, and Water" hypothesis, is Aristotle's explanation. (I renamed it since some of my students are still familiar with the "Earth, Wind, and Fire" rock group; one year I even played a CD from this rock group as students were coming in to class.) The other is Newton's explanation. I had to add a rather elementary explanation of buoyancy.

I then showed the students that I had brought in some pieces of equipment that they could experiment with. I brought beakers, balls of various densities including a superball, water, alcohol, dirt, corks, and other objects. One year I also brought Karo syrup, which is a very high-density liquid in which many objects unexpectedly float. Cleaning up Karo is one of the meanest tasks in teaching. I didn't bring the Karo next year, but was told by a participant that putting the beaker in a spill tray might make the cleanup easier. Students then get together in groups and suggest experiments for me to do to test the Earth, Wind, Fire, and Water hypothesis. A reasonably careful choice of materials will produce a sufficient number of contradictions so that Newton's explanation quickly emerges as the most viable one. For example, superballs float in water and sink in alcohol. A sample of dirt which contains some leaves and sticks will separate in water into floating objects and sinking objects. There usually is enough time in a 75-minute class to deal with the experiments and then talk a bit about what makes an explanation successful enough that we would dignify it with the name of "theory."

Did it work? We did not study subsequent offerings of the "Black Holes" course with the extraordinary depth that was devoted to the 1996 course. In other words, the education research described in this paragraph is the kind of preliminary assessment that can convince a teacher that a particular instructional intervention led to an improvement in student learning. Of the 45 students in 1999 who chose to answer the question "Why do things fall?" 40 (89 %) gave an answer that recognized gravity as an explanation, not just as a phenomenon. In 1996, where this teaching sequence was not in the course, only 5 % of our interview sample gave a similarly acceptable answer. This preliminary assessment of the data is sufficiently positive to demonstrate that this teaching sequence is much more effective than the previous one. I did interview three students from the 1999 course and the findings from these interviews corroborate the results from the exam. Furthermore, our study of theory in the 1996 course (Dagher et al. 2002) indicated that student writings on examinations can be used as a valid source of data. The validity is enhanced by my practice of distributing a set of 5 essay questions in advance. Two of the five are on the exam and students can choose one of those two.

The Mystery Cube: A more general exercise on observation and inference:

One reason that the broader scientific and science education communities have begun paying attention to the nature of science is the continuing controversy in the United States regarding the status of Darwin's Theory of Evolution in the last two decades of the twentieth century. This chapter is not the place to review the controversy itself and the response of the biology community to it (for nice treatments of these topics see Miller 1999, Numbers 1992, Pennock 1999). Rather, my purpose in mentioning the evolution controversy is to provide a small bit of background to a remarkable book which has emerged from it, a National Academy of Sciences publication which contains some thoughtful discussions on evolution, the nature of science, and how to teach both topics (National Academy of Sciences 1998). The activity I describe below is adapted from the National Academy book and is also described in Lederman and Abd-El Khalick (1998). I have no idea whether the inventor was Rodger Bybee, who contributed it to the Academy document, Norm Lederman, someone else in the Lederman group, or someone else entirely. Both descriptions of it add some depth to its possibilities.

I start the activity by distributing a number of cubes around the classroom. The first cube is very similar to an ordinary die, with the numbers 1-6 on each face and arranged so that the numbers on opposite faces add up to 7. I am careful to place every cube on a student desk with the number 5 facing down. The rules of the classroom are similar to the rules of soccer: students are to keep their hands off the cube, and simply observe it. A series of questions guides students through their activities, which basically consist of observing the cube, recording their observations, processing their observations, and predicting what is on the bottom. Every group in the class comes to the firm conclusion that there is a 5 on the bottom. If I have time, I then distribute a more complex cube, where each face has a name and two numbers. Once again, there are patterns which students can recognize through observation and inference, and they can come to reasonable conclusions about what is on the bottom of the cube even if they can't see it. The most important part of the exercise is a concluding conversation, where students are asked to relate what they did with the cube to the nature of science in general.

The purpose of this activity is to convince the class that it is possible for rational people to accept the reality of something even if they can't see it directly. Historical sciences like paleontology, evolutionary biology, and cosmology often run into some criticism from critics who argue "how do you know what was happening millions (or billions) of years ago when someone wasn't there?" The purpose of this activity is to counter the criticism. I have co-presented this activity with some instructors who go so far as to never show the students what is on the bottom of the cube, using the class consensus as an example of how scientific knowledge is generated. I have also thought of manufacturing a cube where the pattern is broken -- where the bottom face reads 13, or π. I have resisted implementing these variations in my own classes, since I think that the need to form a bond of trust with the class outweighs any benefits that can be gained from being cute, clever, or stubborn. The activity is described in more detail in the National Academy of Sciences book, which is available on-line (http://www.nap.edu).

Does it work? The classes that I have used this in are sufficiently filled with other nature-of-science teaching sequences that it is very difficult to isolate the effects of this particular teaching sequence. A variety of student self-reports, including the simple evaluations done at the end of the activity and journal assignments which I collect over the semester, have shown that students believe that the cube activity teaches them about how science is done. My experience has been that these evaluations, while of limited value in determining exactly what is in the students' heads, can at least steer me away from activities which leave the instructor with a warm fuzzy feeling inside about the activity but with no other evidence of any kind that the activity is working. The literature on evaluations of courses by college students does show that when students evaluate an entire course, their evaluations are valid as long as student motivation and class size are adequately controlled or otherwise accounted for (see McKeachie et al. 1994). Consequently, I have found it reasonable to trust student self-reports on the effectiveness of particular teaching sequences, though like any other data source they should not be used blindly.

GENERAL CHARACTERISTICS OF THE TEACHING SEQUENCES

The teaching sequences illustrated above share some general characteristics. These characteristics are also present in the books mentioned in section 7.3 above and those compiled by Duch, Groh, and Allen (2001) and Siebert and McIntosh (2001). The most obvious one is that all of these teaching sequences ask the students to work together in collaborative groups, reflecting my own approach to the implementation of inquiry (see above). There two other common themes in these teaching sequences which emerge from my knowledge of what people at the college level are doing.

Empiricism and Evidence

All three of the activities above address the importance of the use of evidence in testing various explanations or hypotheses. The first activity (Big Bang and Steady State) asks students to compare two different explanations of the evolution of the Universe and use observational data to decide which model is correct. The second activity (Why Things Fall) deals with an experimental rather than an observational scientific situation in which students propose experiments and use the results to choose one explanation over another. In the third activity (the mystery cube), students make inferences from their own observations and come up with a testable prediction about something they don't see.

Evidence matters. One key aspect of the scientific habits of mind is that knowledge is based on data. This is so old in the scientific method B dating at least as far back as Francis Bacon and Galileo B that many philosophers of science do not think that the importance of evidence is even worth mentioning. It probably isn't worth mentioning; if someone is trying to puzzle out how scientists do science, it is probably not worth dwelling on something that was settled three hundred years ago. However, studies of college students, the general public, and certain political figures reveal that evidence is not always used in a consistent way in decision making (Sagan 1995).

Explanations and Theories

A concept that is emerging in the conversations that take place between philosophers of science and science teachers is that of the explanation. An explanation in science can be a very small one, like using the concept of evaporation to explain why a puddle disappeared. It can also be a major core scientific theory like biological evolution, the Big Bang Theory, or gravity. Inquiry learning activities can engage students in the concept of explanation at both levels. Which kind of inquiry is appropriate depends on the kinds of students you have. Disappearing puddles can be a wonderful way to engage first graders or kindergartners about the nature of science. It would take some ingenuity to use disappearing puddles in a course for college students, though it is probably appropriate for a science methods class addressed to future elementary school teachers. We know that the Big Bang Theory is sufficiently remote from everyday experience that it doesn't make much sense to introduce it to young children, but it is much more appropriate for college students.

SOME THOUGHTS ON THE FUTURE

What more do we need to know in order to make the teaching of the nature of science by inquiry techniques even better? This last section of the chapter is primarily addressed to scholars in the field, though practicing K-16 teachers can certainly contribute to these areas of knowledge, especially with regards to the first suggestion.

The most important barrier facing the practicing teacher is finding appropriate activities which require minimal adaptation in order to use them in the classroom. Teachers at all levels are busy people. While many do invent their own activities or extensively adapt those of others, the more user-friendly an activity is, the more likely it is to be used. Both as a practicing teacher and as a scholar of science education, I am struck by the relative shortage of activities which incorporate both active learning and the nature of science. The literature, especially at the elementary and middle school level, contains lots and lots of activities. However, the explanations invoked by these activities are usually pretty narrow. While the evaporation of puddles or the classification of different shapes of seeds is important for first graders, it would be nice to have teaching sequences which go beyond these relatively simple ones.

Another difficulty faced by scholars in this area is determining what aspect of the nature and philosophy of science is really important in instruction. The areas I have chosen in this paper B the importance of evidence, the nature of explanations, and the nature of theories B came largely from personal preference and conversations with my science education colleagues. However, they are also part of the emerging consensus in this area which have been presented at recent meetings of the National Association for Research in Science Teaching and the International History and Philosophy of Science in Science Teaching Group. We should keep in mind that attempting to reduce something complex like the nature of science to a list of a few items is a bit dangerous. We've tossed the five step scientific method overboard, but we don't want to replace it with the six-item list of essentials of the nature of science.

However, practicing teachers and curriculum developers need to start somewhere. It won't work for a science teacher to transform a science course into a philosophy of science course, and let the philosophy of science drive out all of the science content. You have to choose some aspects of the philosophy of science to include and some to leave out. A greater sense of what's really crucial and what is merely important would help. The standards documents are useful and, considering how they were informed from the organized community of scholars investigating the history and philosophy of science, they are a surprisingly good starting point. Their shortcomings, which historians, philosophers, and science educators describe in detail at conferences, usually deal with issues that lie beyond the scope of most science courses as they are taught now.

A third difficulty is few if any scholars have a full appreciation of the diversity of science. Historically, the philosophy of science has generally been the philosophy of physics and astronomy. The Copernican Revolution and the closely related area of Newtonian mechanics are often treated as the paradigm cases B and so even the physics and astronomy aren't particularly recent. There is a growing appreciation among philosophers that there are some important differences between physics and other sciences like biology. Even within biology, I have found that a biochemist's view of nature is often quite different from that presented by someone like Stephen Jay Gould who sees biology at the level of organisms rather than chemical reactions. The diverse contributions of different genders (see, e.g., Biklen and Pollard 1993) are another component of the diverse natures of science.

Even more radical but probably even more important new vistas are opened up when technology is brought into the curricular picture. In this discussion "technology" goes far beyond the use of technology to enhance learning and means human development of devices that make life easier. The science connection is that many, though not all, of these devices were based on our scientific knowledge. We needed to understand electricity to develop a computer that works, for example. Most students in K-16 classes are not going to become scientists and so many of them, even the engineers, should see how science connects with society through technology.

Different cultures see science and technology are seen as being much more closely linked together than is found in Western classrooms. For example, in many cultures a view of the living world is closely linked to their view of nature's plants and creatures as sources of food and of medicines. The multicultural nature of science has been extensively discussed in the science education community but the impact on classrooms, such as it is, has generally been at the level of convincing students that people other than European males can do science. It would be very interesting to develop a well focused activity which could introduce different cultures' views of science into the classroom in a way that would engage students' interest.

Still another factor to consider when deciding "what to teach" are the perspectives provided by the converging areas of developmental psychology, cognitive science, and neuroscience, popularly known as "brain science." There has been considerable discussion of the impact of these different fields on science education considered generally (e.g., Bransford et al. 2000). We don't know, but could in principle understand better, answers to questions like the following: At what age can students really grasp the concept of a scientific theory? When does it make sense to incorporate the notion of overarching core scientific theories like evolution and gravity? At what age can students generate their own meaningful hypotheses as part of a scientific investigation? How can we structure activities so that student-generated hypotheses are meaningful rather than words to fill a space on a storyboard at a science fair?

SUMMARY

The nature of science can be incorporated into a variety of inquiry-based teaching strategies. In this paper I've described a number of teaching strategies, possible sources for science activities, and described three such activities which have been used in college classrooms. A growing number of college instructors are starting to use such activities in their teaching. The times, indeed, are changing.

NOTE

I thank the National Science Foundation (most directly), the Pew Charitable Trusts, and the William and Flora Hewlett Foundation for financial support of various inquiry-based learning projects active at the University of Delaware during the time that I learned how to apply these pedagogies in college classrooms. I also thank my sister, Anne MacFarland, for use of her books, particularly the William James book (James 1935/1902) which I encountered at the Bradley House in Washington, Connecticut. My wife Valerie Bergeron, Larry Flick, and Norm Lederman provided comments on an earlier draft which clarified my thinking and, I hope, my exposition.

REFERENCES

Akerson, V.L., Abd-El-Khalick, F., and Lederman, N.G. (2002). Influence of a reflective explicit activity-based approach on elementary teachers' conceptions of the nature of science, *Journal of Research in Science Teaching,* (37), 295-317.

American Association for the Advancement of Science (AAAS). (1989). *Science For All Americans,* Washington, D.C., Author.

AAAS. (1991). *Benchmarks for Science Literacy*, Washington, D.C., Author.

AAAS. (2001). *Designs for Science Literacy*, Washington, D.C., Author.

Angelo, T., and Cross, K. P. (1993). *Classroom Assessment Techniques*, San Francisco, Jossey-Bass.

Baltas, A., Gavroglu, K., and Kindi, V. (2001). A Discussion with Thomas S. Kuhn. In Kuhn, T., *The Road Since Structure: Philosophical Essays 1970-1993, with an autobiographical interview,* in Conant, J., and Haugeland, J. (eds.), Chicago, IL, University of Chicago Press, 182-183.

Bauer, Karen W. 2001. A Two Major Assessment Efforts Under Way at the University of Delaware. *Assessment Update*, submitted for publication

Biklen, S.K., and Pollard, D., eds. (1993). *Gender and Education,* 92nd Yearbook of the National Society for the Study of Education, Chicago, University of Chicago Press.

Blank, R.K., and Pechman, E.M. (1995). *State Curriculum Frameworks in Mathematics and Science: How are They Changing Across the States?* Washington, D.C., Council of Chief State School Officers.

Boyer, E. (1987). *College: the Undergraduate Experience in America*, New York, Harper and Row, 145-150.

Bransford, J.D., Brown, A.L., and Cocking, R.R. (Eds). (2000). *How People Learn: Brain, Mind, Experience, and School* ,Washington, D.C., National Academy Press.
Brickhouse, N.W. (2001). Embodying Science: A Feminist Perspective on Learning, *Journal for Research in Science Teaching*,(38), 282-295.
Brickhouse, N.W., Dagher, Z., Letts, W.J. IV, and Shipman, H.L. 2000. Diversity of Students' Views about Evidence, Theory, and the Interface between Science and Religion in an Astronomy Course, *Journal of Research in Science Teaching,* (37), 340-362.
Brickhouse, N.W., Dagher, Z., Letts, W.J. IV, and Shipman, H.L. (2002). Warrants for Belief: Student Views of the Relationship between Evidence and Theory in a College Astronomy Course, *Science & Education*, forthcoming.
Cavanaugh. J.C. (2001). Make It So: Administrative Support for Problem-Based Learning, In Duch. B., Groh, S., and Allen, D.E.(2001).*The Power of Problem-Based Learning*, Sterling, VA, Stylus Publishing, 27-38.
Cobern, W., and Loving, C. C. (1998). The Card Exchange: Introducing the Philosophy of Science, In McComas (1998a), 73-82.
Duch. B., Groh, S., and Allen, D.E. (2001). Why Problem-Based Learning? A Case Study of Institutional Change in Undergraduate Education, In Duch. B., Groh, S., and Allen, D. E. (Eds.)., *The Power of Problem-Based Learning,* Sterling, VA, Stylus Publishing, 3-12.
Duschl, R. A. (1990). *Restructuring Science Education: The Importance of Theories and Their Development,* New York, Teachers College Press.
Dylan, B. (1985). The Times, They Are, a Changin', In Dylan, B. *Lyrics, 1962-1985,* New York, Knopf, 82-83. (Song originally written in 1962).
Fraknoi, A. (ed). (1996). *The Universe At Your Fingertips: A Handbook of Astronomy Activities and Experiments* , San Francisco, Astronomical Society of the Pacific.
Groves, F. H. (1995). Science Vocabulary Load of Selected Secondary Science Textbooks, *School Science and Mathematics,* (95), 231-255.
Hake, R.R. (1999). Interactive Engagement versus traditional methods: a six-thousand student survey of mechanics test data for introductory physics courses, *American Journal of Physics*, (66), 64-74.
Harding, S. (1991). *Whose Science? Whose Knowledge? Thinking from Women's Lives,* Ithaca, NY, Cornell University Press.
James, W. (1935/1902). *The Varieties of Religious Experience: A Study in Human Nature,* London, Longmans Green, 38th impression.
Johnson, D.W., Johnson, R.T., and Smith, K. A. (1991). *Cooperative Learning,* Washington, D.C., ASHE-ERIC Higher Education Report No. 4.
Johnson, D.W., Johnson, R.T., and Smith, K. A. (1998, July/August). Cooperative Learning Returns to College. *Change,* 27-35.
Kenny, S.S., Alberts, B., Booth, W., Glaser, M., Glassick, C.E., Ikenberry, S.O., Jamieson, K.H., O'Neil, R.M., Reid-Wallace, C., Tien, C-L., and Yang, C.-N. (1998). *Reinventing Undergraduate Education: a Blueprint for America's Research Universities*, (Published on the world wide web.) http://notes.cc.sunysb.edu/Pres/boyer.nsf
Kuhn, T.S. (1970/1962). *The Structure of Scientific Revolutions,* Chicago, University of Chicago Press.
Lakatos, I. and Feyerabend, P. (1999/1973). *For and Against Method: including Lakatos' s Lectures on Scientific Method and the Lakatos-Feyerabend Correspondence,* M. Motterlini, ed. Chicago, University of Chicago Press. (The lectures were given in 1973 and the correspondence dates from 1967-1974.)
Laudan, L. (1977). *Progress and Its Problems: Towards a Theory of Scientific Growth,* Berkeley, CA, University of California Press.
Lederman, N.G. (1992). Students' and Teachers' Conceptions of the Nature of Science: A Review of the Research. *Journal of Research in Science Teaching,* (29), 331-359.
Lederman, N. (2002). Review of W Cobern, Everyday Thoughts about Nature: A Worldview Investigation of Important Concepts Students Use to Make Sense of Nature with Specific Attention to Science, *Science Education,* (86), 591-593.
Lederman, N., and Abd-El-Khalick, F. (1998). Avoiding De-Natured Science, Activities that Promote Understandings of the Nature of Science, In McComas (1998a), 83-127.
Longino, H.E. (1998). Values and Objectivity, In Curd. M., and Cover, J.A. (1998). *Philosophy of Science: The Central Issues,* New York, W. W. Norton. pp. 170-191.

Matthews. M. (1994). *Science Teaching: the Role of History and Philosophy of Science*, New York, Routledge.

Mather, J.C., and Boslough, J. (1996). *The very first light : the true inside story of the scientific journey back to the dawn of the universe,* New York , Basic Books.

Mayr, E. (1997). *This Is Biology,* Cambridge. MA, Harvard University Press.

McComas, W., ed., (1998a). *The Nature of Science in Science Education: Rationales and Strategies,* Dordrecht, Kluwer.

McComas, W. (1998b). The Principal Elements of the Nature of Science: Dispelling the Myths, In McComas (1998a), pp. 53-72.

McDermott, L.C. (1996). *Physics By Inquiry,* New York, Wiley.

McKeachie. W. J., Chism, N., Menges, R., Svinicki, M., and Weinstein, C.E. (1994). *Teaching Tips: Strategies, Research, and Theory for College and University Teachers,* 9th edition (first edition published in 1951). Lexington, MA, D.C. Heath, chap. 29.

Miller, K.R. (1999). *Finding Darwin's God: a Scientist's Search for Common Ground between God and Evolution,* New York, HarperCollins.

National Academy of Sciences. (1996). *National Science Education Standards,* Washington, D.C., National Academy Press.

National Academy of Sciences. (1998). *Teaching About Evolution and the Nature of Science,* Washington, D.C., National Academy Press.

Narum. J. (2001). Building Natural Science Faculty Communities, In Siebert, E.J., and McIntosh, W.J., eds., *College Pathways to the Science Education Standards,* Arlington, VA, National Science Teachers Association,135-138.

National Science Foundation (1996). *Shaping the Future: New Expectations for Undergraduate Education in Science, Mathematics, Engineering, and Technology,* NSF Publication 96-139, Arlington, VA, Author.

Newcomb, S., and Holden, E.S. (1883). *Astronomy,* New York, Henry Holt.

Numbers. R. L. (1992). *The Creationists: the Evolution of Scientific Creationism,* New York, Alfred A. Knopf.

Paul, R., and Elder, L. (2001). *Critical Thinking: Tools for Taking Charge of your Learning and your Life,* Englewood Cliffs, NJ, Prentice-Hall.

Pennock, R. T. (1999).*Tower of Babel: The Evidence Against the New Creationism,* Cambridge, Mass., MIT Press.

Polman, Joseph L. (2000). *Designing Project-Based Science: Connecting Learners Through Guided Inquiry,* New York, Teachers College Press.

Provencal, J.L., Shipman, H.L., MacDonald, J., and Goodchild, S. (2001). *Twelfth European Conference on White Dwarf Stars,* San Francisco, CA: Astronomical Society of the Pacific Conference Series, vol. 226.

Sadler, P., and Schneps, M. (1989). *A Private Universe,* [Videorecording], available from the Astronomical Society of the Pacific, 390 Ashton Avenue, San Francisco, CA 94122.

Sagan, C. (1995). *The Demon-Haunted World*, New York, Random House.

Saxon, A.H. (1989). *P.T. Barnum: the Legend and the Man,* New York, Columbia University Press. See also _http://www.mccaddon.net/co.sucker.htm

Shipman, H.L. (1980/1976*). Black Holes, Quasars, and the Universe,* Boston, Houghton Mifflin.

Shipman, H.L. (2000). Thomas Kuhn' s Influence on Astronomers, *Science & Education,* (9), 161-171.

Shipman. H.L. (2001). Hands-On Science, 680 Hands at a Time, *Journal of College Science Teaching,* (30), 318-321.

Shipman, H.L. (2002). The Prevalence of Active Learning Strategies at the University of Delaware, Fall 2001. Available on the world wide web at http://www.udel.edu/~harrys/papers/udel.activelearning.classroomsurvey.htm

Shipman, H.L., Brickhouse, N.W., Dagher, Z., and Letts, W.J.,IV. (2002). AChanges in Student Views of Religion and Science in a College Astronomy Course, *Science Education,* (86), 526-547.

Shipman, H.L., and Duch, B. (2001). Problem Based Learning in Large and Very Large Classes, in B. Duch, S. Groh, and D. Allen, eds., *The Power of Problem-Based Learning*, Sterling, Va., Stylus Publishing.

Shulman, L. (2000). Keynote Address to the International Conference on Problem-Based Learning, Samford University, Birmingham, Alabama.

Siebert, E.J., and McIntosh, W.J., eds. (2001). *College Pathways to the Science Education Standards,* Arlington, VA, National Science Teachers Association.
Smith, M.U., and Scharmann, L.C. (1999). Defining versus Describing the Nature of Science: A Pragmatic Analysis for Classroom Teachers and Science Educators, *Science Education,* (83), 493-509.
Snow, T.P. (1988). *The Dynamic Universe: an Introduction to Astronomy,* St. Paul, MN, West Publishing.
Tillery, B.W., Enger, E.D., and Ross, F.C. (2001). *Integrated Science,* Boston, McGraw-Hill.
Toulmin, S. (1972). *Human Understanding. (Vol. I),* Princeton, NJ, Princeton University Press.
Trinkaus, E., and Shipman, P. (1993). *The Neandertals : changing the image of mankind.* New York , Knopf.
Van Cleave, J. (1989). *Physics for Every Kid.* New York, Wiley.
Verner, C., and Dickinson, G. (1967). The lecture: an analysis and review of research, *Adult Education,* (17), 85-105.
Watson, G.H., and Groh, S. (2001). Faculty Mentoring Faculty. The Institute for Transforming Undergraduate Education, In Duch. B., Groh, S., and Allen, D.E. eds., *The Power of Problem-Based Learning,* Sterling, VA, Stylus Publishing, 13-26.
Weiner, J. (1994). *The Beak of the Finch,* New York, Random House.
Yager, R. E. (1983). The importance of terminology in teaching K-12 science, *Journal of Research in Science Teaching* (20), 577-591.
Zeidler, D.L. (2001). Aligning Courses for Standards-Based Teaching, In Siebert, E.J., and McIntosh, W.J., eds., *College Pathways to the Science Education Standards,* Arlington, VA, National Science Teachers Association, 135-138.

CHAPTER 18

FOUAD ABD-EL-KHALICK

OVER AND OVER AGAIN: COLLEGE STUDENTS' VIEWS OF NATURE OF SCIENCE

INTRODUCTION

Helping students develop informed conceptions of nature of science (NOS) has been, and continues to be, a central goal for science education (American Association for the Advancement of Science [AAAS], 1990; Central Association for Science and Mathematics Teachers, 1907; National Research Council [NRC], 1996). Indeed, for the past 50 years, assessing and attempting to enhance K-12 students' and teachers' views of NOS have been the focus of intensive and wide-ranging research efforts (Abd-El-Khalick & Lederman, 2000; Lederman, 1992). However, relative to K-12 students' NOS views, less attention has been accorded to college science students' conceptions, which have been addressed only by a few studies (i.e., Bezzi, 1998; Fleming, 1988; Gilbert, 1991; Ryder, Leach, & Driver, 1999). Also, with the exception of Kimball (1967-68), attention to college students' NOS views has been relatively recent.

Exploring the meanings that college science students ascribe to various aspects of NOS is significant for several reasons. First, an understanding of NOS is a central component of scientific literacy, which is an espoused goal for *all* citizens in participatory democracies (AAAS, 1990; NRC, 1996). Second, as Ryder et al. (1999) noted, college science graduates pursue careers (including teaching K-12 science) that require an understanding of NOS. Third, research on teaching and learning about NOS in K-12 and science teacher education settings indicates that the assumption that students learn about NOS implicitly through engagement in science-based activities is not empirically valid (Abd-El-Khalick & Lederman, 2000). NOS understandings need to be treated as "cognitive" instructional outcomes, which warrant intentional and explicit instruction. Indeed, our research indicates that a conceptual change approach might be needed to help learners, including college students, question their long-held beliefs, and internalize more informed views about NOS (Abd-El-Khalick, 1998; Akerson, Abd-El-Khalick, & Lederman, 2000). Exploring the meanings that learners ascribe to NOS is a crucial first step to such instructional efforts. Thus, the purpose of this study was to explore the meanings that undergraduate and graduate college students ascribe to several aspects of NOS.

L.B. Flick and N.G. Lederman (eds.) Scientific Inquiry and Nature of Science, 389-425.

REVIEW OF THE LITERATURE

Fleming (1988) assessed undergraduate chemistry students' views on the interaction among science, technology, and society. Participants were administered the *Views on Science-Technology-Society* questionnaire (VOSTS) (Aikenhead, Ryan, & Fleming, 1989). Additionally, individual interviews were conducted with selected students. Fleming reported that participant undergraduate students' conceptions of NOS were very similar to the naïve conceptions held by high school students.

Gilbert (1991) used 12 statements to assess the NOS views of undergraduate college students (59% freshmen) enrolled in an introductory biology course. Six statements, which were derived from the *VOSTS* (Aikenhead et al., 1989) and *Nature of Scientific Knowledge Scale* (NSKS) (Rubba & Anderson, 1978), addressed the nature of scientific knowledge and research. The remaining six statements were analogously constructed to address the nature of models and model building. Participants either agreed or disagreed with a statement and provided a short answer to justify their choice. Each of the 687 participants was administered *one* of the 12 statements, resulting in 42-67 responses for each statement. With one exception, participants' views of scientific knowledge and models were congruent. About 67% of participants believed that while scientific models could be artificial, scientific knowledge depicts the way nature "really" is. Moreover, the greater majority of participants: (a) noted that scientific knowledge and models should be free of error in order to be accepted, (b) believed that scientists follow a universal scientific method, and (c) dismissed the role of creativity and social milieu in the generation of scientific knowledge and models.

Bezzi (1998) used the repertory grid method to elicit undergraduate students' images of geosciences. Participants were five first-year students in geology who completed grids that comprised a range of elements representing various disciplines (e.g., physics, biology, and geology) versus a set of bipolar constructs (e.g., inductive/deductive, scientific/humanistic, and objective/subjective). The results indicated that participants ascribed to a stereotypical image of scientific disciplines with an antithesis between physics, which was viewed as objective and rigorous, and geology, which was viewed as subjective and approximate. Also, participants did not appreciate the societal dimensions and issues inherent to the geosciences.

Ryder et al. (1999) examined the images of science held by 11 undergraduate students from four science departments working on a final-year research project. Participants were interviewed using a set of open-ended "stimulus" questions, which aimed to generate extended discussions about participants' views of science and the contexts upon which they drew in attempting to articulate those views. The study indicated that a majority of participants viewed knowledge claims as separate from data. However, most of them believed that scientific claims were "provable" on empirical grounds, and very few noted that knowledge claims actually go beyond the data. Additionally, very few participants admitted a role for social processes and factors in evaluating knowledge claims. Indeed, while many participants recognized the existence of social interactions among scientists, most were unable to explicate why such interactions are useful or important. Finally, many participants did not seem to appreciate the regulatory and directing role of the institutions of science in guiding and funding research, and almost all did not recognize the role of these institutions in validating scientific knowledge.

NOS

Philosophers, historians, and sociologists of science, and science educators are quick to disagree on a specific definition for NOS. The use of the phrase "NOS" throughout this chapter instead of the more stylistically appropriate "the NOS," is intended to reflect the author's lack of belief in the existence of a singular NOS or general agreement on what the phrase specifically means (Abd-El-Khalick, 1998). This lack of agreement should not be disconcerting or surprising given the multifaceted, complex, and dynamic nature of the scientific endeavor. Nonetheless, there is an acceptable level of generality regarding NOS that is crucial to the understandings of scientifically literate individuals and at which virtually no disagreement exists among experts (Abd-El-Khalick, Bell, & Lederman, 1998). Some of the aspects of NOS that fall under this level of generality are that scientific knowledge is tentative (subject to change), empirically-based (based on and/or derived from observations of the natural world), theory-laden, partly the product of human inference, imagination, and creativity (involves the invention of models and explanations), and socially and culturally embedded. Three additional important aspects are the distinction between observation and inference, the functions of, and relationships between theories and laws, and the lack of a single recipe-like method, "The Scientific Method," that guarantees the development of valid scientific knowledge (detailed discussions of these NOS aspects appear in the introductions to the various subsections of the "Results" section below). These NOS aspects, which were adopted and emphasized in the present study, have been emphasized in recent science education reform documents (e.g., AAAS, 1993; NRC, 1996). In this regard it should be emphasized that these NOS aspects are not conceived of as disparate, but rather as integral components of an epistemology in which scientific knowledge is produced through critical, negotiated, and collaborative processes that are propelled by scientists' imaginations and bound by their observations of nature.

PURPOSE

This study aimed to generate a rich, descriptive, and detailed anatomy or map of undergraduate and graduate college students' views of science and scientific knowledge. The specific questions that guided this investigation were: (1) What are participant college students' views of the aforementioned aspects of NOS? (2) Are participants' views related to academic and background variables, including gender, class standing, major, and science credit hours?

METHOD

Participants

Participants were 153 undergraduate and graduate students, 81 male (53%) and 72 female (47%), enrolled in three history of science courses in a mid-sized Western state university. Their ages ranged from 19 to 45 years (M = 24.4 years, SD = 5.1 years). At the time of the study, 1% of the participants were sophomores, 14% juniors, 75% seniors, and 10% graduate students. Most of the participants (79%) majored in one of the biological sciences (60%) or general science (19%). Three students (2%) majored in the geological sciences and two (1%) in the agricultural sciences. The remaining 18% of the participants had other non-science majors including economics, history, education, history of science, international studies, philosophy, political science, and psychology. Participants had completed an average of 67 undergraduate credit hours (SD = 36.5) in various scientific disciplines. Most of these credits (90%) were in the biological (33 credit hours, SD = 24.4) and physical sciences (27.2 credit hours, SD = 16.4). Graduate participants did not differ much in their science background from the undergraduates. Most of the graduates (about 73%) had just begun their graduate work at the time of the study.

Procedure and Instruments

Participants were administered the *Views of Nature of Science Questionnaire–Form C* (VNOS–C) (Abd-El-Khalick, 1998; Abd-El-Khalick, Lederman, Bell, & Schwartz, 2001). Next, a random sample comprising 38 participants (25%) was selected for follow-up semi-structured individual interviews. During these interviews, participants were provided their questionnaires and asked to explain and justify their answers. Follow-up questions were used to probe interviewees' responses in depth and pursue their lines of thinking. Interviews, which were 45 to 60 minutes long, were audiotaped and transcribed verbatim for analysis.

Using the *VNOS–C* in conjunction with follow-up individual interviews to assess participants' NOS views was undertaken with the intent of avoiding the problems inherent to the use of standardized forced-choice or convergent instruments, such as the *Test on Understanding Science* (Cooley & Klopfer, 1961), *Nature of Science Scale* (Kimball, 1967-68), and *NSKS* (Rubba & Anderson, 1978), which have been traditionally employed to assess learners' NOS views. These problems stem from the assumptions underlying the development of these instruments and their format, and cast serious doubt on whether such instruments generate valid assessments of respondents' NOS views. The open-ended nature of the *VNOS–C* items allows respondents to express their own views on issues related to NOS and alleviates concerns related to imposing a particular view of the scientific enterprise on respondents. Moreover, coupled with data from individual interviews, the *VNOS–C* allows the assessment of not only respondents' positions on certain issues related to NOS, but the respondents' reasons for adopting those positions as well (for detailed discussions of the issues associated with assessing learners' views of NOS and the use of open-ended versus convergent type assessment instruments see Abd-El-Khalick et al., 2001; Aikenhead, 1988; Aikenhead, Ryan, & Desautels, 1989; Lederman, Wade, & Bell, 1998).

VNOS–C

The instrument aimed to assess respondents' views of the tentative, empirical, creative, and theory-laden nature of scientific knowledge; the role of social and cultural contexts in science; observation versus inference; and the functions of, and relationship between scientific theories and laws. Four items were adapted from the *VNOS–A* (Lederman & O'Malley, 1990) and *VNOS–B* (Abd-El-Khalick et al., 1998) and five new items were developed by the researcher. Next, a panel of experts examined these items to establish their face validity. This panel consisted of five university professors: three science educators, a historian of science, and a scientist. The panel had some comments and suggestions for improvement and the nine items used in this study were modified accordingly. These items appear in the Appendix.

The *VNOS–C* asks respondents to elaborate/justify their answers and to support them with relevant examples with the intention of assessing the depth of respondents' understandings of the target NOS aspects. The instrument is writing intensive and the planned administration was expected to consume one instructional period were the questionnaires to be completed in the courses in which participants were enrolled. Due to time constraints, the participants completed the *VNOS–C* at home. Administering the instrument outside a controlled environment raised concerns regarding the validity of student responses. To mitigate, though not eliminate, concerns in this regard, participants were assured that there are no "right" or "wrong" answers to the items and that the researcher was mainly interested in their viewpoints on some issues related to science. Moreover, these concerns were further ameliorated by the fact that having no "right" or "wrong" answers, responses to the items did not lend themselves to being "looked-up."

Providing participants with instructions regarding the lack of "right" or "wrong" answers to the *VNOS–C* items necessitates a clarification. It could be argued, and understandably so, that these instructions typify a contradiction: On one hand, participants were assured that "there are no right or wrong answers" to the questions asked of them. On the other hand, it was argued that the NOS framework adopted in the present study represents consensus among experts regarding the target NOS aspects. In a sense, this latter framework will eventually be used to "judge" participants' NOS views. This *seeming* contradiction, however, should not be disconcerting. To start with, the above instructions reflect a methodological heuristic rather than an epistemological position: Participants were less likely to consult references or authorities while completing the *VNOS–C* at home if they were assured that there were no packaged answers to the questions at hand. After all, this study aimed to assess participants' own views on the target NOS issues. More importantly, these instructions were not disingenuous: There is a crucial difference between characterizing the adopted NOS framework as consensus-based versus being considered "right." The former characterization does not necessarily entail the latter.

A consensus-based framework implies tentativeness while simultaneously adhering to a critical, negotiated, and collaborative knowledge-building epistemological stance. According to this latter stance, certain claims about phenomena or events could be judged to be more accurate or valid than others even while fully realizing the lack of access to any final truths. This author is under no illusion that the following discussions regarding participants' "informed" and "uniformed" NOS views will be obsolete at some point in the future. However, this author adopts the position that, presently, some claims regarding the workings of the scientific endeavor are more accurate or informed than others.

Follow-up Interviews

Interviewees were provided their questionnaires and asked to read and explain their answers. Follow-up and probing questions were then used to clarify ambiguities in some responses, and ask for elaborations and justifications. Initially, these questions were not planned. However, following the first few interviews a common set of follow-up, clarification, and probing questions "emerged" and took form. These questions were asked of interviewees either as individual questions or "sets" of interrelated questions. Certain questions or sets of questions were asked following interviewees' explication of their responses to a certain *VNOS–C* item. Alternatively, other questions or sets of questions were *only* asked when interviewees expressed certain ideas regarding NOS.

In what follows, the notation *VNOS–C* #1, *VNOS–C* #2, etc., is used to refer to the first, second, etc., items of the *VNOS–C* (see Appendix). Many interviewees noted, often in response to *VNOS–C* #1, that science is characterized by the scientific method or other sets of logical and orderly steps. Upon expressing this idea an interviewee was asked, "Do all scientists use a specific method, in terms of a certain stepwise procedure, when they do science? Can you elaborate?" In their response to *VNOS–C* #2 many interviewees defined scientific experiments very broadly as "procedures" used to answer scientific questions. In the attempt to clarify such responses interviewees were asked, "Are you thinking of an experiment in the sense of manipulating variables or are you thinking of more general procedures? Can you elaborate?" Also, mostly in response to *VNOS–C* #1 and #2, many interviewees noted that scientific knowledge is "proven" knowledge or that scientific experiments aim to "prove" hypotheses or theories. Interviewees were then asked, "How would you 'prove' a theory or hypothesis?" A typical response was that scientific claims are "proven" by collecting evidence and/or doing experiments. Interviewees were then asked, "How much evidence or how many experiments does it take to 'prove' a scientific claim?" or "How much evidence and/or how many experiments are 'enough' to prove a scientific claim?"

Some interviewees noted, in response to *VNOS–C* #3, that developing scientific knowledge necessarily requires *manipulative* experiments. In an attempt to elucidate how this view relates to the case of "observational" sciences, interviewees were then asked a set of questions. The first question was, "Let's consider a science like astronomy (or anatomy). Can we (or do we) do manipulative experiments in astronomy (or anatomy)?" If interviewees answered in the positive they were asked to explicate their answers and provide examples. This served to further probe interviewees' conceptions of scientific experiments. However, if they answered in the negative, interviewees were then asked, "But we still consider astronomy (or anatomy) a science. What are your ideas about that?"

Other follow-up questions aimed to assess the depth of interviewees' understanding of the theory-laden NOS and the role that scientific theories and theoretical expectations play in guiding scientific research. Two of these questions followed interviewees' explication of their responses to *VNOS–C* #2 that focused on scientific experiments. The questions were, "When scientists perform manipulative experiments they hold certain variables constant and vary others. Do scientists usually have an idea about the outcome of their experiments?" If interviewees believed that scientists have prior expectations, they were then asked, "Some claim that such expectations would bias the results of an experiment. What do you think?" The other two questions followed *VNOS–C* #4 that related to scientific theories. On noting that scientific theories change, interviewees were asked, "The history of science is full with examples of scientific theories that have been discarded or greatly changed. The life spans of scientific theories, if you will, vary greatly, but theories seem to change at one point or another. And there is no reason to believe that the scientific theories we have today will not change in the future. Why do we bother learn about these theories? Why do we invest time and energy to grasp these theories?" The other question was, "Which comes first when scientists conduct scientific investigations, theory or observation?"

A question that followed interviewees' discussion of *VNOS–C* #5 was, "In terms of status and significance as products of science, would you rank scientific theories and laws? And if you choose to rank them, how would you rank them?" Two other questions followed when interviewees' responses to *VNOS–C* #6 on atomic structure were not informative regarding their views of the role of inference and creativity in science. The first question was "Have we ever 'seen' an atom?" If they responded in the negative, interviewees were then asked, "So, where do scientists come up with this elaborate structure of the atom?" Those interviewees who thought that scientists have actually "seen" an atom were asked to elaborate on their answers. Similarly, *VNOS–C* #7 aimed to assess participants' understanding of the role of inference and creativity in science. On noting that scientists were very certain about the notion of species, interviewees were asked, "There are certain species of wolves and dogs that are known to interbreed and produce fertile offspring. How does this fit into the notion of species, knowing that the aforementioned species are 'different' species and have been given different scientific names?"

To assess whether interviewees thought of creativity and imagination in science more as "resourcefulness" and "skillfulness" or as "invention" of explanations, they were asked, "Creativity and imagination also have the connotation of creating something from the mind. Do you think creativity and imagination play a part in science in that sense as well?" Finally, in response to *VNOS–C* #9, many interviewees thought that the dinosaur extinction controversy was justified given that the available evidence supports both hypotheses. In that case, the interviewer noted, "It is very reasonable to say that the data is scarce and that the available evidence supports both hypotheses equally well. However, scientists in the different groups are very adamant about their own position and they publish very pointed papers in this regard. Why is that?"

Data Analysis

The first phase of data analysis aimed to establish the validity of the *VNOS–C* using the interviewed participants' written and verbal responses. The questionnaires and corresponding interview transcripts of these participants were separately analyzed to generate two NOS profiles for the same group of students. Systematic comparisons indicated that the interviewees' NOS profile as generated from analyzing their questionnaire responses was *generally* consistent with the profile derived from analyzing their interview responses. In other words, the researcher's inferences regarding interviewees' NOS views as derived from their *VNOS–C* responses were generally consistent with the views they explicated during individual interviews.

Nonetheless, comparisons between the interview and questionnaire data allowed the researcher to gain some insight into the nuances of meanings that the participants attached to key terms, such as "theory," "prove," and "creativity," which they frequently invoked to convey their views of NOS. These insights helped clarify some ambiguous statements and led the researcher to reinterpret a few inferences regarding participants' NOS views. Additionally, as noted earlier, participants' views were probed in depth during the interviews. Thus, the interview data allowed the researcher to further refine some inferences regarding participants' views. The interplay between the questionnaire and interview data in generating valid representations of participants' NOS views will become evident as both data sources are used in the "Results" section to present the findings of the present study.

In the second phase of data analysis, all participants' questionnaires were examined. Each questionnaire was used to generate a summary of a participant's conceptions of the target NOS aspects. Each summary was coded under the variables relevant to this study, such as gender, class standing, science major, and science credit hours. Using a systematic process, these individual summaries were then analyzed to generate NOS profiles for various groups of participants (all participants; sophomores, juniors, seniors, and graduates; males and females; etc.). During this process, individual summaries were first searched for patterns or categories. The generated categories were then checked against confirmatory or otherwise contradictory evidence in the data and were modified accordingly. Insights gained from analyzing the interview data were constantly used to check the validity of the generated categories. Several rounds of category generation, confirmation, and modification were conducted to satisfactorily reduce and organize the data. Next, these categories were employed to generate a NOS profile for a certain group of participants. Finally, profiles were systematically compared and contrasted to answer the questions of interest.

RESULTS

Participants' NOS views were not related to their gender, class standing, major, or science credit hours. Thus, the following sections present a profile of all participants' views of the target NOS aspects. These aspects included the tentative, empirical, inferential, creative, and theory-laden nature of scientific knowledge, the lack of a single recipe-like method for "doing" science, and the functions of, and relationship between scientific theories and laws. In particular, the explanatory function of theories and their role in guiding research were highlighted. Three additional aspects emerged from analyzing participants' responses to the *VNOS–C* and interview questions. These were the aim and structure of scientific experiments, the logic of hypothesis and theory testing, and the validity of observationally (as opposed to experimentally) based scientific disciplines. Table 1 presents a summary of the major trends in participants' views of the target NOS aspects. Detailed discussions of these views are presented in the following sections.

The Empirical Nature of Scientific Knowledge

Science is, at least partially, based on and/or derived from observations of the natural world, and "sooner or later, the validity of scientific claims is settled by referring to observations of phenomena" (AAAS, 1990, p. 4). However, most natural phenomena are not "directly" accessible to the senses. Observations of the natural world are always filtered through our perceptual apparatus, interpreted from within elaborate theoretical frameworks, and almost always mediated by a host of assumptions that underlie the functioning of "scientific" instruments.

Table 1. A summary of major trends in participants' views of the target NOS aspects

NOS aspect	*% Responses*[1]	
	Naïve	*Informed*
Empirical NOS		
Science has or is distinguished by an "empirical" base	—	13
Science is solely based on "facts" to the exclusion of other personal and social attributes or factors	26	—
Observations, facts, or evidence are used to "prove" scientific claims "right" or "wrong"	14	—
Science relies on evidence, which does not "prove" claims Much in science is based on belief and the unobservable	—	4
The myth of "The Scientific Method"		
Without being prompted (on the *VNOS–C*)	24	0.7
When explicitly asked during individual interviews[2]	85	15
The experimental approach		
General contrived, controlled or manipulative structure	63	11
General aim of validating or detracting from a claim (*vs.* proving a claim right or wrong)	77	11
Role of prior expectations in designing, and interpreting the results of, experiment[2]	72	28
Validity of observationally based scientific disciplines	13	20
Scientific theories		
Theory change is solely based on new technologies or data (*vs.* admitting a role for new ideas, larger milieu)	94	6
Substantiated nature (*vs.* theory as someone's idea!)	82	4
Logic of theory-testing	62	16
Explanatory function	—	36
Function as guiding frameworks for research[2]	85	13
Scientific theories *vs.* laws		
Simplistic, hierarchical relationship	97	3
Laws proven true via repeated testing (inductive fallacy)	90	7
Tentative NOS	90	10
Creativity and imagination in science		
Not used or use is undesirable	5	—
Use limited to the planning and design of investigations	34	—
Refer to inventing explanations, models, or theoretical entities (*vs.* ascribing other meanings to the terms)	77	14
Inference and theoretical entities		
Case of atomic structure	68	30
Case of the concept of species	51	16
Theory-laden NOS	62	17

[1]Percentages do not necessarily add up to 100 since not all participants provided relevant responses and/or some responses were not categorized.
[2]Inferences exclusively derived from participants' interview transcripts.

The responses to *VNOS–C* #1 of a majority of participants (57%) indicated that science is "empirical" or has "empirical" components. Participants used a variety of terms to convey this view. A few participants used the label "empirical." Many others noted that science is based on and/or seeks tangible, concrete, visible, observable, measurable, or physical facts, data, or evidence. This result is rather encouraging. However, a closer examination of the ideas expressed by many participants indicated that the specific meanings they attached to the notion of an "empirical base" might not be consistent with current views of the empirical NOS.

Of 87 participants who indicated that science is "empirical" or relies on observable evidence, 40 (26%) believed that scientific knowledge is solely based on tangible facts to the exclusion of other factors. For 22 of these participants, these factors included human and personal attributes, such as interpretation, speculation, guess, intuition, abstraction, or personal opinions:

> [Scientific knowledge is] things for which there is evidence: The facts as they are finally presented without any kind of, sort of human interpretation. In religion and philosophy we interpret everything and not just take it for how it is plainly right there as we see it. (P 93, interview)

The other 18 participants believed that reliance on facts exonerates science from the burden of subjectivity or social and cultural attributes, such as values and beliefs, since science "deals with the physical world. Religion and philosophy are based on beliefs, values, and traditions" (P 125).

Additionally, 21 participants (14%) believed that science uses observations, facts, or evidence to "prove" its claims "right" or "wrong." These students attributed to observable evidence the sole role in adjudicating between scientific claims. They seemed to believe that absolute truths could be obtained through the use of physical evidence: "Science is different . . . because it uses concrete facts that have been proven/ are observable/ can be repeated . . . to get a right or wrong answer" (P 53).

Twenty-six participants (17%) expressed more informed views: Many noted that science involves the formulation of ideas and then seeking evidence to either support or discount them: "In science we ask questions and seek evidence for our speculations . . . This differs from religion in the sense that . . . scientists seek evidence to support or refute their explanations" (P 1). Unlike their aforementioned counterparts, these participants did not indicate that tangible data could be used to "prove" scientific claims or that science is based on observations to the exclusion of personal, social, or cultural factors. Yet, it can be argued that not expressing a certain view does not guarantee ascribing to an alternative one. As is evident in the last quote, the majority of these 26 students did not voice these ideas. Indeed, holding students to a stringent level of explicitness diminishes the above number to a mere six participants (4%) who noted that even though science relies on observation, there is much in science that is based on belief, conventions, and the non-observable:

> In science . . . typically experiments and observations are done to answer questions of how, why & what to either document its validity or reject it . . . Even though science can be more concrete and observable, this is not always the case when we're talking about magnetic fields or something along those lines. (P 56)

"The Scientific Method"

One of the most widely held naïve ideas about science is the existence of "The Scientific Method": A recipe-like stepwise procedure that all scientists follow and that guarantees developing valid claims about nature. The *National Science Education Standards* (NRC, 1996) and *Benchmarks for Science Literacy* (AAAS, 1993) explicitly debunked this notion. There is no such method (Bauer, 1994). Scientists observe, compare, measure, test, speculate, hypothesize, create ideas and conceptual tools, and construct theories and explanations. Scientists, however, do not follow a cookbook method when they investigate a problem.

The *VNOS–C* makes no references whatsoever to the phrase "The Scientific Method." In their responses to *VNOS–C* #1, nonetheless, 10% of participants indicated that "science differs from religion and philosophy in that it has the scientific method" (P 62). An additional 14% noted that science is typified by a set of orderly steps and rules or a systematic, structured, rigid, standardized, or logical method: "Science is different from other disciplines . . . because there is a very structured and methodical way that scientists follow" (P 140). The particular steps that participants assigned to this "common method," "logical standardized method," "rigid process," or "The Scientific Method," were somewhat different, but all distilled to some set of prescribed steps: "There is . . . a set way to do things. You come up with a question and you go from there to developing a hypothesis, from there you go to testing the hypothesis, and then reach a conclusion and evaluate your hypothesis" (P 138, interview). So, without any prompts, 24% of participants believed that scientists follow a single method in their investigations. Only 1 of 153 participants explicitly indicated that "science has no single method, rather, it relies on the creativity of the investigator to find ways to answer his/her question" (P 1).

While being interviewed, students were explicitly asked whether they thought scientists followed an orderly step-wise procedure. The majority (73%) answered in the positive. An additional 12% noted that "scientists do not use a single method." However, further probing indicated that these participants still believed in a general overarching method. They held that scientific investigations only differ in the types and specifics of the experiments that scientists conduct:

> Probably not all scientists [use the same method], I think there is lots of variations in the experiments . . . But there is some sort of general, you know, they have to come up with an idea in their head, and then think what they can do about it, and then design an experiment and actually do it and collect the results then go on from there. (P 145)

Thus, when specifically asked, the larger majority of participants (85%) seemed to believe in a single scientific method. Only 15% of interviewees believed otherwise. Some believed that there are discrepancies between the way science is portrayed in scientific reports and the way scientific work is actually conducted:

> No [scientists do not use the same method]. You know when you are in sixth grade you learn that here is the scientific method and the first thing you do this, and the second thing you do that and so on so forth. That's how we may say we do science, but there is a difference between the way we say we do science and the way that we actually do science. (P 135, interview)

The Experimental Approach and Observationally Based Disciplines

VNOS–C #2 asks participants to define an "experiment," but does not specify a certain type or class of experiments, such as laboratory, field, natural or crucial experiments (Diamond, 1986). *VNOS–C* #3 asks whether experiments are *required* for the development of scientific knowledge. These two items were to be used in combination to explore whether participants equated scientific investigation with the experimental method, which is often erroneously labeled "The Scientific Method" in many high school and introductory college level science textbooks (e.g., Curtis & Barnes, 1985; Hewitt, 1998; Hill & Petrucci, 1996). Obviously, the meaning that respondents attach to the term "experiment" was crucial to a valid interpretation of their responses to the third item; hence the second item. Participants' views of the experimental approach and the logic of experiments were not the primary focus here. Analysis of participants' responses to these two items, however, unveiled some naïve views in this regard, which were worthwhile exploring.

Commonly, experiments are distinguished from observations (Bernard, 1957; Harre, 1983). Unlike observations, experiments generally involve elements of control and manipulation, and intervention in the course, of the investigated phenomena. "Experimenters describe their activities in terms of the separation and manipulation of dependant and independent variables" (Harre, 1983, p. 15). Some might argue, and rightly so, that not all experiments involve manipulation. However, in the very least, an experiment should involve "*contrived observation, carried out under controlled, reproducible conditions* [italics in original]" (Ziman, 1991, p. 56). Experiments and observations, nonetheless, are similar in that both experimenters and observers must have prior conceptual frameworks with which to make sense of the outcome of their experiments, and perceive and describe their observations.

Participants' characterizations of experiments were mostly general and poorly articulated. Of 153 participants, 29 (19%) noted that an experiment involves observation or the collection of data or information. No reference was made to the contrived or manipulative nature of experiments. Also, these 29 participants either did not articulate a clear aim for experiments or merely noted that an "experiment is a method by which one could test a theory or hypothesis . . . by gathering data" (P 47). Another 52 participants did not mention that experiments involve controlling or manipulating aspects of the investigated phenomena. However, they noted that an experiment is a test, tool, attempt, project, or process "performed in order to prove a proposed theory" (P 151). Ten more participants thought that experiments are intended to decide whether a hypothesis or theory is true or false (or right or wrong).

Six participants noted that experiments aim to test "the validity or invalidity of a hypothesis" (P 141), but did not allude to their manipulative or controlled nature. Eighteen participants (12%) indicated that experiments are studies "in which experimental units are manipulated by the application of a treatment in order to measure the response of the units to the treatment" (P 88). While these participants expressed familiarity with the general structure of an experiment, they failed to articulate an understanding of its aim. This inference is reinforced by the fact that 10 other participants thought that experiments involve manipulation, but indicated that their aim is to "prove" or show that a scientific claim is true or not: "An experiment is a test [done], to prove or disprove a scientific hypothesis . . . A true experiment is manipulative and involves dependent and independent variables" (P 29).

Only a small minority of participants (11%) demonstrated clear understandings of the general intent and structure of experiments. They noted that an experiment is:

> A controlled way to test and manipulate the objects of interest while keeping all other factors the same. When only one factor at a time is changed or manipulated, the observed result can lead the scientist to assume the factor has either a positive or negative or no correlation with the outcome . . . It is the result of an experiment that will lead the scientist to believe his/her theory has or doesn't have validity. (P 55)

It is noteworthy that only two participants indicated that experiments "cannot prove a theory, only disprove the opposite of the theory" (P 24), two others noted that experiments could help uncover cause-effect relationships, and one of the latter two participants noted that an experiment specifically aims to reject the null-hypothesis.

Of the 52 participants who used general terms (e.g., "test," "procedure," "process" or "activity") to characterize experiments, 13 were among those randomly chosen for interviewing. They were specifically asked whether experiments involve elements of control or manipulation. Ten interviewees (77%) indicated that these elements are not crucial in experiments, and were either too inclusive in their definitions of experiments or did not discriminate between observations and experiments: "An experiment doesn't have to be manipulative . . . It is another way of using the senses to gain knowledge" (P 43, interview). The remaining three interviewees noted that experiments involve some control or manipulation and were able to elucidate clear understandings of the difference between observation and experiment and of the significance of manipulating variables during experiments:

> I think an experiment is something where the experimenter actually manipulates the environment. It differs from observation where you just sit there and you observe what is going on. Whereas in an experiment you actually manipulate, you remove something, you add something, you change the environment. (P 30, interview)

Prior expectations and conceptual frameworks are crucial for designing experiments and interpreting their results. Harre (1983) noted that "without some prior idea of what might be there to be found out we would not know what to look for in the results of our experiments, nor would we be able to recognize it when we had found it" (p. 5). Interviewees were specifically asked whether scientists usually have prior ideas about the outcome of manipulative experiments. About 54% of interviewees explicated naïve views in this regard. Many thought that scientists do not usually have an idea about the outcome of an experiment unless similar experiments have been conducted before:

> I think that is depends on what the experiment is. For instance I worked in genetics and molecular techniques and I know that they expect some results. And so that is an area that has been studied, an area that has been known. And I would say that something not as quite well known you would probably have to do the experiment to know what will turn out. (P 133, interview)

Others thought that scientists usually have an idea about these outcomes. They believed, however, that such ideas are undesirable since they tend to bias the results:

> Ideally, I would think that in a scientific experiment to be scientific and valid you should not have any bias or ideas in advance. But there have been experiments that were biased. So, you usually have some sort of idea but, you know, it should not tell what the results are. (P 125, interview)

Another 18% of interviewees indicated that hypotheses are guesses about the expected outcome of experiments. In that sense, scientists usually have an idea about those outcomes. These participants, nonetheless, did not seem to understand the significance of these "guesses" in developing and conducting experiments. Only 28% of interviewees explicated informed understandings in this regard:

> If you are going to organize the experiment you sort of need to know what you are looking for. I always think that they have some kind of idea . . . of where the results would lie, they kind of know that things will be in this area . . . In order to organize an experiment you need to know what is going to come out of it or it wouldn't really be a test method. I don't know how you would organize a test or something if you don't have a general idea about what you are looking for. (P 55, interview)

VNOS–C #3 asks whether experiments are required for developing scientific knowledge with the aim of eliciting respondents' views regarding observationally based scientific disciplines. As noted earlier, 63% of participants did not ascribe elements of control or manipulation to experiments. Thus, it was natural for these participants to respond to this question in the positive. Their responses, however, were not useful for the purpose of this analysis, since these responses were more likely a reflection of a misunderstanding of the nature of experiments than a view regarding the necessity of experiments to the development of scientific knowledge.

Fifty-seven participants indicated that experiments involve elements of control and/or manipulation. Of these, 20 participants believed that experiments are required for developing valid scientific claims noting that observation is simply not enough: "Science would not exist without scientific procedure which is solely based on experiments . . . The development of knowledge can only be attained through precise experiments" (P 116). Sixteen of the 20 did not provide any examples to support this view. The four participants who did, furnished examples related to drug-testing: "Developing scientific knowledge requires experiments . . . Example: To determine if a particular drug is causing a negative reaction within the body, the drug must be tested under many conditions, with a control and a variety of variables" (P 24). The remaining 37 participants indicated that manipulative experiments are not required for developing scientific claims. They provided examples that indicated a clear understanding of the fact that several scientific disciplines are observational in nature and that many powerful scientific theories rest solely on observation: "A good example, I think is Darwin's theory of evolution . . . [which] cannot be directly tested experimentally. Yet, because of observed data, such as fossils and rock formations, it has become virtually the lynchpin of modern biology" (P 3).

Scientific Theories

Scientific theories are well-established, highly substantiated, internally consistent systems of explanation (Suppe, 1977). Theories serve to explain relatively huge sets of seemingly unrelated observations in more than one field of investigation, and play a crucial role in generating research problems and guiding future investigations. Theories often are based on a set of assumptions or axioms and posit the existence of non-observable entities. Thus, they cannot be directly tested. Only indirect evidence can be used to establish the validity of theories. To test theories (or hypotheses), scientists derive specific testable predictions from those theories and check them against empirical evidence. An agreement between the two serves to increase the level of confidence in the tested theory. Like other types of scientific knowledge (e.g., "facts" and laws) theories are tentative. They change as new evidence, made possible through advances in theory and technology, is brought to bear on existing theories, or as old evidence is reinterpreted in the light of new theoretical advances or shifts in the directions of established research programs.

Responses to *VNOS–C* #4 indicated that only 16% of participants believed that theories do not change. For these participants, the original theory might be refined, elaborated or extended. The theory itself, however, does "not change . . . The fundamental view stays the same, but as new information is found new light is brought into the picture and an addendum can be made to the theory" (P 133). The majority of participants (77%) indicated that theories do change, and almost all of them (94%) attributed this change solely to "new information and technological advances which allow increased accuracy in experimentation" (P 141). Only 6% of participants noted that other factors also play a significant role in theory change. Among these factors were the advancement of new ideas, social and cultural changes, and the role of individuals working "out of context":

> Theories change because one person or a group of people act out, they act out of context basically. Often times their ideas come from people, like a new idea in microbiology might come from a person who has never taken a course in microbiology, an outsider or someone who is just remarkable in the way that they do science and are able to act out of context . . . Also, the reasons we accept or reject theories are so much tied to context in a historical and social-political way. So, to a certain extent we are going to accept the theory that harmonizes with that perspective at the time. (P 127, interview)

Of the participants who indicated that theories do change, 54% did not provide examples, and 21% provided inadequate examples, to substantiate their position. These examples were either historically inaccurate or included ideas that could not be accurately labeled as scientific theories: "There was a point in time when scientists believed that the earth was flat. That theory changed as people gained knowledge through exploration and realized that the earth was in fact round" (P 140). Only 25% of participants provided adequate examples. Evolution theory and atomic theory—both mentioned in *VNOS–C* #4, were the two most commonly cited examples of theory change, accounting for 51% and 23% of all examples respectively. Other examples came from geology (e.g., the shift to plate tectonics theory), biology (e.g., the rejection of the theory of spontaneous generation), and astronomy (particularly the shift from a geocentric to a heliocentric cosmology).

Responses to *VNOS–C* #4 revealed other naïve ideas about the nature of theories. About 37% of participants did not recognize that scientific theories are well substantiated, and ascribed to vernacular meanings of the term: "A scientific theory is just that–a theory. It is just a guess as to what might have possibly happened or what might happen" (P 78). Another 45% believed that theories "are still speculations and . . . there's still not enough evidence for them" (P 95). The responses of only 4% of participants indicated informed views of this aspect of theories: "In the vocabulary of a scientist the word theory is used differently . . . It does not mean someone's idea that can't be proven. It is a concept that has considerable evidence behind it and has endured the attempts to disprove it" (P 137).

These views of the nature of theories seem to be rooted in yet another naïve notion. *VNOS–C* responses indicated that 62% of participants had erroneous views of the logic of testing theories, failing to understand that theories are tested indirectly through checking predictions derived from them against empirical evidence. Instead, they insisted that "a theory is something that cannot be tested. For example, to prove the theory of evolution to be true, this would take scientists roughly a million years for speciation to occur. No one has that kind of time" (P 110).

About 36% of participants recognized the explanatory function of theories noting that theories help us explain or are the "best current explanations" for natural phenomena: "I think we learn scientific theories because they explain so much and incorporate so many other theories and info into one package" (P 25). Another 24% of students noted that we strive to learn about scientific theories because "mankind is curious and has been driven by science since the beginning" (P 147). Others (27%) noted that theories serve as building blocks, springboards, stepping-stones or starting points for expanding our knowledge and understanding. Some participants indicated that knowledge of theories allow scientists to refute them or to avoid reinventing the wheel: "We bother to learn scientific theories because they help us to understand things and they act as a stepping stone for new knowledge" (P 99).

At first these latter ideas seemed to reflect an understanding of the role of theories as guiding frameworks for research. However, further probing during the interviews indicated that these responses were more indicative of a "knowledge builds on itself" view. Participants did not seem to understand that theories generate research problems and guide investigations: "You have at least to know the part that is known now and then you can go on . . . The knowledge kind of builds on itself" (P 88, interview). This latter inference was reinforced by interviewees' responses to the question of which comes first when conducting an investigation, theory or observation. Obviously, respondents were not expected to provide a "right" answer. Rather, it was expected that they would, at least, ponder an answer. About 85% of interviewees indicated without hesitation that "observation comes first. You see something and then you try to figure it out some way or the other and then you develop a theory about whatever is that you want to know about" (P 40, interview). A few interviewees (15%) pondered the question for a minute before they noted that it is kind of the "egg and chicken" story. They noted that it could go either way, thus realizing that, at least, investigation could be triggered by scientific theories:

> Well, I don't know if I can answer that. That is almost like the chicken and the egg thing. I mean you can observe something and make a theory from it. Or you think about something and then try to look at the data and see how it fits it. Let's take relativity theory. Theoretical physics often starts with a general principle and then looks at the data rather than record data first. (P 3, interview)

Only 13% of participants seemed to appreciate that theories "set a framework of general explanation upon which specific hypotheses are developed. Theories, even if temporary, advance the pool of knowledge by stimulating hypotheses and research, which may support the current theory or lead to new theories" (P 62).

The Difference and Relationship between Scientific Theories and Laws

Generally speaking, laws are statements or descriptions of the relationships among observable phenomena. Theories, by contrast, are inferred explanations for regularities in these phenomena. Students often hold a simplistic, hierarchical view of the relationship between these constructs whereby theories become laws with the accumulation of supporting evidence. Thus, students believe that laws have a higher status than theories. They fail to recognize that theories are as legitimate a product of science as laws. Scientists do not usually formulate theories in the hope that some day they would acquire the status of "laws." Theories and laws are different kinds of knowledge and one cannot become the other.

The overwhelming majority of participants held naïve views of laws and their relationship with theories. About 90% believed that laws are certain because they are "proven" true through repeated testing. The fallacy of this inductive dictum is well known because "no rule can ever guarantee that a generalization inferred from true observations, however often repeated, is true" (Popper, 1988, p. 25). Additionally, 97% of participants believed in a hierarchical relationship between theories and laws. Their responses indicated that theories are less valid or supported than laws or that theories are merely precursors to laws.

Participants' particular responses were somewhat different but all were consistent with the abovementioned ideas. About 28% of participants explicitly indicated that a "theory can become a law only if that theory can be proven over and over and over again" (P 59). Another 54% noted that theories and laws differ because laws are "proven" to be correct or true while theories are not:

> A theory is just a suggestion. The knowledge it suggests must be tested in order to prove if it is right or wrong. A law has already been proven to be right and is accepted. For example, the theory of evolution is just a suggestion, it has not been proven to be right or wrong. The law of motion is proven and is accepted as true. (P 68, questionnaire)

Still another 8% of participants expressed similar views that were, nonetheless, cast in different terms. They realized that laws cannot be "proven" but still believed them to be "true." Instead of indicating that laws have been proven, they noted that laws have not been disproven. There were no indications that these participants thought laws are apt to change when and if they are disproven: "A scientific law has not been proven wrong for such a long period of time that it is considered to be true" (P 10).

Participants' belief in this hierarchical relationship was also evident during the interviews, during which 53% ranked laws above theories: "I think that laws are ahead of theories. Theory is made, and if it is not disproven over time then it becomes a law" (P 143, interview). Some interviewees (16%) ranked theories above laws noting that "nothing can be done with laws, they will not change." In contrast, theories may lead to new investigations. Other interviewees (31%) chose not to rank theories and laws, but their responses did not reflect more informed views:

> Well, I think that laws are seen as static, factual and they are not going to change. But a theory is close to being factual but it is still apt to change . . . Both of them are important to learn. So I don't know that if you can really rank them. (P 96, interview)

The responses of only 10% of students reflected informed views of laws and/or their relationship with theories. Some (7%) indicated that laws are not certain and believed that theories and laws are not different since both are subject to change: "There is no difference between the two . . . the idea of an absolute is ridiculous. Even Newton's laws have been reduced to principles . . . All laws & theories are just the best explanations currently possible" (P129). Only five participants (3%) explicated clearly informed views:

> Well, I guess a law would be something that you know, something that holds true while describing what is happening. Whereas a theory is why that happens. An original thought. Like the theory of gravitation . . . Actually a law would be much more descriptive, it is something that happens. Whereas a theory explains why it happens. (P 66, interview)

Finally, 50% of students provided adequate examples of laws, of which Newton's laws and laws of thermodynamics constituted about ninety percent. Another 23% did not provide any examples even though they were explicitly asked to do so. Twenty-seven percent of participants provided inadequate examples, and seemed to confuse laws with empirical facts or conventions: "A scientific law has been proven and will always be that way. A methyl group (CH_3) will always be a methyl group" (P 106).

The Tentative Nature of Scientific Knowledge

Scientific knowledge, though reliable, is tentative and never certain. "Facts," laws, and, theories, are subject to change. Tentativeness arises from the fact that scientific knowledge is inferential, creative, and socially and culturally embedded. Also, there are compelling logical arguments that lend credence to the notion of tentativeness in science (see Popper, 1963, 1988). The overwhelming majority of participants (90%) did not seem to have internalized this tentativeness. Sixteen percent of participants explicitly indicated that science differs from other disciplines of inquiry in that scientific knowledge is definitive, correct or "proven" true: "I think what makes science different from other disciplines of inquiry is the fact that it holds universal truths rather than a view of the truth according to certain individuals" (P 76).

As noted earlier, 77% of participants indicated that scientific theories change. This result might be reassuring because it suggests that these participants embrace a tentative view of science. However, 90% of participants noted that laws are absolute and do not change. This view, coupled with the participants' belief in a hierarchical relationship between theories and laws, suggests a lack of belief in the tentativeness of scientific knowledge. Rather, participants seemed to believe that theories are just a stage in the progression toward the "truth." Theories change not because scientific knowledge is dynamic or tentative, but because theories are "just theories" and have yet to attain the status of law or "proven" fact. Only 15 participants (10%) explicitly indicated in their responses to the *VNOS–C* and/or during the interviews that science is dynamic and that scientific knowledge is not absolute.

The Creative and Imaginative Nature of Scientific Knowledge

Science, contrary to common belief, is not a lifeless, rational, and orderly activity. The development of scientific knowledge involves the invention of explanations and theoretical entities, which requires a great deal of creativity and imagination on the part of scientists. Only 3 of 153 participants indicated that scientists do not use imagination and creativity, which they believed are characteristic of the "arts":

> I *don't* think [italics in original] scientific investigation is best characterized by creativity or imagination. I think a composer can be creative, a novelist can be imaginative, etc. . . . Scientific investigations are often tedious and repetitive, with the sole purpose of generating new data on the basis of previous data. (P 124)

Four participants (3%) thought that scientists use these attributes, but indicated that such use is not desirable because these elements are often utilized to bias or "distort" investigations in order to fit scientists' agendas to publish and/or secure funding:

> They use them at all stages of an investigation . . . All of these stages are creatively distorted to make the experiment reflect their preconceived notion as to how the experiment will turn out. They use their imagination to get published in scientific journals and, to receive monetary grants from the government and corporations. (P 36)

Almost all participants (95%) indicated that imagination and creativity are needed in scientific investigation, but differed in their choice of the specific stages in which they thought these attributes were used. Many (44%) thought that they permeate all stages of scientific investigation including planning and design, data gathering, and the stages following data collection:

> Of course they use their imagination and creativity. If not, science would not progress. They must come up with a testable hypothesis then create an experiment to test the hypothesis and collect data. Collecting data can be tricky and therefore imagination plays a big part. One must decide how to interpret the data. (P 131)

Another 34% limited the use of imagination and creativity to the planning and design stages. They believed that using these elements in data collection, data interpretation or in deriving conclusions would result in "incorrect" findings:

> I think scientists use a lot of creativity and imagination during the planning & design of an experiment . . . During and after data collection scientists shouldn't be, and usually aren't, creative as that would cause all value of the experiment to be lost. If scientists are creative with data then incorrect conclusions are made and information perceived as valid is distributed to an audience that is misled. (P 42)

Finally, about 13% of participants indicated that scientists use imagination and creativity in all stages of investigation with the exclusion of data collection. The fact that participants assigned imagination and creativity to different stages of scientific investigation suggested that they attached varying meanings to these terms. Closer examination of their *VNOS–C* responses and further probing during the interviews substantiated this inference. The majority (70%) did not use these terms to refer to inventing explanations, models or theoretical entities, but to refer to "resourcefulness," "skillfulness" or "cleverness":

> I think creativity and imagination come in at all levels like in designing an experiment and the nitty-gritty details of everyday lab work. When I worked we ran into problems all the time and the lab technician was extremely creative. She would like build little things, I mean if something wouldn't work right she would modify it. And she used all sorts of wooden things that she would build, you know just to make gels run right . . . In a sense you need to be an engineer just to get the experiment done. (P 99, interview)

Among the various other meanings that participants included under the label of creative and imaginative activities in science were "being open-minded," "being curious," "maintaining interest," using "appealing" ways to present results, and not "copying" other scientists' designs and experimental procedures: "Some scientists include lots of graphics and big huge metal models of molecules. But sometimes it is just cut and dry and they have it on paper and it is not the same" (P 66, interview). The responses of only 14% of participants indicated that scientists use imagination and creativity in the sense of inventing explanations, models and theories to explain patterns in natural phenomena: "Logic plays a large role in the scientific process, but imagination and creativity are essential for the formulation of novel ideas . . . to explain why the results were observed" (P 88).

Finally, 88% of participants did not provide any examples to support their views concerning the use of imagination and creativity in science. A few (6%) provided examples derived from everyday life situations rather than from scientific practice: "Police investigators try to recreate a crime scene not knowing everything that happened so they have to use their imagination to fill the holes" (P 95). Only nine participants (6%) provided examples that were taken from history of science and that conveyed informed views of this NOS aspect:

> Look at Sir Isaac Newton. He created calculus. That definitely required creativity and imagination. To think of the "great" scientific theories and laws one must be creative and have a large imagination. I don't think that Albert Einstein would be considered uncreative or lacking in imagination after developing the theory of Relativity. (P 40).

Inference and Theoretical Entities in Science

The world of science is inhabited by a multitude of theoretical entities, such as atoms, photons, magnetic fields, and gravitational forces to name only a few. These latter examples come from the physical sciences. Nonetheless, theoretical entities also abound in the biological sciences. For instance, "'species' . . . is a theoretical term embedded in a significant scientific theory" (Hull, 1998, p. 146). Theoretical entities are not directly observable and can only be accessed and/or measured through their effects or manifestations. These entities are inferred explanations or models that aim to account for regularities in the observed behaviors of phenomena.

It follows that inferential entities, such as atoms and species, are functional theoretical models rather than faithful copies of "reality."

In responding to *VNOS–C* #6, albeit for different reasons, the majority of participants (68%) noted that scientists were "certain" about their characterization of atomic structure, thus, demonstrating a lack of understanding of the inferential nature of this construct and of the distinction between observation and inference. First, an alarmingly high percentage of participants (25%) thought that "scientists are pretty sure about the structure of the atom. The evidence they use is microscopic pictures of the actual atoms" (P 132). For these participants, scientists have literally "seen" an atom. Second, 33% of participants indicated a belief in the certainty of atomic structure without indicating whether they thought atoms were or could be "seen." Their responses, however, did not convey an understanding of the nature of the evidence used to infer atomic structure:

> Scientists are 100% certain of the structure of atoms. The evidence they use is how everything is structured by atoms. The world could not exist if the structures were any different. Molecules such as water have to have a specific shape and by knowing this, the structure of the atom must be correct. (P 92)

Third, 10% of participants explicated unfamiliarity with the relevant evidence, and based their belief in the certainty of atomic structure on their faith in scientists and the efforts that were expended to arrive at the presently depicted structure:

> I think scientists are very certain about this structure . . . Unfortunately, I haven't a clue as to what specific evidence scientists used to determine what an atom looks like. So far, in my undergraduate studies we have only been told that this is how an atom is structured, we never learned how scientists proved this. (P 89)

Only 30% of participants held more informed views in this regard. They noted that atoms cannot be directly observed and indirect evidence is used to determine atomic structure. Many indicated that such structure is a model intended to explain observations of the "behavior" and/or "properties" of atoms in reaction to various experimental manipulations: "Models of the structure of atoms are frequently being updated. Current theories of the structure of the atom explain observed phenomena with a fairly high degree of certainty, but only indirect evidence can be used to formulate such theories" (P 21).

Participants' views of the construct of species were not different from their views concerning atomic structure. One-half (51%) believed that scientists are certain about the notion of species. They advanced two major reasons for such belief: (a) Use of a variety of observational evidence, especially DNA sequencing, to determine species membership: "Scientists are certain about their characterization of a species. Scientists use morphology, reproductive cycles, and bone structure to determine a species. [They also use] genetic information obtained from DNA" (P 141). (b) Conducting crossbreeding experiments: "Scientists are very certain about the characterization of what a species is. Years of experiments (interbreeding, etc.) have lead scientists to their conclusions" (P 89). Additionally, circular logic typified the responses of some participants who attempted to defend this position. They noted that scientists are certain that a species is a group of similar organisms that interbreed and produce fertile offspring because only organisms of the same species can interbreed and produce fertile offspring:

> I think that [scientists] are very certain about their characterization. The evidence used was most likely the fact that only animals which have similar characteristics can mate and have fertile offspring. When other animals tried to mate and couldn't, they realized that they obviously couldn't be the same species. (P 140)

Some participants (15%) noted that scientists are not certain about characterizing species. Their justifications included the: (a) Documented disagreements among scientists about the construct itself, (b) existence of gray areas and exceptions that defy classification and blur the lines between certain species due to the abundance of variations among organisms, and (c) ongoing nature of speciation as indicated by evolutionary theory. These participants' responses, nonetheless, did not convey an understanding of the inferential or theoretical nature of the construct:

> Scientists have difficulties when attempting to pigeonhole organisms into a classification of species. The study of genetics reveals a large spectrum of variations between species. Scientists cannot be too certain about classification because of these variations. (P 33, questionnaire)

Only 16% of participants noted that "species" is a human construct, or part of a man-made classification system intended to help scientists bring some order to the enormous variety presented by nature. Like all classification systems, the construct has merits and limitations. On one hand, the construct helps scientists classify, make sense of the relationships between, and communicate about various, organisms. On the other hand, it leaves much to be desired: "A species is a human convention, an 'artificial' concept created to convey and communicate about organisms with others. 'Species' is a very static term for something that is unstable in reality" (P 38). Thus, sharp lines are often difficult to draw between groups of organisms that seem to span the terrain between the blurred lines dividing closely related populations.

The Theory-laden Nature of Scientific Knowledge

Scientists' disciplinary and theoretical commitments, beliefs, training, experiences, and expectations influence their work. These background factors form a mind-set that affects the choice of problems that scientists investigate, how they conduct their investigations, what they observe (and do not observe), and how they make sense of, or interpret their observations. This (sometimes, collective) individuality or mind-set accounts for the role of theory in the production of scientific knowledge.

In response to *VNOS–C* #9, 62% of participants attributed the dinosaur extinction controversy solely to the scarcity of the available "data." They presented this view in three different ways. First, 20% equated data relevant to the extinction question with "seeing what has happened." Not only did these participants misconceive the meaning of "data" or "evidence," they also demonstrated a misunderstanding of the logic of hypothesis testing. They noted that scientists could not go back in time or crash meteorites into inhabited planets, and "scientists were not around when the dinosaurs became extinct, so no one witnessed what happened .

. . The only way to give a satisfactory answer to the extinction of the dinosaurs is to go back in time to witness what happened" (P 89). Thus, scientists can only produce "theories" (in the vernacular sense) about what happened. Second, 33% of participants used the term "data" adequately to refer to artifacts left by either hypothesized event—meteorite impact or volcanic eruptions. They indicated that both hypotheses "are possible since evidence is limited" and not enough to "prove that one of the two hypotheses is correct." The implication here being that the controversy would be resolved if there were "enough" or "complete" data or if such data is obtained in the future: "There is not much proven evidence to support either one at this time. When more data is collected one of the two hypotheses will be proven" (P 13). These participants failed to recognize that data need to be interpreted from within certain theoretical frameworks to acquire any significance as supportive of one scientific claim or another. Third, 9% of participants indicated that reaching different conclusions is possible because of imagination and creativity. However, they did not seem to believe that these attributes are integral to scientists' work. Rather, they thought, "the data is scarce; therefore, scientists are forced to 'fill in the gaps' using their imagination and creativity" (P 21). Again, the implication being that if there were "enough data," the controversy would be non-existent since scientists need only refer to the data to draw conclusions. Indeed, some participants indicated that the using these attributes in cases such as the extinction controversy is undesirable: "This is the danger of creativity and imagination. I think this [reaching different conclusions] happens when scientists put their own ideas and creativity into their research, instead of looking at the available raw data only" (P 69).

Alternatively, 33% of students attributed the controversy to factors other than, or in addition to, the lack of evidence. Their responses could be grouped into three categories. First, 13% noted that scientists arrive at different conclusions because they interpret the data differently, but did not explain why they thought such interpretation occurs: "I believe that the data collected suggested a massive event that caused the extinction of dinosaurs. Both groups suggested large events and interpreted the data in different ways" (P 110). Second, 3% indicated that factors such as fame, prestige, and scientists' egos, and the race to publish and secure funds for research are behind the controversy. These participants understood that science is another human activity that is infused with these humane attributes: "These conclusions are possible because it's profitable to keep the debate going. If they came to the same conclusion in regards to the cause of the extinction they might loose their jobs and big research grants" (P 36). Nonetheless, this latter representative quote appears to convey a "negative" message according to which these human attributes are not only foreign to science but also undesirable factors.

Third, only about 17% of participants explicated an understanding of the theory-laden NOS. They indicated that different scientists derive different conclusions from the same set of data because they interpret or perceive the data from within various frameworks, which often vary with the "particular scientist's background/personal beliefs . . . [and] 'school of thought' the scientist was trained in, his professors' ideas" (P 118). A few participants (3%) noted that disciplinary commitments and preferences might also lead scientists to varying conclusions:

> It is more that one group are more into geology and more into terrestrial things and they are going to come up with the volcanic hypothesis. And the other group is more extraterrestrial, and that group is more into that and they are going to think that it was a meteor. I mean everything is going to have more than one theory, because different people are going to look at it from their own perspective. (P 30, interview)

Additional Attributes of Participants' Views of NOS

Data analyses revealed other noteworthy aspects of participants' NOS views. These include the meanings participants attached to the term "prove" and the fragmented and inconsistent nature of their NOS views.

The "proving" dilemma

The *VNOS–C* makes no references whatsoever to the terms "prove" and "proven." Nonetheless, these terms were frequently used by almost all participants, especially when delineating the differences between science and other disciplines of inquiry, distinguishing between theories and laws, explicating the goals of experiments, or discussing the mass-extinction controversy. As such, it should "prove" worthwhile to explore in depth the meaning(s) that students attach to these terms.

Participants ascribed different meanings to the term "prove." For instance, some equated the term with providing "support" for a hypothesis: "An experiment is a designed test to prove or disprove/support or knock down an existing hypothesis" (P 132). However, many used the term to refer to an "absolute truism": "A law is something that is absolute. It has been proven" (P 79). These latter participants seemed to have used the term "to prove" in one of its most common connotations: "to establish as true; demonstrate to be a fact" (Neufeldt & Guralnik, 1996, p. 1082).

Of 38 interviewees, 28 (74%) had used the terms "prove" or "proven" in their *VNOS–C* responses. They were asked to explicate what they meant by these terms and how they thouyght scientists go about "proving" hypotheses, theories or laws. About 10% equated the term "prove" with collecting evidence or "physical" data to support or "back up" a certain scientific claim. When asked what they meant by the term "prove," they replied, "To have some sort of concrete evidence that the hypothesis could possibly be true" (P 30, interview). Another 39% of interviewees did not perceive "proving" as providing supportive evidence. They either indicated that "proving" a scientific claim is extremely difficult or noted that the term "prove" implies that a claim is "factual," "absolute," "permanent," or "unequivocal":

> To prove something is to give unequivocal evidence of how it is. I mean this is how it is, this is a fact, and no body can disclaim it or provide proof against it. I don't think that there are a lot of things that are proven in science. (P 129, interview)

About 19% of interviewees noted that a scientific claim cannot be "proven" but can only be "disproven." When probed further, about one-half of these students were not able to clearly articulate or defend this position:

> We can't prove why it is true because, I don't know, there is just, in my mind, and this is kind of philosophical, I guess that in my mind it is just, how do we know what we think is the truth. You know there is no, to the best of our ability, how do we know that that is honestly the truth? (P 100, interview)

The other half was able to explicate more accurate justifications:

> Well, you can't even prove a law. In a sense you can say a law is just that, you know it has never been known to fail but maybe sometime down the line they'll do a test and it will fail. You can never be certain because you don't know what might happen in the future. You just don't have infinite knowledge. (P 144, interview)

Another 11% of interviewees indicated that scientists "prove" hypotheses and theories by conducting experiments. When asked how many experiments are needed to achieve that, they indicated that one or a few experiments would suffice if the results support the tested claim. They, nonetheless, did not seem to use the term "prove" in its "absolute" sense: "I guess proving something, well, you have to do experiments first . . . And it depends, you know. If the experiment is done right, then one good experiment can allow you to prove the hypothesis, but it depends" (P 43, interview). Finally, about one-third of interviewees ascribed to the inductive fallacy. They indicated that scientific claims could be proven true through repeated testing. If a hypothesis or theory is tested "over and over and over again," and if the results "came out the same again and again" then the hypothesis or theory is "proven" true:

> To prove a theory basically you have to test it over and over again, and I guess that different scientists have to test it and see that it is replicable. And we have to be able to prove it over and over again and have basically now doubt that it is true. (P 20, interview)

As such, it is evident that the term "prove" means different things to different participants. This multiplicity of meaning and the various ways participants thought scientists go about "proving" claims coupled with their indiscriminate use of the term seem to have resulted in some confusion or difficulty with what will be labeled here as the *proving dilemma*. This dilemma was particularly evident in responses to *VNOS–C* #5 that asked about the differences between scientific theories and laws. Some participants faced difficulties trying to "by-pass" or reconcile the conceptual and logical problems associated with adopting the view that scientific laws are somehow "proven" or "absolute" while at the same time recognizing the implausibility of "proving" any scientific claim:

> A theory is something of interest that is still in the process of being tested. It has not been disproven, but for whatever reasons it has not given good enough correlation to support its validity either. A scientific law . . . also has never been disproven but it has never been disproven more. To the best of all party's knowledge it has been tested to the point that is believed that it can't be disproven. (P 55, questionnaire)

Thus, confusion about the meaning of the term "prove" has left many participants struggling to reconcile the acclaimed certainty of science with its well-documented tentativeness. At best, participants' conceptions of the term "prove" reflect inaccurate use of the term. However, their views might also reflect the belief that scientific claims can achieve an "absolute" status.

The fragmented and fluid nature of participants' NOS views

The views of many students lacked coherence. Their responses were typified by inconsistencies and, at times, outright contradictions. The *VNOS–C* responses of 36% of participants revealed one inconsistency or another, which often were evident when a participant's responses to the various items were compared. For example, in her response to *VNOS–C* #6, this student noted that: "In science nothing is certain. Scientists can't be certain about the structure of the atom" (P 20). However, in her response to the next item that asked about the concept of species, the same participant indicated that: "Scientists are pretty certain [about the characterization of species] . . . Specific evidence: geographical location, genetic similarity, and many more" (P 20). These two items, it should be noted, were consecutive and spatially separated by the flip of a page. Inconsistencies were sometimes evident in participants' responses to the same item. For example, in her response to *VNOS–C* #4 one participant indicated that scientific theories change and are constant in the one and same paragraph: "I believe they [theories] do change because as time goes on, our knowledge grows, so theories must change. We learn theories, for the most part because they are constant. They explain science in a precise way" (P 27).

The fluid nature of participants' views was also apparent when their *VNOS–C* responses were compared to their responses during interviews. Such inconsistencies are noteworthy because participants were provided their questionnaires during the interview and asked to consult their written responses before follow-up questions were asked of them. For example, one participant indicated in his response to *VNOS–C* #3 that experiments "are required to develop scientific knowledge. Without trial and error and creative thinking and doing the experiments to test theory, there is no way to 'prove' scientific truth" (P 117). When asked to elaborate on his response, he indicated that experiments are not required because disciplines such as ecology may only rely on observation:

> [After reading his response in the questionnaire] I am disagreeing with my own answer, that's funny. I originally said yes. But you know, okay I see. I said yes because you do also thought experiments to prove your theory or base your theory off of. And you still get your hypothesis even though your are not actually physically in a lab. In ecology we just observe for example. (P 117, interview)

Responses to probing questions during interviews revealed inconsistencies in the case of 54% of interviewees. One of the most frequently noted inconsistencies was the discrepancy between participants' responses to *VNOS–C* #1 and #8. When delineating science from other disciplines of inquiry in response to the first item, participants often indicated that science is rigid, systematic, or divorced from individuals' opinions, beliefs, or interpretations. However, in their responses to the eighth item, many participants noted that scientists use their imagination and creativity to interpret data. The following dialogue embodies an example of this inconsistency in participants' views. It should be noted that the participant in this example did not seem to recognize the discrepancy:

> Participant: [Explaining his response to item #1] Science isn't so much explaining something . . . like history explains why something happens. Science is the search for knowledge . . . Things for which there is evidence; the facts as they are finally presented without any kind of, sort of human interpretation.

In religion and philosophy we interpret everything and not just take it for how it is plainly right there as we see it . . . .

Researcher: You indicated in your questionnaire that scientists use imagination and creativity. Can you elaborate on that? [Question #8]

Participant: I think they have to. There is no way to get around it. If they went straight with the facts, you can't go from having a set of facts to having a conclusion without, in a way, tying all those facts together. The facts themselves don't directly point and say 1 + 1 = 2. You have to take them and interpret them to come up with something.

Researcher: Let me take you back to the first question . . . [Researcher reiterates the participant's response quoted above.] Now, it seems that you are saying there is a role for human interpretation, imagination and creativity in science. Do you see any tension between these two notions? If yes, how would you reconcile these seemingly opposing elements in science?

Participant: You can still be creative about it if your creativity doesn't interfere with the information you are receiving and the information that you are putting out. I guess I don't see a conflict between the two ideas. (P 93, interview)

In another example, one participant indicated that theories do change and then noted that they do not in her response to the same question:

Participant: I said theories do change . . . like evolution, people come up with theories on how the world evolved. And those theories, I think that theories change because there would be other people who come up with more theories and prove it, try to prove it. I think that theories don't change because the old ones are still there.

Researcher: So do you think that theories change or do not change?

Participants: I just contradicted myself. Well, I guess theories change. (P 113, interview)

It can thus be seen that a majority of participants lacked a coherent framework for their NOS ideas. Their views were "internally" inconsistent, fragmented, and fluid. Participants seemed to hold compartmentalized views regarding various aspects of NOS. Very few or no connections seemed to bridge their conceptions.

DISCUSSION AND IMPLICATIONS

The naïve views of NOS held by college students in this study are consistent with ones reported in a plethora of studies that assessed high school students' NOS views over the past 50 years and the few more recent studies that addressed college students' images of science (see Abd-El-Khalick & BouJaoude, 1997; Aikenhead, 1972, 1973; Bady, 1979; Broadhurst, 1970; Cotham & Smith, 1981; Gilbert, 1991; Horner & Rubba, 1978, 1979; Mackay, 1971; Rubba, Horner, & Smith, 1981; Ryder et al., 1999; Wilson, 1954). The majority of participants held naïve views, or demonstrated inaccurate understandings of: (a) the tentative, empirical, inferential, theory-laden, and imaginative and creative nature of scientific knowledge; (b) the well-supported nature of scientific theories, their role in guiding research, the role of new ideas, and social and cultural factors in theory change, and/or the nature of theory testing; (c) the general structure and/or aim of scientific experiments, and the role of theory and prior expectations in designing and conducting experiments; and (d) the validity of observationally (as opposed to experimentally) based scientific disciplines. Additionally, a majority of participants endorsed a hierarchical view of the relationship between theories and laws, believed that science is characterized by the use of "The Scientific Method" or other sets of orderly steps, and ascribed to the inductive fallacy, indicating that scientific claims could be "proven" through testing them "over and over and over again." Finally, consistent with prior research findings, participants' NOS views were not related to their gender, class standing, and science backgrounds (e.g., Carey & Stauss, 1969; Wood, 1972).

Other attributes also characterized participants' NOS views. First, participants seemed to ascribe a variety of meanings to terms, such as "theory," "creativity," and "prove," that are crucial to assessing their NOS views. The use of the *VNOS–C*, an open-ended instrument, in conjunction with individual interviews in this study was pivotal to accessing these various meanings and the contexts within which they were used, and relating them to participants' NOS views. The multiplicity of meanings attached to such key terms is most likely masked when forced-choice item or convergent paper-and-pencil instruments are solely used to assess learners' NOS views. For example, many participants used the term "theory" in the vernacular sense to refer to "someone's idea about what had happened" instead of a well-substantiated, internally-consistent web of concepts intended to explain a set of natural phenomena.

Participants also ascribed different meanings to the term "creativity" in science. While a few used the term to indicate that science involves the invention of explanations, theories or conceptual models, the majority of participants had other connotations in mind, such as being resourceful, skillful, open-minded, and curious. Similarly, participants attached different meanings to the term "prove." A few students used the term to refer to gathering tangible evidence, while others equated the term with conducting experiments. However, substantially more participants used the term in its more definitive or robust sense. This multiplicity of meaning entailed various perceptions of how participants believed scientists go about "proving" scientific claims and left many struggling with what was dubbed in this study as the *proving dilemma*: attempting to reconcile the acclaimed certainty of science with its well-documented tentativeness.

The variety of meanings that participants attached to these key terms, which often hold more technical meanings for historians, philosophers, and sociologists of science, indicates, at best, inaccurate use of these terms. Such use, however, might also reflect more entrenched naïve views about certain aspects of NOS. Thus, researchers attempting to assess learners' NOS views need to clarify the meanings of the terms that learners frequently use to explicate their views because ambiguities might result from assuming that students ascribe to these terms the same meanings as the researchers.

Second, the NOS views of a majority of participants *seemed* to be fragmented and lacking a coherent framework. These views were compartmentalized with few or no bridging connections. This finding is inconsistent with results reported by some studies, such as Dibbs (1982) and Hodson (1993), in which learners' NOS views were assigned labels, such as inductivist, verificationist or hypothetico-deductivist, indicating that students held coherent and consistent philosophic stances. Such findings were more likely an artifact of the forced-choice type instruments in use than a faithful representation of participants' views. These instruments are often designed with various philosophical stances in mind. As such, irrespective of the choices the respondents made, they often ended up being assigned one "coherent" philosophical position or another. Thus, the present results bring into question some recent calls to develop new standardized forced-choice type paper-and-pencil NOS assessment instruments (e.g., Good et al., 2000). These latter calls stand in contrast with the results of more than 20 years of research on assessing learners' NOS views (Abd-El-Khalick et al., 2001). Not only were participants' NOS views fragmented and inconsistent, these views were often not associated or reconciled with accurate images of scientific knowledge and practice. Many participants either did not provide any examples or provided inaccurate examples from history or practice of science to support or defend their NOS views, especially when discussing experiments, theories and theory change, scientific laws, and the creative nature of scientific knowledge.

This seeming compartmentalization of, and inconsistency in participants' NOS views warrants further attention. What might be considered as inconsistencies from this researcher's standpoint, might comprise from the perspective of any one participant a collection of ideas that "make sense" within a set of varied and personalized images of science that were invoked by the *VNOS-C* items. For example, when responding to *VNOS–C* #3 some participants noted that experiments are required for developing scientific knowledge and cited research on testing new drugs as an instance. It is no surprise then, when in this case the phrase "scientific knowledge" invoked the image of the high-stakes activity of certifying drugs for mass consumption, that some students made an exclusive association between science and rigorous experimentation. Other participants did not make a similar association because the term "scientific knowledge" for them invoked, for example, images of astronomy. It is reasonable to assume that for the *same* participant the term science might have invoked different images, activities, or episodes based in a variety of personal experiences. As noted earlier, many participants did not even perceive inconsistencies in their views or seemed to believe that such consistency was not worth pursuing. Indeed, by virtue of enrolling in a variety of disciplinary science courses and starting to develop more sophisticated knowledge and understandings, college students have available to them a variety of contexts within which to think about science. It could also be assumed that participants might have started to grapple with the realization that science is not necessarily a unified entity.

If the above two assumptions are empirically corroborated, then it might be more useful to approach college students' NOS views as genuinely fluid and locally situated. It follows that the generalized NOS framework advanced in K-12 reform documents (e.g., AAAS, 1990; NRC, 1996) might not be useful for purposes of exploring college students' NOS views. In this latter case, disciplinary nuances and differences, especially those between the biological and physical sciences (see for e.g., Hull & Ruse, 1998), need to be brought into the mix. In other words, researchers who aim to assess college students' NOS views have the task of further delineating and refining the specific realm and NOS aspects they aim to explore, as well as providing students with more specific contexts in their NOS assessments.

That participants held many naïve views of several aspects of NOS, though disconcerting, should not be surprising. During their years of high school and college science, participants were, at best, not accurately informed about NOS. Many participants indicated that they had not thought or were not given opportunities to think about many of the issues asked of them in the present study. Indeed, as noted by one participant who demonstrated informed views of several NOS aspects, students often need to "go out of their way" to learn about NOS:

When I was an undergraduate . . . I took a philosophy of physics class. And I have been doing reading on my own. That seems to be the only way, unless you go out of your way to do this type of thing you never learn anything about the nature of science. (P 56, interview)

Not only were participants not informed about NOS in their high school and college years, they more likely had been misinformed. This study revealed several substantial, naïve patterns in participants' NOS views, which seem to be persistent among high school and college students. It is highly unlikely that such patterns could be attributed to chance or to what students have inadvertently generalized from their experiences with high school and college science. It is more likely that students have been *explicitly* exposed to naïve ideas about NOS. For example, students are often exposed to—if not taught, what Horner and Rubba (1979) dubbed the "laws-are-mature-theories fable." Students encounter in their science textbooks explicit generalizations such as, "A theory that has withstood repeated testing over a period of time becomes elevated to the status of a law" (Curtis & Barnes, 1985, p. 8). Moreover, the myth of the existence of "The Scientific Method" is propagated in many high school and introductory college level science textbooks (e.g., Emiliani, Knight, & Handwerker, 1989; Hewitt, 1998; Hill & Petrucci, 1996). In fact, as one participant noted, "The Scientific Method" is often "drilled" into students' heads. When asked whether she thought scientists follow a certain set of orderly steps in their investigations, this participant replied:

> No. Well, yes. I think yes, maybe. But at the same time, I mean, that is like how the science textbook goes about it. I don't think that everybody would go about it that way . . . So, I don't think that that is the only way you can go about it but that's what has been drilled into my head, the scientific method, this is how the scientists work. (P 17, interview)

The notion of the "absolute" status of scientific knowledge is also bolstered through traditional evaluation practices often presented in science textbooks and laboratory manuals or used by science teachers. Learners are lead to believe that for every question posed about the natural world, scientists will eventually find the correct and final answer. This notion is not implicitly conveyed to learners. Rather, students are explicitly asked and expected to come up with "the" correct answer to end-of-chapter textbook exercises, choose the "one" correct answer on multiple choice tests, and arrive at "the" right conclusion to "cook-book" laboratory activities.

Thus, patterns in participants' naïve views of NOS are likely to be the result of the aforementioned and other "inadequacies" in their science education experiences. Indeed, such inadequacies have, among other things, invited the current wide-ranging efforts to reform science teaching at the pre-college level (AAAS, 1990, 1993; NRC, 1996). Helping students to achieve scientific literacy and an understanding of NOS are probably the most prominent themes in the discourse underlying these reform efforts. As noted earlier, the objective of helping pre-college students and science teachers develop informed views of NOS has been the subject of an extended line of research and associated curricular development activities (Abd-El-Khalick & Lederman, 2000). Achieving such views is considered pivotal to preparing scientifically literate students who, as future citizens, are capable of meaningfully engaging in public discourse about science and making informed decisions regarding science-related personal and societal issues.

But what about the possible contribution of college science teaching to achieving informed views of NOS? Is it not reasonable to assume that college level science courses would convey an accurate understanding of NOS to students? Or, at least, challenge the well-documented naïve NOS ideas that high school students bring into these college courses? Unfortunately, I believe that answers to these questions are likely to be in the negative. Scientific literacy and NOS are not currently prominent in the discourse or goals of the culture of college science education, which I will refer to as the culture of "scientific education." College science programs are still, by and large, preoccupied with preparing students for disciplinary-based science careers. This objective entails a focus on disciplinary content and associated methodologies and processes. Science students typically spend their early years learning the content and discourse of their disciplines through content-specific courses, and later learn the associated processes, methodological commitments, and instrumental preferences through one form of apprenticeship or another. Scientific education rarely, if ever, focuses on learning *about* science as an epistemic and historical endeavor.

Indeed, as Kuhn (1970) suggested, scientific education is both a-philosophical and a-historical. On one hand, Kuhn argued, initiating science students into disciplinary traditions includes having them take the processes and methods of those disciplines, and consequently the underlying ontological and epistemological values and assumptions, for granted. Epistemological and ontological issues put aside, and the conviction that the methods at hand will generate valid and reliable knowledge at bay, students can engage the (normal) activities of their science disciplines and invest the time and energy required to vigorously pursue answers or solutions to specific questions or problems related to some restricted aspect of a minute corner of the natural world. Thus, addressing epistemological issues in science courses, and exposure to, or coursework in philosophy of science is usually not a required part of scientific education. As Medawar (1969) pointed out, if one were to ask a scientist about scientific method, one is likely to get a "solemn and shifty-eyed expression" because the scientist "feels he ought to declare an opinion . . . [and] . . . is wondering how to conceal the fact that he has no opinion to declare" (p. 11). On the other hand, Kuhn continued, science students' exposure to the history of their disciplines is limited to the kind of history often found in scientific textbooks. Such historical narratives or vignettes present history of science "re-constructed by scientists" to convey images of a seamless and logical progression of problems and problem solutions within the discipline, and to celebrate the achievements of the scientist-heroes of that discipline. Such exposure to history is not aimed to help students learn *about* science. Rather it is pedagogically motivated and chiefly aims to promote certain problem-solutions that have proved successful in dealing with what is perceived—in hindsight, to have been the major problems that fraught the development of a certain discipline.

I am not in a position here to question the ways or effectiveness of scientific education as far as the sciences are concerned. Such education seems to be working: The scientific enterprise continues to be successful and disciplinary scientists continue to achieve major breakthroughs in a goodly number of scientific fields. Yet, if we broaden the circle beyond the scientific enterprise itself and concern ourselves with the interface between science and society, we may take issue with several aspects of scientific education. The lack of attention to NOS is of particular interest to the present discussion. The issue here is twofold. First, as Ryder et al. (1999) argued, scientists participate in public life as citizens and they too are faced with science-related personal and societal issues that lie outside their immediate disciplinary specializations. As such, narrow scientific education disadvantages scientists by not preparing them to engage in informed discourse *about* science and science-related public issues. This is especially the case at the present times where the image of scientists as disinterested objective individuals is (slowly) being displaced by more realistic images.

Second, scientific education is generalized within the academy: Disciplinary science departments do not offer genuinely different programs for students who plan to pursue (or end up pursuing) scientific careers and those who do not. The difference between so-called science courses for "majors" and "non-majors" is only a matter of degree and not kind. As a result, students who go through science programs end up with naïve images of the scientific endeavor, as was the case with a majority of participants in the present study. This shortcoming of scientific education is all too well known within the science education community where prospective science teachers (most of whom hold BS degrees) continue to join teacher preparation programs with naïve views of NOS. Thus, in the absence of systemic reforms of science education at the college level, there is a need to supplement scientific education with coursework in the history, philosophy, and sociology of science with the aim of providing college students with opportunities to critically examine their NOS views. Even more, there might be a need to develop courses on public understanding of science, which draw upon history, philosophy, and sociology of science and specifically aim to address college students' NOS views.

REFERENCES

Abd-El-Khalick, Fouad. The Influence of History of Science Courses on Students' Conceptions of the Nature of Science. Unpublished doctoral dissertation, Oregon State University, OR, 1998.

Abd-El-Khalick, Fouad, Randy L. Bell, and Norman G. Lederman. "The Nature of Science and Instructional Practice: Making the Unnatural Natural." Science Education 82.4 (1998): 417–436.

Abd-El-Khalick, Fouad, and Saouma BouJaoude. "An Exploratory Study of the Knowledge Base for Science Teaching." Journal of Research in Science Teaching 34.7 (1997): 673–699.

Abd-El-Khalick, Fouad, and Norman G. Lederman. "Improving Science Teachers' Conceptions of the Nature of Science: A Critical Review of the Literature." International Journal of Science Education 22.7 (2000): 665–701.

Abd-El-Khalick, Fouad, Norman G. Lederman, Randy L. Bell, and Renée Schwartz. "Views of Nature of Science Questionnaire (VNOS): Toward Valid and Meaningful Assessment of Learners' Conceptions of Nature of Science." Paper presented at the annual meeting of the Association for the Education of Teachers in Science, Costa Mesa, CA, January, 2001.

Aikenhead, Glen. The Measurement of Knowledge about Science and Scientists: An Investigation into the Development of Instruments for Formative Evaluation. Dissertations Abstracts International 33 (1972): 6590A. (University Microfilms No. 72-21, 423).

Aikenhead, Glen. "The Measurement of High School Students' Knowledge About Science and Scientists." Science Education 57.4 (1973): 539–549.

Aikenhead, Glen. "An Analysis of Four Ways of Assessing Student Beliefs About STS Topics." Journal of Research in Science Teaching 25.8 (1998): 607–629.

Aikenhead, Glen, Ryan A., and J. Desautels. "Monitoring Student Views on Science-Technology-Society Issues: The Development of Multiple-Choice Items." Paper presented at the annual meeting of the National Association for Research in Science Teaching, San Francisco, CA, April, 1989.

Aikenhead, Glen, A. Ryan, and R. Fleming. Views on Science-Technology-Society (from CDN.mc.5). Saskatoon, Canada: Department of Curriculum Studies, University of Saskatchewan, 1989.

Akerson, Valarie L., Fouad Abd-El-Khalick, and Norman G. Lederman. "Influence of a Reflective Explicit Activity-Based Approach on Elementary Teachers' Conceptions of Nature of Science." Journal of Research in Science Teaching 37.4 (2000): 295–317.

American Association for the Advancement of Science. Science for All Americans. New York: Oxford University Press, 1990.

American Association for the Advancement of Science. Benchmarks for Science Literacy: A Project 2061 Report. New York: Oxford University Press, 1993.

Bady, R. A. (1979). "Students' Understanding of the Logic of Hypothesis Testing." Journal of Research in Science Teaching 16.1 (1979): 61–65.

Bauer, Henry H. Scientific Literacy and the Myth of the Scientific Method. Champaign, IL: University of Illinois Press, 1994.

Bernard, Claude. An Introduction to the Study of Experimental Medicine (H. C. Greene, Trans.). New York: Dover, 1957. (Original work published 1865)

Bezzi, Alfredo. "What is this Thing Called Geoscience? Epistemological Dimensions Elicited with the Repertory Grid and their Implications for Scientific Literacy." Science Education 83 (1999): 675–700.

Broadhurst, N. A. "A Study of Selected Learning Outcomes of Graduating High School Students in South Australian Schools." Science Education 54.1 (1970): 17–21.

Carey, R. L., and N. G. Stauss. "An Analysis of the Relationship between Prospective Science Teachers' Understanding of the Nature of Science and Certain Academic Variables. Bulletin of the Georgia Academy of Science 27.3 (1969): 148–158.

Central Association for Science and Mathematics Teachers. "A Consideration of the Principles that Should Determine the Courses in Biology in Secondary Schools." School Science and Mathematics 7 (1907): 241–247.

Cooley, W. W., and Leopold E. Klopfer. Test on Understanding Science. Princeton, NJ: Educational Testing Service, 1961.

Cotham, J, and E Smith. Development and Validation of the Conceptions of Scientific Theories Test. Journal of Research in Science Teaching 18.5 (1981): 387–396.

Curtis, Helena, and Sue N. Barnes. Invitation to Biology. 4th ed. New York: Worth Publishers, 1985.

Diamond, Jared M. Overview: Laboratory Experiments, Field Experiments, and Natural Experiments. In Community ecology, edited by Jared M. Diamond, and Ted J. Case. New York: Harper & Row, 1986.

Dibbs, D. An Investigation into the Nature and Consequences of Teachers' Implicit Philosophies of Science. Unpublished doctoral dissertation, University of Aston, England, 1982.

Emiliani, Cesare, L. B. Knight, and Mark Handwerker. Earth Science. Chicago: Harcourt Brace Jovanovich, 1989.
Fleming, R. "Undergraduate Science Students' Views on the Relationship between Science, Technology and Society." International Journal of Science Education 10 (1988): 449–463.
Gilbert, Steven W. "Model Building and a Definition of Science." Journal of Research in Science Teaching 28.1 (1991): 73–80.
Good, Ronald et al. "Guidelines for Nature of Science (NOS) Researchers." Symposium conducted at the annual meeting of the National Association for Research in Science Teaching, New Orleans, LA, April, 2000.
Harre, Rome. Great Scientific Experiments. New York: Oxford University Press, 1983.
Hewitt, Paul G. Conceptual Physics. 8th ed. Menlo Park, CA: Addison-Wesley, 1998.
Hill, John W., and Petrucci, Ralph H. General Chemistry. Upper Saddle River, NJ: Prentice Hall, 1996.
Hodson, Derrick. "Philosophic Stance of Secondary School Science Teachers, Curriculum Experiences, and Children's Understanding of Science: Some Preliminary Findings." Interchange 24 (1993): 41–52.
Hoffmann, Ronald. For the First Time, You Can See Atoms. American Scientist 81 (1993): 11–12.
Horner, J, and Peter Rubba. "The Laws are Mature Theories Fable." The Science Teacher 45.2 (1978): 31.
Horner, J, and Peter Rubba. (1979). The myth of absolute truth. The Science Teacher 45.1 (1979): 29–30.
Hull, David, and Michael Ruse. Eds. The Philosophy of Biology. Oxford: Oxford University Press, 1998.
Kimball, M. E. "Understanding the Nature of Science: A Comparison of Scientists and Science Teachers." Journal of Research in Science Teaching 5 (1967-68): 110–120.
Kuhn, Thomas S. The Structure of Scientific Revolutions. 2nd ed. Chicago: The University of Chicago Press, 1970.
Lederman, Norman G. "Students' and Teachers' Conceptions of the Nature of Science: A Review of the Research." Journal of Research in Science Teaching 29.4 (1992): 331-359.
Lederman, Norman G., Philip D. Wade, and Randy L. Bell. Assessing Understanding of the Nature of science: A historical perspective. In The nature of science in science education: Rationales and strategies, edited by William McComas. Dordrecht, The Netherlands: Kluwer Academic Publishers, 1998.
Lederman, Norman G., and Molly O'Malley. "Students' Perceptions of Tentativeness in Science: Development, Use, and Sources of Change. Science Education 74 (1990): 225–239.
Mackay, L. D. Development of Understanding about the Nature of Science. Journal of Research in Science Teaching 8.1 (1971): 57–66.
Medawar, Peter B. Induction and Intuition in Scientific Thought. Philadelphia, PA: American Philosophical Society, 1969.
National Research Council. National Science Education Standards. Washington, DC: National Academic Press, 1996.
National Science Foundation. Shaping the Future: New Expectations for Undergraduate Education in Science, Mathematics, Engineering, and Technology. Arlington, VA: Author, 1996.
Neufeldt, Victoria, and David Guralnik. Eds. Webster's New World College Dictionary. 3rd ed. New York: Macmillan, 1996.
Popper, Karl R. Conjectures and Refutations: The Growth of Scientific Knowledge. London: Routledge, 1963.
Popper, Karl R. The Open Universe: An Argument for Indeterminism. London: Routledge, 1988.
Rubba, Peter A., and H. Andersen. "Development of an Instrument to Assess Secondary School Students' Understanding of the Nature of Scientific Knowledge." Science Education 62.4(1978), 449-458.
Rubba, Peter A., J. Horner, and J. M. Smith. "A Study of Two Misconceptions about the Nature of Science among Junior High School Students." School Science and Mathematics 81 (1981): 221-226.
Ryder, Jim, John Leach, and Rosalind Driver. "Undergraduate Science Students' Images of Science." Journal of Research in Science Teaching 36.2 (1999): 201–219.
Suppe, Frederick. The Structure of Scientific Theories. 2nd ed. Chicago: University of Illinois Press, 1977.
Wilson, L. (1954). "A Study of Opinions related to the Nature of Science and its Purpose in Society." Science Education 38.2 (1954): 159–164.
Wood, R. L. "University Education Student's Understanding of the Nature and Processes of Science." School Science and Mathematics 72.1 (1972): 73–79.

Ziman, John. Reliable Knowledge: An Exploration of the Grounds of Belief in Science. Cambridge: Cambridge University Press, 1991.

APPENDIX: *VNOS–FORM C*

(1) What, in your view, is science? What makes science (or a scientific discipline such as physics, biology, etc.) different from other disciplines of inquiry (e.g., religion, philosophy)?

(2) What is an experiment?

(3) Does the development of scientific knowledge *require* experiments?

* If yes, explain why. Give an example to defend your position.

* If no, explain why. Give an example to defend your position.

(4) After scientists have developed a scientific theory (e.g., atomic theory, evolution theory), does the theory ever change?

* If you believe that scientific theories do not change, explain why. Defend your answer with examples.

* If you believe that scientific theories do change: (a) Explain why theories change? (b) Explain why we bother to learn scientific theories? Defend your answer with examples.

(5) Is there a difference between a scientific theory and a scientific law? Illustrate your answer with an example.

(6) Science textbooks often represent the atom as a central nucleus composed of protons (positively charged particles) and neutrons (neutral particles) with electrons (negatively charged particles) orbiting that nucleus. How certain are scientists about the structure of the atom? What specific evidence *do you think* scientists used to determine what an atom looks like?

(7) Science textbooks often define a species as a group of organisms that share similar characteristics and can interbreed with one another to produce fertile offspring. How certain are scientists about their characterization of what a species is? What specific evidence *do you think* scientists used to determine what a species is?

(8) Scientists perform experiments/investigations when trying to find answers to the questions they put forth. Do scientists use their creativity and imagination during their investigations?

* If yes, then at which stages of the investigations you believe scientists use their imagination and creativity: planning and design, data collection, after data collection? Please explain why scientists use imagination and creativity. Provide examples if appropriate.

* If you believe that scientists do not use imagination and creativity, please explain why. Provide examples if appropriate.

(9) It is believed that about 65 million years ago the dinosaurs became extinct. Of the hypotheses formulated by scientists to explain the extinction, two enjoy wide support. The first, formulated by one group of scientists, suggests that a huge meteorite hit the earth 65 million years ago and led to a series of events that caused the extinction. The second hypothesis, formulated by another group of scientists, suggests that massive and violent volcanic eruptions were responsible for the extinction. How are these *different conclusions* possible if scientists in both groups have access to and use the *same set of data* to derive their conclusions?

CHAPTER 19

RANDY L. BELL

PERUSING PANDORA'S BOX

Exploring the What, When, and How of Nature of Science Instruction

In his presentation at the 1968 annual meeting of the National Association for Research in Science Teaching, Marshall Herron (1969) noted that the latest trend among curriculum planners was to promote the nature of science as a curricular objective of "paramount importance." Yet he doubted that most who used the term had reached a "sound and detailed understanding" of its educational ramifications. He feared that *nature of science* was in danger of becoming another trite educational cliché to be bandied about in discussions among science educators and ignored by classroom teachers.

While Herron believed an adequate understanding of the nature of science to be "critically necessary to science curriculum building and to the training of science teachers," (p. 107), he labeled the pursuit of this goal a "Pandora's box." Along with discussions about the precise meaning of *nature of science* came a host of other questions. Which point of view is most appropriate for schools? Is there a single view appropriate for all students? How much detail should be supplied? How much should the concepts be simplified?

More than 30 years after Herron presented these concerns, few would say that these issues have been settled. Philosophers of science still spar with one another over specific aspects of the nature of science. Scientists offer a variety of perspectives on the nature of science that are often at odds with those of the philosophers. And science educators constantly debate the nuances of the construct and question whether the nature of science can be taught and learned in absence of a universally accepted definition. Yet, since Herron's time, consensus has emerged on several aspects of the nature of science, which appear to be relevant and developmentally appropriate for school-aged children. Additionally, three decades of research has provided much insight into what K-12 students can learn and when they can learn it. Studies have found that, when taught purposefully, students *can* grasp important nature of science concepts. And research in developmental psychology has provided evidence in favor of children's capabilities for abstract thinking about epistemological issues at various ages.

In this chapter, I will address some of the contents of Herron's Pandora's Box by first describing a view of the nature of science that is supported by current science

L.B. Flick and N.G. Lederman (eds.) Scientific Inquiry and Nature of Science 427-446.

education reform documents. Next, I will discuss student conceptions of the nature of science – what they believe apart from purposeful instruction, how the typical school science experience informs (or mis-informs) their views, and whether altering their views through instruction is even possible. Finally I discuss the issue of developmental appropriateness for teaching contemporary understandings of the nature of science.

WHOSE NATURE OF SCIENCE?

As Herron emphasized in his presentation, science educators have been much more successful at prescribing nature of science instruction than describing what this instruction should look like, or what the nature of science is, for that matter. Certainly, the complexity of the nature of science construct is part of the problem. Philosophers and historians of science have found it difficult to describe the scientific enterprise in ways that are neither too inclusive (identifying as science enterprises that are clearly non-science) nor too restrictive (excluding endeavors that clearly fall within the realm of science). Furthermore, as Lederman (1992) pointed out, the nature of science construct is a moving target. Accepted views of science have changed in the past and will likely change in the future as philosophers, historians, and sociologists continue their efforts to characterize the scientific enterprise. To complicate matters further, these changes in philosophic stance should not be viewed as progressive, since views that are currently *in vogue* are not necessarily any closer to the "truth" than those of the past (Driver, Leach, Millar, & Scott, 1996). Thus, rather than narrowing our view of what science is, contemporary studies of the practice and epistemology of science have tended to emphasize its varied and idiosyncratic nature.

One unfortunate result of this state of affairs is lack of consensus among stakeholders regarding a specific definition or list of characteristics of the scientific enterprise. The complexity and lack of agreement regarding the tenets of the nature of science have been recognized by both science philosophers (e.g., Feyerabend, 1993; Kuhn, 1996; Latour, 1987) and science educators (e.g., Duschl, 1994; Hodson, 1991; Lederman, 1992; Loving, 1997; Matthews, 1994, Welch, 1984, Ziman, 1988). On one extreme, some have argued that lack of consensus is a critical impediment to nature of science instruction and assessment (Alters, 1997; Good et al., 2000). After all, how can students be expected to develop sophisticated understandings of a construct on which philosophers of science do not agree? Others have argued that while stakeholders do not agree on specific tenets of the nature of science, consensus does exist at more general levels (Driver et al., 1996; Matthews, 1994; McComas & Olson, 1998; Smith, Lederman, Bell, McComas, & Clough, 1997). Furthermore, it is at these more general levels that the nature of science is accessible and relevant to students (Abd-El-Khalick, Bell, & Lederman, 1998; Matthews, 1994), and where we see clear connections between students'/citizens' knowledge about science and their decisions made regarding scientific claims (Bell, in press). This latter point is critical, considering the science education community's stated and implied goals for nature of science instruction (Driver, et al., 1996, McComas, Clough, & Almazroa, 1998).

When considering these issues, it is important to remember that this lack of consensus on specific definitions and aspects is not unique to nature of science instruction. Many commonly addressed concepts in biology, for example, are difficult to define and present challenges to those who are uncomfortable with disagreement and uncertainty. Consider the lack of consensus among the knowledgeable over operational definitions of the biological concepts of life, species, and genes, just to name a few. Neither is such disagreement over basic concepts unique to the biological sciences. As recently as 1998, members of the American Physical Society failed to agree upon a satisfactory definition of their discipline (Trocco, 2002). And, despite over a century of efforts to delineate the boundary between science and non-science, scholars have been unable to reach agreement on what science is (Macilwain, 1998). Yet, in light of this debate, biology teachers still manage to instruct students on the nature of life, species, and genes. Physics instructors offer definitions of their discipline, and teachers of science in general are being challenged to address the issue of what science is and how it differs from other ways of knowing.

WHY TEACH THE NATURE OF SCIENCE?

Early admonitions to include nature of science topics in the science curriculum appeared in print nearly 100 years ago (see Central Association for Science and Mathematics Teachers, 1909). Yet, despite decades of curriculum development and teacher education, research has shown that students rarely achieve desired conceptions of the nature of science (Lederman, 1992). Given the apparent difficulty involved in learning about the nature of science, why should teachers and students bother?

Science educators have promoted a variety of justifications for teaching about the nature of science. For example, Matthews (1997) has argued that the nature of science is inherent to many critical issues in science education, including the evolution/creationism debate, the relationship between science and religion, and delineation of the boundaries between science and non-science. Others have related teaching about the nature of science to increased student interest (Lederman, 1999; Meyling, 1997) and the development of cultural awareness of the increasingly important impacts of science in society (Driver et al., 1996). Perhaps the most frequently cited justification for including nature of science instruction in the K-12 science curriculum is enhanced learning of science content (Cleminson, 1990; Songer & Linn, 1991). For more detailed examinations of the rationales for nature of science instruction, see Driver et al. (1996) and McComas et al. (1998).

It is important to note that while many of these outcomes for nature of science instruction are intuitive, there is little evidence that learning the nature of science produces the desired results. For example, another of the commonly stated benefits of nature of science instruction is its potential for facilitating decision making on science and technology based issues. Driver et al. (1996) referred to this justification as the "democratic argument" for nature of science instruction:

> The democratic argument for promoting public understanding of science focuses on the understandings needed to participate in the debates surrounding [socioscientific issues] and in the decision-making process itself...an understanding of the issues requires not just knowledge of science content, but also an understanding of the nature of science and scientific knowledge. (p.18)

While compelling, the "democratic argument" merits closer scrutiny in light of research suggesting that understandings of the nature of science may not play a primary role in decision making after all (Bell, in press; Fleming, 1986a, 1986b; Lederman & O'Malley, 1990; Zeidler, Walker, Ackett, & Simmons, 2002).

Current science education reform documents promote the nature of science as a critical aspect of scientific literacy (American Association for the Advancement of Science [AAAS], 1993; National Research Council [NRC], 1996). As such, understanding the nature of science can be seen as a worthy goal in itself—just as we deem the more traditional concepts of photosynthesis, evolution, and kinetic molecular theory as worth knowing because they contribute to scientific literacy.

In the typical classroom, instruction has focused almost exclusively on the well-established products of science and cookbook approaches to laboratory exercises, using authoritarian teaching modes. Under these conditions, science instruction has inadvertently and, in some cases intentionally, fostered absolute notions of scientific knowledge. Students who experience this form of science instruction are in danger of viewing science as irrelevant and ineffectual when confronted with the real-world "messiness" of the science surrounding science and technology based issues (Bell, 1999; Collins & Pinch, 1998).

If science educators are to meet the goal of scientific literacy, science instruction must promote a more realistic, less absolutist view of scientific knowledge. This assertion is supported in the US science education reform documents (AAAS, 1993; NRC, 1996) and the writings of a number of science educators (Cleminson, 1990; Duschl, 1990; Hodson, 1988; Lawson, 1995; Lederman, 1992; Matthews, 1994; Smith et al., 1997). McComas and Olson (1998) found further support for this assertion in their comparison of the frequency of various aspects of the nature of science promoted by eight international science education reform documents. The analysis revealed 17 separate philosophical statements appearing in two or more of the documents, with "Scientific knowledge is tentative" as the only philosophical statement common to all eight documents.

At the most general level, the development of scientific literacy implies an instructional approach emphasizing the tentative nature of scientific knowledge. But what makes for a sophisticated understanding of the tentative nature of science?

THE TENTATIVE NATURE OF SCIENCE

In an effort to combat the absolute views that K-16 students typically develop, Lederman and his colleagues have compiled a set of critical assertions designed to guide instruction and assessment supporting a more accurate view of the nature of science (Lederman, Abd-El-Khalick, Bell, Schwartz, & Akerson, 2001; Lederman & O'Malley, 1990; Smith, Lederman, Bell, McComas, & Clough, 1997)[1]. These assertions include understandings that scientific knowledge is

a) Tentative (subject to change and revision).
b) Empirically based (derived from observations of the natural world).
c) Partially based on human inference.
d) Partially based on human imagination and creativity.
e) Subjective (data is collected and interpreted in light of current scientific perspectives. as well as the experiences and values of individual scientists).
f) Socially and culturally embedded.

In general, the last five tenets should be understood as supporting the first tenet that scientific knowledge is subject to change. In other words, the last five tenets provide explanations for why scientific knowledge is tentative. Thus, all the tenets are also interrelated. For instance, the roles of idiosyncrasy and creativity in the construction of scientific knowledge contribute to the subjective nature of science, which in turn, contributes to its tentativeness—see Abd-El-Khalick & Lederman (2000a) for a more detailed treatment of the interrelationships among these aspects. What follows is a discussion of each of these tenets and how they relate to the tentative nature of science.

First and foremost among the aspects of the nature of science, scientific knowledge should not be viewed as absolute—rather, it is characterized as tentative and revisionary. This characterization of science as tentative includes all forms of scientific claims, including facts, models, theories and laws. The tentative nature of scientific knowledge is derived from several factors. First among these are compelling logical arguments disputing the common belief that scientific hypotheses, theories, and laws can be absolutely "proven." As Popper (1988) asserted, "proof" in the absolute sense lies outside the scope of science, irrespective of the amount of supporting empirical evidence. For example, to be "proven," a particular scientific law would have to account for every single instance of the phenomenon it claims to describe at all times. Since our limited view of the universe can never adequately assess all possible instances, we can never be certain that the law is "true" in any absolute sense.

Science philosopher Nancy Cartwright (1983, 1988) elaborated on this concept, pointing out that scientific laws do not propose generalizations that are universally true. Rather, they hold only under special, usually ideal conditions. We then assume that they apply in normal, less than ideal conditions. The Ideal Gas Law is a classic example of this characteristic of scientific laws, in that it is used to describe and predict the behavior of gases in everyday situations, even though it was written to apply only to a hypothetical "ideal" gas. This ideal gas is one in which molecules have no volume and experience no inter-molecular forces. It is thus an imaginary standard to which we assume the behavior of known gases is related. In actual practice, this assumption is well-founded in that most gases under ordinary conditions conform quite closely to the ideal standard. Even so, the law as written cannot be seen as universally true because it applies perfectly only to a non-existent ideal gas.

The empirical nature of scientific knowledge also contributes to the tentativeness and revisionary nature of scientific claims. As used here, "empirical" refers to knowledge claims that are based ultimately on observations of the natural world. The trouble (from an absolutist viewpoint) is that new evidence is constantly being imposed on existing claims. Science abounds with the revision of facts, theories, and laws as new evidence is accumulated. A fairly recent example taken from headlines is the discovery of water on the lunar surface by the unmanned spacecraft *Clementine*. The old "fact" that there is no water on the moon has been replaced with a new estimate of as much as 10 billion tons of water at the lunar poles (Recer, 1998). Additionally, the possibility of measurement errors and perceptual illusions give rise to a degree of uncertainty even to data based upon direct observation. And as we have seen, because no one can ever be certain of observing all instances of a phenomenon (there is always the possibility of a black swan), scientific claims based upon observation are also tentative (see Popper, 1963).

However, the accumulation of new evidence is not the only way scientific knowledge changes. Occasionally, new ways of looking at current knowledge come to light as a result of paradigm shifts (Kitcher, 1982; Kuhn, 1996). For example the shift from a geocentric to heliocentric model of the solar system did not occur simply due to the accumulation of new data (although data from the newly developed telescope did play a role), but was also the result of a change in the mindset of scientists and philosophers. This new mindset, or paradigm, was influenced by social and cultural changes, such as religious and political views, and resulted in an entirely new way of looking at existing data (Kuhn, 1957).

When considering its empirical nature, it is important to remember that scientific knowledge is the product of both observation <u>and</u> inference. Observations constitute the empirical basis of scientific knowledge and are descriptions of natural phenomena that may be directly perceived by the senses (or instrumental extensions of the senses). For example, astronomers might observe that one star is brighter than another. Inferences, on the other hand, are conjectures that go beyond what is directly accessible to the senses. The claim that the brighter of two stars is significantly closer to earth is usually an inference, since the distances to stars are typically too great to be directly measured. Rather, estimates of stellar distances are commonly based upon the inferred relationship between a star's spectral class and intrinsic luminosity. Subsequent comparison of the inferred intrinsic luminosity and apparent (observed) brightness yields the star's distance using the inverse square law of light. Since inferences are conjectures beyond observable data, these claims are inherently tentative. In such cases we can be quite certain, if our inferences are supported by much data, but never absolutely positive

The distinction between observations and inference in science is significant because these two types of scientific knowledge give rise to different kinds of scientific claims. For example, part of the work of science is to recognize/create patterns and relationships among observations. Useful patterns and relationships that withstand the scrutiny of scientists and the test of time may become known as scientific laws. Charles' law states that at constant pressure, the volume of a gas is directly proportional to its Kelvin temperature. In other words, as temperature changes, the volume of a gas will change proportionally, providing that the pressure remains the same. The law expresses a relationship that can describe *what* happens under specific conditions, but offers no explanation for *why* it happens. Explanations for why the relationship expressed in Charles' law exists must go beyond what is observable, and are, therefore, inferential in nature. Over time, as these inferential explanations are elaborated upon, supported and substantiated, they may become known as scientific theories (Suppe, 1977). For example, the kinetic molecular theory explains Charles' law in terms of the inherent motion of the molecular particles that make up gases.

Two critical misconceptions are dispelled by understanding the respective observational and inferential basis of laws and theories. Because scientific laws and theories are based on different kinds of evidence, the common conception that laws are proven theories is unfounded (Campbell, 1953; Dunbar, 1995; Rhodes & Schaible, 1989; Horner & Rubba, 1979). Theories and laws represent different kinds of knowledge and play different roles in science. As such, one can never "morph' into the other. Additionally, because both laws and theories are based on tentative knowledge (i.e., observations and inferences), neither is absolute. Laws are no more "proven" than theories. Both are subject to change (Horner & Rubba, 1979), albeit scientific laws are less often revised due, in part, to their relative simplicity and their lower reliance on conjecture and inference.

Nonetheless, scientific laws can change in light of new evidence, as explicated in the March 31, 2000 issue of the journal *Science*. The article recounts how a 3M team of research scientists found evidence calling for the modification of Brewster's Law (Weber, Stover, Gilbert, Nevitt, & Ouderkirk, 2000). This nearly 200-year-old axiom of optics predicts lower reflectivity for p-polarized light at material interfaces as incidence angle increases. A film developed by the 3M team maintains high reflectivity of select wavelengths of light regardless of the angle of incidence, in clear "violation" of Brewster's Law. The 3M corporation hopes to use the film in a wide variety of applications from improving laptop displays to making eye-catching clothing and accessories. Thus within a few years, we may be seeing daily reminders of the fallibility of a scientific law.

Another reason for the inherent tentativeness of scientific knowledge is its basis in human imagination and creativity. Science is not simply the product of logic and rationality. There is no single "scientific method" that scientists follow in order to produce scientific knowledge. Rather, science incorporates a host of methodologies and involves the creative invention of explanations and the imaginative construction of patterns and relationships. Darwin's synthesis of the theory of natural selection required the creative work of pulling together data and ideas from several diverse sources, including observations and samples from his tenure on the *H.M.S. Beagle*, Lyell's geologic principles, and Malthus' theory of populations. Consider also the intellectual creativity that was required for Niels Bohr to construct his elaborate model of the atom from the available observational data of atomic spectral lines. Thus, creativity and inventiveness are at the heart of scientific process, making the endeavor more personal than many suppose.

The role of creativity and imagination in the development of scientific knowledge also has implications for the supposed objectiveness of science—even so-called "objective facts" in science are not really free from subjectivity.

Furthermore, scientists' backgrounds, theoretical and disciplinary commitments, and expectations combine to provide strong influences on their work. These factors produce a mindset that affects what scientists investigate, how they conduct their investigations, and how they interpret their observations. Contrary to common belief, science seldom (if ever) starts with neutral observations. Rather, observations are typically motivated by, guided by, and acquire meaning in the context of specific theoretical frameworks (Hanson, 1958; Kuhn, 1996; Popper, 1934).

Consider the early sketches of Galileo and others who first used the telescope to observe the planet Saturn (Figure 1). As Sheehan (1988) characterized the historical episode, early observers did not know what to make of their observations of the planet and its varying appearance. At times, Saturn appeared to be flanked by appendages (which Galileo incorrectly interpreted as moons). At other times, to his dismay, the appendages appeared to have disappeared. Subsequent observers were similarly confused by the varying appearance of Saturn. Not until 1659, when Christiaan Huygens published a theory (or model) of the Saturnian system (Figure 2) did the observations of Saturn finally begin to make sense. It is significant that following publication of this model, no observer of Saturn reported seeing anything but rings. As R. L. Gregory later noted, such drawings do not depict the appearance of an object through a telescope at a specific time. Rather, they are the synthesis of many observations and reflect a belief in what the object is "really" like (Gregory, 1970). Before the explanation provided by Huygens' model of Saturn, observers had no perceptual framework with which to interpret the unfamiliar data provided by their senses.

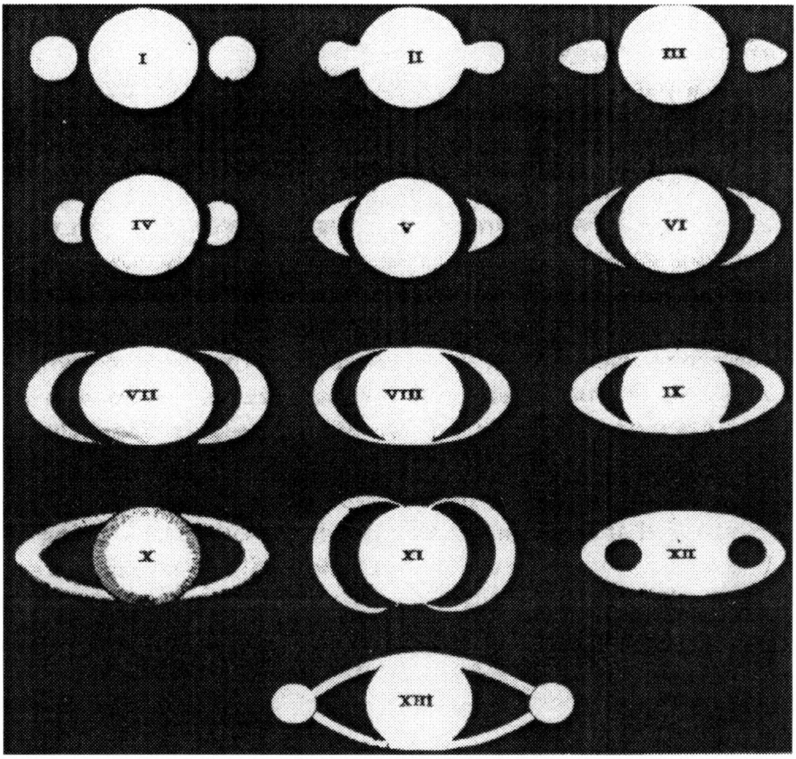

Figure 1. Telescopic views of Saturn as drawn by Galileo, Gassendi, and other early observers and compiled by Christiaan Huygens in Systema Saturnium (1659

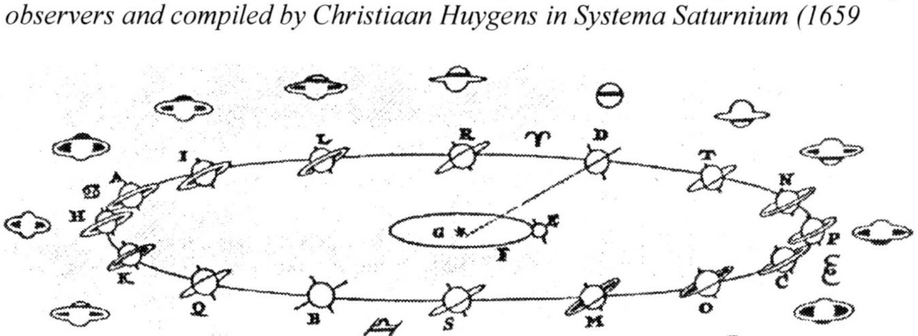

Figure 2. Huygens' diagram in Systema Saturnium illustrating a ringed and inclined model of Saturn that could account for its varied telescopic appearances.

Finally, cultural influences within and outside the scientific enterprise contribute to the tentative nature of scientific claims. Cultural influences operating within the scientific enterprise include such forces as accepted paradigms and theoretical frameworks, peer review, and collegial rivalries, to name a few. For example, whether the mass extinction of dinosaurs and other animals during the late Cretaceous Period is attributed to (1) meteorite impact, (2) increased volcanic activity, or (3) something as mundane as epidemic disease, depends largely on the background of the researcher (see Bakker, 1986 for an excellent discussion of the competing theories explaining the mass extinctions of the Cretaceous Period). Cultural influences operating outside the scientific enterprise include such elements as politics, socioeconomic factors, and religion. Each of these cultural influences has the potential to impact what research is done, how scientific findings are reported and received, and even the conclusions of scientific investigations. One need only consider such well-known episodes as the Catholic Church's suppression of Galileo's discovery that the moons of Jupiter revolve around the planet or Darwin's 17-year long delay in publishing his theory of natural selection to illustrate the major impact that society can have on science.

STUDENT CONCEPTIONS OF THE NATURE OF SCIENCE

Given that an informed, tentative view of the nature of science is desirable and may be delineated, the next questions that come to mind are "Do students possess such understandings?" and if not, "Can their understandings be altered through instruction to become more consistent with this informed, tentative view?" The typical classroom approach of focusing on the products of science, cookbook laboratory exercises, and authoritarian teaching modes was referred to earlier in this chapter. It appears unlikely that students would develop the complex understandings inherent to an informed view of the tentative nature of science in such an educational environment, a point that has not been overlooked by science educators (Duschl, 1988; Gallagher, 1991; Hodson, 1988, 1991; Matthews, 1994; Nadeau & Desautels, 1984). But what has research to say about the views of nature of science students develop as they progress from grade school through college?

Nearly five decades of assessments have consistently shown that students do not develop adequate conceptions of the tentative nature of science from their educational experiences. Whether the research has targeted the understandings of elementary students (Driver et al., 1996; Smith, Maclin, Houghton, & Hennessey, 2000), secondary students (Aikenhead, 1987; Bady, 1979; Klopfer & Cooley, 1963; Larochelle & Desautels, 1991; Lederman & O'Malley, 1990; Rubba & Anderson, 1978; Wilson, 1954), college students (Cotham & Smith, 1981; Gilbert, 1991), or international students (Broadhurst, 1970; Griffiths & Barman, 1995; Mackay, 1971) the result has been the same—typical science instruction focusing on the products of science with little attention to the values and assumptions inherent to the development of scientific knowledge tends to produce naively absolute views of the nature of science.

Research exploring school-aged children's understandings of the construction of knowledge (both general and scientific) has likewise found that students typically view knowledge as arising directly from nature and that absolute truth can be obtained through diligent observation. For example, researchers have identified a "common sense" epistemology, in which children see knowledge in general as arising directly from observation, and view bodies of knowledge as collections of (absolutely) true beliefs (King & Kitchener, 1994; Kitchener & King, 1981; Kuhn, Amsel, & O'Loughlin, 1988). When applied to science, the result is an essentially inductivist or empiricist view, placing the origin of scientific knowledge completely within the realm of observation (Carey & Smith, 1993; Hodson, 1985, 1988; Nadeau & Desautels, 1984; Strike & Posner, 1985). Direct assessments of 7th- and 11th-grade students' epistemologies of science have yielded similar results: Students typically leave school with absolutist views of the construction of scientific knowledge (Carey, Evans, Honda, Jay, & Unger, 1989; Grosslight, Unger, Jay, & Smith, 1991).

What is the source of this widespread and undesirable outcome? Studies reviewed by Carey and Smith (1993) and Smith et al. (2000) suggested three viable candidates for the source of students' limited scientific epistemologies: (a) limitations of science curricular materials and instructional approaches, (b) interference from students' everyday epistemologies, and (c) developmentally based constraints on reasoning (stage theory). As previously mentioned, curricular approaches emphasizing memorization of facts and knowledge with little opportunity for students to engage in the "messiness" of scientific processes afford few opportunities for students to develop complex understandings of the construction of scientific knowledge (Driver et al., 1996; Duschl, 1988; Matthews, 1994, among many others).

Recent research has directed attention to the relative effectiveness of implicit vs. explicit nature of science instruction. For example, some have argued that science teachers and educators have often assumed that the nature of science may be taught effectively through implicit approaches focusing on science processes and/or "constructivist" instructional practices (Bell, Abd-El-Khalick, & Lederman, 1998; Bell, Lederman, & Abd-El-Khalick, 2000). However, a recent review of empirical studies on improving science teachers' understandings of the nature of science concludes that of the three general approaches reported in the literature (historical, implicit, and explicit), the explicit approach consistently effected the most significant conceptual change (Abd-El-Khalick & Lederman, 2000b). Underlying the explicit approach to nature of science instruction (not to be confused with direct instruction) is the philosophy that teaching is a purposeful act, and that, to maximize learning, instruction must be purposive and goal-driven. The conceptual change literature adds further support, indicating that explicit, purposive instruction is necessary to address the misconceptions students develop both implicitly and explicitly (Butts, Hoffman, & Anderson, 1993; Joshua & Dupin, 1987; Strike & Posner, 1992). In light of these issues, it is possible that reliance on implicit approaches to nature of science instruction may well be a factor in students' limited views of science epistemology.

Further exacerbating this problem are the textbooks and other curricular materials rife with explicit references to misconceptions about the nature of the scientific enterprise. Take the issue of the single approach to doing science, commonly referred to as the Scientific Method. Nearly every popular textbook refers to this narrow five or six-step, algorithmic view of scientific process (although some recent editions include vague references to other ways of doing science). At best, this "Scientific Method" is a general characterization of the process of experimentation and reflects the format in which scientific investigations are typically reported. Even so, it leaves out the complex, cyclic relationship between theory and observation and fails to convey the creativity, imagination, and serendipity that permeate the art of doing science. At worst, it paints a picture of scientists as following a checklist, with little or no room for ingenuity, creative thought, or imagination.

Another explicit misconception commonly found in secondary and college-level textbooks concerns the supposed hierarchical relationship between scientific theories and laws, where theories become laws when "enough" evidence has accumulated for them to be proven (see, for example, Curtis & Barnes, 1985). This despite the reality that no theory has ever transformed into law, and that scientific laws themselves are subject to modification as new information comes to light (Horner & Rubba, 1979, McComas, 1996).

It appears then, that the curricular issue goes beyond students' lack of opportunities to engage in and reflect upon authentic science processes. Even when educators have had the nature of science on their agenda, they have too often inappropriately assumed their students would be able to make the necessary conceptual changes through implicit instruction. This is especially unlikely, given the tenacity of students' absolute views of science and the reality that popular curricular materials explicitly teach misconceptions about the nature of science. Like any instruction designed to achieve conceptual change, effective nature of science lessons should provide opportunities for students to consider and express their current views, challenge these views when inadequate, and work together with their teacher to construct more appropriate understandings.

Common sense epistemology

The literature on students' understandings of knowledge in general, sometimes referred to as "common sense epistemology" is both rich and complex. Over the past 30 years, a number of researchers have probed students' and adults' views of the nature of knowledge, its source, and justification (see Carey & Smith, 1993). In general, these studies have supported a developmental progression of students' epistemologies that begins with young children viewing knowledge as absolute and arising directly from authority and/or sensory experiences, and ending with adult college students' more sophisticated view of how different perspectives influence one's interpretation of experience (e.g., Chandler, 1987; Kitchener & King, 1981; Kuhn, et al., 1988). It is easy to see that the endpoints of this developmental progression are closely related to the absolute vs. tentative views of the nature of science discussed earlier. Less clear is to what extent the development of students' general epistemologies constrains their abilities to develop accurate conceptions of the nature of scientific knowledge.

Stage theory

If these developmental frameworks are biologically based in the Piagetian sense, then the stages through which students' abilities progress pose barriers to their understandings of the nature of science. For example, in the Piagetian framework children's absolutist views of the nature of science are the necessary result of concrete operational thought. Thus, in elementary school, nature of science instructional practice informed by such a framework would emphasize hands-on experiences focusing on the construction of concrete, straightforward concepts. Students would be engaged in the processes of science, such as observing, classifying, and inferring, in order for them to build a foundation upon which more abstract ideas about the construction of scientific knowledge could later be added. Once students reach the formal operations stage, they become capable of more complex reasoning and entertaining a variety of viewpoints. Only then would they be prepared for more abstract concepts of the nature of science, such as the realization that knowledge is dependent upon perspective and is, therefore, not absolute.

Cognition Theory

Recent commentary and investigative studies have suggested that alternative (e.g., absolute) epistemologies of science may not stem from biologically based stages of development. Rather, science epistemology may be viewed as a part of a particular cognitive domain (i.e., an intuitive theory of mind) with a specific developmental history (Carey & Smith, 1993). The fact that even students in age groups consistent with formal operations typically maintain absolute views of the nature of science indicates that acquisition of formal operations does not automatically lead to more tentative views, at least in regard to scientific knowledge. Furthermore, recent work in developmental psychology has demonstrated that young children are capable of greater intellectual sophistication than once was assumed (Metz, 1995). For example, investigations into the beliefs of six-year-olds have shown that elementary school children can recognize inference as a source of knowledge (Sodian & Wimmer, 1987) and that background knowledge can influence interpretation of visual stimuli (Taylor, 1988; Taylor, Cartwright, & Bowden, 1991). The picture that is unfolding points to a more flexible view of cognitive development that, in many cases, may respond to targeted instruction. As Metz (1995, p. 156) cautioned,

> Science educators cannot assume that age characteristics are simply a function of development in the sense of immutable cognitive characteristics of the stage. While some age-correlated weaknesses may be fairly robust at a particular stage and readily ameliorated at a subsequent stage, other weaknesses may to varying degrees respond to instruction, and still others may constitute an enduring challenge at all ages and all levels of expertise.

TEACHING THE NATURE OF SCIENCE

It appears, then that some of the basic assumptions guiding curriculum development in the past may have set unnecessary limits on children's abilities to learn the more abstract concepts of scientific reasoning and the nature of science. In fact, the limitations set by rigid adherence to developmental stage theories may, in themselves, unintentionally contribute to misconceptions about the scientific enterprise. Having elementary school children participate in disjointed activities that focus on the processes of science, such as observation and classification, apart from any explicit overriding purpose, may serve to reinforce their already absolute views. For example, a third grade teacher aware of the constraints implied by stage theory might have her students classify leaves in a typical hands-on lesson in a biology unit. An innovative teacher might well have students come up with their own classification schemes for leaves. However, without explicit attention to the concept that classification schemes are human constructs, that they cannot be read directly from the book of nature, and that multiple classifications may well be equally viable, students are likely to walk away from such activities with reinforced views of the single "right answer" and scientific knowledge as absolute.

As argued throughout this chapter, one goal of science instruction should be to combat such absolute views of scientific knowledge. In the earliest years, the approach to addressing students' absolute views might be modest. For example, students might focus on observation and inference early on, with explicit attention to the differences in these types of knowledge and the possibility that many valid inferences may stem from any particular set of observations. Later, as students are encouraged to continue to make observations and inferences, they may be taught to assess competing inferences, with value placed on evidence and logic. Eventually, students could be taught through explicit reflection on their own learning processes how observations and inferences complement one another in the construction of knowledge and how this is true for science as well. The ultimate goal would be for students to view inferences as human constructs with inherent limitations on the one hand, but with the potential to be robust and informative when based on evidence and logic. Therefore, even if science cannot infer "Truth" in an absolute sense, it can produce knowledge useful for everyday decisions.

On the other side of the issue, there is danger in introducing abstract nature of science concepts too soon. From the stage theory perspective, age-correlated constraints may cause particular misconceptions to be resilient at early stages but readily ameliorated at later stages. From the cognitive domain perspective, students may not be prepared to handle certain abstract concepts of the nature of science until sufficient groundwork has been laid. Either way, science educators must consider the appropriate timing of nature of science instruction, especially in terms of the more abstract issues. Thus, there are aspects of the nature of science (as in the previously cited observation/inference example) that are appropriate to introduce even in the primary grade levels. Other aspects may be inappropriate for elementary school science but perfectly reasonable at the secondary level. Still other issues, such as debates over the existence of reality independent of the observer and the influence of these debates on the interpretation of scientific knowledge, may only be appropriate postsecondary instruction.

Even with the caveats of careful planning and explicit instruction, some have expressed concern about the unintended consequences of challenging students' (absolutist) presuppositions of science. For example, over 20 years of teaching philosophy of science to science teachers led Winchester (1993) to question the wisdom of teaching students to doubt the straightforward, absolute views of the nature of science they take for granted:

> Should we challenge systematically the presuppositions which our students are picking up by their scientific training at schools and universities?... Are we likely to affect their citizenship in the community of scientists for the better by suggesting, for example, that it is doubtful that science aims at truth? Or by suggesting that there is no sense to the notion that a scientific generalization can be "confirmed" although it may sometimes be disapproved? Do we do them a service by suggesting that science may have limitations and listing them?...Someone who believed this, if these presuppositions really do lie behind their being able to function as scientists and to maintain their faith in Science, would see no reason to continue with science or the scientific enterprise. (Winchester, 1993, pp. 196-7)

Winchester's concerns are well founded and should be considered by all who strive for conceptual change in students' understandings of the nature of science. However, before adopting the most pessimistic viewpoint, one must consider two critical points. First, the alternative to challenging students' absolutist presuppositions about science is to leave them with a propensity toward a simplistic view that science is an almost superhuman endeavor, able to wholly compensate for human flaws and biases, and able, with diligent application, to produce absolute truth. This view, which ignores the history and philosophy of science, is but a caricature of the scientific enterprise. Considered in this way, the question becomes do we leave our students with a "safe" caricature or help them gain a more realistic, if less comforting, understanding of the scientific endeavor? On grounds of intellectual honesty alone, the answer should be obvious.

The second point concerns whom we are targeting with instruction about the nature of science. Most of Winchester's concerns target the effect of disillusionment on the next generation of scientists and engineers (even though his students were predominantly teachers). The defining goal of current U.S. science education reforms is science for ALL Americans. This goal recognizes the need for better understandings of the scientific enterprise and scientific inquiry in order to build a scientifically literate society, presumably a society in which individuals are able to make reasoned decisions on socio-scientific issues. As Behnke asserted, "Widespread public understanding of the nature of scientific endeavor is essential for a healthy society" (Behnke, 1961, p.206). Even if one embraces Winchester's concern for the potential of explicit nature of science instruction to disillusion the scientists and engineers of the future, these are not the principle targets of nature of science instruction. Rather, the target is everyday citizens who are asked to make personal and public decisions on science and technology based issues.

CONCLUSION

Herron (1969) aptly labeled the pursuit of the nature of science when he cautioned that it may turn out to be "less the panacea and much more the Pandora's Box that we are pursuing (p. 107)." In mythology, when Pandora opened the forbidden box, out swarmed all the calamities of mankind, from earthquakes to bunions. Despite her best efforts and intentions, Pandora could not prevent them from spreading out across the world once they had been released. Certainly there are parallels to the science education community's "opening the box" to nature of science instruction. As Heron explained, opening the box has released a swarm of issues, not unlike the calamities that swarmed out of Pandora's forbidden box:

- Are some viewpoints of the nature of science more appropriate than others for different curricular purposes?
- Can we agree upon a single particular point of view appropriate for students in all science classes?
- Are some points of view appropriate to the biological sciences and others to physics or chemistry?
- To what extent does one falsify such viewpoints for the sake of "simplification" to increase the probability of their comprehension by those at a relatively unsophisticated level?
- How much time can we devote to such questions when there is so little time to do what we would like to do now?

In a less-familiar aspect of the mythological story, Pandora managed to shut the lid of the box in time to prevent the escape of its final occupant. This was Elpis (hope). Thus, no matter how bad things appear, hope remains. After 30 years of research and commentary on the issue of nature of science instruction and learning, we have begun to assimilate answers to many of the questions that have plagued our community since we first opened Pandora's box. Not every question and concern has been addressed to the satisfaction of all. However, we have made significant progress toward understanding the whats, whens, and hows of nature of science instruction. There remains much work to be done before our task is complete (if it ever will be), but this should not be seen in a negative light. "Quite the contrary," as Heron (1969) concluded, "I think it is a task worth every effort we can expend if we are ever to put an end to what appears to be a wide disparity between what we are saying in science education, and what we are, in fact, doing" (p.107).

REFERENCES

Abd-El-Khalick, F., Bell, R. L., & Lederman, N. G. (1998). The nature of science and instructional practice: Making the unnatural natural. *Science Education, 82,* 417-436.

Abd-El-Khalick, F., & Lederman, N. G. (2000a). The influence of history of science courses on students' views of nature of science. *Journal of Research in Science Teaching, 37,* 1057-1095.

Abd-El-Khalick, F., & Lederman, N. G. (2000b). Improving science teachers' conceptions of the nature of science: A critical review of the literature. *International Journal of Science Education, 22,* 665-701.

Aikenhead, G. (1987). High school graduates' beliefs about science-technology-society: Characteristics and limitations of science knowledge. *Science Education, 71,* 459-487.

Alters, B.J. (1997). Whose nature of science? *Journal of Research in Science Teaching, 34,* 39-55.
American Association for the Advancement of Science. (1993). *Benchmarks for science literacy: A Project 2061 report.* New York: Oxford University Press.
Bady, R.A. (1979). Students' understanding of the logic of hypothesis testing. *Journal of Research in Science Teaching, 16,* 61-65.
Bakker, R.T. (1986) The dinosaur heresies: New theories unlocking the mystery of the dinosaurs and their extinction. New York: Kensington Publishing.
Behnke, F.L. (1961). Reactions of scientists and science teachers to statements bearing on certain aspects of science and science teaching. *School Science and Mathematics, 61,* 193-207.
Bell, R. L. (in press). Understandings of the nature of science and decision making on science and technology based issues. *Science Education.*
Bell, R. L. (1999). *Understandings of the nature of science and decision making on science and technology based issues.* Unpublished doctoral dissertation, Oregon State University, Oregon.
Bell, R. L., Abd-El-Khalick, F., & Lederman, N. G. (1998). Implicit versus explicit nature of science instruction: An explicit response to Palmquist and Finley. *Journal of Research in Science Teaching, 35,* 1057-1061.
Bell, R. L., Lederman, N. G., & Abd-El-Khalick, F. (2000). Developing and acting upon one's conception of the nature of science: A follow-up study. *Journal of Research in Science Teaching, 37,* 563-581.
Broadhurst, N. A. (1970). A study of selected learning outcomes of graduating high school students in South Australian schools. *Science Education*, 54, 17-21.
Butts, D. P., Hoffman, H. M., & Anderson, M. (1993). Is hands-on experience enough? A study of young children's views of sinking and floating objects. *Journal of Elementary Science Education, 5*(1), 50-64.
Campbell, N. (1953). *What is science?* New York: Dover Publications.
Carey, S., Evans, R., Honda, M., Jay, E., Unger, C. (1989). "An experiment is when you try it and see if it works": A study of grade 7 students understanding of the construction of scientific knowledge. *International Journal of Science Education, 11,* 514-529.
Carey, S., & Smith, C. (1993). On understanding the nature of scientific knowledge. *Educational Psychologist, 28,* 235-251.
Cartwright, N. (1983). *How the laws of physics lie.* Oxford: Oxford University Press.
Cartwright, N. (1988). The truth doesn't explain much. In E.D. Klemke, R. Hollinger, and D. Kline (Eds.). *Introductory readings in the philosophy of science* (pp. 129-136). Buffalo, New York: Prometheus Books.
Central Association for Science and Mathematics Teachers. (1909). A consideration of the principles that should determine the courses in biology in secondary schools. *School Science and Mathematics, 9,* 241-247.
Chandler, M. (1987). The Othello effect: Essay on the emergence and eclipse of skeptical doubt. *Human Development, 30,* 137-159.
Cleminson, A. (1990). Establishing an epistemological base for science teaching in the light of contemporary notions of the nature of science and of how children learn science. *Journal of Research in Science Teaching 27,* 429-445.
Collins, H.M. & Pinch, T. (1998) *The golem: What you should know about science.* (2nd ed.). Cambridge, MA: Cambridge University Press.
Cotham, J., & Smith, E. (1981). Development and validation of the conceptions of scientific theories test. *Journal of Research in Science Teaching, 18,* 387-396.
Curtis, H., & Barnes, N. S. (1985). *Invitation to biology*, 4th edition. New York: Worth Publishers.
Driver, R., Leach, J., Millar, R., & Scott, P. (1996). *Young people's images of science.* Philadelphia: Open University Press.
Dunbar, R. (1995). *The trouble with science.* Cambridge, MA: Harvard University Press.
Duschl, R.A. (1988). Abandoning the scientistic legacy of science education. *Science Education, 72,* 51-62.
Duschl, R. A. (1990). *Restructuring science education: The importance of theories and their development.* New York: Teachers College Press.
Duschl, R. A. (1994). Research on the history and philosophy of science. In D. L. Gabel (Ed.), *Handbook of research on science teaching and learning* (pp. 443-465). New York: Macmillan.
Feyerabend, P. (1993). *Against method.* New York: Verso.

Fleming, R. (1986a). Adolescent reasoning in socio-scientific issues, Part I: Social cognition. *Journal of Research in Science Teaching, 23,* 677-687.

Fleming, R. (1986b). Adolescent reasoning in socio-scientific issues, Part II: Nonsocial cognition. *Journal of Research in Science Teaching, 23,* 689-698.

Gallagher, J. J. (1991). Perspective and practicing secondary school science teachers' knowledge and beliefs about the philosophy of science. *Science Education, 75,* 121-134.

Gilbert, S.W. (1991). Model building and a definition of science. *Journal of Research in Science Teaching, 28,* 73-78.

Good, R. et al. (2000, April). *Guidelines for nature of science (NOS) researchers.* Symposium conducted at the annual meeting of the National Association for Research in Science Teaching, New Orleans, LA.

Gregory, R.L. (1970). *The intelligent eye.* New York: Simon and Schuster.

Griffiths, A.K. & Barman, C.R. (1995). High school students' views about the nature of science: results from three countries. *School Science and Mathematics* 95, 248-255.

Grosslight, L., Unger, C.M., Jay, E., & Smith, C. (1991). Understanding models and their use in science: Conceptions of middle and high school students and experts. *Journal of Research in Science Teaching, 28,* 799-822.

Hanson, (1958). *Patterns of discovery.* Cambridge: Cambridge University Press.

Herron, M.D. (1969). Nature of science: Panacea or Pandora's box. *Journal of Research in Science Teaching, 6,* 105-107.

Hodson, D. (1985). Philosophy of science, science and science education. *Studies in Science Education, 12,* 25-57.

Hodson, D. (1988). Toward a philosophically more valid science curriculum. *Science Education,* 72, 19-40.

Hodson, D. (1991). The role of philosophy in science teaching. In M. R. Matthews (Ed.), *History, philosophy, and science teaching: selected reading* (pp. 19-32). New York: Teachers College Press.

Horner, J. K., & Rubba, P. A. (1978). The myth of absolute truth. *The Science Teacher, 45,* 29-30.

Horner, J. K., & Rubba, P. A. (1979). The laws are mature theories fable. *The Science Teacher, 45,* 31.

Joshua, S., & Dupin, J. J. (1987). Taking into account student conceptions in instructional strategy: An example in physics. *Cognition and Instruction, 4,* 117-135.

King, P.M . & Kitchener, K.S. (1994). *Developing reflective judgment: understanding and promoting intellectual growth and critical thinking in adolescents and adults.* San Francisco: Jossey-Bass

Kitchener, K.S., & King, P. M. (1981). Reflective judgement: Concepts of justification and their relationship to age and education. *Journal of Applied Developmental Psychology,* 2, 89-116.

Kitcher, P. (1982). *Abusing science: The case against creationism.* Cambridge, MA: MIT Press.

Klopfer, L. E., & Cooley, W. W. (1963). The history of science cases for high schools in the development of student understanding of science and scientists. *Journal of Research for Science Teaching,* 1, 33-47.

Kuhn, D., Amsel, E., & O'Loughlin, M. (1988). *The development of scientific thinking skills.* Orlando, Fl: Academic.

Kuhn, T.S. (1957). *The Copernican revolution.* Chicago: The University of Chicago Press.

Kuhn, T. S. (1996). *The structure of scientific revolutions.* 3rd Edition. Chicago: The University of Chicago Press.

Larochelle, M., & Desautels, J. (1991). "Of course, it's just obvious": Adolescents' ideas of scientific knowledge. *International Journal of Science Education,* 13, 373-389.

Latour, B. (1987). *Science in action: How to follow scientists and engineers through society.* Cambridge, MA: Harvard University Press.

Lawson, A.E. (1995). *Science teaching and the development of thinking.* Belmont, CA, Wadsworth Publishing.

Lederman, N. G. (1992). Students' and teachers' conceptions of the nature of science: A review of the research. *Journal of Research in Science Teaching, 29,* 331-359.

Lederman, N.G. (1999). Teachers' understanding of the nature of science and classroom practice: Factors that facilitate or impede the relationship. *Journal of Research in Science Teaching, 36,* 916-929

Lederman, N.L., Abd-El-Khalick, F. S., Bell, R. L., Schwartz, R., & Akerson, V. L. (2001, January). *Assessing the "un-assessable": Views of nature of science questionnaire (VNOS).* A paper presented at the Annual Meeting of the National Association for Research in Science Teaching, St. Louis, MO.

Lederman, N. G., & O'Malley, M. (1990). Students' perceptions of tentativeness in science: Development, use, and sources of change. *Science Education, 74,* 225-239.
Loving, C. C. (1997). From the summit of truth to its slippery slopes: Science education's journey through positivist-postmodern territory. *American Educational Research Journal, 34,* 421-452.
Mackay, L. (1971). Development of understanding about the nature of science. *Journal of Research in Science Teaching, 8,* 57-66.
Macilwain, C. (1998). Physicists seek definition of "science." *Nature, 392,* 849.
Matthews, M. R. (1994). *Science teaching: The role of history and philosophy of science.* New York: Routledge.
Matthews, M. R. (1997). Editorial, *Science & Education, 6,* 3232-329.
McComas, W. F. (1996). Ten myths of science: Reexamining what we think about the nature of science. *School Science and Mathematics,* 96, 10-16.
McComas, W.F., Clough, M.P., & Almzroa, H. (1998). The role and character of the nature of science in science education. In W. McComas (Ed.), *The nature of science in science education: Rationales and strategies* (pp. 3-39). The Netherlands: Kluwer Academic Publishers.
McComas, W.F., & Olson, J.K. (1998). The nature of science in international science education standards documents. In W. F. McComas (Ed.), *The nature of science in science education: Rationale s and strategies* (pp. 41-52). Boston: Kluwer Academic Publishers.
Meyling, H. (1997). How to change students' conceptions of the epistemology of science. *Science & Education, 6,* 397-416.
Metz, K. (1995). Reassessment of developmental constraints on children's science instruction. *Review of Educational Research, 65,* 93-127.
Nadeau, R., & Desautels, J. (1984). *Epistemology and the teaching of science.* Science Council of Canada, Ottawa.
National Research Council. (1996). *National science education standards.* Washington, DC: National Academic Press.
Popper, K. R. (1934). The logic of scientific discovery. London: Routledge. (Reprint edition, 1992).
Popper, K. R. (1963). *Conjectures and refutations: The growth of scientific knowledge.* London: Routledge.
Popper, K. R. (1988). Science: Conjectures and refutations. In E. D. Klemke, R. Hollinger, & A. D. Kline (Eds.), *Introductory readings on the philosophy of science* (Rev. ed., pp. 19-27). Buffalo: Prometheus Books..
Recer, P. (1998, September 4). New evidence of moon water found. *Corvallis Gazette-Times,* pA9.
Rhodes, G., & Schaible, R. (1989). Fact, law, and theory: Ways of thinking in science and literature. *Journal of College Science Teaching, 18,* 228-232.
Rubba, P.A., & Anderson, H.O. (1978). Development of an instrument to assess secondary school students' understanding of the nature of science. *Science Education, 62,* 449-458.
Sheehan, W.S. (1988). *Planets and perception: Telescopic views and interpretations, 1609-1909.* Tucson, AZ: University of Arizona Press.
Smith, M.U., Lederman, N. G., Bell, R. L., McComas, W.F., & Clough, M.P. (1997). How great is the disagreement about the nature of science? A response to Alters. *Journal of Research in Science Teaching 34,* 1101-1103.
Smith, C.L., Maclin, M., Houghton, C., & Hennessey, M.G. (2000). Sixth-grade students' epistemologies of science: The impact of school science experiences on epistemological development. *Cognition and Instruction 18,* 349-422.
Smith, M. U., & Scharmann, L. C. (1999). Defining versus describing the nature of science: A pragmatic analysis for classroom teachers and science educators. *Science Education, 83,* 493-509.
Sodian, N. B., & Wimmer, H. (1987). Children's understanding of inference as a source of knowledge. *Child Development, 58,* 424-33.
Songer, N., & Linn, M. (1991). How do students' views of science influence knowledge integration? *Journal of Research in Science Teaching, 28,* 761-784.
Strike, K. A., & Posner, G. J. (1985). A conceptual change view of learning and understanding. In L. West & A. Pines (Eds.), *Cognitive structure and conceptual change* (pp. 211-231). New York: Academic.
Strike, K. A., & Posner, G. J. (1992). A revisionist theory of conceptual change. In R. A. Duschl and R. J. Hamilton (Eds.), *Philosophy of science, cognitive psychology, and educational theory and practice* (pp. 147-176). New York: State University of New York Press.

Suppe, F. (1977). *The structure of scientific theories* (2nd ed.). Chicago: University of Illinois Press.

Taylor, M. (1988). Conceptual perspective-taking: Children's ability to distinguish what they know from what they see. *Child Development, 59*, 703-718.

Taylor, M., Cartwright, B., & Bowden, T. (1991). Perspective taking and theory of mind: Do children predict interpretive diversity as a function of differences in observers' knowledge? *Child Development, 62*, 1334-1351.

Trocco, F. (2002). On the fringes of credibility: The boundary question between science and non-science. *Skeptic, 9*, 32-39.

Weber, M.F., Stover, C.A., Gilbert, L.R., Nevitt, T.J., & Ouderkirk, A.J. (2000). Giant birefringent optics in multilayer polymer mirrors. *Science*, 287, 2451-2456.

Welch, W. W. (1984). A science-based approach to science learning. In D. Holdzkom & Lutz (Eds.). *Research within reach: Science education.* Washington, DC: National Science Teachers Association.

Wilson, L. (1954). A study of opinions related to the nature of science and its purpose in society. *Science Education, 38*, 159-164.

Winchester, I. (1993). "Science is dead. We have killed it, you and I"—How attacking the presuppositional structures of our scientific age can doom the interrogation of nature. *Interchange, 24*, 191-198.

Zeidler, D. L., Walker, K. A., Ackett, W. A., & Simmons, M. L. (2002). Tangled up in views: Beliefs in the nature of science and responses to socioscientific dilemmas. *Science Education, 86*, 343-367.

Ziman, J. (1988). What is science? In E.D. Klemke, R. Hollinger, and D. Kline (Eds.). *Introductory readings in the philosophy of science* (pp. 28-33). Buffalo, New York: Prometheus Books.

---

[1] There is, of course, no single representation that is inclusive of all aspects and issues related to the nature of science. Therefore, this representation should be viewed as one framework among many that can be used as a guide for nature of science instruction. However, it should be noted that this framework is consistent with others that have been described in the science education literature (e.g., Carey & Smith, 1993; Smith & Scharmann, 1999), as well as in the current science education reform documents. In addition, the generalizations presented here are to be understood in the context of K-16 science education, rather than in the context of educating graduate students in history or philosophy of science. Obviously, these nature of science aspects should be approached at different levels of depth and complexity, depending on the background and grade level of students.

INDEX

ABOUT THE EDITORS

Dr. Lawrence B. Flick is chair of Science and Mathematics Education at Oregon State University. He has a B.S. in Electrical Engineering from Purdue University, an MAT from Northwestern University, and a Ph.D. in Science and Environmental Education from Indiana University. He worked in the communications industry before entering the teaching profession where he taught middle school science in both public and private schools.

His research is in the area of student conceptual development and the application of social cognition theory to classroom teaching practices. His work has investigated strategies for developing strategic thinking skills that support student understanding of science as inquiry.

He has been president of the Association for the Education of Teachers of Science and is currently co-editor, with Norm Lederman, of School Science and Mathematics. Dr. Flick has served on the board of directors of NSTA, AETS, Council for Elementary Science International, and the School Science and Mathematics Association. He has served on the editorial boards of the Journal for Research in Science Teaching, the Journal of Science Teacher Education, and School Science and Mathematics. He has been a national consultant for numerous educational projects such as WGBH Teaching High School Science, PBS SCIENCELINE, Westinghouse Hanford Company, Environmental Education, and the AT&T Distance Learning Network.

Dr. Norman G. Lederman is Chair and Professor of Mathematics and Science Education at the Illinois Institute of Technology. Dr. Lederman received his Ph.D. in Science Education and he possesses MS degrees in both Biology and Secondary Education. Prior to his 20 + years in science teacher education, Dr. Lederman was a high school teacher of biology and chemistry for 10 years.

He received the Illinois Outstanding Biology Teacher Award (1979), the Burlington Resources Foundation Faculty Achievement Award for Excellence in Teaching and Research (1992), and the AETS Outstanding Mentor Award (2000).

Dr. Lederman is internationally known for his research and scholarship on the development of students' and teachers' conceptions of the nature of science and scientific inquiry. He has been author or editor of 10 books, written 15 book chapters, published over 150 articles in professional journals, and made over 500 presentations at professional conferences around the world.

Dr. Lederman is a former President of the National Association for Research in Science Teaching (NARST) and the Association for the Education of Teachers in Science (AETS). He has also served as Director of Teacher Education for the National Science Teachers Association (NSTA), and has served on the Board of Directors of NSTA, AETS, NARST, and the School Science and Mathematics Association.

Science & Technology Education Library

Series editor: William W. Cobern, *Western Michigan University, Kalamazoo, U.S.A.*

Publications

1. W.-M. Roth: *Authentic School Science.* Knowing and Learning in Open-Inquiry Science Laboratories. 1995 ISBN 0-7923-3088-9; Pb: 0-7923-3307-1
2. L.H. Parker, L.J. Rennie and B.J. Fraser (eds.): *Gender, Science and Mathematics.* Shortening the Shadow. 1996 ISBN 0-7923-3535-X; Pb: 0-7923-3582-1
3. W.-M. Roth: *Designing Communities.* 1997 ISBN 0-7923-4703-X; Pb: 0-7923-4704-8
4. W.W. Cobern (ed.): *Socio-Cultural Perspectives on Science Education.* An International Dialogue. 1998 ISBN 0-7923-4987-3; Pb: 0-7923-4988-1
5. W.F. McComas (ed.): *The Nature of Science in Science Education.* Rationales and Strategies. 1998 ISBN 0-7923-5080-4
6. J. Gess-Newsome and N.C. Lederman (eds.): *Examining Pedagogical Content Knowledge.* The Construct and its Implications for Science Education. 1999 ISBN 0-7923-5903-8
7. J. Wallace and W. Louden: *Teacher's Learning.* Stories of Science Education. 2000 ISBN 0-7923-6259-4; Pb: 0-7923-6260-8
8. D. Shorrocks-Taylor and E.W. Jenkins (eds.): *Learning from Others.* International Comparisons in Education. 2000 ISBN 0-7923-6343-4
9. W.W. Cobern: *Everyday Thoughts about Nature.* A Worldview Investigation of Important Concepts Students Use to Make Sense of Nature with Specific Attention to Science. 2000 ISBN 0-7923-6344-2; Pb: 0-7923-6345-0
10. S.K. Abell (ed.): *Science Teacher Education.* An International Perspective. 2000 ISBN 0-7923-6455-4
11. K.M. Fisher, J.H. Wandersee and D.E. Moody: *Mapping Biology Knowledge.* 2000 ISBN 0-7923-6575-5
12. B. Bell and B. Cowie: *Formative Assessment and Science Education.* 2001 ISBN 0-7923-6768-5; Pb: 0-7923-6769-3
13. D.R. Lavoie and W.-M. Roth (eds.): *Models of Science Teacher Preparation.* Theory into Practice. 2001 ISBN 0-7923-7129-1
14. S.M. Stocklmayer, M.M. Gore and C. Bryant (eds.): *Science Communication in Theory and Practice.* 2001 ISBN 1-4020-0130-4; Pb: 1-4020-0131-2
15. V.J. Mayer (ed.): *Global Science Literacy.* 2002 ISBN 1-4020-0514-8
16. D. Psillos and H. Niedderer (eds.): *Teaching and Learning in the Science Laboratory.* 2002 ISBN 1-4020-1018-4
17. J.K. Gilbert, O. De Jong, R. Justi, D.F. Treagust and J.H. Van Driel (eds.): *Chemical Education: Towards Research-based Practice.* 2003 ISBN 1-4020-1112-1
18. A.E. Lawson: *The Neurological Basis of Learning, Development and Discovery.* Implications for Science and Mathematics Instruction. 2003 ISBN 1-4020-1180-6
19. D.L. Zeidler (ed.): *The Role of Moral Reasoning on Socioscientific Issues and Discourse in Scientific Education.* 2003 ISBN 1-4020-1411-2

Science & Technology Education Library

Series editor: William W. Cobern, *Western Michigan University, Kalamazoo, U.S.A.*

20. P.J. Fensham: *Defining an Identity. The Evolution of Science Education as a Field of Research.* 2003 ISBN 1-4020-1467-8
21. D. Geelan: *Weaving Narrative Nets to Capture Classrooms.* Multimethod Qualitative Approaches for Educational Research. 2003 ISBN 1-4020-1776-6; Pb: 1-4020-1468-7
22. A. Zohar: *Higher Order Thinking in Science Classrooms: Students' Learning and Teachers' Professional Development.* 2004 ISBN 1-4020-1852-5; Pb: 1-4020-1853-3
23. C.S. Wallace, B. Hand, V. Prain: *Writing and Learning in the Science Classroom.* 2004 ISBN 1-4020-2017-1
24. I.A. Halloun: *Modeling Theory in Science Education.* 2004 ISBN 1-4020-2139-9
25. L.B. Flick and N.G. Lederman (eds.): *Scientific Inquiry and the Nature of Science.* Implications for Teaching, Learning, and Teacher Education. 2004 ISBN 1-4020-2671-4

KLUWER ACADEMIC PUBLISHERS – DORDRECHT / BOSTON / LONDON

LaVergne, TN USA
11 April 2010
178816LV00001B/37/A
9 781402 026713